PERENNIAL SOLUTIONS

PERENNIAL SOLUTIONS

A GROWER'S GUIDE TO PERENNIAL PRODUCTION

Paul Pilon

Ball Publishing
Batavia, Illinois, U.S.A.

Ball Publishing
Post Office Box 9
335 North River Street
Batavia, IL 60510
www.ballpublishing.com

Library of Congress Cataloging-in-Publication Data

Pilon, Paul.
Perennial solutions : a grower's guide to perennial production / Paul Pilon.
p. cm.
ISBN-13: 978-1-883052-47-8 (hardcover : alk. paper)
ISBN-10: 1-883052-47-5 (hardcover : alk. paper)
1. Perennials. 2. Floriculture. I. Title.

SB434.P53 2006
635.9'32--dc22

2006007484

ISBN: 10:1-883052-47-5
ISBN: 13:978-1-883052-47-8
Printed and bound in China by Imago.

12 11 10 09 08 07 06 1 2 3 4 5 6 7 8 9

Without the love and support of my wife, Jennifer, and daughters, Kaylan and Carissa, this book would never have been possible.

"When you look at your life, the greatest happinesses are family happinesses."

—Joyce Brothers

Contents

Preface

Back in the late 1990s when I made the transition from a greenhouse grower of annuals and potted holiday crops to a perennial grower, there was a great disconnect between the vast array of cultural information available for the crops I have become accustomed to growing and the seemingly lack of information regarding commercially produced perennials.

I've spent years trying to understand the many aspects of perennial crop production. The learning process has involved an awareness of the perennial industry (past, present, and future), a network of peers who contributed to my education, the ability to be observant, the composition of numerous databases, hundreds of experiments and in-house grower trials, the use of past experiences and information to design or modify production plans and systems, and the desire to not be complacent. By default and situation, I wanted to learn and fill the gaps between available information and successful crop production.

Today, with the popularity of perennials, there has been great efforts made and research conducted on perennial crops. With more emphasis on commercial production practices, cultural information is more readily available. Perennial producers are much better positioned in this sector than in the past; however, growers still often have difficulty pulling it all together. Perennial Solutions is my attempt to compile the many aspects of commercial perennial production into one easy-to-use reference. I hope it proves useful to your business, and I'm proud to be a part of your success.

Acknowledgments

I have been fortunate to have a great network of peers who believed in me, helped create opportunities, and encouraged me to grow professionally. I would like to thank my colleagues and friends, as well as all of those in the industry and at universities whose previous endeavors and willingness to share their expertise helped me form a knowledge base and foundation for my personal awareness and growth. The content presented in this book is the accumulation of my own experiences combined with information provided and/or acquired from numerous sources. A book of this magnitude is the compilation of many hard-working individuals.

My parents, Doug and Vel Pilon, provided an exceptional amount of support and encouragement as my horticultural interests developed at a very early age, and they have played a significant role in who I have become today. My mother and my wife, Jennifer, contributed countless hours of proof reading and editing; I want to express my gratitude to them for their hard work and patience.

1

An Introduction to Perennials

The last two decades have seen considerable growth in the popularity and sale of containerized perennials. Perennials provide a great value to the consumer, as they can survive for many years in the landscape. Perennials consist of numerous plant species with a wide range of foliage and flower characteristics, and these plants have adapted to nearly every growing environment or landscape situation. With increased interest and demand from consumers, more commercial growers have added perennials to their production plans to capitalize on the opportunities at hand.

The current demand for perennials is strong, and this market continues to expand. Unfortunately, the economy, increased competition, and the costs of energy have created a tough business environment for many growers to maintain profitability. Perennial growers essentially have to produce the highest quality crops as efficiently as possible to retain a competitive advantage over other producers. *Perennial Solutions* focuses on the necessary cultural information and other topics specific to commercially producing perennials to help growers successfully grow these crops. The information provided should be used as a guideline and often needs to be modified to fit the production system. With this information, growers can increase their success and take advantage of the numerous opportunities available with perennial crops.

Although the focus of this book is commercial perennial production, it is still important to review the general concepts of what perennials are. As producers, we need to understand where perennials fit in the marketplace and the general terminology used by both our customers and the end users, the consumers. This general knowledge lays the foundation for much of the more detailed cultural information found in the upcoming chapters.

Commercially, there are several categories of plants that are marketed. Each category has its own distinguishing characteristics, such as being classified as a tree, a bedding plant, or a tropical houseplant. These are general, roughly defined groupings of plants. In fact, many of these groupings do not have clear boundaries, with many plants falling into several categories. The probability of being promoted across several markets increases the opportunity for some of these plants to be produced in larger quantities or be grown during different times of the year, which allows for an extended growing season and perhaps more turns of the production space. Basically, regardless of how they are marketed, plants belong to one of three groups: annuals, biennials, and perennials.

Annuals are non-woody (herbaceous) plants that live for only one year, or a single growing season. Typically, annuals are produced from seed. Whether they are commercially produced or grown naturally, the life cycle of an annual usually begins in the spring of each year. Most bedding plants are classified as annuals. Examples of bedding plants include geraniums, impatiens, marigolds, and petunias. Vegetative annuals are currently making waves in the marketplace. Many of these vegetative varieties are actually perennials in their natural environments and are marketed as annuals in colder parts of the world, where they grow or perform as annuals.

Biennials are plants that mature and die after two years. These plants usually remain vegetative during the first growing season, overwinter, resume growth in the spring, and flower sometime during the second growing season. After biennials flower and produce seed, their life cycle is complete and they die. Many perennials are actually classified as biennials. Many of the biennials that are often classified and marketed as perennials actually self-sow, or reseed themselves, giving the effect of being perennial.

Perennials are typically defined as plants that live for three or more years. They include woody plants, such as trees and shrubs, and a diverse array of herbaceous plant species. A herbaceous plant usually dies back to the ground in the fall or winter and resumes growth again in the spring. There is some crossover in plant categories within this group. For example, several varieties of woody, shrub-type plants, such as *Buddleia,* are loosely produced and marketed as perennials. Or some perennial plants that grow well in northern climates cannot tolerate the heat and humidity of southern climates and must be grown as annuals. Today there are at least five thousand species of perennials in commercial production. Unless otherwise specified, the remainder of this book will focus on these herbaceous, non-woody, perennial plants.

USDA Hardiness Zones

A useful way to classify perennials is to group them into regions, or zones, where they can survive the environmental elements. Hardiness is a plant's ability to survive in a specific geographic area, based on the ability to withstand minimum winter temperatures. The U.S. Department of Agriculture (USDA) has collected environmental information from five thousand recording stations since 1990. Using this information, they have updated the previous standard Hardiness Map, which was originally developed in 1960. The new hardiness zones were updated and released during the later part of 2001. The plant hardiness zones divide the United States and Canada into 11 areas (zones) based on a 10° F (approx.

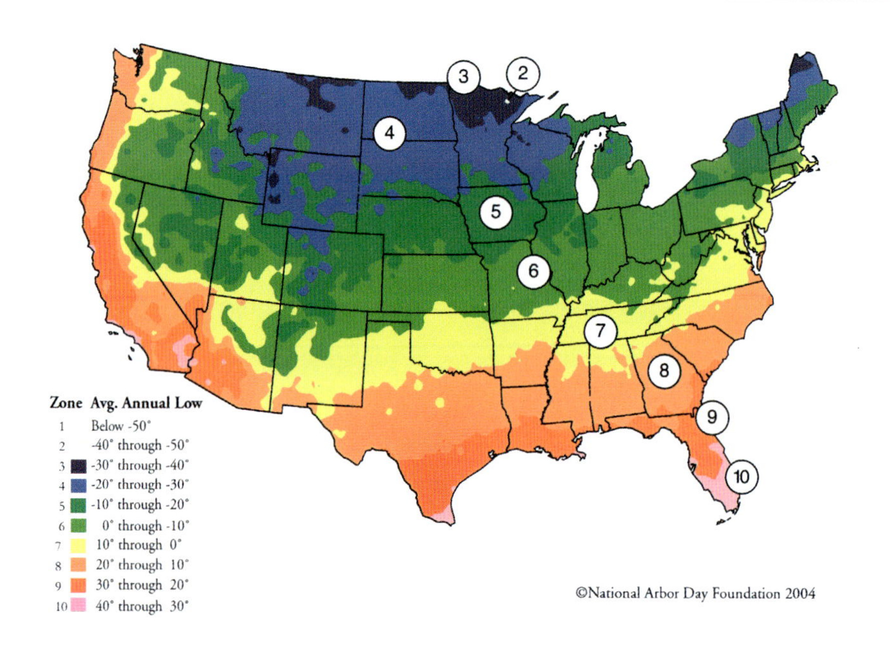

Figure 1.1. **Hardiness Zone Map**

6° C) difference in the average annual minimum temperature. For example, the average annual minimum temperatures in Zone 7 is 0–10° F (-18–-12°C); Zone 6 is –10–0° F (-23–-18° C), etc. These zones roughly show a plant's ability to survive cold temperatures and should be used only as a guide. In fact, the precise hardiness of many perennials has not been determined.

Certainly, there are exceptions and borderline cases, but generally the zones recommended for any given plant are the zones where that plant is expected to grow and survive. There are many factors that affect the hardiness of perennials. Tolerance to cold, heat, moisture, soil type, wind, plant size and age, light intensity, and other conditions might affect the viability of certain plant species or even specific cultivars. While a plant may survive in one area, for a number of reasons it may not be as successful in another. Again, the hardiness zones represent average conditions; plant injury and death may still occur because of extreme weather conditions or other factors. With good culture and winter protection, many perennials can survive in somewhat warmer or colder zones than listed.

Most perennial suppliers list the hardiness zones in their catalogs. It is not uncommon to find slight differences in the hardiness zones between suppliers. These differences likely indicate that some suppliers have updated their catalogs using the new hardiness zones, while others are using the hardiness zones developed in 1960. And as new cultivars hit the marketplace, sometimes there is not adequate time to conduct full hardiness trials (usually conducted by universities) across the country; therefore, growers distribute the information they have available at that time.

Take care when interpreting hardiness zone maps or when reading a plant's recommended zones. If you live in Zone 5 and a plant in your landscape is listed as a Zone 5 plant, that does not guarantee that that particular plant will survive an unusually cold winter. If you have Zone 3 or 4 plants in your Zone 5 landscape, chances are high that they will survive an unusually cold year. The same is true for the higher recommended zones. A Zone 6 plant might be able to survive in Zone 5 under certain conditions, but if extreme weather happens, chances are high that plant survival will be in jeopardy. Also keep in mind that these hardiness zones were derived from plants growing in the ground, not in containers above ground. Containerized plants will be more sensitive to cold temperatures and need to be treated differently.

For more information on the USDA Hardiness Zone Map, including a search by ZIP Code, visit the National Arbor Day Society's Web site, www.arborday.org.

AHS Heat Zones

In 1997, under the direction of Dr. Marc Cathey, the American Horticulture Society (AHS) introduced the Plant Heat-Zone Map. Up until then, growers and consumers used the USDA Plant Hardiness Zone Map (described above) to identify the coldest temperature zone a specific plant could survive in, based on minimum winter temperatures. The AHS Plant Heat-Zone Map determines a plant's ability to survive a specific location based on heat rather than cold.

The AHS Plant Heat-Zone Map uses summer temperatures to demonstrate a specific plant's ability to survive or handle the stress of high summer temperatures. The map has twelve zones. Each zone is derived from the number of "heat days," or the number of days the temperatures are 86° F (30° C) or higher. This temperature is significant because many plant proteins become damaged at temperatures above 86° F (30° C). Each zone is identified by the average number of "heat days" it experiences in a year.

Currently, heat zone ratings are expressed with the highest (or warmest) heat zones listed first, followed by the lowest. For example, the heat zone rating for *Platycodon* is listed as follows: AHS Heat Zone: 8–1. This basically illustrates that *Platycodon* can tolerate and survive the summer heat in the zones 8 through 1. With some plants, lower heat zones

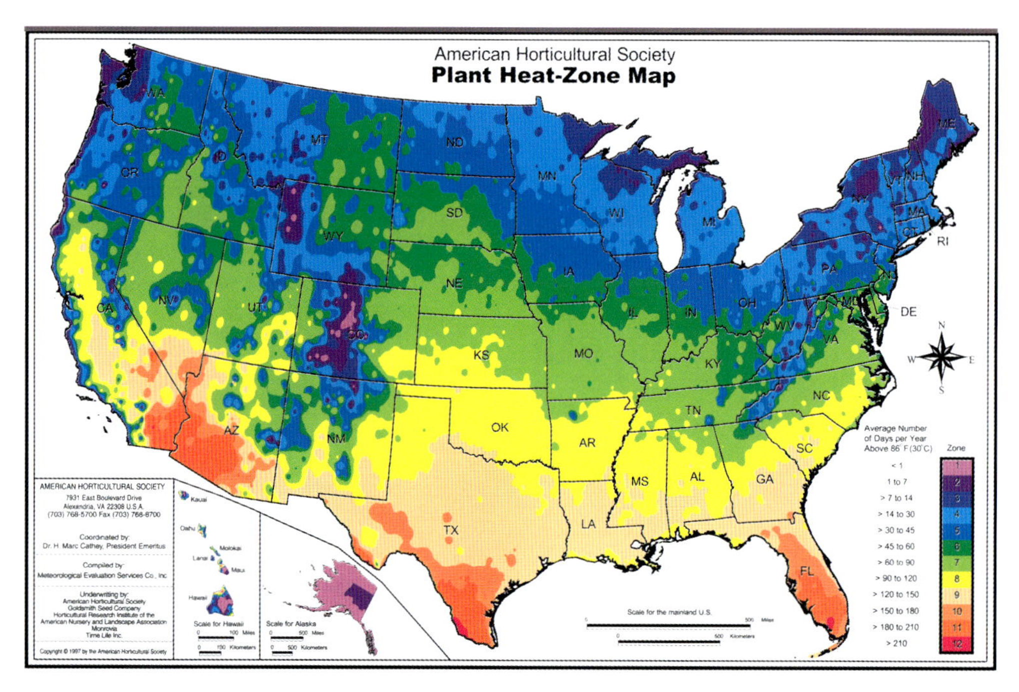

Figure 1.2. Heat Zone Map.

may not be listed. *Gaura,* for example, is listed as AHS Heat Zone: 9–5. In Zones 1 to 4, the growing season may be too short or the summer temperatures not high enough to produce a flowering plant.

There are several factors that can skew the accuracy of a particular heat zone. The factors affecting heat zones include water, light, day length, oxygen, air movement, and pH. Of these factors, water is the most critical. Plants lacking water for even short durations can become more susceptible to injury from heat; heat injury is always linked to an insufficient amount of water being available to the plant.

The Heat Zone Map was not developed to replace the Hardiness Zone Map, but to supplement it, allowing growers to choose plants based on their ability to survive both temperature extremes in their region. Growers should first determine the USDA hardiness zone they live in, and then determine the AHS heat zone. Now using both zone maps, growers can better select plant varieties suitable to both the winter and summer extremes at their location.

The AHS Heat-Zone Map currently is not as widely used as the USDA Hardiness Zone Map. With the relatively recent introduction of this map, it may take several more years for growers to become familiar with it. In time, the AHS Heat Zone Map should prove to be a useful tool for growers and become more widely used. Currently, until further trials are conducted, it is not possible to provide heat zone listings for every perennial being grown. Many perennial plant suppliers are beginning to provide the heat zones in the perennial listings of their catalogs.

For more information on the AHS Heat-Zone Map, visit www.ahs.org.

How Are Perennials Named?

Perennials usually are listed and marketed by their botanical (scientific) names. Scientific plant names are derived from the taxonomic hierarchy using a binomial system of nomenclature. The first portion of the scientific name is called the genus. A genus is a closely related grouping of plants consisting of one or more species. The second portion of a scientific name is referred to as the species (actually called a specific epithet). The species is a grouping of plants that are capable of interbreeding and have distinctly different characteristics from other species in the same genus. Many perennials have a cultivar name listed after the genus and species. A cultivar (cultivated variety) is the name given to plants that have resulted from hybridization, random mutation, or plant selection, and retain their distinguishing characteristics when reproduced.

The following is an example of the correct taxonomic hierarchy for the perennial *Sedum spectibile* 'Brilliant'

Kingdom	Plant
Division	Spermatophyta
Class	Angiosperm
Subclass	Dicotyledon
Order	Rosales
Family	Crassulaceae
Genus	*Sedum*
Species	*spectibile*
Cultivar	'Brilliant'

Each tier of the hierarchy essentially classifies plants with similar growth and flowering characteristics together. The genus, species, and cultivar are generally the only names used to identify perennials in our trade.

The general public usually knows perennials by their common names. These common names are generally easier to remember, pronounce, and use. These user-friendly names are derived from various aspects of the plant including the leaves, the flower, the history, the medicinal properties, the origin, the discoverer, and other plant uses. Quite frequently a plant will have several common names. For example, the groundcover *Vinca minor* is often referred to as myrtle or periwinkle. It is also common for two different perennials to share the same common name. For instance, plants in the genus *Rudbeckia* and *Echinacea* are both often referred to as coneflowers. It is also not uncommon for a perennial to have different common names in different parts of the country. For these reasons, it is not practical to rely on the use of common names when classifying plants. Most distributors of perennials include both the common and scientific names when marketing their perennials.

Where Do They Grow?

Most perennial references indicate the amount of light each perennial requires in the landscape by describing the amount of sunlight or shade it can tolerate. There are usually various ways of describing the amount of sunlight, many of which have the same meaning. For example, one meteorologist might say, "Today will be partly cloudy," and another one may declare, "Today will be mostly sunny." For the most part, aren't they both giving the same forecast? With sunlight, there are generally three different categories: full sun, partial shade, and full shade. Perennials that can be grown under full sun do not require any shade whatsoever during the day. Some plants require partial shade; these plants generally require shade or indirect sunlight during some portion of the day. A full shade requirement means that plants need to be grown with no direct sunlight.

Of course, it is not possible to classify the exposure of sun in every possible landscape situation. There often is a combination of conditions throughout the day. For example, it may be possible to have a location that receives morning shade, full sun in the afternoon, partial shade in the late afternoon, and full sun again throughout the rest of the evening. Most plants can tolerate conditions slightly different from the recommendations listed in various resources. A full sun plant can tolerate partial shade to some degree; a full shade plant can tolerate some partial shade conditions; and a partial shade plant can tolerate some full sun or some full shade. It is not uncommon to see a plant listed as both partial shade and full shade. This indicates that the plant may be able to tolerate small amounts of full sun exposure but will perform best under shady conditions. Conversely, a plant may be listed as both partial shade and full sun. This plant will perform best when grown under sunny conditions but will tolerate some shade.

The Past, Present, and Future of Perennials

Perennials are the fastest growing sector in floriculture. It is still difficult to measure what the growth of the perennial industry has been in recent years. The USDA just recently separated annuals from perennials in their Floriculture Crop Summary surveys, so it is difficult to precisely establish the real trends at this time. According to the 1998 Census of Horticultural Specialties, perennials consisted of 26.6% of the entire bedding/garden plant category. This means that roughly $1 of every $4 spent in the bedding/garden category is spent on perennials.

The interest to produce and market perennials is definitely high. Each year more operations are looking into or are including perennials in their production cycles, and more research is conducted to learn about perennial breeding, culture, and marketing. In many ways, the perennial industry is young; it can be compared to where the bedding plant industry was twenty years ago. But it will not take twenty years to catch up, as far as knowledge and production practices are concerned. Sales of perennials may never surpass sales of annuals, but there is still room to take more market share away from bedding plants.

Initially, perennials were produced by nurserymen, whose main crops were trees and shrubs. These crops historically have not required sophisticated facilities or many specific or specialized practices. Due to the nature of the crops, controlling inventory, or turning the production space, was not a priority. "If it doesn't sell this year, it can be held over and sold next year" was a general philosophy many nurserymen shared.

Then there was the evolution of the perennial specialist. These perennial producers were often former nurserymen who decided to focus on producing only perennial crops. They made slight changes in their growing practices, providing better facilities and using cultural practices more suitable to these herbaceous crops. But most often they still held onto the nurserymen philosophy, which rarely considered attention to detail or turnover of production space. These specialists did play an important role in bringing perennials to the marketplace and in laying the foundation for the many opportunities available today.

Today, bedding plant growers are moving into perennial production. With more greenhouse operations producing perennials, there is a new standard being set for all perennial producers. One of the main differences is the concept of turning the greenhouse space, each turn adding to the overall profitability of the operation. Greenhouse growers are often able to deliver an entire garden plant product line—including bedding plants, patio containers, hanging baskets, and now perennials—to their customers. This one-stop shopping is something a perennial specialist has not been able to compete with. Many traditional greenhouse companies have developed new marketing programs to promote perennial sales. Greenhouse production practices have improved the overall quality and consistency of the perennials being sold today. With this new focus on marketing, consistency, turning of production space, and quality, the pressure is on all perennial producers to raise the bar, manage their production, and find ways of remaining competitive.

The future of perennials is yet to be written, but it appears to be bright. Refined growing practices, focus on producing flowering plants, improved plant quality, new marketing strategies, and increased competition will all fuel the growth of perennials. Most operations will experience some sort of learning curve. The traditional greenhouse operations will have to learn all of the complexities and special needs of perennials, such as vernalization or the required cold period to induce flowering. Nurserymen and perennial specialists will have to adjust their philosophies and production practices to improve plant quality and to remain competitive in this industry. Successful production of perennials will require a combination of both sound production practices and growing philosophies, creating a hybrid grower.

This manual is intended to help fill the information gap in the perennial industry, help growers produce high quality crops, and smooth the transition between nursery and greenhouse perennial production. Producing plants of any type is a fragile business; things can change immediately for the better or worse. With the complexity of crop production and the seemingly endless amount of information needed to produce consistent, high-quality perennials, it is easy to become overwhelmed or discouraged. When production practices seems too complex, go back to the basics. In many cases, growers are so focused on the specific details of growing perennials; they overlook the general principles necessary for crop production.

Regardless of the size of operation, education levels, or personal goals, remember why you chose this industry: passion. The business side of crop production is necessary, allowing us to make a living and provide for our families. But do not let the need to make a profit or the competitive nature of producing perennials drive who you are or how you function. Instead, allow your passion and desires to provide you with a competitive edge—the profits will follow. Dr. David Viscott, one of the best-known psychiatrists in this country, said, "There is some place where your specialties can shine. Somewhere that difference can be expressed. It's up to you to find it, and you can."

References

Cathey, H. Marc. "Zone Map Heats Up." *Greenhouse Grower.* May 2000.

Perry, Leonard P. *Herbaceous Perennials Production: A Guide From Propagation to Marketing.* NRAES-93. 1998.

United States Department of Agriculture. *1998 Census of Horticultural Specialties.* National Agriculture Statistics.

2

Properties of Growing Mixes

All growers producing plants in containers use some type of growing medium or potting substrate to produce their crops. This medium is the foundation to successful crop development. Ideally, a container mix should have the following characteristics: provides a good balance between air and water to the roots; allows for adequate drainage; provides a base level of nutrients, including acceptable beginning ranges for pH and soluble salts; is free of weeds and diseases; and fits into the grower's crop management style.

Substrates are used by many, but understood by only a few. If the production of perennials is so dependent on the growing medium, shouldn't growers learn all they can about this aspect of crop production? Or should I say, if growers had a better understanding of potting substrates, could they grow higher quality crops, increase sales, and acquire greater profits?

A potting mix provides a structure for the plant to anchor itself with its root system, to hold water and nutrients necessary for plant growth, and to allow air exchange to and from the roots. Besides providing anchorage, the growing medium must also have sufficient weight so plants do not fall over from lack of support. The root medium should have an adequate water-holding capacity, allowing for easy absorption of water by the roots. The growing mix should also have a good cation exchange capacity (CEC), or ability to hold nutrients, making them available for roots to absorb. The ability to provide air to the root zone is important, as the roots are actively respiring, taking in oxygen and giving off carbon dioxide.

There are a number of components commercial growers use, most often blended together, and each individual component contributes its own unique characteristics and properties to the mix. The components used by individual growers are largely dependent on regional availability and cost. Each component affects the mix's physical and biological characteristics in various ways. Rather than focusing on the components of the blend individually, growers should focus on the desired properties of the medium as a whole. Most commonly, components are used to create a better root zone environment to support plant growth and development. The basic properties growers are looking for are adequate porosity, water holding ability, and nutrient-holding ability.

Media Components

Peat

Many greenhouse operations utilize Canadian sphagnum peat moss as a major component of their potting medium. Peat is formed by the partial decomposition of plant materials in acidic, waterlogged environments. Sphagnum peat moss is harvested from peat bogs located throughout Canada. There will be some variability in the physical and chemical properties of Canadian sphagnum peat moss, as it is often harvested from different locations. Peat moss is lightweight and available in a variety of grades ranging from fine to course. Compared to other components, peat is relatively expensive.

Canadian sphagnum peat moss typically has high water- and nutrient-holding capacities, a relatively low bulk density, low soluble salts, and pH ranging from 3.0–5.0. Peat moss is mostly free of insects, weed seeds, and disease organisms. The low pH provides some level of suppression of disease-causing pathogens. It is usually necessary to add dolomitic limestone to mixes containing large percentages of peat in order to raise and maintain the pH levels during crop production.

The water-holding capacity varies with the age of the peat. For example, relatively undecomposed fibrous peat moss can hold twenty to thirty times its own weight, whereas peat moss that is greatly decomposed only holds four to eight times its weight of water. Due to its waxy cutin layer, once peat moss dries out, it will actually repel water and is, therefore, difficult to rewet. This characteristic has brought about the need to use wetting agents, which help the mixes to absorb and evenly distribute water. Once the moisture level of peat moss becomes less than 30 percent, it does not wet well, even when a wetting agent has been used.

Bark

Currently, growers are using a lot of softwood barks as a major component of their growing mixes. These softwood barks are usually derived from pine, fir, spruce, or hemlock. Pine bark is mostly utilized in the eastern half of the United States, while the western half of the U.S. mostly uses barks derived from fir. Bark from pines and firs are easily processed because the bark is large and platy (flat and chunky), which allows for easy separation from the tree. Hardwood barks are used to a lesser extent because they tend to contain excessive amounts of certain minor elements and other potentially toxic compounds.

Bark is widely available, lightweight, and has a relatively low cost. In fact, it is the second most utilized component of container mixes, second only to sphagnum peat moss. Barks add weight and physical stability to a mix. They do not compress or shrink significantly in a pot over time, making it a good component for long term or outdoor crops. For use in containers, such as 1 gal. (3.8 L) perennial production, the particle size is usually less than half an inch. Another potential benefit is the fact that composted bark is biologically active and will provide some disease suppression to such pathogens as *Fusarium*, *Pythium*, *Phytophthora*, and *Rhizoctonia*.

The most widely used practice is to compost the bark in a controlled manner, which involves frequently turning the piles, adding appropriate amounts of nitrogen, monitoring temperatures, and adding moisture when necessary. The composting process produces a usable, stable product in less time than does the aging process. Aging bark is a passive process, which simply entails storing the bark in piles until it becomes biologically stable. A drawback of using aged barks is the potential to receive bark that has gone anaerobic and may contain acetic acid, which is toxic to plants.

If either the aging or composting process is incomplete, saprophytic fungi will often become present in the container, which may lead to the formation of a layer, or mat, that is difficult for irrigation water to pass through. Whether the bark is aged or composted, the key to success is stability. Typically, stable barks have a low carbon-nitrogen ratio (C:N ratio) of 30:1, a pH range between 4.0–6.5, and an electrical conductivity (EC) lower than one.

Bark most often increases the bulk density, slightly increases the air space, and decreases the water-holding capacity of the mix. Barks are usually associated with low relative soluble salts, and a high cation exchange capacity. Compared to sphagnum peat moss, bark is less resistant to physical breakdown, less acidic, provides better drainage, and has some nutrient-holding capacity. These characteristics make the use of bark compelling for many greenhouse and nursery operations.

As the percentage of bark within a mix increases, there is a greater demand for nitrogen, due to the nitrogen draw from the bark. Barks that are not properly composted or aged may contain excessive amounts of fresh (green) bark, which will tie up the nitrogen within the mix, which ultimately may lead to nitrogen deficiency in the plants being produced. This is commonly referred to as nitrogen immobilization, or drawdown, and occurs when the microbes responsible for the decomposition of the bark compete with the plants for available nitrogen within the growing mix. Even the best aged or composted barks will cause some nitrogen immobilization. Most often, the company composting the bark will usually add a nitrogen source such as Nitroform (urea formaldehyde) into the mix to satisfy the bark's need for nitrogen.

Perlite

Perlite is made from naturally occurring siliceous volcanic rock that is expanded in a furnace, forming a white, lightweight, sterile aggregate. It is used to increase air space and provide drainage in the medium; it has no water-holding ability. The pH of perlite is near neutral, and the cation exchange capacity is negligible. Being essentially inert, perlite has virtually no effect on the pH or fertility of a growing mix. There are a number of grades of perlite available, the most common being medium, or horticultural grade. Unless it is mishandled prior to filling the containers, perlite is stable and does not decompose readily. Compared to other commonly used components, perlite is relatively expensive. Although not commonly observed, poor-quality perlite can contain potentially harmful levels of fluoride, which can be toxic to some plants at high pH levels.

Vermiculite

Another commonly used component is vermiculite. This aggregate is made from a micaceous mineral that when heated to approximately 1,800° F (982° C) expands to form a lightweight material. It provides a growing mix with air space, water-holding capacity, and a high cation exchange capacity. Vermiculite tends to be slightly basic to neutral and provides small amounts of potassium, magnesium, and calcium to the mix. It is a fragile material that breaks down with improper handling or over time due to irrigation practices and moisture levels within the medium. Compression of moist vermiculite from handling or weight compression within the container causes the expanded particle to collapse. As it breaks down or becomes compacted when wet, the desirable porous structure is destroyed, resulting in a reduction in drainage. Like perlite, it is also relatively expensive and available in a variety of grades.

Coir

In recent years, growers have taken great interest in using coconut coir as a component in their growing mixes. The coir is the "dust" or byproduct from the coconut fiber industry. It originates and is produced in several tropical regions of the world, including Sri Lanka, the Philippines, Indonesia, Mexico, and South America. Coir is used in a similar manner as Canadian sphagnum peat moss and looks like peat, only it is more granular in nature and has slightly different characteristics.

Coir has very high air- and water-holding capacities. The pH range is 5.2–6.5, and the soluble salts are slightly higher than both peat and bark. It has lower nutrient-holding capacity than peat. Coir does not have any waxy cutin; therefore, it does not repel water, as peat moss does, allowing it to wet and rewet easier. Due to the lignins in coir, it is more resistant to microbial breakdown, there is less shrinkage and loss of aeration over time, and it is resistant to degradation from improper handling procedures. Some sources of coir may contain high levels of sodium and chloride. It is essential that processors properly age and leach coir to remove these elements before using it in growing mixes. Growers using coir-based mixes find that they need to modify their fertility and irrigation programs to compensate for the coir's ability to hold water and inability to hold nutrients. Growing mixes usually do not contain more than 10 to 15% coconut coir.

Rice hulls

Many growers are incorporating various forms of rice hulls into their growing media. Some growers use a hammer-milled rice hull, which is chopped, while others are using the unchopped hull. Most growers use an aged or composted form of rice hull. Fresh hulls are occasionally used but increase the likelihood of weed seeds (rice seedlings). The primary function of rice hulls is to provide drainage, and they are often used as a substitute for sphagnum peat and/or bark. Other characteristics include low water-holding capacity, low cation exchange capacity, and they have a pH of 5.5–7.5. Rice hulls do not decompose readily; therefore, they do not cause shrinkage of the medium over time.

Peanut hulls

Peanut hulls are a byproduct of the shelled peanut industry and are becoming a popular component in container mixes. The hulls are generally hammer-milled, or chopped, to produce a range of particle sizes, and they are composted before being used as a media component. Peanut hulls are used as a replacement for pine bark, which is becoming more difficult to obtain each year. Composted hulls are not currently used to completely replace bark, but as a substitute for up to 50% of the bark in the growing medium. They supply nitrogen and potassium, among other nutrients, into the growing medium. Hulls provide a mix with increased total porosity without decreasing the water-holding capacity, resulting in increased air space with a lower bulk density.

Other components

Over the years, there have been several different types of aggregates growers have incorporated into their mixes, many of which are still being used today. Some of the products used are materials that are only available in certain regions or in limited supplies.

Many growers blend sand into their container mixes to improve drainage and add weight to the mix. Sand is available in a variety of grades, and most often clean (washed) sand or course builders' sand is used. Sand is heavy, and when used in large proportions or if fine grades are used, it may plug up the drainage space of a mix. Typically, sand accounts for no more than 20% of a production mix. Other characteristics include a low water-holding capacity, good drainage, low to moderate amounts of aeration, and negligible cation exchange capacity. It does not decompose readily, and depending on the source, it may have a high pH. Besides the potential to plug up pore spaces, sand can also bring plant pathogens into the medium.

Composted yard wastes are also commonly used as a component of growing mixes. Yard compost is inexpensive and provides good water-holding abilities with moderate air space. Depending on the source of the compost and seasonal variations, there is great variability with the quality in regard to nutrients, pH, and particle size. Mixes containing large percentages of composted yard waste often tend to shrink over time as the material continues to decompose and the fines (small particles) settle, causing the mix to become plugged. This settling of the mix causes the water-holding ability to increase and the aeration to decrease, both of which are undesirable characteristics. Compost has also been shown to suppress diseases such as *Pythium* and *Phytophthora*.

Components such as mineral soil, topsoil, and sand are nonsterile and must be pasteurized prior to blending in a container medium. These nonsterile components may contain unwanted disease pathogens or herbicides, which may cause damage to a crop. Using these components, growers can establish pH stability, add micronutrients, and increase the nutrient-holding capacity of the mix. Using large percentages of these aggregates within a mix also reduces the air space and increases the water-holding abilities of the medium. This characteristic could lead to slower plant development and overly wet growing conditions, making the plant more susceptible to disease problems. Growers using mineral soil or topsoil as a component in a container medium typically use no more than 10%; higher levels tend to impede drainage and affect crop development.

Other components often used in container mixes include but are not limited to calcined clay, hydrophilic gels, pumice, sawdust, polystyrene beads, and rock wool. Growers are constantly searching for new and affordable aggregates to utilize in their production systems. Future components will most likely utilize excessive raw materials from certain industries or possibly consist of completely synthetic growing substrates. The characteristics of future components must be predictable, consistent, and cost effective.

Blending Your Own Media

Many growers purchase individual components, blending them together at different proportions to suit their production needs. For the experienced grower, there certainly can be several advantages, such as the ability to change the medium to fit the crop's production requirements. But more often, it is the savings that can be achieved (typically 20–40%) that leads many growers to blend their own container mixes. However, although it may seem financially attractive, it takes a great amount of technical savvy to properly manage the raw materials and the blending process. Don't get me wrong—many growers are very successful at mixing their own media, but more growers just "get by." Inadequate handling of materials or inconsistent blending is often to blame for insufficient crop development and, possibly, crop losses.

Pad-and-skid mixing is one of the oldest and most widely used methods of blending container mixes. It is accomplished by first layering the desired components on a concrete pad, then using a front-end loader, turning the materials over and over, causing the raw materials to become mixed with each other. This type of mixing procedure is best used when growers are only blending two or three components together. It does not provide very consistent results when blending small amounts of items such as fertilizers, where perhaps only 1 lb. (454 g) of product needs to be incorporated in each cubic yard (0.76 m^3) of medium. Using these small-volume amendments make it next to impossible to achieve uniform distribution and will cause each batch to vary slightly.

Batch mixers, which are designed to blend soils, do an adequate job, provided they are used properly. There are several types: ribbon blenders, tumbling devices, and even converted cement mixers. Premeasured volumes of the individual components are added into the hoppers or catch basins, and then they are mechanically mixed together. This type of mixing usually provides consistent results and is easy to do.

Many do-it-yourself growers use cement mixers or a tractor and bucket. These tools may appear to uniformly blend the components, but most often they do not blend the media uniformly enough, especially when filling small containers. With these methods, it is very difficult, nearly impossible, to evenly blend the components used in small proportions. There is also a greater likelihood for components to become damaged during the mixing process.

The most uniform and consistent results are achieved when using continuous mixing systems. These systems entail loading the raw materials and amendments into individual material hoppers or feeder bins and then layering the components onto a moving conveyor belt in known volumes. After the components are layered, they become blended either by a tumbling tube or a mixing head. This type of mixing equipment provides very quick, consistent, and uniform distribution of the aggregates. Depending on the style and the manufacturer, continuous mixing systems have the capacity to blend 10–400 cu. yd. (7.6–306 m^3) of medium per hour.

Mixing components for extended periods breaks down their structure, forming many small particles called fines. These fines clog the air pores within a medium, reducing the water drainage and limiting the air exchange for the roots. This reduction in oxygen to the roots will cause root and shoot development to slow down. While filling the containers, it is also important to avoid compaction. Compaction is caused by compressing (forcing) excess

Figure 2.1. **A continuous mixing system is the most uniform option available.**

medium into a container. Like mixes with too many fines, a compacted medium has less air space and increased water-holding ability, often causing overly wet growing conditions and inadequate plant development.

Be sure when blending the components that they are mixed thoroughly. The physical properties such as the total porosity, drainage, and the water-holding capacity can vary significantly when a medium is mixed improperly. These changes in physical properties will often result in uneven drying of the medium in the greenhouse, making the management of irrigation practices more difficult. Thorough mixing is also necessary to ensure uniform distribution of amendments such as fertilizers or lime. For example, uneven mixing of lime could result in varying pH levels among the containers.

Usually the decision to blend your own medium has more to do with controlling costs than it does with obtaining a higher quality, more consistent growing medium. This decision really depends on grower preference and the degree of control desired. With control of the media comes complete responsibility. Blending your own media should be a function of preference and specific needs and should never be based on price alone.

Growers who blend their own mixes often overlook quality control and the additional, less obvious costs associated with blending. In addition to the raw materials, the costs may include: storage facilities for the components, expensive equipment to provide accuracy, overhead, labor—including a technical manager—nutrients, and wetting agents. There is also a cost associated with the quality of the product that is made. If the mix is inconsistent, uneven quality and plant losses will definitely reduce the value of the crops being produced. With poor mix quality, even with just one batch, the potential reduction in value from crop losses and reduced quality might cost more than the savings achieved by mixing your own. If you are currently mixing your own potting mixes or considering doing so, do not cut corners and focus on making quality control your priority.

Storage of Container Media

You should use container mixes as soon as possible after they are blended or received. Container mixes stored longer than three months often experience changes that are not advantageous to crop production. The longer a mix is in storage, the greater the tendency is for it to dry out, making it more difficult—seemingly impossible—to rewet after the containers are filled. Storing bulk materials for extended periods often leads to contamination from diseases, insects, and weed seeds. Prepackaged mixes may develop a slime mold between the growing mix and the plastic bag, where condensation usually occurs. These molds are saprophytic in nature and do not harm plant development; they usually disappear within a couple weeks.

Storage will also affect the chemical nature of the mix, often causing the pH to rise while the EC and the nitrogen levels fall (EC levels will increase if controlled-release fertilizers have been incorporated). Always test both the chemical and the physical properties of a mix to determine if any changes have occurred during storage and how to best compensate for them.

Pallets of growing media stored outside are not impervious to water, freezing temperatures, or exposure to sunlight, even if they are shrink-wrapped and have pallet caps. These conditions can lead to several undesirable drawbacks: excessive drying, waterlogged conditions with anaerobic pockets, and damaged packaging providing the entry points for insects, pathogens, or weed seeds. Changes to the chemical properties, such as increased salt levels and activation of lime affecting the pH, are also very likely to occur during storage.

Depending on the circumstances, it may be necessary to handle media that have been stored differently than media that are used fresh. It is often necessary to leach the media after planting to remove the excessive salt levels that have built up. In some cases, the fertility programs need to be adjusted to compensate for changes in the pH, wetting agents may need to be reapplied, or fungicide drenches should be made to reduce root zone pathogens.

It is best, if possible, to store growing media or components inside a building or a protected area. If these materials must be stored outside, store them on the original shipping pallets with the shrink-wrap intact. Additional protection can be achieved by placing a tarp over the products. Tarps provide protection from intrusion of water, chemical contaminants, weed seeds, and disease organisms. Bulk materials are best stored on concrete pads or in bunkers to prevent contamination of weed seeds and pathogen spores associated with bare soil. To avoid contamination from runoff, all storage, mixing, and handling areas should be located at higher elevations than the growing areas. When possible, manage the growing media inventories by rotating the existing inventories so that the oldest media is always used first and as soon as possible.

Figure 2.2. **Outdoor storage of mini-bulk bags at a minimum should have plastic pallet caps and be shrinkwrapped as shown.**

Figure 2.3. **Bulk media storage is best when contained inside of a covered building.**

Handle with Care

The selection of what components to use, and at what ratios, is only part of what makes a mix perform successfully. How a mix is blended and handled has a significant impact on a medium's physical properties. Every time a container mix is handled, there is a slight change in the physical properties, which alters the aeration and the water retention capabilities of the mix. Changes to the physical nature of a growing mix can dramatically impact root development and the efficacy of irrigation and/or fertility programs. Mixers, screw conveyors, flat fillers, potting machines, and other handling systems are all commonly used and cause damage to the growing medium.

Most of the damage to the mix is a result of the breakdown of larger particles into smaller ones. This breakdown results in an increase in the mix's water-holding abilities and a decrease in the available air space. Certain components, such as perlite and vermiculite, break down more easily than other components, such as peat moss and barks. Growers should recognize that any handling of growing mixes will cause slight damage to the physical properties, but excessive handling may cause permanent and even detrimental injury to the mix, most likely decreasing plant development and quality.

Prior to filling containers or plug flats, it is very important to obtain the proper moisture content of the medium. Adding water to dry components, such as peat, causes them to hydrate and swell, increasing aeration by reducing the particles' tendency to intertwine with one another. Small containers contain less air naturally by their size and height. (This principle is described in the porosity section of this chapter.) The smaller the container size, the more water that should be added prior to filling to increase the amount of air in the medium. I recommend the moisture content of the growing mix should be approximately 50–60% moisture when filling the containers just prior to planting.

If the containers are filled with a dry medium, it is often very difficult to add water, or to wet the medium, as the water will often channel through the medium and drain out the bottom of the pot. The moisture content is important to maximize the pore space of the medium, reduce shrinkage, and allow any amendments to react within the medium. If your mix has an excessive amount of shrinkage after the initial watering, increasing the moisture content of the mix before filling will help to reduce it.

When filling containers, they should be filled firmly with little compaction. Avoid excessive firming or compacting of the growing medium during transplanting, as this alters the large pore spaces, reducing the ability of the mix to hold air. Containers should be lightly filled, with the excess medium brushed off. Compaction of the growing medium can severely reduce or eliminate air space. Loosely filled containers will also lead to disappointing performance of your plants. Initially, loosely filled containers will provide an excessive amount of drainage and settling of the substrate in the containers. This settling will result in a change of the physical properties of the mix by adversely changing its ability to hold air and water.

Try to avoid stacking filled pots on top of each other, which will compress the medium. This compression causes an undesirable reduction of the air-filled pore space and drainage, which results in uneven drying of the medium in the greenhouse. Researchers have found that compaction from stacking trays directly over one another will reduce the air space in the medium by at least 50%. If stacking of the containers is necessary, use dividers between each layer of the containers, or stack them so that the rims of the pots are supporting the weight.

Many growers pre-fill their pots in advance of starting the crop. In many cases, pre-filling is done to save time when the planting does occur. It is an activity that can be done between shipping orders or to keep employees productive and to help get ahead of the next crop cycle. When containers are filled far in advance, there is a tendency for them to become extremely dry and resistant to rewetting. To reduce the difficulty of rewetting, keep the medium moist with an occasional light watering, cover the filled pots, add a wetting agent to the water when rewetting, and irrigate them thoroughly.

Figure 2.4. **Prefilled pots should be stacked so no pressure is on the potting mix.**

Understanding Chemical Properties

pH

The pH of the soil solution has a great impact on the plant roots' ability to acquire and utilize fertilizer nutrients. The pH of the soil is a measure of the hydrogen ion (H^+) concentration, which indicates the relative acidity or basicity of the soil solution. It is measured on a logarithmic scale ranging from 0 to 14, with values of 0 to <7 being acidic, 7 being neutral, and >7 to 14 being basic. Every perennial species has its own optimum pH range suitable for plant growth. Plants are able to tolerate a range in pH values and slight changes in the soil solution. Extreme changes in the pH, or extended periods of time outside of the optimum pH range, can have negative effects on plant growth.

Most perennials perform well at 5.5–6.5 pH. Some require slightly higher or lower pH values to maintain optimum growth. Greenhouse growers have adopted the practice of placing their plants into pH groups, using a separate potting medium, starter charge, or fertility program for each group of plants. The pH of the growing medium is influenced by several factors: the medium's components, the type and amount of agricultural limestone added, the alkalinity of the irrigation supply, fertilizer type, and the root activity.

Depending on the components used in a growing medium and the production needs of the grower, it is a common practice to incorporate lime into the mix, which acts as a pH buffer and a source of calcium and/or magnesium. Dolomitic lime (calcium and magnesium carbonate) and calcitic lime (calcium carbonate) are both commonly incorporated in container mixes to increase the pH. Dolomitic lime is more commonly used and provides a longer-term effect on the medium's pH and also provides a source of magnesium for plant uptake. Calcitic lime tends to increase the medium's pH faster than dolomitic lime does, but it provides a shorter-term effect. The particle size of the lime greatly affects how quickly the medium's pH will change. Lime with a smaller particle size will have a more immediate, but shorter term, effect on the media pH as compared to a lime source with a larger particle size.

Dolomitic limestone comes in numerous grades and particle size distributions. Both have a great impact on how quickly the pH changes and how long the pH levels can be maintained. The finer grades tend to raise the pH quickly but do not provide much residual control. Courser grades do not raise the pH quickly, but they remain active for extended periods, providing some residual activity. Different grades of limestone require different lengths of time (from twenty-four hours to seven days) before they react and raise the pH of the potting medium to the desired level.

Since dolomitic limestone can take up to seven days to react, many growers incorporate a small amount (1–2 lb. per cubic yard of mix [454–907 g/0.76 m^3]) of hydrated lime (calcium hydroxide) to raise the pH quickly before the dolomitic limestone becomes active. Hydrated lime is very water soluble, which allows it to raise the pH quickly, but it does not last a long time because it usually gets leached from the soil profile.

To determine the amount of dolomitic lime to add to a potting mix, growers must take into consideration the initial pH of the medium's components, the crops being grown, the alkalinity of the water source, and the fertility program.

The best method to determine the amount of dolomitic limestone to add is to mix all of the components and starter charges, except for the limestone, and have an institution or private laboratory test the pH. Most growers will need one or two crop cycles to identify and implement the necessary changes to get the mix the way they need it.

Growers can determine the quantity of dolomitic limestone to add by mixing several batches of media, each with a different limestone rate. Fill pots with each batch and label them well to avoid confusing them with one another. Water the media with distilled water and measure the pH after five to seven days. After recording the results, determine if additional batches need to be mixed to find the rate necessary to achieve the desired pH range. The general rule of thumb is to incorporate 1 lb. dolomitic limestone per cubic yard (454 g/0.76 m^3) to raise the medium's pH by 0.1 units. If a grower wishes to raise the pH from 5.0 to 5.5, that would be 0.5 pH unit increase and would require the incorporation of 5 lb. (2.3 kg) of dolomitic limestone per cubic yard. Generally, growers and media suppliers incorporate 5–15 lb. of dolomitic limestone per cubic yard (2.3–6.8 kg/0.76 m^3) of mix.

Some growers find that with the components they are using, the crops they are growing, and their water quality, they do not need to add limestone to raise the pH of the potting media. These growers often would like to include a source of calcium and magnesium that does not change the pH. The calcium source they typically add is gypsum (calcium sulfate). The rate of gypsum incorporated varies by the grower; a commonly used rate is 5 lb. per cubic yard (2.3 kg/0.76 m^3). If a starter charge of magnesium is necessary, they will often apply Epsom salts by mixing 0.5 lb. (227 g) in sufficient water to apply to 1 cu. yd. (0.76 m^3). Epsom salts are very water soluble and leach quickly, not providing any residual activity. When applying Epsom salts in this manner, the crop must be monitored closely to determine if reapplication is necessary.

When using calcitic limestone, it may be beneficial to provide a magnesium source, such as Epsom salts, in the medium. This source of magnesium is short term, and additional magnesium should be added during production, either through liquid fertilizer programs or as part of a controlled release fertilizer.

Currently, there are no standards for limestone particle sizes. It is usually best to work with one reputable limestone supplier and find a rate that gives you the desired results. Changing brands will most likely result in a different particle size distribution and will change the affect on the medium's pH. Always purchase ground limestone; avoid purchasing pulverized limestone, as it is ground very fine, raising the pH too quickly and providing no residual results.

Another method of adjusting the pH of the growing medium is to use the characteristics of the water source and the fertility regime. For example, growers can raise the pH of a mix by applying water with high alkalinity or by injecting high-nitrate fertilizers into the irrigation water. Conversely, injecting high ammoniacal fertilizers or an acid, such as phosphoric or sulfuric, into the irrigation water can reduce the pH of a potting mix. Depending on the current pH levels in the medium, a grower can usually maintain the pH within an acceptable range simply by choosing or alternating between different sources of fertilizer.

EC

Growers also need to monitor the soluble salts or the electrical conductivity (EC) of a growing medium. The EC is a measure of all salts and does not necessarily provide a clear image of the nitrate levels or other fertilizers, which are usually beneficial for plant development. It is one of the easiest and quickest measurements to obtain through in- house testing, and it is the most commonly used indicator of the fertility levels in the root zone. The water source, beginning nutrient charge, fertilizers added by incorporation or through liquid feed programs, container size, and the medium's pH all have an effect on a growing mix's EC at any given time.

Upon blending or receiving your medium, it is a good idea to conduct an in-house test of the medium's EC and pH, to determine if they are within acceptable ranges (0.7–2.0 mS/cm) and to establish what the initial watering should consist of, clear water or fertilizer. A high EC at planting will most likely delay plant establishment by inhibiting and stunting root growth; it should be leached out of the medium by watering heavily with clear water. Low EC levels may cause chlorotic foliage, formation of smaller leaves, and increased stem elongation. If a mix is within the acceptable

EC range, then normal watering and fertility procedures should be followed. In-house testing methods are described in more detail in chapter 3, Fertility for Perennials.

Nitrogen

Any organic component in a growing mix continues to break down over time. The decomposition process involves microorganisms that require nitrogen to break down organic products. The nitrogen required for the microorganisms and the decomposition of materials is referred to as nitrogen draw. If the components are stable, there will be little breakdown and a smaller nitrogen draw. Conversely, if the components are less stable, more breaking down will result, requiring a larger nitrogen draw.

A carbon-nitrogen ratio (C:N ratio) of 80:1 provides a mix that is stable enough for container production. Peat moss is more stable and has a better nutrient buffering capacity than bark or coir. Due to the larger nitrogen draw, bark based mixes require 10–15% more fertility than peat-based mixes. It is a common practice of media companies to blend in a nitrogen source, usually urea formaldehyde (Nitroform), to offset the nitrogen draw. Nitroform generally lasts up to six weeks within a medium. This practice reduces the fertility requirement of a bark-based mix, making it more comparable to a peat-based mix.

Growers using bark-based mixes should understand the importance of knowing the C:N ratio, the age of the bark, and how to offset the nitrogen draw. Bark quality is not constant and changes over time, requiring growers to monitor and adjust their practices accordingly.

Nitrogen and potassium

Growers often include a starter charge in the potting mix, consisting of nitrogen and potassium, to provide sufficient fertility to support rapid plant growth. The starter charge used to supply nitrogen and potassium usually consists of calcium nitrate and potassium nitrate at a rate of 1 lb. per cubic yard (454 g/0.76 m^3) of potting mix. At this rate, there will be sufficient amounts of these nutrients available for the first ten to fourteen days of crop production. The rates may vary from grower to grower, depending on the crops grown and the container size. For example, a plug grower would typically reduce the initial starter charge by about 50% from what a finished container grower would begin with. Incorporating nitrogen and potassium in the growing medium is optional. Growers with liquid fertilizer capabilities might not incorporate these elements into the potting mix but would apply balanced liquid fertilizers to the medium during the blending process.

Phosphorus

Phosphorus is one of the macronutrients that growers often add to the growing medium when blending the components together. The two sources of phosphorus usually include superphosphate (0-20-0) or treble superphosphate (0-45-0). Superphosphate is most widely used because it is ground to a particle size (fine powder) that is conducive to incorporating in potting mixes. Treble superphosphate is a course pellet material that is more slowly available in the growing mix and often provides growers with inconsistent results. Superphosphate also contains gypsum, which is a good source of calcium and sulfur.

Growers often incorporate 2.5–4.5 lb. of superphosphate per cubic yard (1.1–2.0 kg/0.76 m^3), depending on the needs of the crop. When using treble superphosphate, they often incorporate 2.25 lb. per cubic yard (1.0 kg/0.76 m^3) plus 1.5 lb. (680 g) of gypsum to provide the calcium and sulfur.

Many growers do not incorporate a phosphorus source into the potting medium prior to potting. Instead, they rely on liquid fertilizer programs to deliver the desired amount of phosphorus to the crop. Some growers have found that even with incorporating phosphorus, it is often necessary to apply it at reduced amounts in their liquid fertilizer programs.

Micronutrients

Growers have found it beneficial to incorporate micronutrients into the potting medium prior to planting. There are a couple of commercial micronutrient products designed for this purpose. These products usually contain boron, copper, iron, manganese, molybdenum, and zinc. These micronutrient products generally use water-soluble micronutrient salts that are either impregnated in clay granules or on fine particles of glass (fritted) to slow the potential leaching and facilitate mixing with the medium. Preplant micronutrient products will supply adequate amounts of micronutrients to last about three to four months.

A few growers apply micronutrients as a post-planting drench, using a water-soluble micronutrient product at a high rate. This type of application can also supply micronutrients for three to four months. There is the possibility of overapplying these elements, which could lead to crop injury caused from micronutrient toxicity, so be careful.

Quality Control

To help maximize the performance of a growing medium, it is beneficial to inspect the medium and conduct some simple on-site testing to determine if there are any potential problems. You are looking for any inconsistencies that might affect the chemical or physical properties of the mix. Tears or holes in the packaging may allow for the intrusion of moisture and outside contaminants. Try to locate the batch number or quality monitoring number provided by the manufacturer, as this can be used for tracking shipments, stock rotation, and resolving potential problems.

Conducting on-site pH and EC tests will provide a quick indication if there are any potential problems regarding the chemical properties. A visual inspection is also recommended to verify the mix looks the same from batch to batch. Besides visual inspection, conducting periodic on-site porosity tests or sending samples to a laboratory for a sieve analysis will provide an indication of potential problems related to the physical properties of the growing mix. I also advise to set aside a small amount of growing medium (1 cu. ft. [28.3 dm^3] should suffice) from each delivery. This medium, called retains, should be kept for at least as long as the crop is being produced in the medium with the corresponding delivery date. Storing retains is a good method of providing proof of soil-related problems to the manufacturer, if cultural problems are detected during crop production.

Many growers reuse old containers for the production of new crops. These old containers often are contaminated with old medium and disease-causing organisms. If you must reuse old containers, sanitize them with a disinfectant such as Greenshield, Triathlon, or a 10% bleach solution. Prior to disinfecting, remove all of the old medium from the containers, as the organic matter will tie up the active ingredients of the sanitizers. Completely immerse the containers in the disinfectant for a minimum of ten minutes. Check the labels of these products to verify the concentration of the solution and the correct soaking time.

Understanding Physical Properties

The physical properties are what give a growing mix the characteristics desirable for plant growth. These characteristics largely include the mix's total porosity (air space, available water-holding capacity, and unavailable water content), the size and distribution of the particles used in a medium, bulk density, and longevity. The physical properties of media will change slightly from batch to batch, even when mixing the components in the same ratios. It is a good idea to periodically conduct in-house or laboratory tests to verify that the physical properties are within acceptable ranges for your operation. Changes in the physical properties will mean that your handling and management practices will have to change in order to get the same level of performance from the mix. Ideally, you want a mix to fit in with your production practices and management style; you should not have to adjust your systems to fit the medium.

The growing mixes used by perennial growers consist of solids, liquid, and air. Between the media aggregates (solids) are pores that are occupied by water and/or air. Porosity is the volume of substrate available for air, water, and roots. The total porosity measures the volume of water and air in the mix that is not occupied by solids. Four factors affect the air and water of a growing medium used for containerized crops: the substrate, the height and shape of container, substrate handling procedures, and watering practices.

Air porosity is the percent volume of medium filled with air after the mix has been saturated and allowed to drain. The amount of air in the growing medium affects plant growth and the irrigation practices. Mixes with low air porosity remain wet for extended periods, suppress plant growth, and increase the likelihood of root rot pathogens. Conversely, growing mixes with high air porosity require frequent irrigation due to their high drainage abilities and are more prone to root injury from insufficient moisture levels. After irrigation, enough water should drain from the container to allow for 15–25% aeration. Growing mixes containing at least 15% air space have fewer problems related to overwatering.

The water retention porosity, often referred to as water-holding capacity, is the percent volume of a medium filled with water after it has been saturated and allowed to drain. Some of the water is available for plants to uptake while some is unavailable. The individual components used in the growing medium have differing characteristics, which affect the mix's water-holding capacity. Organic components, such as peat moss and pine barks, have complex internal structures that hold water, making it unavailable to plant roots. For example, the water-holding capacity of peat moss is made up of approximately 25% unavailable water and 75% available water. After a crop has been irrigated and natural drainage has occurred, the mix should consist of 30–40% by weight of water. Retention of less water will require more frequent irrigation, and holding more water can result in waterlogged conditions.

Both the size and the number of pores in the medium make up its total porosity. Small pores (micropores) hold more water and less air as compared to larger pores (macropores). After a crop is irrigated, the water quickly drains out of the macropores and is replaced by air. Media components that are large, such as pine bark, and spongy components, such as coir and peat moss, tend to pack loosely in the finished mix, creating a large amount of macropores. These large pores allow the downward movement of water through them and are responsible for drainage and aeration, which the roots require for exchanging oxygen and carbon dioxide. Fine media particles, from composts or the fines from bark and peat moss, create micropores, which increases the water holding capacity as water does not readily drain from them. These small pores are also responsible for the upward capillary movement of water through the root zone against the force of gravity. Both pore sizes are necessary and should be present in balanced amounts to provide the proper aeration, drainage, and water holding capacity suitable for plant roots.

It is possible to have two mixes with different pore sizes, similar porosities, but very different water-holding capacities. A mix's ability to hold the right balance of air and water is critical to crop production. All growing mixes contain some fines; however, if they contain excessive amounts of them, the fine particles plug up the macropores, increasing the water-holding capacity, reducing the air porosity, suffocating roots, and inhibiting plant growth.

The particle size of the components greatly affects the pore size, porosity, and water-holding abilities of a mix. Excessive amounts of fines or small particles (usually less than 0.1 in. [2.5 mm]), will dramatically increase the water-holding ability and decrease the air space of a mix. Organic components such as peat, coir, or bark may contain a large percentage of fines, often holding too much water for proper root growth and plant development. Very fine particles, less than .03 in. (7.6 mm), will silt out, or move to the bottom of the container,

Figure 2.5. **Sieve screens (left) are useful in determining the particle sizes of a growing mix.**

blocking drainage, which may lead to water-logged conditions. Course particles help provide aeration and drainage to a mix, thus reducing the water-holding ability of a mix. The key is to find a medium that provides good aeration while providing enough available water to the plant between irrigation applications.

The availability of water for plant growth is largely determined by how tightly it is held by the solid components in the medium. The closer a molecule of water is to a component, the more tightly it is held by the forces of adhesion and cohesion. Fine mixes may hold more water than course mixes, but a higher percentage of the water in the mix is unavailable to the plant due to the number and size of the media components. Course mixes cannot hold onto water molecules as well, since the water is not as close to the solid particles. Growing mixes with large components and poor spaces tend to have more water available to the plant and also drain faster than mixes containing smaller particles and pore spaces.

The chemical components of a mix, such as lime or fertilizers, generally do not affect the air space or water-holding capacity of a mix. Wetting agents are the exception and are commonly used to enhance the wetting properties of peat- and bark-based mixes.

The size of the container greatly affects the physical properties of a mix. The shorter the container, the higher the water-holding ability will be. Drainage of water from the medium is greatly affected by the pore size and the length of the column. Smaller pore sizes and shorter water columns will retain more water than larger pore sizes and taller water columns. On a percentage basis (not volume), shorter containers will hold more water than larger containers when using the same medium. For example, if two containers hold the same volume of growing medium and one is short and the other is tall, the tall container has a greater force of gravity on the water occupying the pore spaces. This will cause the water to drain more, increasing the air occupying the pores. Conversely, the short container will not drain as readily, holding more water and less air in the pore spaces.

To demonstrate that the height of a column has a great affect on the water-holding ability of any given medium, perform a sponge demonstration. Take a 2 x 4 x 6 in. (5 x 10 x 15 cm) sponge and immerse it in a tub of water until it is fully saturated. Remove the sponge from the water, holding it flat, so it has a 2 in. (5 cm) height (resembles a 72-cell plug height), some water drains and is replaced with air. When the water stops draining, turn the sponge on its side to create a height of 4 in. (10 cm; resembles a 1 qt. [1 L]

Table 2.1. Sieve Sizes Commonly Used

Sieve size
1/2 inch (12.6 mm)
3/8 inch (9.5 mm)
1/4 inch (6.3 mm)
Number 4 (4.8 mm)
Number 6 (3.35 mm)
Number 8 (2.4 mm)
Number 12 (1.7 mm)
Number 18 (1.0 mm)
Number 20 (0.85 mm)
Number 40 (0.42 mm)
Number 100 (0.15 mm)
Pan*

* Pan size varies with each testing facility and is often consider to be the fines measuring less than 0.5 mm.

Table 2.2. Particle Size Distribution Profile (Sieve Analysis)
Example sieve analysis from a bark-based perennial growing mix*

Sieve Size	Weight Distribution (lbs per yard)	Percent	By Size
3/8 inch (9.5 mm)	95	12%	21%
1/4 inch (6.3 mm)	75	9%	
Number 4 (4.8 mm)	55	7%	24%
Number 8 (2.4 mm)	135	17%	
Number 18 (1.0 mm)	180	23%	23%
Pan (<1.0 mm)	255	32%	32%

To provide the proper physical properties, allowing the mix to wet, drain, and rewet, all particle sizes are important. Obtaining a mix with fairly equal proportions of each particle size grouping as shown above (By Size column) will help growers maintain adequate air and water within the root zone, encourage good root growth, and allow for easy water management.

*Results determined and calculated and based on dry weight in grams

container). Notice that additional water drains from the sponge, demonstrating that a 4 in. container contains more air than a 72-cell plug. Again, once the water stops draining, turn the sponge so the height is now 6 in. (15 cm; resembles a 1 gal. [3.8 L] container), and note that even more water drains from the sponge. In this example, the sponge is the growing medium and the size or volume of the sponge remained constant. Only the orientation or the height of the column changed, allowing gravity to pull additional water from the sponge as the height increased. This demonstration shows that the height of the container dictates the total air space of the substrate after drainage of excess water.

Bulk density is a physical property referring to the weight of the growing medium per cubic foot. More specifically, the bulk density represents the weight per volume of medium and is expressed as pounds per cubic foot, or grams per liter. Growing mixes usually weigh 8–18 lb. per cubit foot (3.6–8.2 kg/28.3 dm^3) when dry and 40–65 lb. per cubic foot (18.1–29.5 kg/28.3 dm^3) when they are wet. Bulk density is important for providing the proper anchorage and stability to the plant. Growers often consider the bulk density or, more specifically, the weight of the container for moving and shipping purposes.

The formula for bulk density is:

Bulk density = weight of soil/volume of soil

Organic components tend to break down over time, causing the physical properties to change during the production cycle. Longevity or stability implies that the mix does not change or decompose dramatically during crop production. Growers optimally would like to see the medium in the pot remain at the same level in the container from the start of the crop to the finish. Coir and bark do not break down as rapidly as peat, allowing the physical properties to remain stable for longer periods of time. All components are subject to breaking down if mishandled during blending or filling.

The settling of a medium within a pot is referred to as shrinkage, resulting in compaction of medium in the bottom of the container, reduced aeration, and poor plant health. There are two options for addressing containers with excessive shrink: replanting the pots with shrinkage or lifting the plant out of the container, placing fresh growing medium in the bottom of the pot, and then placing the plant back into the container. Some growers try to solve shrinkage by placing more growing medium on top of the settled medium surface. This practice only worsens the effects of the compaction and further reduces root aeration conditions because the orginal growing medium contains small particles that pull water from the larger particles of the newly applied medium, causing it to reach saturation before the water can drain out.

To maximize the porosity, it is important that the components of the medium are blended properly. Containers filled with a poorly mixed medium can have different physical properties; the total pore space, aeration porosity, and water-holding capacity can all vary significantly. Variations in the physical properties can cause water content and drainage differences between the containers, resulting in uneven drying of the medium in the production site, retail area, and in the customer setting. Thoroughly blended mixes also ensure the chemical properties are maximized by properly incorporating amendments such as fertilizer, limestone, or micronutrients.

Initially, it is important to establish the proper moisture relationship between the medium and the plant roots. If the mix is too wet, there is the potential for compaction in the container, which causes a reduction in porosity (air space). If the mix is too dry, the components might excessively swell, which also reduces the mix's porosity. Mixes that are difficult to wet often require excessive amounts of water and frequently the results are undesired compaction or some of the medium floating out of the container, reducing the medium volume and porosity of the container.

To promote optimum plant growth, all container substrates must provide a proper balance between air space and water availability. Many benefits are observed when this balance is achieved. Growers usually produce their crops in less time, decrease overall plant losses, increase total sales, and decrease production costs. By providing the proper characteristics in the container medium, growers improve plant quality and their overall profitability.

It is inevitable for mixes to change from load to load and from season to season, as they contain a large percentage of organic materials with variable quality characteristics, which change by availability, source, and time. To achieve consistency, it is necessary to observe and test the medium for physical and chemical properties as part of a routine integrated health management program.

There are a couple of in-house testing procedures growers can use to determine physical properties such as the overall porosity, the air porosity, and the water-holding capacity of a mix. There are a number of independent or university laboratories that will conduct these and/or other tests that are helpful in determining the physical characteristics of potting media. Frequent testing and monitoring can provide growers with specific information, allowing them to properly manage the growing medium while avoiding or addressing any issues that should arise.

Physical Property Testing

A simple and free testing method is called the moisture squeeze test. This test provides an indication of the relative moisture of the mix. To conduct this test, simply pick up a handful of potting medium and squeeze it to see if any water comes out between your fingers. If the mix crumbles when you open your hand, this indicates that there is not enough moisture in the mix. Conversely, if moisture is easily squeezed out of the mix and runs between your fingers, then the mix is too wet. At the proper moisture level, you will feel moisture when you squeeze the mix, but no dripping will occur, and when you open your hand the medium should easily break apart.

Another simple in-house test can be done to provide an indication of a mix's wetting ability, or how well the water will distribute through the medium at the initial watering. Fill a container, such as a 1 gal. (3.8L) pot with the potting mix. The pot should be filled as if it were going to be planted, meaning that the same volume and compaction of soil should be used. Slowly add a known volume of water to the medium until it begins to drip out from the container. Place your hand on the top of the mix surface, tip the pot upside down and slowly remove the pot from the mix while your hand supports the potting mix. Observe whether the mix has wet through completely; there should not be any dry areas. The volume of water applied roughly indicates the water-holding capacity of the mix.

Porosity Testing

Checking the porosity of a potting medium is another on-site test growers can conduct to provide an indication to what the total pore space is and whether it consists of air or water. The best method I have found is to use empty coffee cans with drainage holes punched in the bottom, simulating the drainage rate of a pot. Ideally, you would like the same number and size of drainage holes as found on the growing container. Place the cover on the bottom of the can to prevent water from leaking out when you are conducting the test.

Fill the can with water, up to the level you intend the medium to reach. With a marker, draw a line on the inside of the can to demonstrate where to fill the container with media. Pour this water into a measuring cup; the amount of water represents the total volume of the container. Now fill the container with potting medium up to the line or the level the water was in the previous step. The can should be filled in a similar manner (similar firmness) as a pot with a plant in it. After filling the can to the desired level, tap it several times on a countertop, causing it to settle, add some more medium, if necessary, and re-tap the can. Be sure the medium is level with the line drawn, as this a known volume that will be used to determine the results later on. Do not over-tap or compact the medium; over-filling or compacting will greatly alter the results.

Slowly add known volumes of water to the potting mix until it is saturated. At the surface, water will just be visible, or glistening. Do not fill past the line or completely cover the surface with water, or your results will not be accurate. If the medium's surface is rising or floating, you have added too much water and should conduct this test again using new medium. The total amount of water added represents the total pore volume. Let the can set for about an hour to ensure that the components become saturated, then remove the lid from the bottom of the can, allowing the water to drain out and into a collection container. Be sure to keep the can vertical; do not tip it at an angle or tilt it sideways. You are not trying to force the water out of the can, but trying to simulate drainage as it might occur on a production bench in the greenhouse. The water that drains out represents the aeration pore volume or the air space in the medium.

To recap, the container volume is the total volume of the container, the pore volume represents the total pore space (air and water) and is measured as the volume of water added to the medium to reach saturation, and the aeration pore volume is the total air space in the medium after the water is drained out. To calculate the percent of total porosity, divide the total pore volume by the container volume and multiply by 100. The percent aeration porosity is determined by dividing the aeration pore volume by the container volume then multiplying by 100. To figure the percent water retention porosity, subtract the percent aeration porosity from the percent total porosity.

$$\% \text{ Total Porosity} = \frac{\text{total pore volume}}{\text{container volume}} \times 100$$

$$\% \text{ Aeration Porosity} = \frac{\text{aeration pore volume}}{\text{container volume}} \times 100$$

$$\text{Water Retention Porosity} = \% \text{ Total Porosity} - \% \text{ Aeration Porosity}$$

So now that you have all of these numbers, how well does your mix measure up? Conducting the porosity tests in this manner, for 1 gal. (3.8 L) perennial production, I try to obtain mixes that have a total porosity of 50–65%, aeration porosity ranging from 20–25%, and the water retention porosity of 30–40%. There will be slight variations from test to test and from load to load. Observe the trends, paying attention to drastic changes in the test results. If you are not happy with your porosity results, work with your media supplier to make the appropriate changes to your potting mixes.

By knowing how each component contributes to the macro- and micropore spaces, growers can determine how best to amend the growing mixes with too high or too low water or air porosities, moving the porosity in the desired direction. Mixes with poor drainage or low air porosities can be corrected by increasing the size of the pore spaces in the growing medium. The larger pore spaces can be accomplished by adding a component with a courser particle size. Mixes with high air porosities and little water-holding capacities can be altered by adding components with finer particle sizes.

Growing mixes need particles of all the various sizes to create macropores to hold air and micropores to hold water. Using different components and particle sizes, growers can achieve the necessary pores and growing environment necessary for healthy root and plant development.

Avoiding Media Problems with Culture

Adjusting the growing environment and removing the favorable conditions for disease development can avoid many of the disease problems growers face within the growing mix.

Ventilation

By periodically opening the vents, growers can exchange the humid air inside the greenhouse with fresh air with a lower relative humidity. Controlling the humidity in a greenhouse reduces the amount of condensation and allows the medium to dry out quicker and more uniformly. The potting mix surface can also dry quickly with adequate airflow within a greenhouse. To increase airflow and to provide more uniform temperatures in a greenhouse, growers use horizontal airflow (HAF) fans. With a drier medium surface, there will be less algae growth, fewer shore flies and fungus gnats present, and less activity from the disease pathogens *Pythium* and *Phytophthora*.

Figure 2.6. **Cold frame with ventilation fan.**

Watering

Water, in combination with the physical properties of the medium, has a great impact on the growth of the plant, as well as creating a favorable environment for disease organisms. Applying water to crops aggressively, such as at high water pressures, can compact the medium, reducing the air spaces in the root zone. Irrigation should be applied in a non-aggressive, uniform manner.

Every mix will have its own water-holding characteristics, and the grower must learn how to manage irrigation practices to fit the properties of the growing medium. Here are a couple general rules. When irrigating, water thoroughly to be sure each container reaches a similar moisture level. Watering in this manner will also leach out some of the salts, limiting the potential for salt build up, injury to plant roots, and the presence of disease organisms. Another recommendation is to perform the irrigation early in the day, allowing the foliage to dry and the temperature of the medium to rise back up to normal before the end of the day. Refrain from watering in the afternoon, unless your plants are wilting.

Light green foliage and chlorotic growth usually indicate poor nutrition, inadequate pH, low temperatures, onset of root diseases, or that the plants are over-watered. These symptoms are often expressed as a combination of these factors. For example, high salt levels may have occurred, causing injury to the root system, and this root injury led to the onset of a root rot problem. The root rot limited the plants' ability to take up nutrients, causing the plants to become malnourished and appear chlorotic. Plants that are grown too dry will generally have a dark green or gray-green appearance and will generally be shorter than plants grown with an adequate supply of water. Nutritional and disease problems are also likely to occur on plants grown under water stress.

Growing Medium during Cold Seasons

Cool growing conditions often result in reduced water loss and consumption from the growing medium, as the rate of evaporation and plant uptake is reduced. Water loss through evaporation is reduced with cool soil temperatures and high relative humidity levels near the surface. During these conditions, it is important to manage the crop differently in order to prevent additional cultural problems.

When the production environment is cool, it is important to reduce the irrigation frequency. It is common for growers to irrigate too frequently or to not adjust their normal practices during periods where the production temperatures and light levels have been dramatically reduced. It is beneficial to let the potting mix surface dry out slightly between irrigations, which also helps to reduce algae growth and populations of fungus gnats and shore flies.

The growing medium dries out quicker when the airflow can be increased and the humidity levels reduced. Increasing the airflow using HAF fans mixes the air, providing more uniform temperatures and reducing the humidity level near the plant canopy by moving the humid air away from the plants and replacing it with drier air. Humidity levels can also be reduced by purging the facility or moving the humid air out of the facility by means of venting or using exhaust fans and replacing it with drier air from the outside.

Growers can also select growing mixes with higher air porosity and less water-holding capacity. These mixes often contain bark, or other components, with large particle sizes that create large macropores (air spaces) in the mix, allowing for sufficient drainage. These mixes work well during cool production seasons but tend to dry out rapidly when used during warmer seasons, requiring the need for frequent irrigations.

Uneven drying of the medium is another common problem growers face when growing plants in cool environments. Inconsistent moisture levels are often the result of uneven container filling, variation in compaction from pot to pot, improper mixing and handling of the medium, and uneven water applications. In many cases, growers do not apply enough water initially after planting. Instead, they water each container individually, which results in various quantities of water being applied to each pot, causing the medium to dry out at different rates and requiring water at differing times. To remedy uneven drying, growers should move away from applying small amounts of water to each pot and should water each pot to saturation. The excess water will run through the pore spaces and out of the pot with gravity.

Growers should also allow the containers to dry out more between irrigations. Generally, you cannot overwater a crop by applying too much water at one time, but you can overwater by applying water too frequently. Less frequent, thorough watering will reduce the moisture on the medium's surface, reducing the environment necessary for dampening off pathogens and the build up of undesirable salts such as sodium and chloride. When uneven drying is observed, growers should spot water the areas that are dry and then come back and thoroughly water the entire crop. Spot watering alone will only cause more uneven drying.

Root rot pathogens *Pythium, Rhizoctonia*, and *Thielaviopsis* thrive under cool and wet conditions. These diseases often attack plants at the medium's surface, causing them to fall over and die. Managing the moisture level on the surface is important and can dramatically reduce the occurrence of these diseases when plants are grown under cool conditions. The key to reducing these pathogens is to increase the airflow, reduce the humidity level, and allow the surface of the growing medium to dry.

Nutritional problems are likely to occur during cool periods. In many cases, growers do not fertilize during the first few weeks of production to prevent the plants from stretching. At some point, the starter charge in the growing medium runs out, and there just aren't adequate amounts of nutrients available to support plant development. Some nutrients, such as phosphorus, become less available to plants when the medium's temperatures are low. It is usually best to provide some nutrients within the first couple weeks of production, even if they are applied at reduced rates.

Summary

The selection and management of the growing medium is one of the most important tasks a grower has. The growing medium provides the foundation on which all future plant growth will be built. A weak foundation leads to weak crops. The properties of growing medium and the resulting plant performance should be the primary consideration in the selection of container mixes. Understanding the chemical and physical properties of a container medium will help growers to produce more consistent, high-quality crops.

Growers obviously need to be cost conscious when formulating and purchasing the growing mixes used at their facilities. However, cost alone should not be the determining factor. Factors such as availability, consistency, and physical properties should be the primary considerations. Other considerations before looking at the price should include pH, salts, and stability of the medium. The costs incurred for shrink, plant replacement, and management of plants being grown in a low-quality growing medium are far more costly in the long run than paying a little more for a high-quality growing mix up front.

References

Bailey, Douglas A., William C. Fonteno, and Paul V. Nelson. "Bedding Plant Substrates and Fertilization." *PPGA News*. July 1996.

Behe, Bridget, Diane Brown-Rytlewski, et. al. *Management Practices for Michigan Wholesale Nurseries*. East Lansing, Mich.: Michigan State University. 2004.

Bloodnick, Ed and Troy Bluechel, "Media Matters: Physical Properties of Media are Important and Their Effects on Your Crops Are Often Overlooked." *Greenhouse Grower*. July 2004.

Bloodnick, Ed, Troy Buechel, et. al. "Pay Attention to Details: Maximize the Performance of a Growing Medium by Paying Attention to What Happens to It from the Time It Arrives until the Time It Leaves with the Finished Crop." *GMPro*. January 2003.

Carpenter, Tim. "Mixing up Media Success." *Greenhouse Grower*. July 1995.

Evans, Michael R., and William C. Fonteno. "Get a Handle on Your Growing Media: Even the Best Medium Can Perform Poorly if Mixed or Handled Improperly." *GMPro*. September 1999.

Fernandez, Tom. "Don't Settle for Pot Luck: Know Your Container Substrate Physical Properties." *The Voice*. January/February 2000.

Ferry, Shannen, Ron Adams, et. al. "Soilless Media: Practices Make Profit, Part I." *Greenhouse Grower*. July 1998.

————. "Soilless Media: Practices Make Profit, Part II." *Greenhouse Grower*. August 1998.

————. "Soilless Media: Practices Make Profit, Part III." *Greenhouse Grower*. September 1998.

————. "Soilless Media: Practices Make Profit, Part IV." *Greenhouse Grower*. October 1998.

Jacques, Daniel, and Al Toops. "Growing Mix Component Effects: Find Out the Impact of Growing Medium Components on Plant Growth When They Are Incorporated into a Mix." *GMPro*. January 2001.

Jacques, Daniel J., Nancy Morgan, et. al. "Regional Components Could Meet Your Growing Needs." *GMPro*. September 2003.

Jacques Daniel J., Ron Adams, et. al. "Your Growing Media: Determining the Physical Properties Is Important." *GMPro*. April 1999.

Powell, Charles C. "Media Blitz: Start at the Root of the Problem by Knowing the Properties and Management Techniques for Your Container Media." *GrowerTalks*. October 2004.

3

Fertility for Perennials

Managing the nutrition of perennial crops can be difficult, due to such factors as the great diversity of plant species being produced, a wide range of production times, and many different types of growing facilities, such as greenhouses, shade houses, and outdoor sites. Containerized perennials in greenhouses and nurseries are often grown at what is called luxury fertility levels. The fertility needs of a plant grown in a pot are very different than one grown in the landscape. Most potting mixes do not contain sufficient amounts of nutrients, and the environments these plants are grown in provide super-optimal conditions for plant growth. In addition to the complexity of managing nutrients during crop production, providing adequate fertilization has often been inefficient, inadequate, and threatening to our environment.

Developing, implementing, and monitoring a fertility program is critical to successful perennial production. Most growers do not understand fertilizers well enough or tend to oversimplify them without understanding the impact they may have on plant growth under various growing conditions. There is often confusion regarding what types of nutrients and how much of them are needed. In order to properly design a fertility program for your perennial crops, growers should first learn what nutrients, if any, will be supplied from the irrigation water or the potting medium.

Perennials, like all plants, have eighteen chemical elements essential for healthy plant growth. These elements are carbon, hydrogen, oxygen, nitrogen, phosphorus, potassium, calcium, magnesium, sulfur, iron, manganese, nickel, copper, boron, sodium, zinc, molybdenum, and chlorine. Without these elements, plants may not develop properly or carry out certain plant processes. Other elements may be taken up and improve plant growth, but plants can survive without them.

Growers supply nutrition to their crops using a variety of methods, such as incorporating them in the growing medium, using liquid feed programs, or applying controlled-release fertilizers. Each of these methods will be discussed in detail later.

Macronutrients

The six elements required by plants in substantial amounts are called macronutrients. The macronutrients are required to build plant cells and to form the compounds that carry out plant processes. The primary macronutrients are nitrogen (N), phosphorus (P), and potassium (K). These primary elements are often depleted from the growing medium and need to be replenished with some frequency. The secondary macronutrients consist of calcium (Ca), magnesium (Mg), and sulfur (S). Like the primary elements, the secondary macronutrients are required in relatively large amounts and also may need to be replenished, or supplemented, in the growing medium.

Nitrogen

Nitrogen is the primary macronutrient that is used by plants in the largest amounts. It plays a major role in growth and reproduction of the plant. Nitrogen is involved in the structure of amino acids, enzymes, plant hormones, proteins, nucleic acids, and chlorophyll.

When nitrogen levels become deficient, most plants turn pale green to yellow in color and generally become unhealthy, with reduced growth rates. Nitrogen is a mobile element, so deficiencies typically show up in the older foliage. If nitrogen is excessive, plants tend to have weak or soft growth and have less resistance to diseases. High nitrogen levels contribute to increasing soluble salts, which may lead to fertilizer burn in severe cases. Nitrogen is commonly taken into the plant in the form of nitrates, which are easily leached from the root zone and most often need to be replaced using various fertilizers.

There are three types, or forms, of nitrogen fertilizers used by growers: ammonium, nitrate, and urea. Plants respond differently to each of these forms. Plant responses to ammonium and urea are similar because the urea must be converted into ammonium in the soil before plants can absorb it into their roots. Fertilizers with high levels of ammonium and/or urea typically stimulate rapid leaf expansion and internode elongation while suppressing flower and root development. Fertilizers consisting of high levels of nitrates tend to produce more compact plants with less internode elongation while improving flowering and root development.

Phosphorus

The primary macronutrient phosphorus serves many functions that are critical to plant development. Phosphorus is a component of membrane structure, nucleic acids, enzymes, RNA, and DNA, and is a key element in many energy transfer reactions. Optimum levels of phosphorus will influence protein formation, seed germination, photosynthesis, root development, rapid plant growth, maturity, flower development, and they aid in the use of other elements.

Phosphorus bonds to the soil particles and does not leach readily from the root zone. Due to these tight bonds, it might be in the root zone in seemingly adequate amounts, but may not actually be available for plants to uptake and use. It is difficult to determine when the phosphorus supply in the soil runs out. Phosphorus may leach out of the root zone when the pH falls to extremely low levels. Historically, growers apply too much phosphorus, with the excessive amounts ending up in the groundwater.

Phosphorus deficiencies may cause plants to appear stunted, dark green, and/or cause a purple or bronze cast to the foliage. In most cases, deficiency symptoms appear on the older leaves first. An excessive amount of phosphorus in the potting medium reduces plant growth and flower size and causes some plants to elongate or grow taller. High levels of phosphorus can block the uptake of other elements and has been known to cause nutrient deficiencies of copper, iron, manganese, and zinc.

Temperature plays an important role with the uptake of phosphorus by many plants. *Clematis* and *Rudbeckia* are two perennials that have difficulty taking up phosphorus when the temperatures are below 55° F (13° C). At these low temperatures, they often express purpling of the older foliage as the predominant symptom of phosphorus deficiency.

Potassium

Potassium serves many functions as a macronutrient for plant growth. It is necessary for the formation of carbohydrates, sugars, and starches and is involved in protein synthesis and cell division. Potassium plays an important role for the movement of food and nutrients within the plant (ionic balance), photosynthesis, enzyme activity, maintaining overall vigor, and promoting the formation of roots. It also plays a significant role with plant/water relations (opening and closing of stomata), stem rigidity, and cold hardiness. Potassium is not needed in the same quantities as nitrogen, but at times it may become depleted and need to be replenished.

Potassium deficiency symptoms appear as reduced growth and/or necrosis along the leaf margins. The primary consequence from excessive potassium levels is the potential for calcium and magnesium uptake to be reduced.

Calcium and magnesium

Calcium is a very important nutrient for sufficient plant growth. It activates enzymes for cell mitosis, cell division, and elongation. Calcium also is a structural component of cell walls, maintains cell wall integrity, maintains membrane permeability, and enhances pollen germination and growth. Shortages of calcium are usually expressed as browning of leaf tips, often referred to as tip burn. It is very immobile, causing deficiencies to occur in the new leaves. Excessive calcium is rare but does cause reduced uptake of potassium and/or magnesium.

Magnesium is a critical structural component of chlorophyll and is necessary for functioning of plant enzymes and the production of carbohydrates, sugars, and fats. It is mobile within the plant and exhibits the classic deficiency symptom of interveinal chlorosis of the older leaves when magnesium levels become low. Magnesium uptake can be reduced in soils with high pH levels.

The primary source of calcium and magnesium in the potting substrate is usually dolomitic limestone, which is mainly used to adjust the medium's pH. Irrigation water is also a common source of these elements. Many growers observe that dolomitic limestone does not always release fast enough for some crops or may even run out for crops that are around for long periods of time. It is recommended to test the water source and the potting mix to determine how much of these nutrients are present and to help determine what type of fertility program to set up.

Not all water-soluble fertilizers contain these nutrients. Generally, acidic fertilizers do not contain any—or only small amounts of—calcium and magnesium. Basic high-nitrate fertilizers often contain calcium but may not contain magnesium. To provide adequate amounts of these nutrients, growers often rotate between these types of fertilizers (acidic and basic) and/or supplement magnesium using Epsom salts.

Sulfur

Sulfur is an essential structural component of amino acids, proteins, and enzymes. Sulfur is also critical for the production of chlorophyll. Deficiency symptoms are similar to those of nitrogen deficiency, except they are expressed in the whole plant as opposed to the older foliage. When excessive sulfur is present, there may be premature leaf drop. Usually, excessive sulfur decreases the pH below optimum levels, resulting in micronutrient imbalances or an increased uptake of aluminum, which is toxic.

Micronutrients

Nine of the eighteen essential elements critical for plant growth are often referred to as micronutrients. These elements—boron (B), chlorine (Cl), copper (Cu), iron (Fe), manganese (Mn), molybdenum (Mo), nickel (Ni), sodium (Na), and zinc (Zn)—are needed in relatively small amounts by plants. With the exception of sodium and chlorine, which are sufficiently available in the growing medium, irrigation water, or fertilizer contaminants, the other six micronutrients must generally be supplemented with fertilizer applications.

Micronutrients, although needed in small quantities, often serve large roles in respect to plant development. For example, iron has the central role in the chlorophyll structure. Some micronutrients, like copper, which plays a role in plant respiration, do not necessarily become part of the plant tissues but are essential for certain biochemical reactions necessary for plant survival.

With a majority of containerized perennials being produced in soilless mixes, micronutrient deficiencies have become more problematic for growers. Growing media based on natural soils usually had sufficient amounts of these elements to sustain plant growth and functions. Soilless mixes require an application of minor elements prior to using and may require additional applications throughout the growing season. A contributing factor to needing additional amounts of minors in the growing medium is that organic components compete for micronutrients, tying them up or making them unavailable for plants to uptake.

Boron

Some of the cellular functions of boron include cell division, differentiation, membrane integrity, maturation, development, cell wall formation, and growth. It is needed for photosynthesis and activation of several plant enzymes. Boron is involved in several plant functions including flowering, pollen germination, and fruiting.

Chlorine

Chlorine plays a role with water relations, nutrient uptake, and photosynthesis. It is involved in the movement of water or solutes in cells (osmosis) and the ionic balance necessary for plants to take up nutrients.

Copper

The immobile micronutrient copper is involved in many plant functions, including photosynthesis, activation of enzymes, and the metabolism of nitrogen, carbohydrates, and proteins.

Iron

Iron is a very important nutrient for plant growth, as it is necessary for the synthesis of chlorophyll. It is also a component of many enzymes, proteins, and electron transport systems.

Manganese

Manganese is involved in many cellular functions within plant cells, including nitrogen assimilation, photosynthesis, respiration, enzyme activity, and pollen germination.

Molybdenum

Molybdenum is a structural component of enzymes that reduces nitrates to ammonia. This process is critical; without it the synthesis of proteins is blocked and plant growth ceases.

Nickel

Nickel has just recently been acknowledged as an essential micronutrient. It is a component of the protein urease, which is required for the breakdown of urea into the nitrogen form usable by plants. Nickel is also needed for iron absorption and seed germination.

Sodium

Sodium plays a similar role in plants as the micronutrient chlorine; it is involved in the movement of water and the ionic balance in plants.

Zinc

Zinc is an immobile micronutrient that is a component and an activator of enzymes. It is essential to carbohydrate metabolism, protein synthesis, and cell elongation.

Beneficial micronutrients

Cobalt

Cobalt helps plants to absorb ammonium in root nodules and aids nitrogen fixation in legumes.

Silicon

Silicon is a component of cell walls, making them stronger, helping to improve the plant's ability to withstand heat and drought. It also helps increase the plant's resistance to insect and disease infections.

Nutrient Ratios

Research has shown that optimum plant development can be achieved when certain ratios are maintained between individual elements. There are exceptions to these ratios as many plant species have ratios specific to them. Below is a brief description of different nutrient ratios growers should be aware of in order to maximize plant development.

A fertilizer ratio is typically expressed with a number, the atomic symbol of the first element followed by a colon (:), then a number and the atomic symbol of the next element. For example, 3N:2K would read three parts nitrogen (N) to two parts potassium (K). Many times growers would refer to the above ratio as the N:K ratio, which is three to two.

When discussing these ratios, there are frequent references to reduced uptake of other elements. The presence of certain nutrients can interfere with the uptake of others; this interference is referred to as antagonism. For example, high levels of potassium can interfere with the uptake of calcium and magnesium. Deficiencies of these elements can occur even though soil analysis indicates adequate levels are present.

Nitrogen-potassium ratio

Most perennials perform well when there is a fairly equal balance between nitrogen (N) and potassium (K). This can be expressed as a 1-to-1 N:K ratio or as 1N:1K. Excessive amounts of potassium can reduce the uptake of ammonium, calcium, magnesium, manganese, and zinc. To reduce competition for uptake, or antagonism, among these nutrients, growers should maintain a 4K:2Ca:1Mg ratio.

Phosphorus-nitrogen ratio

It is difficult to determine the exact amount of phosphorus needed for satisfactory plant growth. Although phosphorus (P) deficiencies can and do occur, many growers typically apply more than what is needed, causing undesirable plant stretch and a reduction of flower size. A good guideline is to apply no more than half the concentration of the rate of nitrogen. Expressed as a ratio, it would appear like this: 1P:2N, or one part phosphorus to two parts nitrogen. High levels of phosphorus also can reduce the uptake of other nutrients. Maintaining the following ratio in the potting soil will reduce the competition between these elements for uptake by the plant: 10N:10K:10Ca:5Mg:2P.

Calcium-magnesium ratio

Many growers have heard of the calcium (Ca)-magnesium (Mg) ratio and often use it as a benchmark to maintaining crop quality. There is a range that growers often function within, usually between a 2Ca:1Mg ratio and a 5Ca:1Mg ratio. Perhaps the most common goal for most growers is to maintain a 3Ca:1Mg ratio.

It is difficult to make blanket nutrient ratio recommendations that meet the needs of all perennials being grown. Maintaining ratios such as the ones mentioned above helps sustain nutrient balance. Balancing nutrients available to the plant is important; this allows growers to avoid nutrient deficiencies, toxicities, or nutrient antagonism (competition between nutrients during uptake). Most perennials will perform quite well when the following ratio is maintained: 13N:1P:13K:7Ca:2Mg:2S.

Nutrition from Irrigation Water

The North American water supply has great variability in regard to its quality characteristics and how growers must manage crop nutrition. Growers commonly use different water sources depending on availability, water quality, and cost. These sources include: pond, river, well, and municipal. As described below, the quality characteristics vary dramatically from source to source and location to location. Some irrigation sources are almost pure as rainwater, while others have high mineral content. Additional quality characteristics such as soluble salts, pH, and alkalinity vary widely and affect perennial crop production. The only real way of knowing the quality characteristics of the irrigation source is to test the water regularly.

There are several characteristics of the water supply that growers should understand prior to producing a crop. The three most critical characteristics to consider are the nutrient content, the effect on the potting medium's pH, and the amount of soluble salts.

Irrigation water from some sources, especially shallow wells or surface water such as ponds or creeks, may contain high levels of one or several dissolved nutrients (typically N, Ca, Mg, or S) that may contribute to the total necessary for plant growth and possibly would need to be subtracted from subsequent fertilizer applications. Many sources of irrigation water, particularly shallow wells, may be alkaline, gradually causing the potting media pH to increase, often to the point nutrient deficiencies, poor plant growth, or plant damage occurs. Some sources of water might contain dissolved solids, such as sodium bicarbonate or sodium chloride that, together with other nutrients, could contribute to high soluble salt levels that may be detrimental to plant growth.

Water quality does not remain constant and can change often. It is important to test the water prior to building any type of production facility and before each crop cycle. It is recommended to check the water as many variables come into play. Such variables include seasonal use, average rainfall versus little rainfall, unusual droughts, and city water that comes from multiple sources.

Nutrition from Potting Medium

Many growers have found it both convenient and cost effective to supply both macronutrients and micronutrients as additives to potting media. To determine which nutrients are available from their media, I recommend growers take samples of unused media (right from the bag or bulk delivery) and send them to a soil-testing laboratory for testing. Many potting media suppliers will provide these analyses free of charge if you request them.

The most common approach is to blend various nutrients into the potting medium, to adjust the medium's pH, and to provide a modest amount of fertilizer (often referred to as a starter charge) so plants have sufficient nutrients available to begin growing quickly. Perennial and nursery growers often incorporate these starter charges, combined with controlled-release fertilizers, to provide sufficient amounts of these nutrients to last the entire duration of the crop.

The starter charge of most growing media usually contains low amounts of most nutrients and usually has a soluble salt reading of 0.75–2.0 mS/cm. It is very common for growing mixes to contain lime and dolomitic lime, which provide pH buffering to the mix. These types of lime and gypsum are valuable sources of calcium, which are often lacking in fertility programs.

For more information regarding the starter charges commonly found in potting media, please refer to chapter 2, Properties of Growing Mixes.

pH

The pH refers to the acidic or basic properties of the growing medium or a water solution. It is the relative amount of hydrogen (H+) or hydroxide ions (OH-) of a solution. A pH reading is a measurement of the hydrogen ion concentration of a solution. The pH of a solution is expressed on a scale from 0–14. When a solution contains more hydrogen ions than hydroxide ions, the solution is acidic (pH less than 7.0). A solution containing more hydroxide ions than hydrogen ions is considered basic (pH greater than 7.0). A balance of hydrogen and hydroxyl ions yields a neutral pH solution (pH equal to 7.0).

Although a pH reading of 7.0 is considered to be neutral, 7.0 is not the optimal pH for irrigation sources or of the root zone. The acceptable substrate and irrigation pH varies widely with the crops being grown, generally ranging from 5.2–6.8 for irrigation water and 5.4–6.3 for the substrate solution. Most perennials grow best in slightly acid growing conditions, in the range of pH 5.8–6.2. Producing perennials within these ranges creates the greatest average level of availability for all essential plant nutrients. Fluctuations above or below these optimal ranges can cause deficiency or toxicity of nutrients.

The growing medium contains a water solution that has the nutrients needed for plant growth. When growers measure the pH of the potting mix, they are actually measuring the pH of the water solution in the growing medium. This solution contains mineral elements dissolved in ionic form. The acidic properties of the soil solution have a marked effect on the availability of mineral elements to plant roots.

The pH of the growing medium is by far more important than the pH of the irrigation water or fertilizer solution since the pH of the growing medium determines how available certain nutrients are. The availability of the macronutrients (N, P, K, S, Mg, and Ca) is not greatly affected by the pH of the growing medium. However, the medium's pH does determine the availability of most micronutrients. The pH of the root zone has a great effect on the availability of micronutrients such as iron, manganese, zinc, copper, and boron. At high pH levels, some nutrients that are essential for plant growth become unavailable, causing nutrient deficiencies to occur. The most common deficiency associated with high pH levels is iron deficiency. Many micronutrients (B, Cu, Fe, Mn) are not mobile within the plant and cannot be redistributed to other plant parts. Deficiency symptoms can develop quickly when the pH of the growing medium is not within the optimum range for these elements.

Growers should note that pH-related deficiencies occur because plants are unable to take up these nutrients from the growing medium, not because the nutrients are not present. In many cases, growers are instructed to add more micronutrients. Adding more micronutrients will not solve deficiencies related to the pH—only correcting the pH will make these nutrients more available. Providing additional micronutrients could prove detrimental, leading to micronutrient toxicities, once the pH is corrected.

Growers should become familiar with the primary four factors that affect the pH of the growing medium: (1) amount of lime in the medium; (2) alkalinity of the irrigation water; (3) type of fertilizer; and (4) the plant roots themselves. These factors are discussed throughout this chapter.

Alkalinity

The ability of a solution, namely water, to neutralize acids is referred to as alkalinity. It is the concentration of the soluble alkalis in a solution. Do not confuse the term alkaline with alkalinity. (Alkaline describes conditions where the pH levels exceed 7.0.) The major chemicals that contribute to water's alkalinity are the dissolved bicarbonates—such as calcium bicarbonate ($Ca[HCO_3]_2$), sodium bicarbonate ($NaHCO_3$), and magnesium bicarbonate ($Mg[HCO_3]_2$)—and the carbonates—such as calcium carbonate ($CaCO_3$). Other chemicals such as ammonia, borates, dissolved hydroxides, organic bases, phosphates, and silicates can also contribute to the alkalinity of a solution but are relatively insignificant compared to the effects that carbonates and bicarbonates have.

For the most part, the total carbonates (carbonates plus bicarbonates) equals alkalinity. Many testing laboratories refer to alkalinity in this manner. Alkalinity is often expressed as milligrams per liter, or parts per million of calcium carbonate (mg/L or ppm $CaCO_3$), or as milliequivalents per liter of calcium carbonate (meq/L $CaCO_3$). To convert between these units, multiply the milliequivalent value by 50.04 to determine the milligrams per liter $CaCO_3$. Alkalinity values expressed as parts per million $CaCO_3$ and mg/L $CaCO_3$ are equivalent: 50 ppm $CaCO_3$ is the same as 50 mg/L $CaCO_3$. There are a few laboratories that report alkalinity as it were derived solely from bicarbonates using either mg/L or ppm HCO_3^- or meq/L HCO_3^-. To convert between these units, multiply the milliequivalent value by 61 to determine the milligrams per liter HCO_3^-.

Alkalinity establishes the buffering capacity of the water, and it affects how much acid is required to change the pH; it makes irrigation water resist changes in pH. Alkalinity can be thought of as the amount of limestone found naturally in the water. In theory, with high-alkalinity water sources, lime is being added to the growing medium with each irrigation, raising the pH of the medium over time. The alkalinity of a solution has more of an effect on the availability of nutrients than does the actual pH of the solution. Growers can grow with high pH readings as long as the alkalinity is low.

The alkalinity of irrigation water used by growers ranges across the country depending on the source of the water (surface water versus well water) and their geographic location, from 0–400 ppm $CaCO_3$. Acceptable alkalinity ranges vary with the plant species, growing medium, irrigation methods, and fertility programs.

Problems are likely to arise with the availability of nutrients when the alkalinity of the irrigation water is above 100 ppm $CaCO_3$.

These moderate alkalinity levels may cause the medium's pH to increase over time. Container growers should maintain alkalinity levels of 80–120 ppm. Plug growers often experience problems when the alkalinity is above 75 ppm $CaCO_3$ because the small volume of substrate provides little buffering against a rise in pH; they should maintain alkalinity levels of 60–80 ppm. Alkalinity levels below 50 ppm are too low and generally push the pH levels down below the desirable range.

Growers whose irrigation water has an alkalinity of greater than 150 ppm normally have to acidify the water before irrigating the crops to neutralize the bicarbonates (alkalinity) present. Acid sources commonly used to reduce the alkalinity of the irrigation water include nitric, phosphoric, or sulfuric acids.

Growers can measure their alkalinity using a titration test kit. The alkalinity is measured by titrating a water sample with an acid to an endpoint pH of about 4.6. A pH indicator dye is added to a known volume of water, and then acid is added until the solution changes color. When shopping for alkalinity test kits, growers should look for kits that can detect the naturally occurring alkalinity levels of most sources of irrigation water (0–400 ppm). It is usually acceptable for test kits to have ±20 ppm $CaCO_3$ level of accuracy, though more precise kits are available. Testing kits are available from $35–150 for one hundred tests. The more expensive kits tend to have twice the accuracy and often contain a digital titrator that can be used to test other solution parameters, such as water hardness and chlorine concentrations.

Lime

Many commercial growing mixes containing pine bark or peat moss often contain lime due to extremely low pH levels of these components. Lime is added to raise the pH level of the medium to a level more suitable for plant growth (5.5–6.5). There are three common sources used for this purpose, each with its own characteristics: hydrated lime, calcitic lime, and dolomitic lime. Hydrated lime reacts very quickly but does not last long. It is useful only for short-term adjustments to the growing medium. Calcitic lime reacts slower than hydrated lime and lasts longer. Dolomitic lime is a slow-reacting, long-lasting lime that provides residual control of the pH and is also used as a source of magnesium. Gypsum (calcium sulfate) is commonly used as a source of calcium but does not alter the pH of the growing medium. The type, amount, particle size, and the combination of limes used will determine how fast the medium's pH will increase, to what level, and for what duration.

For more information regarding the effects of lime in potting medium, please refer to chapter 2, Properties of Growing Mixes.

Soluble salts

All of the organic and inorganic components in a solution (soil or water) that conduct electricity are called soluble salts. They are the total dissolved salts in a solution at any given time. The total dissolved salts in a solution can be determined by measuring how well a solution conducts electricity, referred to as electrical conductivity (EC). For example, table salt (sodium chloride) dissolved in water separates into its two individual charged components, sodium (Na+) with a positive charge and chloride (Cl-) with a negative charge. The saltwater contains water and lots of individual ions (Na+ and Cl-). Once the table salt has been dissolved in water, the conductivity can be measured, showing an increase in electrical conductivity as compared to water without table salt.

Soluble salts provide an indication of the nutrients available to a crop. Soluble salts do not indicate which elements are contributing to the measurement or are available, but rather provide growers with a general sense of the availability or level of nutrients in solution. Irrigation sources often contain mineral salts that contribute to the soluble salts. Common mineral salts in irrigation water include calcium bicarbonate, magnesium bicarbonate, and sodium chloride.

Growers use soluble salts to provide an indicator of the presence of macronutrients in the growing medium. The growing medium in a container contains individual components such as bark and perlite, water, and air. Nutrients are dissolved in the water portion of the growing mix, often referred to as the media solution, or they become fixed or attached to the medium's components. Plant roots come into contact with the media solution where they can absorb water and nutrients.

Soluble salts in the potting medium, or media solution, commonly come from many sources such as the medium's components, preplant amendments, water-soluble fertilizers, controlled-release fertilizers, and the irrigation water. Some of the most common fertilizer components contributing to the total soluble salts include ammonium (NH_4), bicarbonate (HCO_3), calcium (Ca), chlorides (Cl), magnesium (Mg), nitrate (NO_3), phosphates (PO_4), potassium (K), sodium (Na), and sulfates (SO_4).

All fertilizers are salts that when dissolved in water disassociate into individual ions. For example, calcium nitrate disassociates into calcium (Ca^{+2}) and nitrate (NO_3^-). Plants do not absorb inorganic nutrient sources as whole chemicals such as calcium nitrate, but as individual charged Ca^{+2} and NO_3^- ions.

Excessive soluble salts in the growing medium can be detrimental to crop production. Not only will plants be growing under stressful conditions, they will most likely have damaged root systems, which open the door to root rot pathogens. Plants that are wilting during the middle of bright, sunny days often identify high salt levels. These plants partially or fully recover in the evening or during the night, as the temperature and light levels have decreased. Typically, overall growth is reduced and the plants appear stunted. In some instances, high soluble salts can cause the leaf margins and tips to become necrotic; this is especially prevalent on the lower leaves.

Cation Exchange Capacity

Cations are positively charged ions in the soil, and anions are negatively charged ions. Why are cations and anions important? Plants do not take nutrients up in their elemental form; instead they are taken up in their charged or ionic form. Nitrogen, for example, is taken up as nitrate (NO_3^{-1}) or ammonium (NH_4^{+1}). In most cases, positively charged ions (cations) are held longer in the soil than negatively charged ions (anions). See table 3.1 below for a listing of the cations and anions plants absorb.

Media components have fixed negative charges. These charges will attract and hold positively charged cations. Ammonium (NH_4^+), calcium(Ca^{2+}), copper (Cu^{2+}), hydrogen (H^+), iron (Fe^{2+}), magnesium (Mg^{2+}), manganese (Mn^{2+}), potassium(K^+), and zinc (Zn^{2+}) are all positively charged fertilizer cations that can be held by the medium's components. The cation exchange capacity (CEC) can be thought of as the soil's ability to retain nutrients. CEC measures the magnitude of fixed negative charges and is expressed as milliequivalents per 100 cubic centimeters (meq/100 cc) of dry growing medium. Perennial mixes generally have CEC measurements of 6–15 meq/100 cc. Soils with low CEC will require more frequent fertilization of cations such as potassium and magnesium. Higher CED measurements of the growing medium, although not very common, are desirable.

Monitoring Fertility

Nutritional problems can arise to even the most carefully thought out and monitored fertility programs. Properly identifying nutritional problems involves a combination of experience, observational skills, and knowledge regarding plant nutrition. Keeping up-to-date records regarding all the events of crop production—including production temperatures, light levels, irrigation practices, and the fertility regime—will remove some of the uncertainty in identifying nutritional problems.

There are a number of steps and tools such as visual diagnosis, graphical tracking of soluble salts and pH levels, soil and water testing, and tissue analysis, that are helpful in determining the nutritional status of crops. The two most important and practical measurements that growers can collect at their facil-

Table 3.1. Nutrient Forms Taken up by Perennials

	Chemical Symbol	Uptake Form
Macronutrients		
Carbon	C	CO_2 (air)
Hydrogen	H	H_2O (water), H^+
Oxygen	O	CO_2, CO_3^-, O^-, OH^-, SO_4^-
Nitrogen	N	NH_4^+, NO_3^-
Phosphorus	P	$H_2PO_4^-$
Potassium	K	K^+
Calcium	Ca	Ca^{+2}
Magnesium	Mg	Mg^{+2}
Sulfur	S	SO_4^-
Micronutrients		
Iron	Fe	Fe^{+2}, Fe^{+3}
Manganese	Mn	Mn^{+2}, Mn^{+4}
Zinc	Zn	Zn^{+2}
Copper	Cu	Cu^{+2}
Boron	B	$H_3BO_3^-$, $B(OH)_4^-$
Molybdenum	Mo	MoO_4^-
Chloride	Cl	Cl^-

ities are the pH and EC of the growing medium. With regular monitoring of the growing medium, most potential problems associated with fertilization can be identified and eliminated. Most growers do not have a proactive plan to prevent nutritional problems; instead, they respond to problems as they arise. Unfortunately, when using a reactive strategy, it may be impossible to undo crop injury that has already occurred. Monitoring the nutritional levels of the growing medium is a preventative approach, minimizing crop damage.

Visual diagnosis

Although identifying nutritional problems visually is often the first indication a problem exists, it is not a reliable method of monitoring crop fertility. Properly identifying the nutri-

Table 3.2. Cation Exchange Capacity

Soil Texture Field/Landscape	CEC Ranges (meq/100 cc soil)
Organic	>50
Fine (clays)	25–50
Medium (silts)	8–30
Coarse (sands)	2–15
Greenhouse mixes	6–15

tional status of a crop visually is problematic because several nutritional disorders can occur simultaneously and can be confused with other cultural problems, such as plant pathogens. Often, by the time growers observe and can properly identify nutritional problems, crop quality may already be reduced or the timing of the crop is delayed. When this occurs late in the production cycle, the crop may never reach the quality level expected, even after the problem has been corrected.

pH meters

I'm amazed that all growers don't have the proper nutritional testing equipment. These tools are relatively inexpensive (often less than \$300 each) compared with the value of the crops we produce. They are cheap insurance policies and can provide an indication that a serious problem is developing, often long before it becomes severe.

There are a number of inexpensive pH meters (\$50–300) available to growers that are accurate enough for greenhouse and nursery applications. Cheap pH meters are good enough for greenhouse monitoring but tend to have a short life span, needing to be replaced in as little as one year. Meters in the \$200–300 range should last considerably longer. The pH probe (not the meter) is not very durable and usually needs to be replaced every couple of years.

When shopping for pH meters, there are a few specifications you should look for. The range of the meter should always be 1–14, with an accuracy of ± 0.1 pH unit. Any meter you purchase should come with a calibration feature, allowing you to reset the equipment in order to take the most accurate measurements. To calibrate these meters, you will also need standard pH solutions, usually pH 4.0 and pH 7.0. These meters need to be calibrated before their initial use and at the manufacturer's recommendation thereafter.

Figure 3.1. pH testing meter in use.

Testing for soluble salts

Measuring soluble salts is a fairly accurate method for growers to determine the nutrient status of a crop, provided the majority of the salts are coming from the intended fertilizers and not some other source. Remember, the soluble salt reading measures all the salts in a solution, not the individual fertilizer nutrients. To determine and quantify the individual nutrient components, growers will have to send samples to a testing facility.

The unit of measure or the term used to quantify the soluble salts of a solution is electrical conductivity (EC). The electrical conductivity of a sample solution is measured using an EC meter. These monitoring devices can be obtained from many reputable greenhouse supply companies. A good, reliable EC meter can be purchased for less than \$200 and will last for several years. To properly calibrate these meters, you will also need to purchase the appropriate EC standards (see the manufacturer's recommendations and carefully follow their instructions, as each meter is different).

Even though growers refer to soluble salts as EC, the unit of measurement for EC is often confusing. The most common unit used by growers to measure electrical conductivity is mmhos/cm (mmhos/centimeter or millimhos/centimeter). Recently, industry professionals have been using the unit dS/m (deciSiemen/meter) to measure EC. Many universities and testing laboratories use S/cm (Siemen/centimeter) or mS/cm (milliSiemen/centimeter) as the unit for measuring electrical conductivity. The units mmhos/cm, mS/cm, and dS/m are synonymous and can be used interchangeably, as they all are EC x 10^{-3} (1 dS/m = 1 mS/cm = 1

Figure 3.2. **Electrical conductivity meters provide quick, reliable measurements of soluble salts of solutions.**

mmhos/cm). Sometimes EC is reported using the unit µmhos/cm (micromhos/centimeter), which is EC x 10^{-6}. Converting µmhos/cm to mmhos/cm is relatively simple and involves moving the decimal three places to the left. For example, 1,000 µmhos/cm is equal to 1.0 mmhos/cm.

Conductivity meters can be used to directly measure the soluble salts of a water or fertilizer solution. Some laboratories will express the salts in a solution as total dissolved solids (TDS) or ppm salts. Growers can convert the EC value into the approximate ppm TDS by multiplying the EC by 640 (EC x 640 = ppm TDS) or convert the ppm TDS into an approximate EC value by dividing the ppm TDS by 640 (ppm TDS/640 = EC).

Remember that measuring the EC only determines the total amount of soluble salts and does not provide any information about the presence or absence of individual nutrients. It is possible to obtain an EC reading that is within the acceptable ranges and still have certain nutrients not present in sufficient amounts. Therefore, measuring EC is a valuable method used to provide an indication of the nutritional status of a crop, but it should only be used as a guideline or reference point. To determine the presence of specific nutrients, samples should be sent periodically to a testing laboratory.

One of the best methods of tracking the fertility status of a crop is to conduct nutrient analysis of the potting medium at periodic intervals during the production cycle. Medium testing helps growers determine, from a nutritional perspective, where their crops are and where they are likely to go. The results can be used to determine if the crop is on target or if modifications to the fertility program need to be made.

To keep the fertility program on track, it is recommended to conduct medium testing using in-house methods on a weekly basis and to periodically send samples to a commercial or university laboratory for nutritional analysis. In many cases, growers only conduct these tests if a nutritional problem is apparent. Routine testing can help growers track nutrition, diagnose nutritional shortfalls or overages, and correct any minor problems before they develop into big ones that could lead to potential crop losses or unsalable product.

It is useful to test unused growing medium out of the bag or hopper before any amendments are added and before any crops are sown or planted. Samples taken a couple weeks after the crops are started are also helpful and provide an indication of the true effect of the lime in the medium. Any key crops or crops suspected of having nutritional problems should be tested on a weekly basis.

Collect potting mix samples from eight to ten pots of the same plant. Each sample is usually collected from the outer edge of the root zone; a top-to-bottom profile is taken, excluding the top 0.5–1 in. (12–25 mm) of the medium, where salts are likely to be concentrated. The samples from each pot can be combined with the other samples to form a representative sample (a composite) from the entire crop. Laboratories usually require 1–2 cups (227–454 g) of medium sent in a well-marked zipper-lock bag.

The EC of the growing medium cannot be measured directly. To obtain an EC of the root zone, the salts must be extracted from the medium and then measured. There are a few sample extraction methods used by universities, laboratories, and growers to measure the EC values of the growing medium. These methods include the 2:1 extraction method, the saturated media extraction, and the pour-through method. Laboratories, both institutional and private, use the saturated media extraction (SME) method, while growers use the 2:1 extraction and the pour through methods for soil testing. Laboratory tests usually run about $30, but are often conducted free of charge through various commercial fertilizer and media suppliers. Consult your supplier to see if this service is available for you.

The SME method is the most accurate technique to quantify many plant nutrients, but involves expensive equipment that most growers do not have. However, growers can use the SME method to affordably measure the pH and soluble salts in-house. An advantage to using the SME method is the results will more closely resemble those obtained through a professional laboratory.

The 2:1 extraction method entails combining 2 parts distilled water to 1 part air-dried medium. How to collect media samples and conduct this procedure is outlined in the box below. A few growers use the 5:1 extraction method to monitor soluble salts.

Many growers often only check the nutritional status of their crops when problems are apparent. Scheduled nutritional monitoring and measuring of the soluble salts will allow growers to track the trends over time, allowing them to make minor corrections throughout the growing cycle before problems arise.

I recommend that growers sample a minimum of five to ten plants weekly from a

Saturated Paste or SME

1. Collect about 8 oz. (237 ml) of growing medium for each sample.
2. Place the sample in a clean container. Fill the container firmly to represent the consistency of the medium as it was in the pot. Do not lightly fill or heavily pack the medium while filling the measuring cup.
3. Add distilled water to the medium until the sample just glistens at the surface. Stir the sample and let sit for thirty minutes.
4. Pour the slurry through a cheesecloth or paper towel to filter out the medium's components and collect the solution in a clean container.
5. Place the pH or EC electrodes into the solution (one at a time), record the value after about one minute or until the number stabilizes.

1:1 and 2:1 Extraction Method

1. Collect samples from the lower two-thirds of the soil profile. Collect medium from the middle to the edge of the pot from several containers of the selected crop. The sample should be taken before the plants are irrigated.
2. Mix all the samples to form a representative sample of the entire crop.
3. Measure a known volume of medium, such as 4 oz. (118 ml). Fill the cup firmly to represent the consistency of the medium as it was in the pot. Do not heavily pack or lightly fill the measuring cup.
4. Place into a large cup. Add one equal volume (same volume as media) of distilled water into the cup. Mix and wait at least fifteen minutes. Measure the pH by placing the electrode directly into the slurry. This is a 1:1 extraction method.
5. After the pH measurement has been taken, add an additional equal volume (equal to the original medium volume) into the large cup. This is the 2:1 extraction method. Mix and wait a few minutes. Measure the EC by placing the electrode directly into the slurry.

crop, measure the EC (pH and nitrate levels should also be monitored), and average the results of the testing procedures. These results should be recorded on a chart or a spreadsheet so they can be easily compared with the target ranges for that crop. Target ranges are simply where plant growth can be maximized; all growers should try to maintain levels between the maximum and minimum target ranges. When readings are below the minimum target range, action should be taken, such as a corrective liquid feed, to bring the EC back up to the desired levels. Conversely, when the EC reading exceeds the maximum target range, growers should take action to reduce the EC back down into the target range.

The pour-through method has become one of the most popular methods growers use to monitor the fertility levels of their crops. There are several advantages of using the pour-through method over other testing procedures. The results obtained closely resemble the actual soil solution that plant roots are growing in, providing a clearer picture of the availability of plant nutrients. The pour-through method requires less time to conduct and does not require sophisticated equipment to provide accurate results. It is a nondestructive testing procedure that does not disturb the root systems of healthy plant materials. Pour-through extracts can still be sent to laboratories for thorough nutrient analysis.

The pour-through method of nutrient extraction involves pouring distilled water onto the surface of the medium and collecting the extract, often called leachate, from the bottom of the container. The leachate collected results from the displacement of soil solution by the distilled water. Some dilution of the extract will occur, but the readings collected from the sample should be fairly close to the soil solution surrounding the roots.

Using water sources besides distilled water to conduct pour-through testing is acceptable. The pH of the leachate collected may be altered slightly, especially if the alkalinity of the alternative water applied is over 150 ppm. If your water source has an alkalinity of over 150 ppm, it is best to use distilled water to conduct this test.

It is important to water the pots to be tested thoroughly—using the same water source, or fertilizer solution, that is normally applied—one to two hours before conducting this test. Failure to water the pots thoroughly before testing causes an improper displacement of the existing medium solution. If this occurs, it is possible for the leachate collected to closely resemble the EC and pH of the water you poured on the pots instead of the pH and EC of the growing medium.

Conduct pour-through tests on at least five pots from every crop. Each pot should be tested individually, but the results of the five pots should be averaged to determine the nutritional status of the crop and if any fertility adjustments must be made. The sample pots should originate from the center of the block;

Optimum Growing Ranges for Perennials Grown in 1 Gal. Containers Using the Pour Through Method

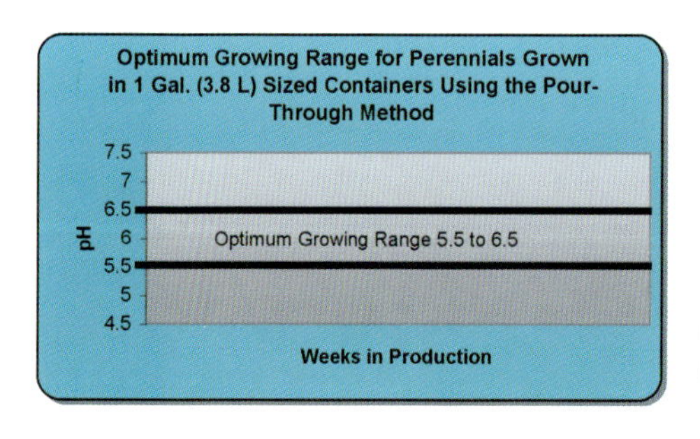

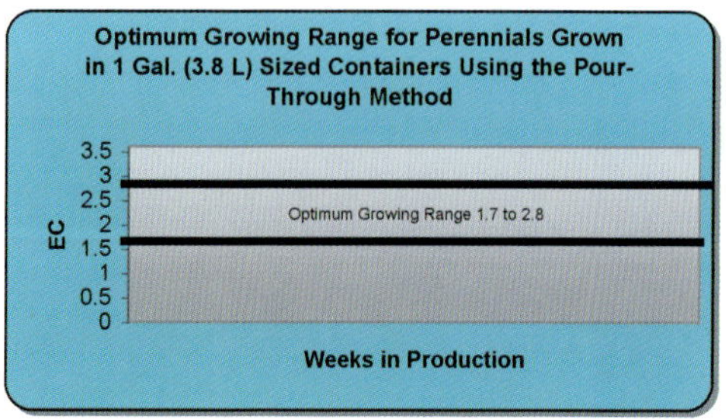

Pour-Through Method

1. The pots to be sampled should be near container capacity. Water them with the normal water or fertilizer source one to two hours before conducting this test.
2. Begin by placing a saucer under the container being tested to collect the leachate from the drainage holes in the container. If the bottom of the container does not have raised edges, place a 1 in. (2.5 cm) section of PVC pipe under the pots so the water drains from the holes unobstructed.
3. Add enough distilled water to the surface of the medium to displace around 2 oz. (59 ml) of solution (leachate) out of the pot and into the saucer. The quantity of distilled water required will vary with the container size. For example, 6 oz. (177 ml) of distilled water may be needed for a 4 in. (10 cm) pot, while 10 oz. [296 ml] may be needed for a 6 in. 9(15 cm) pot to obtain the appropriate quantity of leachate. Always try to collect 2 oz. (59 ml) of leachate regardless of the container size. Collecting more leachate will dilute the extract and provide misleading results.
4. Measure the EC and pH of the leachate.
5. Take multiple samples from each crop. Do not combine the samples before taking measurements. Measure each leachate separately, and then average the results to indicate the crop's current fertility levels.

samples taken from the outside rows are usually not reflective of the true nutritional status of the crop, as they dry out faster and may have been irrigated differently. To obtain the most representative results, pick sample plants randomly; do not select the best or the worst looking plants.

Regardless of the size of the pot being produced, apply enough water to the top of the growing medium to collect 2 oz. (59 ml) of leachate. Note that only the first 2 oz. of leachate should be collected, collecting more than 2 oz. will result in diluted results. It is best to use the least amount of water applied to the top of the pot as possible to collect the 2 oz. The pH and EC are usually measured from the solution collected, though some growers are even testing the nitrates of the leachate solution. The pH of the leachate should be between 5.4–6.0 and the EC between 2.0–3.5 mS/cm for most crops. Salt-sensitive plants should be grown with lower EC levels, 1.0–2.6 mS/cm. Heavy feeders prefer EC levels from 2.6–4.6 mS/cm.

Tissue analysis

A reliable testing method used to determine the concentration of nutrients found within the plant is a tissue analysis. These tests involve taking representative plant parts, namely leaves, and sending them to a university or commercial laboratory for analysis. Normally, recently matured leaves from near the top of the plant are tested. Leaf samples should not come from plants that are wilted or diseased, as they are likely to suffer other problems that will confound the tissue analysis data. To acquire an accurate interpretation of the tissue test results, it is important to take samples from the appropriate plant tissues at the proper time or stage of development. Contact your testing laboratory for the correct sample collection and handling procedures.

When submitting tissue samples for testing, I recommend also submitting media samples from the corresponding plants. Media testing provides an indication of what the plants *could* take up while tissue tests indicate what the plant *did* take up. Frequently, growers will observe that adequate amounts were present in the soil, while insufficient amounts were found in plant tissues due to secondary factors, such as media pH levels. This dual testing will more accurately diagnose and verify that nutritional problems are present.

The elemental concentrations from plant tissue testing are expressed in two ways: macronutrients are expressed in percent dry weight, and micronutrients are expressed in parts per million (ppm). For help interpreting your tissue test results, contact the testing facility or your extension service.

Currently, there are not many nutritional standards developed for perennial crops. It is recommended to submit tissue samples from symptomatic (problem) plants and non-symptomatic (healthy) plants so the results can be compared. In addition to demonstrating the nutrients that have been accumulated in plant tissues, tissue analysis can also demonstrate if any toxic elements have been taken up into the plant.

Testing summary

With any type of testing procedure, it is important to establish consistency over time. Designate one person to conduct these tests; this will reduce the likelihood of variability and improve consistency in the testing procedures. The tests should be conducted in the same manner every time, using the same calibrated equipment and procedures. Each of the extraction methods will provide a different EC value; when noting the results, it is helpful to list the extraction method. Table 3.3 compares the results from each of the above-mentioned testing procedures.

Testing results can be graphed to provide a picture of what is happening over time. Use the graphs to observe the general trend. Are the pH and EC levels increasing or decreasing over time? What effect has a change in the fertility rate applied had on the growing medium? Does the appearance of the plants (are they happy?) agree with your testing results?

The quality and characteristics of the water supply affect the interpretation of growing medium EC values. The salts from the irrigation water often accumulate in the growing

Table 3.3. EC Sufficiency Ranges for Several Extraction Methods

5:1 Media Extraction Method	2:1 Media Extraction Method	Saturated Media Extract (SME)	Pour-Through Method	Indication	Interpretation
0–0.11	0–0.25	0–0.8	0–1.0	Very low	Deficiency symptoms are often apparent.
0.12–0.35	0.25–0.75	0.8–2.0	1.0–2.5	Low	Acceptable for seedlings and salt sensitive plants.
0.35–0.65	0.76–1.25	2.0–3.5	2.6–4.6	Normal	Established plants.
0.65–0.9	1.26–1.75	3.5–5.0	4.7–6.5	High	May begin to reduce plant growth of sensitive varieties.
0.9–1.1	1.76–2.25	5.0–6.0	6.6–7.8	Very high	May cause marginal burn. Do not allow to dry out.
>1.1	>2.25	>6.0	>6.5	Extreme	High potential for burn, root damage, stunting, and wilt.

medium. Growers can adjust for this by measuring the EC value of the irrigation water (with no fertilizers or acids) and subtract this from the EC of the growing medium as measured by the pour-through method. When the EC of the irrigation water is above 1 mS/cm, growers should maintain an adjusted EC value (growing medium EC minus irrigation water EC) toward the lower side of the recommended EC range.

Nutrient Disorders

Plant health, vigor, and performance are greatly influenced by the availability of the necessary macro- and micronutrients. Every plant species, and even plant cultivar, requires differing amounts of these elements. The stage of plant development also plays a role in which nutrients are most important. There is often a fine line between providing adequate amounts of these elements and delivering either excessive or inadequate amounts of them. When excessive amounts are delivered, it often causes side effects, abnormalities, or even severe plant injury; this is called nutrient toxicity. Conversely, when insufficient amounts of nutrients are delivered, it too may cause side effects, abnormalities, or severe plant injury; this is referred to as nutrient deficiency.

There are a number of deficiency and toxicity symptoms that growers can observe to help determine if nutrient-oriented problems are occurring. Many times these symptoms take the form of slow growth, abnormal leaf shape, or changes in leaf coloration. The general function for plant growth and typical disorder symptoms of each element will be described in more detail later in this chapter.

One of the major factors leading to nutrient deficiencies in perennials is the pH of the growing medium. It is generally recommended to maintain the pH of the substrate at a range of 5.6–6.2. When the pH rises above 6.5, certain micronutrients, particularly boron, copper, iron, manganese, and zinc, become difficult for plants to uptake (not available). With low pH levels (less than 5.4), calcium and magnesium become less available.

Plants obtain nutrients with their root systems from the soil solution. These nutrients are moved into the plant for use by plant cells to build plant tissues, produce more cells, and for photosynthesis. When nutrients are less available to a plant, from general lack of fertilizers or due to root rot problems, the plant tries to provide nutrients from the older tissues so that the younger, newer regions can continue to develop. Many elements such as nitrogen, phosphorus, potassium, and magnesium are mobile and can be translocated, or moved, from older tissues to newer ones. Some nutrients such as calcium, sulfur, iron, and manganese are relatively immobile and cannot be translocated within the plant.

Understanding the mobility of nutrients, or their ability to be translocated to other plant parts, will help growers to narrow down which type of deficiency is present, based on the location of the symptoms. Nutrient deficiencies of mobile elements will be expressed in older leaves, while deficiencies of immobile elements will be expressed in the new growth. To understand why the symptoms show up in this manner, think of the growing points of perennials as "sinks" for nutrients because they are rapidly expanding, creating a high demand for nutrients. Mobile elements can be translocated from mature leaves to meet the demand for the new growth, creating a deficiency in the older leaves. Immobile elements cannot be moved from the old leaves to the new growth, so the new growth will show the deficiencies.

Nitrogen is involved in many of the cellular functions, and it is a significant component in the structures of all amino acids, proteins, and many enzymes. Nitrogen does have the ability to be moved from the older, lower leaves to the younger, actively growing shoot tips. The first indication of a nitrogen deficiency, as the result of this translocation, results in the older leaves turning yellow.

Calcium moves up the plant in the water solution as the plant transpires. Calcium only moves through the water-conducting xylem tissues and may not be transported efficiently when excessive air humidity is present. When water is unavailable for transpiration, or when the plant is transpiring at low rates, such as during cloudy weather, calcium uptake is limited, and deficiencies could occur. Calcium is immobile once in plant tissues and cannot move from one plant part to another. Signs of calcium deficiencies in some plants, such as *Rudbeckia,* resemble an upward cupping of the leaves. Other plants, such as some *Leucanthemum* or *Coreopsis* cultivars, show symptoms as wavy or rippled leaves.

There are numerous occasions where growers have provided excessive amounts of certain nutrients, which in turn caused uptake of other nutrients to be hindered. This is referred to as nutrient antagonism. There are many minerals researchers have found that behave antagonistically in regard to the availability and uptake of other nutrients by plant roots. Excess amounts of certain nutrients can decrease the uptake of other elements. For example, excess amounts of phosphorus can inhibit the uptake of copper, iron, and zinc. To avoid antagonism between elements, growers have established certain ratios that when maintained will prevent most antagonism from occurring. These ratios are described in the Nutrient Ratios section of this chapter.

To minimize the likelihood of nutrient deficiencies, growers should take steps to ensure that the injection equipment (proportioner) is calibrated and working properly. When the injection equipment is not working properly, growers may experience problems with several nutrients that may make proper diagnosis and control methods difficult to determine.

Water management practices can also affect the availability of micronutrients. The root growth and uptake of nutrients by plants that are either overwatered or remain constantly saturated for extended periods of time is dramatically reduced or inhibited by low oxygen levels in the root zone. Saturated conditions can lead to inefficient uptake of iron or phosphorus by the inactive root systems. Without adequate transportation of water from the roots to the shoots, plants may also experience calcium deficiency symptoms. Whether root rot pathogens begin as a result of saturated medium conditions or other factors, perennials with root rots quite commonly show deficiency symptoms to various macro- and micronutrients.

Plant nutrients should be monitored during crop production. Simple in-house soil tests can be used to determine if the soil conditions are adequate for plant growth. More detailed results regarding the presence of all nutrients can be obtained by sending samples to a reputable soil testing company. To better determine the levels of elements actually present in a plant, growers can submit tissue samples to laboratories to help illustrate fertilizer imbalances.

One or more of these factors often causes nutrient deficiencies: inadequate fertility, excessive leaching, and poor root health. Insufficient fertility levels occur as the result of inappropriate fertilizer selection, using rates lower than required, and injector malfunctions. Excessive leaching often occurs when the irrigation is run too frequently and for long durations, or in outdoor production sites where growers cannot control or limit the amount of rainfall. Poor root health from plant diseases or low oxygen levels in waterlogged potting mixes often reduce nutrient uptake. To correct deficiency problems, growers should check each of these factors to determine if they can be adjusted to improve the nutritional situation. In many cases, it is necessary to make additional corrections specific to the nutrients that are undersupplied before returning to normal fertility practices.

Correcting Nutrient Problems

Nutritional problems are very common with production of containerized perennials and may go undetected for prolonged periods. Both over- and under-fertilization can lead to reduced plant vigor and increased susceptibility to other cultural problems, such as insects and diseases. Below are several nutritionally related problems growers often face and how these situations can be overcome.

Low pH

Potting substrates occasionally reach pH levels that are too low for certain crops. The primary reasons for low pH levels are consistent use of acids or acidic fertilizers, an insufficient amount of limestone had been incorporated into the potting medium, the limestone incorporated was too course, or the limestone ran out toward the end of a long crop cycle.

Where low media pH conditions exist, there is an increased availability of micronutrients, which possibly could lead to toxicity symptoms from iron and manganese. Low media pH levels could also lead to deficiencies from calcium and magnesium, which could increase the likelihood of ammonium toxicity. There are numerous symptoms crops will exhibit when grown with low media pH. The lower leaves get necrotic spots, stippling, or have marginal burn often caused by micronutrient toxicities from iron, manganese, copper, and zinc. Additionally, there is sometimes a general yellowing and stunting of the crop.

Growers who continuously acidify the irrigation water occasionally observe the pH of the growing medium falling below the desired levels. These growers can simply stop using acid for short periods, allowing the pH to rise, using the alkalinity or the liming effects of their water source. Some growers with extremely low pH levels leach the crops with clear water to increase the pH back into the optimum range.

Some growers apply limestone onto the surface of the growing medium using approximately 1 lb./cu. yd. (454 g/0.76 m^3) for every 0.1 unit increase in pH desired. This is equivalent to ⅛ tsp. (0.6 g) per standard 6 in. (15 cm) pot. Limestone reacts slowly, but when applied to the surface of the growing mix, it will react even slower, which will add to the time it will take to have its full effect on the medium's pH. This type of limestone application takes up to two weeks to begin having an effect on the pH and up to six weeks for the effect to be significant.

Over time, growers can make small adjustments in the pH level, usually 0.5–1.0 units, by switching from acidic, ammoniacal nitrogen fertilizers (ammonium nitrate, ammonium sulfate, urea) to basic, nitrate nitrogen fertilizers (calcium nitrate, potassium nitrate). Using fertilizers to correct the pH level is effective but occurs over several weeks.

More rapid, and larger, adjustments to the medium's pH can be accomplished by mixing 1 lb. of hydrated lime with 100 gal. of water (454 g/379 L), allowing it to sit overnight. Drench the plants with the clear portion of this mixture; do not use the residue that has settled to the bottom. After the application, rinse the foliage to remove any residues.

A similar method to quickly raise the medium's pH entails mixing 1 lb. (454 g) of hydrated lime with 3–5 gal. (11–19 L) of water for five minutes, allowing this mixture to sit overnight. Using only the clear portion of this mixture, inject it into the water supply at a ratio of 1:100 and apply to the crop. One quart of solution should be applied for every square foot of production space, or approximately 10 oz. (296 ml) per 1 gal. (3.8 L) container. The growing medium should be moist prior to applying. After the application has been made, rinse the foliage to remove the residue.

Applying a drench of potassium bicarbonate at a rate of 13.4 oz. per 1,000 gal. (396 ml/3,785 L) will also raise the pH of the growing medium. Using potassium bicarbonate in this manner will increase the alkalinity by 1 meq/L, or 50 ppm, and supplies 39 ppm of potassium. Other growers apply potassium bicarbonate at 2 lb./100 gal. (907 g/379 L). Subsequent fertilizer applications should be adjusted to account for the additional amount of potassium being added using this method. It is also recommended to apply a calcium-containing fertilizer within a few days of application.

Commercially available flowable limestone is also a viable option for growers who need to raise the pH levels of a crop. Flowable limestone reacts faster than surface applications of ground limestone. These products are slightly more expensive than using hydrated lime, but they offer growers more safety, as they are less corrosive. Follow the recommendations on the label to determine the precise rates based on existing and desired pH levels. A general corrective rate recommendation is to apply liquid lime at 4 qt. per 100 gal. (3.8 L/379 L). The leaves should be rinsed following the application to move any residues from the lime.

High pH

Growers occasionally observe media pH levels that are too high for certain crops. High pH levels may be caused by using irrigation water with high alkalinity, consistent use of basic fertilizers, excessive amounts of limestone incorporated into the potting medium, or the limestone used was ground too fine. Perennials grown at high pH levels are prone to micronutrient deficiencies particularly from iron; other deficiencies can occur with boron, copper, manganese, and zinc.

Growers can gradually adjust the pH by small amounts (0.5–1.0 units) by consistently using acidic fertilizers. This requires them to switch from using high nitrate fertilizers to using formulations with high ammonium or urea levels.

To reduce the pH of the growing medium by large amounts quickly, growers can inject acids into the irrigation water to neutralize about 80% of the alkalinity. Many growers have water sources that have high pH and/or alkalinity levels, which require acidification in order to lower these components down to the desirable ranges. The first recommendation is to lower pH down to 5.8. If the alkalinity level is still above the desired range (80–100 ppm), then inject more acid, lowering the pH to about 5.1. For corrective measures, I have successfully injected acid into the water lowering the pH to 4.0 for short durations (a couple of irrigations); do not irrigate with pH levels this low for extended periods of time.

Occasionally, growers use iron sulfate or aluminum sulfate drenches to lower the pH level. They drench either of these products individually, not together, using 1.0–2.5 lb. aluminum sulfate per 100 gal. (454–1,114 g/379 L) or 2–3 lb. iron sulfate per 100 gal. (0.9–1.4 kg/379 L). It is important to rinse off the foliage after the application has been made to prevent possible foliar damage. Both of these products will raise the EC in the growing medium and may lead to micronutrient toxicity.

When growers know the pH and/or alkalinity levels of the water source are high, they can take several steps to help not make matters worse, once a crop is being produced. Future problems can be reduced or eliminated by reducing the lime charge in the growing medium, avoiding high nitrate fertilizers, and acidifying the irrigation water. In many cases, it is recommended to use several of these methods simultaneously.

Low soluble salts

Growers often experience low soluble salts (EC) levels while growing perennial crops. Low EC values indicate that the overall fertilizer levels are too low. Growers often experience low fertility levels as a result of excessive leaching during irrigation, too many clear water irrigations, by applying fertilizers infrequently, or at rates that are too low for the crops being grown.

Common symptoms expressed by plants grown under low EC conditions include slowed growth rate, lower leaf yellowing caused by lack of nitrogen, lower leaf purpling caused by too little phosphorus, and lower leaf interveinal chlorosis from too little magnesium. These conditions, as well as other nutritional deficiencies, are likely to occur when the EC values fall below 0.75 mS/cm using the saturated media extract method.

Growers can correct low soluble salts by increasing the frequency of the fertilizer applications or by increasing the rates that are applied. They should also look at reducing the leach fraction if necessary, allowing no more than a 10% leach fraction with each irrigation. Other methods to increase the soluble salts include increasing the nutrient-holding capacity of the soil and using controlled-release fertilizers. To remedy these situations, it is usually beneficial to use a combination of these methods to increase soluble salt levels.

High soluble salts

Fertilizer levels occasionally rise to concentrations not conducive to plant growth and development. If the EC levels have risen too high, it is usually the result of fertilizing at high rates, or applying fertilizers too frequently. The best method to avoid high EC levels is to apply the proper amount of fertilizer. It may be necessary to decrease the fertilizer rate and the frequency in which they are applied. Salt levels also build up when growers do not allow a leach fraction with each irrigation. It is recommended, whenever irrigating, to always allow for a 10% leach fraction, which should help maintain soluble salts at acceptable levels.

Allowing crops with high soluble salts to dry down too far causes the salts around the roots to increase three to four times the concentration they are when ambient moisture is present. Dry growers should take measures to avoid high salts.

The quickest method for reducing soluble salt levels is to leach the crop. Leaching entails overhead irrigating the crop with sufficient amounts of clear water (no fertilizer) to flush the nutrients inside of the pot out. A common practice is to double leach the crop, leaching it two times, allowing about an hour between each application of clear water. After leaching, let the growing medium dry to the normal levels and test the medium to determine if the soluble salts have been reduced to an acceptable level.

In many cases of high EC, the root system has been damaged, possibly to the point where root rot pathogens have set in. Examine the root system and apply any necessary fungicides after the salt levels have been reduced. Do not apply fertilizers until the root system has begun to grow again.

Besides increasing the leach fraction, reducing the fertilizer rates, and decreasing the application frequency, growers can reduce the nutrient-holding capacity of the growing medium, change fertilizer formulations, and switch water sources as options for managing salt problems.

Ammonium toxicity

Many growers have either experienced or expressed concerns that in greenhouse environments there are certain conditions that may cause ammonium fertilizers to become toxic to plants. Toxicity is most likely to occur when the following conditions exist: cloudy, cool weather; low media pH levels; and a nearly saturated medium. Under normal growing conditions, nitrifying bacteria in the potting substrate converts urea and ammonium to nitrate. This process occurs best when the pH levels are about 7.0 and temperatures are above 60° F (16° C). When the growing medium is cool (below 60° F [16° C]), and the pH is low (below 6.0), these bacteria are less active, causing more potential for the ammonium to turn into a gas that may lead to dangerously toxic levels within enclosed structures.

Ammonium toxicity symptoms look like interveinal V-shaped chlorosis with green veins, and the younger leaves cup upwards or downwards, depending on the plant species. The margins of older leaves may also curl up or down, develop chlorosis, and then turn necrotic. Fewer roots are formed on plants with ammonium toxicities, and the root tips often become necrotic, often with an orange to brown coloration.

The best approach for preventing ammonium toxicity is to use fertilizers consisting of no more than 40% of the total nitrogen being ammonium or urea, or some combination of them, particularly during the winter months. During warmer, brighter months of the year, using ammonium or urea as nitrogen sources is beneficial to promote vegetative growth. Ammonium and urea are not bad sources of nitrogen. It is just important to understand the potential risk when using these products under certain growing conditions and manage the fertility program to minimize the potential for this problem to occur.

I am not familiar with any toxicity cases where growers were using urea-based controlled-release fertilizers within enclosed structures. Time-release fertilizers release relatively small amounts of urea at any given time, so toxicity problems should not occur.

Nitrogen deficiency

Nitrogen deficiency is perhaps the easiest nutrient deficiency for growers to identify. The most predominant symptom is a general change in coloration of the entire plant, from dark green to pale green or yellow. Nitrogen is mobile inside the plant and can be moved from healthy, older plant tissues to newer deficient leaves. Severe deficiencies cause the lower leaves to turn yellow, as the nitrogen is moved to the new growth, and eventually fall off. In many cases, the plants are stunted, have weak stems, and are of lower quality than healthy plants. With slight deficiencies, flowering will occur earlier. Some plant species will express nitrogen deficiencies with a purple to red coloration of the older leaves.

Growers can easily correct nitrogen deficiencies by applying drenches using high rates of nitrate or ammonium-based fertilizers. One or two applications of these fertilizers at rates from 200–400 ppm nitrogen will normally return the color of the leaves back to the normal green color within one or two weeks. Lower leaves that have turned yellow will not return to a green coloration and will usually continue to senesce and fall off of the stem. The rates listed here are for corrective purposes and should not be used for extended periods of time, as this would be an over-application and high soluble salts may result. It is worthwhile to note that high amounts of potassium in the growing medium will create an antagonistic affect with nitrogen, reducing its uptake. If high levels of potassium are present, do not correct nitrogen deficiencies using fertilizers containing potassium.

Phosphorus deficiency

The most noticeable symptom of phosphorus deficiency is a purple pigmentation of older leaves. In many cases the foliage turns to a dark green, sometimes appearing black-green. Phosphorus is mobile within the plant and can move to the newer growth, if necessary. Advanced stages of phosphorus deficiency produce necrotic patches on the leaf margins and cause a general stunting of the plant growth. Older leaves are likely to develop chlorosis, followed by necrosis. Young plants may bypass the chlorotic stage and exhibit only the necrosis. The internodes become short and the plants appear stunted. Deficient plants produce fewer roots, but they are much longer than healthy root systems. Cool temperatures (below 60° F [16° C]) affect the uptake of phosphorus by the plant, even when adequate amounts are available in the growing medium.

Growers can correct phosphorus deficiencies by making drench applications of liquid fertilizers such as 20-10-20 or 20-20-20. Applying these fertilizers at 200 ppm nitrogen will result in 44 ppm and 88 ppm phosphorus, respectively. These fertilizers contain phosphate (P_2O_5); it is recommended to use fertilizers with the percent phosphorus at 50% or more of the nitrogen concentration for corrective situations. Apply complete fertilizers using the standard rates of nitrogen until the deficiency is corrected. Another option is for growers to apply monoammonium phosphate (MAP), injected into the water at 40–80 ppm phosphorus. When deficiencies are present and adequate amounts of nitrogen are already in the root zone, I would consider using the MAP, as it contains only a small amount of nitrogen as compared with the other sources of phosphorus.

Potassium deficiency

Potassium is mobile and can move from plant part to plant part, and deficiencies are harder to identify than nitrogen and phosphorus deficiencies. Early signs that potassium shortages are present include compact and deeper green growth. The predominant symptoms are marginal chlorosis on older leaves, slow growth, and scorched leaves in the advanced stages. The chlorosis may occur at the tips and margins of the older leaves. Necrosis rapidly kills older leaves, starting either at the tips or as scattered spots. Potassium uptake into the plant may also be reduced by antagonism from high substrate levels of ammonium, calcium, magnesium, sodium, or any combination of these.

Applications of potassium nitrate 13-0-44 (KNO_3) or a balanced formulation of a commercial water-soluble fertilizer containing K_2O—such as 20-10-20, 17-5-24, or 15-2-20—using rates of 300–400 ppm potassium can be made for corrective situations. Using these rates of potassium, the normal leaf coloration will return within one to two weeks, except where necrosis has occurred.

Calcium deficiency

Calcium is immobile once in plant tissues and cannot move from one plant part to another. Calcium deficiencies appear near the top of the plant and often occur without growers recognizing them. Deficiencies first appear in the newest growth, or the growing points of the plant, causing poor leaf expansion and cupping to occur. Moderate calcium deficiencies are more subtle and cause leaves to cup, crinkle, or curl; take on a rippled or wavy appearance; form leaves that are strap-like; or cause leaves to be distorted or unfold improperly. Symptoms with severe calcium deficiencies appear as necrotic leaf margins on young leaves and eventual death of terminal buds and root tips. Growers often refer to calcium deficiencies as tip burn, where necrosis appears on terminal shoots. Deficiencies during flower formation may cause the flower stems and petals to collapse and/or bud abortion to occur. Calcium deficiencies cause the formation of short, thick, and densely branched roots.

It is very difficult, and in many cases impossible, to reverse the damage caused from calcium deficiencies. Corrective actions should occur immediately after the problem has been identified. With many crops, calcium deficiencies occur frequently and may require a more preventative approach to managing this nutrient. Calcium deficiencies are also associated with low pH levels, so test the medium's pH and make adjustments if necessary.

Besides inadequate application of calcium, deficiencies can occur as the result of inadequate preplant incorporation of limestone in the growing medium. In some cases, calcium uptake is reduced due to antagonisms from high levels of magnesium, potassium, ammonium, sodium, or any combination of these in the substrate.

Corrective measures most commonly involve drenching crops using fertilizers with high calcium levels, such as calcium nitrate or 15-0-15, using 200 ppm nitrogen. Within two weeks of this application, visible improvements to the new growth should be evident. The leaves with severe symptoms, such as cupping or necrosis, will not return back to a normal, healthy appearance. Growers experiencing low calcium levels and having low pH can apply a flowable dolomite limestone or hydrated lime to correct both shortages.

To prevent anticipated calcium deficiencies, preventative applications of calcium chloride can be applied weekly using 200–400 ppm. I have seen some benefit using calcium chloride sprays on such crops as *Coreopsis, Leucanthemum,* oriental lily, and *Rudbeckia.* For future crop cycles, growers experiencing calcium deficiencies should consider making adjustments to the dolomitic limestone, which is incorporated into the potting medium.

Magnesium deficiency

Symptoms of magnesium deficiencies appear as marginal, interveinal chlorosis of the older, lower leaves. The chlorosis starts along the terminal margins of leaves, moving inward between the veins. The symptoms gradually advance inward on the leaves, followed by marginal necrosis, then defoliation. Young plants may express magnesium deficiencies as leaf curl. Some plant species may take on a pink, red, or purple coloration of the older leaves, followed by the onset of chlorosis.

In most cases, high available levels of calcium relative to magnesium cause magnesium deficiencies. Calcium becomes antagonistic due to the use of calcitic rather than dolomitic limestone in the substrate, high levels of calcium in the irrigation water relative to magnesium, or the continued use of fertilizers containing calcium but little or no magnesium.

The best corrective measure is to make supplemental drench applications of magnesium sulfate (Epsom salts) at a rate of 1–2 lb./100 gal. of water (454–907 g/379 L). The chlorotic symptoms can generally be corrected, returning the leaves back to their normal green coloration within one to two weeks, but the marginal necrosis will remain on the injured leaves.

For future crop cycles, growers experiencing magnesium deficiencies should consider making adjustments to the dolomitic limestone, which is incorporated into the potting medium.

Sulfur deficiency

Sulfur deficiency symptoms can easily be confused with those of nitrogen, as plants appear to take on a lighter green coloration and have a slow growth rate. In most cases, sulfur deficiencies occur over the entire plant simultaneously, whereas nitrogen deficiency often begins with the lower leaves.

Corrective applications of magnesium sulfate (Epsom salts) can be applied. Drenches using 1 lb./100 gal. water (454 g/379 L) will generally return plants to normal color within two weeks. An alternate method worth considering is to acidify the irrigation water using sulfuric acid. Acids should only be used if the pH of the crop is already above the optimum level during the time of the deficiency. Applying acids when the crop is below the optimum level could create additional nutritional problems.

Micronutrient disorders

Micronutrients are commonly applied in combination with macronutrients during each fertilizer application throughout the crop cycle. Many commercial formulations have a complete micronutrient package as a component of the formulation. Several growers use commercial micronutrient formulations, such as STEM and Compound 111, separate from the macronutrient fertilizer formulations. Some growers formulate their own micronutrients, which they inject into the irrigation water using individual micronutrient components.

The first step to correcting problems associated with micronutrients is to check the pH of the growing medium and make sure it is within the recommended range. Before adding any micronutrients, take steps to correct the pH. If a micronutrient problem still persists, consider making additions of the deficient nutrient or take the necessary action to alleviate toxicities.

The second method used to correct micronutrient deficiencies involves applying high concentrations of the deficient element in a single drench application. The drench application should resemble the same volume of solution as would be applied during a normal watering exercise.

Foliar applications of various micronutrients can be made to crops to reverse micronutrient shortages. Spray applications have more risk of phytotoxicity (namely leaf burn) than does correcting the medium's pH or making soil applications of micronutrients. The most effective time to apply these foliar applications is just after sunrise, when the leaves can remain wet for a long time, enhancing nutrient uptake. Also, nutrient uptake through plant leaves is greater during the light period of the day than during the dark night, making morning applications more advantageous than evening sprays. For optimum plant uptake and to prevent phytotoxicity, avoid spraying micronutrients during the midday heat. Using a spreader/sticker with micronutrient sprays will provide more effective coverage.

Before taking any corrective measures regarding any specific deficiency or toxicity problems, growers should make an effort to diagnose the status of all micronutrients. The presence of one nutrient being deficient or in excess does not indicate that all microelements will be the same. Micronutrient disorders can involve one or more nutrients and may consist of several deficiencies or toxicities

Table 3.4. Micronutrient Sources and Rates

These corrective procedures are to be applied only once. Additional applications could prove detrimental to the crop and should only be made after soil and tissue tests indicate the need.

Deficient Nutrient	Fertilizer Source	Rate of Application—Substrate Drench*
Mg	Magnesium sulfate (Epsom salts): 10% Mg	16–32 oz. (454–907 g)
S	Magnesium sulfate (Epsom salts): 13% S	32 oz. (907 g)
	Or switch N or K source to ammonium sulfate (24% S) or potassium sulfate (18% S) for a few weeks. Or if lowering the pH is also necessary, consider using sulfuric acid.	
Fe	Iron chelate: 10% Fe or ferrous sulfate: 20% Fe	4.0 oz. (113 g)
	Or foliar spray iron chelate or ferrous sulfate	4.0 oz. (113 g)
Mn	Manganese sulfate: 28% Mn	0.5–2.0 oz. (12–57 g)
	Or foliar spray manganese sulfate	2.0 oz. (57 g)
Zn	Zinc sulfate: 36% Zn	0.5–2.0 oz. (12–57 g)
Cu	Copper Sulfate: 25% Cu	0.5–2.0 oz. (12–57 g)
	Or foliar spray tribasic copper sulfate: 53% Cu	4.0 oz. (113 g)
B	Borax: 11% B	0.75 oz. (21 g)
	Or Solubor: 20% B	0.4 oz. (11 g)
Mo	Sodium molybdate: 38% Mo	2.7 oz. (77 g)
	Or ammonium molybdate: 54% Mo	1.9 oz. (54 g)
	Or foliar spray ammonium or sodium molybdate with a spreader sticker.	2.0 oz. (57 g)

*All applications are intended to be applied as a media drench unless foliar sprays have been indicated. All rates listed for foliar and drench applications are to be mixed with 100 gal. (379 L) of water.

simultaneously. Taking action without this understanding will, in many cases, increase the problem.

Visual observation is the first step to identifying a micronutrient disorder, but identifying problems visually is subjective and provides no quantifiable results. The best diagnostic tool is to submit leaf samples to a testing facility to conduct a nutrient analysis. Micronutrient values from foliar analysis can be used to compare against standards that have been determined for many plants. The critical foliar micronutrient levels vary slightly from crop to crop, but are fairly consistent across a broad range of plant species (see table 3.5).

Many growers submit soil samples that coincide with the tissue samples for nutrient testing to help determine, or verify, the true nutritional status of the crop. Soil tests alone do not provide a good indication of the micronutrient status, as many of the micronutrients are not included in the test results.

Deficiencies and toxicities caused by excessive micronutrients

Many micronutrient problems occur as the result of too many micronutrients in the growing medium. These excessive amounts commonly lead to micronutrient toxicities, which are the first logical side effect, but most growers don't realize that micronutrient deficiencies are just as common from exposure to high levels of micronutrients. These deficiencies are due to the antagonisms between micronutrients during plant uptake. When two nutrients are antagonistic (micro- or macronutrients), an excessive concentration of

Table 3.5. Suggested Nutrient Ranges and Limits

Nutrient	Irrigation Water	Substrate Leachate	Saturated Media Extract	Plant Tissue
pH	5.4–7.0	5.2–6.3	5.2–6.5	—
Conductivity	0.2–2.0	0.5–2.0	0.75–3.0	—
Bicarbonate	<1,000.0 ppm	—	—	—
Alkalinity	40–160 ppm	—	—	—
Total dissolved salts	<1,000.0 ppm	<1,400.0 ppm	—	—
N	—	100–150 ppm	—	2.5–3.5%
NO_3-N	10.0 ppm	50.0 ppm	35–180 ppm	—
NH_4-N	2.0–10.0 ppm	50.0 ppm	0–30 ppm	—
P	<1.0 ppm	3–15 ppm	5–50 ppm	0.2–0.5%
K	<10 ppm	<100 ppm	50–200 ppm	1.5–3.0%
Ca	<60 ppm	40–200 ppm	40–200 ppm	0.6–1.5%
Mg	<6–24 ppm	10–50 ppm	25–100 ppm	0.5–1.0%
S	<24 ppm	75–125 ppm	50–200 ppm	0.2–0.7%
Fe	0.2–4.0 ppm	0.3–3.0 ppm	0.3–3.0 ppm	35.0–150 ppm
Mn	<0.5–2.0 ppm	0.02–3.0 ppm	0.03–3.0 ppm	30.0–90 ppm
Zn	<0.3 ppm	0.3–3.0 ppm	0.3–3.0 ppm	20–55 ppm
Cu	<0.2 ppm	0.01–0.5 ppm	0.001–0.6 ppm	4–10 ppm
B	<0.5 ppm	0.5–3.0 ppm	0.05–0.5 ppm	25–70 ppm
Mo	<0.1 ppm	0.0–1.0 ppm	0.02–0.15 ppm	0.5–1.0 ppm
Al	0.05–0.5 ppm	0.0–3.0 ppm	—	<300 ppm
Fl	<1.0 ppm	—	—	—
Na	<50.0 ppm	<50.0 ppm	<75 ppm	.01–0.1 ppm
Chlorides	<70.0 ppm	<70.0 ppm	<75 ppm	100-500 ppm

one nutrient in the growing medium will suppress the plant's uptake of the other nutrient.

Several examples of micronutrient deficiencies caused by high levels of other micronutrients are listed below. High levels of manganese in the growing medium will commonly cause iron deficiencies. Conversely, high levels of iron in the root zone will cause manganese deficiencies. Zinc deficiencies may result when either excessive amounts of manganese or iron are present. It is possible for these deficiencies to occur even when adequate amounts of the deficient element are present.

Excessive amounts of micronutrients usually occur when multiple sources of them have been applied. For example it is very common for the growing mixes used by perennial growers to have micronutrients already incorporated in them. Many commercially available fertilizers, both water soluble and controlled release, contain a standard micronutrient package in them. Many growers will inject micronutrients into the irrigation water without knowing that their media supplier has incorporated them into the growing medium already. As the pH of the growing medium decreases, the availability of micronutrients (except molybdenum) increases. Any of these factors, or various combinations of them, can lead to increased micronutrient availability.

Unfortunately, when micronutrient deficiencies occur as the result of antagonistic relationships with an excessive nutrient, most growers do not recognize that the deficiency is the result of an *excess*. The natural reaction is to make a correction by supplementing with a micronutrient package. This action makes the situation worse as the excessive nutrient increases in concentration, continuing to block the uptake of the deficient nutrient.

Whether correcting deficiency or toxicity problems, it is difficult to reverse the effects that excessive amounts of micronutrients may have caused. There are several strategies growers can implement to correct excessive micronutrient situations. The first would be to stop the application of micronutrients. If fertilizers are necessary, use formulations without micronutrients. The second method is to raise the pH of the growing medium, which will decrease the availability of most micronutrients. For example, when raising the pH by one unit, the availability of iron is decreased tenfold. When dealing with antagonistic pairs of micronutrients, growers can apply the nutrient that is deficient to counteract the one that is in excess. For example, when growers have an iron deficiency due to excessive manganese levels, they could apply iron to suppress manganese uptake while increasing iron uptake, thus alleviating both problems. The final method would be to leach the growing medium using clear water, removing the available micronutrients from the soil solution, reapplying them after the problem has been corrected.

Boron

Boron, like several of the micronutrients, is an immobile nutrient that has many roles within the plant. Deficient levels of boron are caused when the pH levels are above 6.5 or excess levels of calcium and potassium are present.

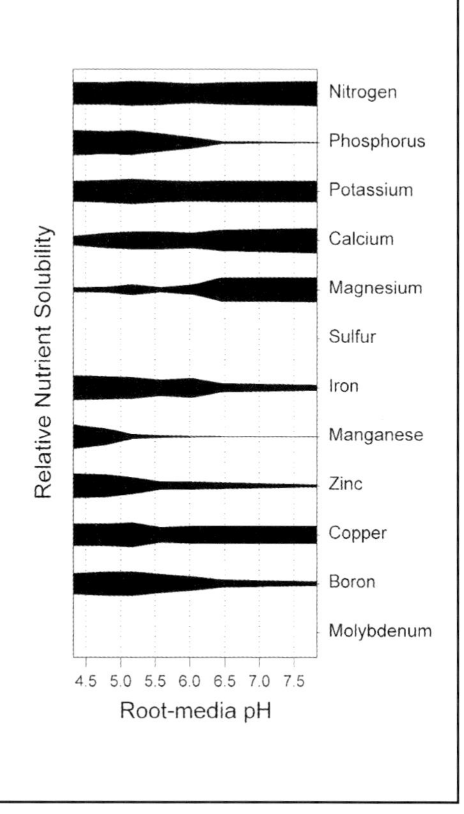

Chart 3.2. **Relative Solubility**
Relative solubility of nutrients at different pH levels in one peat-based medium (graph based on work by John Peterson). Wide sections of each bar indicate greater relative solubility than narrow sections.

Symptoms of boron deficiency appear as clubby, distorted young leaves and stems, followed by necrosis or death of the shoot tip and early maturing lateral shoots. The internodes become short at the top of the plant, and increased branching is likely. The leaves often appear thick and crinkled, and corking may become present. The stems and leaves are often brittle with necrotic spots (black and sunken) just below the nodes. Similar to calcium deficiency symptoms, the roots become short, thick, and densely branched; the root tips may eventually die. It is difficult to reverse the injury caused by boron deficiencies, making early detection and immediate action critical.

Boron deficiency symptoms of the flowering parts can include incomplete formation of the flower parts, fewer petals, small petals, sudden wilting, notches of tissue missing in flower stems, and collapse of flower stems; bud abortion is also likely. Death of the flower bud promotes branching, followed by death of the new buds; this proliferation of shoots is often called witches broom.

Target ranges for boron are 0.25–0.60 ppm in the potting medium and 60–80 ppm in plant tissues. To correct boron deficiencies, apply a drench using 0.5–0.75 oz. of borax or 0.25–0.43 oz. of Solubor per 100 gal. of water (7–12 g/379 L). Applying these products using the high rates will provide 6.25 ppm of boron.

Boron toxicity symptoms are also occasionally observed. The symptoms are found on the older leaves and appear as necrotic leaf margins with a characteristic reddish-brown coloration. Necrotic spots can also be observed across the older leaves but tend to be concentrated near the margins.

Chlorine

Chlorine plays a role with water relations, nutrient uptake, and photosynthesis. It is involved in the movement of water or solutes in cells (osmosis) and the ionic balance necessary for plants to take up nutrients. Chlorine deficiency symptoms include wilting, chlorosis, bronzing, and stubby roots.

Cobalt

Cobalt helps plants to absorb ammonium in root nodules and for nitrogen fixation in legumes. Cobalt deficiencies result in symptoms resembling nitrogen deficiencies.

Copper

Copper becomes deficient when the pH rises or by antagonism with high levels of iron, manganese, or zinc in the growing medium. When this immobile micronutrient becomes deficient, the plants will have distorted younger leaves, necrosis of the apical meristem, and reduced or stunted growth. The young leaves curl, twist, and may become chlorotic. The leaves often take on a bluish green, dark olive, or blackish coloration. Flowering is suppressed and may stop altogether; if flowers develop, they are often lighter in color. When copper is deficient, there is often sudden necrosis of the young, fully expanded leaves that spreads rapidly and resembles desiccation.

To correct copper deficiency, growers can apply copper sulfate as a drench to the growing medium using 0.5 oz./100 gal. of water (14 g/379 L; 9.3 ppm copper). Care should be taken to not overapply. Corrective foliar applications can be made using tribasic copper sulfate at a rate of 4 oz./100 gal. (113 g/379 L), which provides 159 ppm of copper. Tribasic copper sulfate is available as bacteria/fungicides under the trade names Top Cop and Cuprofix (these chemicals are not labeled for use on perennials as of press time). Too much copper can cause toxicity.

Iron

Symptoms of iron deficiency appear as interveinal chlorosis on the young leaves that can progress over the entire plant. Iron is immobile and cannot be translocated to other parts of the plant. In severe cases the young leaves to appear bleached (white), then turn necrotic, leading to shoot tip death.

Iron deficiencies can occur when the medium's pH rises above 6.5, the potting medium is overwatered, low root zone temperatures are present, the root system is damaged, root rots are present, or excessive amounts of limestone, copper, manganese, or phosphorus are applied. The uptake of iron is antagonized by high levels of manganese, zinc, and copper (particularly manganese) in the growing mix. When the medium's pH is high, there may be abundant amounts of iron in the root zone, but it is unavailable for plants to take up.

The most common and effective means of correcting iron deficiencies is to lower the pH to 6.2–6.4 for optimum availability of iron. Foliar applications of iron sulfate (ferrous sulfate) at a rate of 4 oz./100 gal. of water (113 g/379 L; 62 ppm iron) have worked well for growers. It is beneficial to apply foliar applications of iron in the morning, as foliar uptake is enhanced with slower spray drying times. Sprays with iron products are effective but often leave unsightly residues and can cause some foliar damage.

If the medium's pH is within the desirable range for the crop, iron sulfate can be applied as a drench to the crop using 4–6 oz./100 gal. of water (113–170 g/379 L; 62–93 ppm iron). Using chelated iron products such as Sprint 330 (DTPA) as foliar sprays or media drenches has provided growers with some success reversing deficiency symptoms. Apply drench applications of chelates using 4–7 oz./100 gal. of water (113–198 g/379 L). Corrective drench applications should only be applied once to prevent the likelihood of going from deficient to toxic levels of iron in the root zone.

Chelated iron products or complete micronutrient packages including iron are often injected as part of many growers' liquid feed program. The amount of iron added with each application is much lower than the rates applied for corrective situations. Supplementing iron in this manner and adjusting the pH as needed will go a long way to prevent the occurrence of iron deficiencies.

Under certain circumstances, iron toxicity may become prevalent. The symptoms first appear as bronzing on the recently matured leaves. The bronzing usually consists of numerous pinpoint spots that begin yellow and quickly turn bronze. The affected leaves become necrotic. Some plants develop slightly different toxicity symptoms, beginning as numerous pinpoint necrotic spots across the leaf, often black in color. As these spots enlarge, they turn necrotic until the entire leaf is covered and dies.

Manganese

Deficiency symptoms of manganese appear as mottled, interveinal chlorosis on young leaves that progresses into necrosis on leaf tips and margins. The chlorosis may affect the entire plant but is more pronounced in the young leaves. Many times, deficiency symptoms on young leaves are described as a network of green veins on a light green background. As it advances, the light green portion turns white, and the leaves fall off of the plant. Sometimes growers confuse manganese and iron deficiencies. Symptoms in the late stages may also appear as tan flecking in the chlorotic areas between the veins of young leaves. Excess levels of calcium, iron, and magnesium often induce symptoms of manganese deficiency. It is difficult to reverse the injury caused by manganese deficiencies, so early detection and immediate action is critical.

Manganese is immobile and does not translocate once inside of the plant. It is recommended to maintain a 2:1 iron-manganese ratio in the potting mix to limit the competition for uptake between these nutrients. High substrate levels of iron, zinc, or copper create antagonistic conditions, reducing the uptake of manganese into the roots. Deficiencies often coincide with adversely high pH in the medium.

To correct manganese deficiencies, apply a foliar spray of manganese sulfate at a rate of 2 oz./100 gal. of water (57 g/379 L). Foliar applications applied at this rate will supply 40 ppm manganese. Apply manganese sulfate early in the morning to allow the plant leaves to remain wet for long periods, which will enhance uptake into the plant. Manganese sulfate can also be applied to the soil as a drench using 0.5 oz./100 gal. of water (14 g/379 L). Drench applications should be applied using the same volume of water applied to the crop as used in the normal watering practices. Drench applications will take slightly longer to see results from compared to foliar applications. Do not over-apply manganese sulfate with either foliar or drench applications. Fungicides containing mancozeb, such as Protect T/O, can also be applied to meet the manganese needs of the plant in addition to controlling plant pathogens.

Under certain circumstances, manganese toxicity may become prevalent. Toxicity symptoms begin with interveinal chlorosis of young leaves due to iron deficiencies, causing a high manganese antagonism of iron uptake. Eventually, the toxicity takes the form of necrotic tips and margins or reddish-brown spots on the older leaves. The spots are usually scattered over the entire leaf; as they become more numerous, they eventually coalesce into patches.

Molybdenum

Molybdenum is a structural component of enzymes that reduces nitrates to ammonia. This process is critical; without it the synthesis of proteins is blocked and plant growth ceases. Plants with the ability to fix nitrogen (root nodule bacteria) require molybdenum. Deficiencies are rare and appear as pale green leaves with rolled or cupped margins. The margins of the leaves at the middle of the plant become chlorotic, quickly turning necrotic. The symptoms move down the plant.

Nickel

Nickel has just recently been acknowledged as an essential micronutrient. It is a component of the protein urease, which is required for the breakdown of urea into the nitrogen form usable by plants. Nickel is also needed for iron absorption and seed germination. Plants deficient in nickel tend to mature and reach the reproductive stage quicker than normal; however, they fail to produce viable seed.

Silicon

Silicon is a component of cell walls, making them stronger, helping to improve the plant's ability to withstand heat and drought, and increasing resistance to insect and disease infections. The presence of silicon in plants can offset potential nutrient toxicities from manganese, iron, phosphorus, and aluminum as well as zinc deficiency. Silicon has not historically been considered an essential element, but it often composes up to 10% of the dry weight of plant tissue. The complete role of silicon is not fully understood, and there are no known deficiency or toxicity symptoms at this time.

Sodium

Sodium plays a similar role in plants as chlorine does; it is involved in the movement of water and the ionic balance in plants. Under normal growing conditions, it is uncommon for growers to observe any signs of deficiencies or toxicities.

Zinc

Zinc becomes deficient when the pH rises or by antagonism with high levels of iron, manganese, or copper in the growing medium. Deficiency symptoms include curling of the upper new leaves, mottled leaves with chlorosis in the interveinal areas, short internodes forming a rosette appearance, and leaves and flowers that fall off of the stem. Other symptoms observed include abnormally small young leaves, curled leaf margins that often form spoon-shaped leaves, and necrosis on recently matured leaves. Zinc deficiencies often lead to iron deficiencies.

Growers can take corrective action by drenching zinc sulfate to the growing medium using 0.5 oz./100 gal. (14 g/379 L). Foliar applications can also be applied using zinc sulfate at a rate of 2 oz./100 gal. (57 g/379 L), which will provide 56 ppm of zinc. It is important to not overapply zinc sulfate when making corrective applications. Fungicides containing mancozeb, such as Protect T/O, can be applied to meet the zinc needs of the plant in addition to controlling plant pathogens. Lowering the pH of the growing medium can cause zinc to become more available, possibly leading to zinc toxicity.

Acidification

It is often necessary for growers to inject acid into the irrigation water to lower the alkalinity and/or the pH. Acidification reduces the quantity of carbonates and bicarbonates in the water. Essentially, a reaction occurs between the hydrogen ions (H^+) from the acid and the bicarbonates (HCO_3^-) in the water to form carbon dioxide gas (CO_2) and water (H_2O).

The acids used to neutralize water alkalinity include sulfuric (H_2SO_4), phosphoric (H_3PO_4), nitric (H_2NO_3), and citric ($H_3C_6H_5O_7$). There are several factors to consider when deciding which source of acid to use. These factors include: ease of use, safety, relative cost, availability, and which nutrients are being injected (nitrogen, phosphorus, or sulfur).

The amount of acid growers must add to the water depends both on the pH and the alkalinity of the water source. Water with high alkalinity usually requires more acid than water with low alkalinity, even if the initial pH of the water source is lower.

Handling

Acids can be dangerous to use if the proper precautions aren't followed or if they are misused. Weaker acids such as citric, 75% phosphoric, or 35% sulfuric are relatively safe to work with as compared to 67% nitric or 93% sulfuric acid. Nitric acid is very caustic and can cause serious injury to exposed tissues, especially the eyes. Nitric acid also can create fumes during handling procedures, and steps should be taken to avoid breathing these fumes.

When handling any acid, take *all* the precautionary steps you can to avoid exposure to your skin, eyes, and lungs. Use acid-resistant eyewear (preferably a full-shield face mask), gloves, and coveralls. Acids are corrosive and will cause damage instantly to your clothing and body if you are not properly protected. Always keep some baking soda around your acid mixing or injection facilities. If you come into contact with an acid, apply baking soda immediately (before flushing with water). The baking soda will neutralize the acid, greatly reducing the amount of injury it can cause to you. If you flush only with water, a reaction between the acid and the water occurs that can actually cause more harm to your body. Neutralize the acid first, and then flush it off of your body.

Many growers have injection equipment that does not require them to mix or handle the acids. If mixing is necessary, follow these guidelines. When mixing stock solutions of acid, always add the acid to the water. Never add water to the concentrated acid. Improperly mixing acids will increase the risks of this procedure, causing these components to react with each other, creating heat and fumes that could lead to unnecessary exposure. The heat formed from incorrect mixing practices weakens the supply lines and injection equipment, which could rupture, creating a spill or emergency situation.

Cost

Generally, sulfuric and nitric acid are less expensive that citric and phosphoric acid. The prices, of course, will vary throughout the country depending on availability, supplier, and the quantity being purchased.

It may be cost prohibitive to use citric acid to neutralize the alkalinity of irrigation water. It is an expensive and weak acid that is not well suited for applications where high alkalinity levels in the irrigation water exist. Growers do commonly use citric acid to neutralize the water used in pesticide spray tanks and fertilizer stock solutions.

Nutrients

Most acids (citric acid is an exception) used for acidification also supply plant nutrients in conjunction with the hydrogen ions that neutralize the bicarbonates. These nutrients are sulfur from sulfuric acid, phosphorus from phosphoric acid, and nitrogen from nitric acid. In many cases, these elements are beneficial to plant development, provided they are not supplied in excess. But they can also react with other fertilizer salts, reducing these elements from being available to the plant.

Growers should take into consideration the source of the acid they are using, get an understanding of how it will affect their fertility programs, and make any necessary adjustments prior to irrigating with acids and fertilizers. For example, growers using phosphoric acid should reduce the source of the phosphorus supplied by the fertilizer by the anticipated amount of phosphorus being supplied by the acid.

It is somewhat difficult for growers to calculate the precise amounts of these nutrients, or even the total amount of acid, being injected at any given time. If you have a good handle on your water quality and know your alkalinity, you can mathematically determine how much acid would be required to bring the alkalinity down to an acceptable range.

If your alkalinity is 300 ppm and you are using phosphoric acid to neutralize the alkalinity, 126 ppm of phosphorus (280 ppm P_2O_5) is supplied at each irrigation. The maximum amount of phosphorus recommended to growers is usually around 55 ppm phosphorus (125 ppm P_2O_5). This example shows that using phosphoric acid to reduce high alkalinity sources of irrigation water is not feasible and could lead to other nutritional problems. Alternate sources of acid should be considered to treat high alkalinity water sources.

Using 67% nitric acid to neutralize 300 ppm alkalinity would supply 67 ppm nitrogen with each irrigation. This is a significant amount of nitrogen and may in itself be adequate nitrogen for plant growth. Or it may become problematic if nitrogen levels were already optimal or on the high side.

Sticking with the same scenario of reducing irrigation water with 300 ppm alkalinity, using sulfuric acid would supply 75 ppm of sulfur. Generally, it is recommended to have 20–30 ppm sulfur for most floricultural crops. Excessive amounts of sulfur have a limited effect on crops and may be the best choice to reduce high alkalinity irrigation sources.

Which acid and how much?

Each grower seems to have his own preferences or circumstances that lead to using a particular source of acid. From a safety standpoint, sulfuric acid is better than nitric acid, not only to human safety, but plant safety. It would be really unusual to experience a problem due to excessive amounts of sulfur being applied, but too much nitrogen or phosphorus could have a negative affect on plant growth and overall crop quality. Sulfuric acid is also a better choice when dealing with very alkaline water sources.

To determine the proper amount of acid to inject into the irrigation water, both the pH and alkalinity of the water source need to be taken into consideration. Table 3.6 demonstrates how much of different sources of acids must be added to the water source for every 50 ppm of alkalinity that is present.

To more precisely calculate the necessary amount of acid to inject and the costs of doing so, researchers from North Carolina State University have developed an alkalinity calculator that growers can use to answer these questions. Growers simply input their water pH and alkalinity, the source and cost of acid being used, and the target pH and alkalinity. The spreadsheet will then calculate the nutrient additions from the acid injection and the costs of doing so. This is an invaluable tool and is available over the Internet at http://floricultureinfo.com.

Be sure that the manufacturer approves any injection equipment used to inject acids for this type of application. The manufacturer should be consulted to select the best injection option for your circumstances. Failure to use the right equipment may cause permanent damage to the injector, void any warranties, create an emergency situation, and may cause unnecessary exposure of workers to acid.

It is recommended to not mix fertilizer stock solutions with acid stock solutions. Depending on the acid-fertilizer combination, fertilizer salts could precipitate out of solution. Always use separate injectors for each type of solution.

When using acid in greenhouses with galvanized piping, check the integrity of the plumbing periodically, as acidic water will corrode them.

Fertilizer Programs

Fast-growing perennials, especially under greenhouse conditions, tend to have a high demand for nutrients. Growers commonly deliver nutrients to their crops under these growing conditions by means of time-release fertilizers or frequent fertilizations using water-soluble sources of nutrients. A good fertility program provides adequate nutrients to perennial crops at all times. If nutrients in the growing medium fall below or above the optimal range for any given plant, plant health will be compromised.

It would be ideal to deliver nutrients only to the plant when they need them and in the exact proportions required; however, fertilizers are not delivered in this manner. In many instances, fertilizers contain inadequate ratios of nutrients, excessive amounts, and undesirable elements. Many of the nutrients from the salts in the growing medium, irrigation water, and fertilizers are not used by the plant and are often left behind. If these nutrients are not leached out of the root zone regularly, they will accumulate and could become problematic.

Table 3.6. Properties of Acids

Acid	Formulation	Amount of Acid Per 1,000 Gal. (3,785 L) to Reduce Each 50 ppm of Alkalinity Present*	Concentration of Nutrient Added by 1 oz. Acid Per 1,000 Gal. (3,785 L) Water**	Approx Cost***	Safety Comments****
Citric ($H_3C_6H_5O_7$)	99.5% granular	9.1 oz. (269 ml)	None	$0.71	May cause minor skin and eye irritation.
Citric ($H_3C_6H_5O_7$)	50% liquid	14.5 oz. (429 ml)	None	$0.99	May cause minor skin and eye irritation.
Nitric (H_2NO_3)	67% liquid	6.6 oz. (195 ml)	1.64 ppm N	$0.28	Very caustic and dangerous. Use extreme caution. Avoid contact with the acid and its fumes.
Phosphoric (H_3PO_4)	75% liquid	8.1 oz. (240 ml)	2.88 ppm P	$0.47	Slightly caustic and dangerous. Use with caution. May cause skin and eye irritation.
Sulfuric (H_2SO_4)	35% liquid	11.0 oz. (325 ml)	1.14 ppm S	$0.17	Slightly caustic and dangerous. Use with caution. May cause skin and eye irritation.

* Add this amount for each 50 ppm reduction of alkalinity desired. For example, if your alkalinity is 200 ppm and you wish to lower it to 50 ppm, then you would need to add 33 oz. (976 ml) of 35% sulfuric acid per 1,000 gal. (3,785 L) of water ([11.oz/1,000 gal.] x [150 ppm/50 ppm]) = 33 oz. 35% sulfuric acid/1,000 gal.

** Using the above example, you would be supplying 37.6 ppm sulfur at each irrigation (33 oz. x 1.14 ppm sulfur/1 oz. [30 ml] sulfuric acid).

*** Cost will vary. These costs we derived from the following: $1.25/lb. for 99.5% citric acid, $8.75/gal. for 50% citric acid, $5.35/gal. for 67% nitric acid, $7.45/gal. for 75% phosphoric acid, and $1.90/gal. of 35% sulfuric acid.

**** Use caution with all acids.

High salts make interpretation of soluble salts from in-house testing less meaningful, as the grower cannot distinguish whether the salts are from useful or undesirable elements. Accumulated salts from undesirable elements, such as sodium and chloride, can also lead to plant stress and increased disease susceptibility.

Successful plant nutrient management requires a thorough understanding of the relationships between irrigation practices, fertilization programs, and growing medium selection. These three cultural disciplines are interconnected. Even the slightest change in one of these cultural practices can greatly affect plant nutrition. Understanding, recognizing, and manipulating these relationships can help growers provide and maintain successful fertility programs. There are numerous factors, changing from crop to crop, season to season, and year to year, that impact plant nutrition and the need to adjust nutritional programs constantly.

The availability of nutrients is controlled by the frequency of fertilizer applications, the cation exchange capacity (CEC) of the growing medium, and the medium's pH. Using low rates of water-soluble fertilizers is desirable to maintain nutrient availability since the CEC of most growing mixes is low. The growth rate of a plant often corresponds with its water use and nutrient demands—as the plant grows, there is a higher demand for nutrients and water. Liquid fertilizer programs work quite well. With each required irrigation, more nutrients are supplied, keeping up with the nutrient demands of the plant.

The environment plays an important role in regard to the EC of the medium. During cool conditions, less watering usually occurs, often causing the soluble salts to build up in the container. Periods of low light levels, cloudy conditions, and high humidity will cause the same effect.

Factors affecting nutrient management programs

- Any change to the potting mix, large or small
- Starter charges of the growing medium
- Moisture content of the medium prior to filling containers
- The porosity of the growing medium
- Source of irrigation water: pond, well, municipal
- The alkalinity of the water source
- The irrigation method: overhead versus subapplication
- The irrigation frequency and quantity applied
- New water technicians
- The specific nutritional needs of each crop
- The form of nitrogen in the fertilizer: $NH4+$ or $NO3-$
- Frequency and rates of fertilizers applied
- The acidity or basicity of the fertilizers applied
- The growth response from the nitrogen source: lush versus toned
- Understanding the nutrient contributions of the water, media, and fertilizer
- Knowing where each needed nutrient is coming from: water, fertilizer, or medium
- Weather conditions
- Container size
- Age of the crop
- The occurrence of insects and diseases

When putting together a fertility program, it is important to look at the ratios of certain chemicals to each other, in addition to looking at the actual values in the medium. Many nutrients interfere with or reduce the uptake of other nutrients. Each perennial will have its own optimum nutrient ratios, but some general ratios have been established to provide growers an indication of which ratios generally produce normal plant growth. For example, a general recommendation would be to supply a nitrogen-potassium ratio of 1:1, which generally promotes normal growth and height development. When this ratio sways one way or the other, the rate of growth and quality characteristics will be altered.

The maturity or the stage of development plays a role in the types of nutrients required or what ratios they are delivered to the plants. Typically, young, actively growing plants tend to require relatively more calcium and nitrogen than do flowering ones, which usually require more phosphorus and potassium.

Forms of fertilizers

There are several forms of fertilizer available for perennial growers to apply to their crops. These include:

- Liquid sources: must be diluted and injected into the irrigation water
- Soluble crystals or dry salts: must be dissolved in water prior to injecting
- Tablets: easy-to-measure soluble sources that dissolve over time
- Slow release: granules or plastic-coated forms that can release over extended periods of time
- Organic: materials that release low rates of nutrients over long periods of time

Growers can deliver nutrients using one or several of the above forms. The form used is not important, as long as the required nutrients are provided. There are several factors growers consider when selecting a particular type of fertilizer: the form of nitrogen, solubility, chemical purity, its basic or acidic properties, ease of handling, longevity, personal experience, and price.

There are a number of commercially available water-soluble and controlled-release fertilizers in various formulations. All commercially available fertilizers list the percent of nitrogen, phosphorus, and potassium in a standard format that looks like this: 20-20-20. The first number always refers to the percent nitrogen, the second expresses the percent phosphorus in the oxide form (P_2O_5), and the last number always represents the percentage of the oxide form of potassium (K_2O). In the above example, 20% of the compound is nitrogen, 20% is phosphorus, and 20% is potassium. Another example is a 15-5-10 formulation, which contains 15% nitrogen, 5% phosphorus, and 10% potassium.

To show the relationship between the percentages and the weight of these materials, we can take the 15-5-10 formulation, as demonstrated above, and for simplicity use 10 lb. (4.5 kg) as the size of the package being used by the grower. This provides 1.5 lb. (680 g) of nitrogen (10 lb. package x 15%), 0.5 lb. (227 g) phosphorus, and 1 lb. (454 g) potassium for each 10 lb. (4.5 kg) bag. The remaining portion is filler used to complete the mixture, providing the desired percentages.

Selecting a fertilizer is also dependent on what type of growth a grower is trying to achieve. Fertilizers containing high levels of nitrate nitrogen provide toned growth (compact growth of small leaves and short internodes), while fertilizers with high percentages of ammoniacal nitrogen promote soft, lush growth (large leaves and long internodes). Phosphorus promotes cell elongation and can be used by growers to regulate plant height. For example, during periods of low light, where plant stretch is more likely to occur, growers can reduce the phosphorus applied to the crop to help reduce stem elongation.

Growers should also take into consideration the nutrient charges of the growing medium and the water supply. Many commercial mixes contain calcitic or dolomitic limestone, which contain calcium and/or magnesium. With these elements premixed in the growing medium, the amount of these elements in the fertility program can often be reduced or adjusted.

Water-soluble fertilizers can be somewhat misleading. For example, many of them are marketed as "complete" fertilizers, implying that they contain all of the elements necessary for plants. The reality is that most complete fertilizers only contain the primary macronutrients (nitrogen, phosphorus, and potassium), not all nutrients as the name suggests. Many of these commercially available fertilizers contain trace elements. Some of them contain calcium and magnesium, while many of them do not. It is often necessary to rotate fertilizer sources, or supplement with certain components, to complete the fertility needs of various crops.

Fertilizer rates

The amount of fertilizer to apply varies greatly, depending on the plants being produced, growing conditions, light levels, fertilizer form, application frequency, irrigation practices, water quality, and the predetermined need. The rate of fertilizer to apply varies depending on the plant species; some perennials are light feeders, while others are considered heavy feeders. The age of a plant also affects the fertility rate; young seedlings require less fertilizer than do older plants. It is generally recommended to apply low application rates on a regular basis, which is effective at maintaining plant health without encouraging excessive growth.

The fertilizer rates should be adjusted to coincide with the light levels. Perennials grown under high light conditions require higher levels of fertilizer. The rates applied in early spring need to be increased by 50–100%, corresponding with the increased light levels during the late spring and early summer. During periods of low light, the fertility rates can be decreased, as the growth rate is reduced. An exception to lowering the fertility rates with the light levels occurs when the medium doesn't dry out very quickly, during periods of low light and/or cool conditions; fertilizer rates may need to be increased since the frequency of application is often decreased.

Growers should conduct routine soil tests to determine the current nutritional status of the crop. When these tests indicate fertilization is required, a nutrient form, formulation, and rate can be determined.

Fertilizers are soluble nutrient salts. These salts, when dissolved in water, compete with root hairs for moisture. As long as the chemical concentrations are higher inside of the roots than outside, water can move into the plants. Once fertility levels rise in the root zone, water fails to move into the roots as easily, causing a fertilizer burn (leaves often lose color and turn brown) to occur. Excessive nutrient levels can occur from a single application, multiple applications, improper watering practices, poor water quality, poor drainage, or overly dry conditions.

The influence of air on plant nutrition

Often overlooked by growers, the air and certain characteristics about the air in the growing environment have a great effect on plant growth and plant nutrition. The humidity and carbon dioxide (CO_2) levels are the two most important characteristics that affect plant development.

High humidity levels can increase the incidence of many foliar diseases and often cause physiological disorders such as edema and guttation. Edema appears as callused corky spots on the undersides of plant leaves and stems. Guttation occurs as seepage of cellular fluids from leaf margins.

The movement of calcium through a plant is greatly influenced by the humidity level of the growing environment. Calcium is only transported through the xylem, the water-conducting tissues inside of the plant. Calcium is only translocated through transpiration. Under high humidities, transpiration is greatly reduced, decreasing the amount of calcium that is taken up by the roots and moved to the growing points, where it is needed. Once inside of the plant, calcium does not redistribute, or translocate, from one plant part to another. In contrast, nitrogen can easily move from older leaves to newer plant growth when inadequate amounts are available to be taken up by the roots.

During extended periods of cloudy, rainy weather, adequate amounts of calcium cannot be taken up by the plant, causing deficiencies to develop with many plants. *Rudbeckia fulgida* 'Goldsturm' exhibits calcium deficiency by developing leaves that are cupped upward. Several perennials, such as *Coreopsis* and *Leucanthemum* cultivars, form leaves that appear to be rippled or misshaped, as they do not unfold and develop properly. To prevent these types of calcium/humidity related disorders, growers can apply sprays of calcium chloride during periods when humidity levels are expected to be high for several days.

Carbon dioxide is essential to plant growth and development. Next to water, CO_2 is the chemical compound used in the largest quantities by the plant. When the level of CO_2 in the air drops below 200 ppm, plants will stop growing. Natural CO_2 levels vary from 200–400 ppm; the average level in most locations is around 340 ppm. Under normal growing conditions, sufficient amounts of carbon dioxide are present in the air to sustain normal, healthy plant growth. Some growers supplement the CO_2 levels in their facilities by running carbon dioxide burners to increase the level of CO_2 in the air.

Liquid fertilization plans

The most common means of addressing the fertility needs of perennial crops is to provide liquid fertilization using water-soluble fertilizers. Water-soluble fertilizers (WSFs) offer growers a great amount of flexibility. The rates applied can be dialed up or down depending on the needs of the crop. WSF is most economical and environmentally friendly when used through subirrigation systems, drip irrigation, or to closely spaced plant materials.

It is important to supply constant amounts of certain nutrients such as nitrogen and potassium because they are easily leached from the medium. Liquid fertility programs should provide adequate amounts of both the macro- and micronutrients and maintain the pH within an acceptable range. Note that not all commercial liquid fertilizer blends offer calcium, magnesium, or micronutrients; refer to the analysis on the fertilizer bag to check the total elemental composition.

During low light conditions, plants are more prone to stretching, requiring growers to adjust the amount of fertilizer they apply to their crops. When these conditions are prevalent, there are a number of strategies growers can implement to minimize undesirable plant stretch. They can fertilize less often, use lower rates, or use a fertility program that contains lower amounts of ammonium, urea, and phosphorus.

When high light conditions occur, plants tend to grow more rapidly and use nutrients more quickly, especially when grown outdoors or in retractable-roof greenhouses. Under these conditions, growers should fertilize more frequently, use higher rates, or use fertility programs with more ammonium, urea, and phosphorus.

It is important when following a liquid fertility program to have good irrigation practices. Whether growers are considered to be wet growers or dry growers, it is important that when they water, they water thoroughly. Thorough watering establishes a healthy root environment by dissolving excess soluble salts and moving them out of the root zone while drawing air into the medium from the surface.

Flushing soluble salts out of the medium, referred to as leaching, can be done with or without using fertilizers. Leaching is important to keep soluble salts, such as sodium and chloride, from accumulating to levels where injury to the roots could occur. The leach percentage refers to the amount of irrigation that drains out of a container after it has been applied. With each irrigation, it is recommended to have 10–15% of the volume of water applied come out of the bottom of the pot. For example, if 10 oz. (296 ml) of water or fertilizer solution is applied to a 4 in. (10 cm) container, then 1.0–1.5 oz. (30–45 ml) should be flushed out of the bottom of the pot.

It is important to not have excessive leach fractions, which just wastes water and fertilizer. The leach percentage has a great influence on the crop's fertility requirement; higher leach percentages require higher fertilizer rates. Growers with higher leach percentages remove more fertilizer from the root zone and often require higher fertility rates. Conversely, growers using no or small leach percentages are removing very few nutrients from the growing medium and can usually fertilize with lower rates.

Fertilization frequency

Growers using water-soluble fertilizers deliver nutrition to their crops using a number of different methods. The method to apply fertility is often determined by the crops being grown, grower preference, labor inputs, and the availability of injection equipment.

Constant liquid fertilizer programs involve applying a diluted solution of fertilizer at each watering. This type of fertilization method ensures growers that the nutrients in the soil are always available for plants to uptake with little fluctuation. When growing conditions are optimal, the chance of fertilizer levels building up is reduced because only low concentrations are applied with each irrigation. Most growers using constant liquid feed programs have adopted the practice of irrigating with clear water every third or fourth irrigation to prevent high salt levels from accumulating.

To some extent, constant feed programs automatically adjust to the needs of the plant. For example, during sunny, warm periods, plant growth increases, requiring more nutrients to support plant growth. These conditions also cause plants to use more water, as they tend to dry out faster, which creates the need to irrigate more frequently. Irrigating more frequently also allows the plant to receive fertilizer with each irrigation. Conversely, during cool, cloudy weather, the rate of growth is reduced, as well as the need for water, requiring less irrigation and, therefore, less fertilizer is applied.

Applying fertilizer constantly is slightly easier to manage than using other methods, such as scheduled applications, because the workers do not have to make fertility decisions and two processes are occurring simultaneously.

Scheduled fertilizer applications consist of using a more concentrated fertilizer solution that is applied at periodic intervals, such as weekly. This type of fertilization leads to variability, as the nutrients available for uptake will fluctuate greatly between the intervals, ranging from very high concentrations to very low levels within a short period of time.

A third method involves determining when fertility is necessary and applying it on an as-needed basis. Providing nutrients in this manner can be very effective for those growers using routine testing but very challenging for growers who are only guessing at when to apply plant nutrients. This strategy, unless meticulously managed, could lead to wide swings in nutrient availability (excessive to deficient) and plant health.

Water-soluble fertilizer rates

The fertilizer rates of water-soluble fertilizers are usually expressed as parts per million (ppm) of nitrogen in the solution being applied. The ppm that growers use today is typically lower than they were ten years ago. For example, a decade ago growers using scheduled fertigations applied 250–700 ppm nitrogen on a weekly basis. Today, scheduled applications range from 150–400 ppm nitrogen. Growers using constant liquid-feed programs of 100 to 250 ppm nitrogen with every watering in the past are using 50–150 ppm today. The reduction in the rates being applied is a result of research showing that additional amounts of fertilizer do not always increase plant growth or improve plant quality. In many instances, many cultural problems arise due to overfertilizing crops.

Figure 3.3. **This Anderson injection system is just one of the numerous injector systems growers use to accurately inject fertilizers into the irrigation water.**

When discussing fertilizer rates, growers often use ranges such as the ones listed above to refer to their typical targets and production practices. Growers tend to use the lower end of the range during periods when plant growth is slow and during clouding weather. When optimal conditions for plant growth exist, such as sunny periods, growers often deliver fertilizer concentrations at the upper end of the range to provide adequate nutrition during periods of rapid growth.

Acidic or basic properties of fertilizers

Most water-soluble fertilizers have some effect on the pH of the potting medium after application. Each fertilizer is categorized by its acidity or basicity. It is very important for growers to know and understand the affects acidic or basic fertilizers have on the medium's pH and crop production.

A common practice of fertilizer suppliers is to list the potential acidity or basicity of the fertilizer on their bags, labels, or technical

Table 3.7. Potential Acidity or Basicity of Common Water Soluble Formulations*

Formulation	Potential Acidity**	Formulation	Potential Acidity**	Formulation	Potential Acidity**	Formulation	Potential Basicity***
17-6-6	1,800	20-20-20	597	7-40-17	350	15-5-25	37
21-7-7	1,700	18-18-18	597	15-20-25	298	5-11-26	60
21-7-7	1,556	25-5-20	585	20-7-19	273	16-4-12	73
30-10-10	1,035	20-20-20	583	15-20-25	270	15-3-20	75
25-10-10	1,039	25-5-20	570	15-15-15	261	17-0-17	75
28-14-14	1,023	18-18-18	570	4-25-35	250	15-5-15	135
15-45-5	1,000	20-20-20	560	20-15-25	243	7-11-27	188
9-45-15	960	20-20-20	526	17-17-17	218	14-4-14	200
20-30-10	950	24-10-20	525	15-16-17	215	13-2-13	200
9-45-15	940	20-9-20	510	15-16-17	190	5-11-26	210
10-52-10	940	20-2-20	508	15-16-17	165	14-0-14	220
27-15-12	940	20-20-20	474	15-15-18	170	13-2-13	220
11-41-8	920	15-17-17	465	20-5-30	153	15-2-20	241
28-14-14	900	21-8-18	449	17-5-24	125	15-0-14	281
24-8-16	853	16-17-17	440	21-1-20	122	15-0-15	300
20-2-20	825	15-17-17	440	20-5-30	118	14-5-38	307
5-50-17	800	10-30-20	435	20-5-30	100	15-0-15	319
20-2-20	800	4-25-35	433	15-11-29	91	13-4-20	332
25-15-15	788	20-10-20	425	15-15-30	80	14-0-14	337
24-8-16	730	10-30-20	425	15-5-25	78	15-0-15	338
9-45-15	720	21-5-20	418	15-10-30	76	13-2-13	342
20-19-18	710	7-40-17	414	10-15-20	65	13-2-13	380
20-18-18	710	20-10-20	410	18-3-18	50	15-0-15	380
24-8-16	691	25-0-25	405	15-11-29	50	15.5-0-0	400
12-36-14	680	20-10-20	404	20-0-20	40	12-2-14	400
15-30-15	680	21-5-20	389	15-10-30	40	14-0-14	410
18-3-18	670	17-5-19	385	19-6-12	30	15-0-15	418
15-30-15	660	21-7-7	369	8-20-30	28	12-4-12	420
25-5-15	640	20-5-19	380	21-0-20	10	12-0-43	452
20-19-18	640	20-7-19	375	17-5-17	0	13-0-44	460
24-7-15	612	20-8-20	350	17-2-17	0	12-0-44	508
20-19-18	610	20-5-19	350	20-0-20	0	13-0-44	520
20-18-20	610	10-30-20	350				

Acidic	200–500 lb./ton*
Neutral	0–200 lb./ton*/**
Basic	200–500 lb./ton**

*Variances in formulations with identical analysis is due to the source of nutrients the fertilizer is derived from and differences between manufacturers.

**Pounds of calcium carbonate limestone required to neutralize the acidity caused by using 1 ton of the specified fertilizer.

***Application of 1 ton of the specified fertilizer is equivalent to adding this many pounds of calcium carbonate limestone.

sheets. The measurement used (expressed as pounds of calcium carbonate per ton) represents the amount of calcium carbonate required to neutralize a ton of fertilizer. The potential acidity or basicity is calculated using calcium carbonate as a benchmark. (See table 3.7.) In both cases, the greater the number, the more acidic or basic the fertilizer is. For example, a fertilizer with the potential acidity of 535 is more acidic than one with the potential acidity of 230. Conversely, a fertilizer with the potential basicity of 420 is more basic than one with the potential basicity of 200.

Acidic fertilizers tend to contain higher amounts of ammonium nitrogen, while basic fertilizers contain much less or no ammonium nitrogen. Ammonium- and urea-based fertilizers tend to lower the medium's pH through nitrification; as bacteria in the soil convert ammonium (NH_4^+) into nitrates (NO_3^-), hydrogen ions (H^+) are released into the medium, lowering the pH.

Acidic fertilizers can be used to lower the medium's pH over time, while basic formulations gradually increase the pH. The higher the potential acidity value, the more of an effect it will have on lowering the pH. Conversely, the lower the potential acidity value, a fertilizer will have less effect on altering the medium's pH. In some cases, water quality does not allow growers to solely use fertility as a pH management tool; they most often will need to rely on injecting acids for sufficient control of the pH.

Basic fertilizers usually contain high levels of nitrate nitrogen from such sources as calcium nitrate and potassium nitrate. These fertilizers tend to increase the pH of the growing medium over time.

Growers can often manage the medium's pH primarily by using the acidic or basic properties of the fertilizers to their advantage. Extreme pH modification cannot be corrected using fertilizers, but maintaining reasonable pH levels using fertility is a viable option for many growers. Some growers alternate between acidic and basic fertilizers to maintain desirable pH levels. Another option for growers is to use "neutral" fertilizers such as 17-5-17, which have very little effect on the medium's pH.

Growers should keep in mind the types of plant growth that ammonium-based fertilizers (soft and lush growth) and nitrate-based formulations (toned growth) provide. They should consider the current growing conditions to decide if the type of growth from these formulations is desirable or if different pH management and fertility plans should be followed.

The ions plants absorb also affect the pH of the growing medium. When plants absorb positively charged ions such as ammonium, the roots release positively charged hydrogen ions into the medium, thus lowering the pH. Conversely, as plants absorb negatively charged ions such as nitrates, they release negatively charged hydroxide ions; the hydroxide binds with hydrogen ions to form water, removing water from the medium, thus increasing the media pH.

Salt index

Fertilizer labels also provide information regarding the salt index of the product. Each fertilizer affects the soluble salt levels in the potting substrate differently. The salt index refers to the product's relative drought-inducing effect it has compared to an equal weight of sodium nitrate. In addition to selecting a fertilizer based on the nutrient content of the formulation, growers generally should select fertilizers with a low salt index.

Fertilizer mixing calculator

North Carolina State University has developed software called FERTCALC to help growers easily calculate fertilizer stock solutions. Growers enter some basic information regarding the fertilizer source, analysis, and the desired ppm of nitrogen, phosphorus, or potassium. Based on the information entered, the calculator will determine the amount of fertilizer needed, the ppm of the nutrients supplied, and the cost of the fertilizer solution. When using this calculator, growers can limit the misapplication of fertilizers due to incorrectly calculating the rates or the amounts to measure when making the stock solution. Growers can obtain a free download of the software at www.floricultureinfo.com. The program is designed in Microsoft Excel 5.0 and can be downloaded for both Windows and Mac systems.

Controlled-release Fertilizers

Controlled-release fertilizers (CRFs), also commonly called time-release fertilizers, are fertilizer salts designed to release nutrients at a controlled rate over a period of time. These long-lasting products are coated with materials such as acrylic resins, polyethylene, polyurethane, or sulfur, which aid in the release of the nutrients over time. Temperature is the most important factor determining the rate of release of CRFs. Higher temperatures increase the rate of release, while lower temperatures slow the fertilizer release. Other factors such as moisture levels and thickness of the coating will also influence the release of nutrients from these products.

Controlled-release fertilizers usually contain the three major macronutrients (nitrogen, phosphorus, and potassium) and in many cases include other elements, such as micronutrients. Many of the minor elements are not coated but should provide from several weeks to months of availability. Depending on the type of coating, the source of nutrients, and various environmental factors, the release of nutrients occurs over a period of time, perhaps through the entire crop cycle.

Most time-release fertilizers are activated by soil moisture, not free water. Water vapor enters the prills, dissolving the fertilizer inside of the protective coating. Growing mixes always contain moisture, 40–60% moisture by weight. The speed the CRF prills are activated depends on the moisture level of the medium. The actual release of the dissolved nutrients is then a function temperature. Microbial activity and the medium's pH are additional factors that may determine the release of nutrients from the coated fertilizer.

Time-release fertilizers are either incorporated into the potting substrate prior to planting or topdressed on the surface of the growing medium after planting or during crop production. Both of these application methods require using predetermined rates specific to each application.

Many perennial growers have found it useful to use controlled-release fertilizers as their primary source of nutrients, which provide a base level of fertility, but using supplemental liquid fertilization as needed to complete the fertility program. These growers report that they achieve better results than when using water-soluble or controlled-release fertilizers alone.

Growers who use CRFs experience several advantages over traditional overhead liquid feeding. These advantages include: low initial fertilizer levels in the growing medium, availability of nutrients during the entire growing season, reduction in the amount of nutrients lost from runoff and leaching, and a more efficient and localized supply of nutrients. In the past, greenhouse growers have primarily relied on overhead liquid fertilization programs to deliver nutrients to their crops. Liquid feed programs offer the grower great control and flexibility in choosing the type of nutrients to be used, rates, and time of application. In spite of its popularity, overhead liquid feeding is very inefficient. Depending on crop spacing, plant canopy, and the volume of irrigation water applied, as much as 80% of the water and nutrients that are applied to the crop is lost to leaching and runoff. By shifting fertilization practices toward a greater reliance upon time-released fertilizers, growers can more effectively add nutrients to their crops while limiting the potential of contaminating water sources.

Characteristics of common trade formulations

There are several controlled-release fertilizers (CRFs) available for growers to use. The most commonly used trade names include Multicote, Nutricote, Osmocote, Plantacote, and Polyon. Under each trade name, there may be several formulations with different characteristics, such as the types of nutrients provided, various lengths of release, and differing release patterns. Each trade formulation usually has its own unique properties, such as the type of coating, prill size, and release mechanism.

Perhaps the most recognized CRF is Osmocote. As water passes through the resin coating, the nutrients become dissolved and internal pressure disrupts the coating, allowing for the osmotic release of nutrients. The thickness of the coating varies from one pellet to another, causing nutrients to be released at different times from separate pellets. As previously stated, the release rate is dependent on temperature, moisture, and thickness of the coating.

With Nutricote, the duration of the nutrient release is controlled by the composition of the resin and quality of a special chemical release agent added to the resin. When applied to the soil, the water enters the granule through microscopic pores, dissolving the nutrient elements inside the granule. The nutrients will then be released steadily through the same pores based on the amount of the special release agent. The release rate is influenced only by soil temperature, not by soil moisture, pH, or microbial decomposition. Nutricote has uniform granule size, shape, and thickness of coating, which allows for easy and precise mechanical distribution.

Plantacote releases nutrients through a natural membrane process, which is controlled by temperature. The release time is not affected by substrate type, pH, microorganisms, or soil moisture. The pattern of release is somewhat attuned to the physiological requirements of the crop, releasing nitrogen early for plant establishment and potassium during the later stages. It consists of a tough elastic polymer coating, which reduces mechanical damage during mixing and ensures crop safety during periods of frost and hot weather.

Multicote and Polyon contain an ultrathin reactive-layer coating that bonds a polymer to nutrients, creating a durable granule for 100% controlled release. The polymer coating used on Multicote is a thermoset plastic, and Polyon uses durable polyurethane. Water vapor is absorbed into the prill, which dissolves the water-soluble core, forming a soluble nutrient solution inside of the prill. Nutrients are released by osmosis, allowing for a constant, gradual diffusion of nutrients through the polymer coating, regardless of the amount of moisture in the soil. Soil type, pH, and microbes do not affect the release of nutrients; temperature is the only factor that affects the rate of release.

Other common sources of CRFs

Sulfur-coated urea (SCU) is another form of coated fertilizer; it is manufactured by coating urea with sulfur and sealing it with polyethylene oil or microcrystalline wax. Nitrogen is released when the sealant is broken or by diffusion through pores in the coating. The rate of release is dependent on the thickness of the coating. Microbes, moisture, and abrasion break down SCU. Typically, sulfur-coated urea releases over eight to twelve weeks. This type of CRF is best utilized as a topdress application rather than by incorporating it into the soil.

Some of the older time-release formulations often contain sulfur-coated compounds, which are prone to cracking due to the wetting and drying cycles of crop production. Once the coating has been cracked, the nutrients become immediately available and the controlled-release characteristics are gone.

Sulfur coatings are brittle and full of cracks, which lead to the unpredictable release of nutrients. To address the fragile nature of sulfur-coated products, researchers have developed plastic-coated, sulfur-coated urea (PCSCU). This type of coating can be applied to additional nutrient sources besides urea. The manufacturing process involves spraying molten sulfur on various nutrient sources and then applying a thin plastic film over the sulfur shell. Many of the newer formulations use resin or polymer coatings, which are much less prone to cracking.

Some CRFs are chemically altered to create a portion that is insoluble. For example, urea formaldehyde is a chemically modified fertilizer that has 38% nitrogen, with 70% of that being water insoluble. This form of nitrogen is released by microbial activity in the soil.

Nitroform is a methylene-urea nitrogen source produced by reacting urea with formaldehyde to provide slow-release nitrogen chains called ureaforms, which under greenhouse conditions could potentially last up to two months in the growing medium. One-third of the nitrogen is immediately available to the plants through water solubility within the first four to six weeks, and the remaining two-thirds is gradually released by microbial activity. Nitroform is more resistant to leaching and runoff due to the need for microbial activity to release the nitrogen.

IBDU (isobutylidene diurea) is similar to the ureaforms but consists of 31% nitrogen, of which is 90% is insoluble. IBDUs are not dependent on microbial activity, as the release of nitrogen occurs as water moves across the granule. The size of the IBDU particle or the surface area in contact with water greatly affects the release rate of nutrients and the longevity of the product. With temperature having no effect on the release of nutrients, IBDU fertilizers are ideal to use during the spring or fall, when temperatures are not conducive to releasing nutrients from most CRF sources.

Urea from both urea formaldehydes and IBDU products must go through a conversion process before plants can take up the nitrogen. Urea is first converted to ammoniacal nitrogen via an enzyme, and then the ammonium is converted to nitrate nitrogen by bacteria.

Considerations for using CRFs

Every operation has a unique set of needs. So when using controlled-release fertilizers, growers must determine what their own growing conditions and production needs are. They'll need to ask:

- What are the nutritional requirements of the crops being grown?
- How long is the crop being grown?
- What temperatures are the crops going to be grown under?
- What release pattern is needed (steady release throughout the crop cycle or more release as the crop develops)?
- Should the nutritional benefits carry over to the consumer, or the retail location? If so, for how long?
- Are rates needed for each specific crop, or can an average rate be used for a broad range of varieties?
- What is the water quality?
- What are the current irrigation practices?
- What is the container size?
- Will CRF be the sole source of nutrients or will supplemental liquid feed be implemented?
- Is reducing runoff from the facility an issue?

Putting CRFs into production

The two most common methods of using controlled release fertilizers are topdressing and incorporation. Topdressing entails adding a predetermined amount of fertilizer on the top of the growing medium of each container. Nutrients released from the topdressed material have to travel throughout the soil profile and have a good chance of being retained by the soil or taken up by the plant. When topdressing CRFs, fewer nutrients are lost from leaching than with incorporation. With the entire amount of fertilizer on the soil surface, the nutrients, once released, have to move through the entire soil profile before leaching out of the container. As the nutrients move through the growing medium, some become attached to the medium components, are dispersed in the water solution, or are taken up by the roots. Topdressing is more labor intensive, as it is usually applied to one container at a time after the crop is in the production area. Also, some of the nutrients can be lost if the pots are tipped or blown over.

The release of nutrients from topdressing is also not as constant compared to other application methods. The prills on the surface tend to

Figure 3.4. **With topdressing, the fertilizer is spread on top of the growing medium, as with this clematis.**

dry out between irrigations, slightly reducing the release process. Surface temperatures can fluctuate more throughout the day and night than the temperatures within the growing medium. With temperature and moisture being the two predominate factors determining the release of nutrients from CRFs, it stands to reason that the variability in these factors on the medium's surface causes these products to release slightly slower than other application methods where these factors remain more constant.

With incorporation, a predetermined amount of fertilizer is evenly distributed within the growing medium (prior to planting) and becomes available throughout the soil profile. Time-release fertilizers incorporated into the growing medium are exposed to more constant moisture levels and usually temperatures higher than those found on the substrate surface. These conditions cause the nutrients to release more rapidly and consistently than fertilizers applied to the surface. With incorporation, the fertilizer is distributed fairly evenly, throughout the container (top to bottom), and is usually more prone to leaching since the distance to the drainage holes is often small. Growing mixes with CRFs incorporated prior to planting should be used within fourteen days following incorporation, and they should be watered thoroughly after planting. These precautions are recommended since CRFs become activated by water vapor and may start releasing prior to planting.

Dibbling is a method used primarily by tree and shrub nurseries that entails placing a predetermined amount of fertilizer directly below the liner while potting. Dibbling is a method used to provide fertilizer to the crop while controlling weeds from developing on the surface. With a lack of nutrients at the surface, the small germinating weed seeds become starved and in many cases cannot survive. It is important if dibbling CRFs to use products with long-release patterns (eight months minimum); using short-term products (four months or less) will cause too many nutrients to be released and plant injury is likely to occur as a result of high salts.

During and after periods of high temperatures (85° F [29° C] and above), careful monitoring of the soluble salts is essential, as excessive levels of fertilizer may have been released. Since CRF coatings vary, each formulation may respond differently to temperature. If excessive amounts of fertilizer have been released, the salts could burn the roots and may lead to root rot problems. To resolve these situations, heavy irrigations of clear water may be necessary to leach out the high feed levels in the medium. During warmer weather, growers need to water more frequently and may be leaching a portion of these high salt levels out of the pot with each watering. On the other end of the spectrum, if temperatures are too cool (below 55° F [13°C]), there may be insufficient amounts of nutrients being released, often requiring nutrients be added using water-soluble fertilizers.

The simplest way to monitor nutrient release is through regular conductivity testing using a conductivity meter. The pour-through and 2:1 extraction procedures are both effective tools for monitoring electrical conductivity. At minimum, growers should know what the conductivity of their irrigation water is and what the maximum allowable conductivity of the growing medium is before plant injury occurs.

The irrigation practices play an important role in the availability of nutrients from CRFs. The release of nutrients is a function of temperature, but the availability of nutrients is greatly affected by the irrigation practices. How plants are watered, the frequency, and the quantity of irrigation applied all affect the performance of controlled release fertilizers. Irrigation can be used to increase, decrease, or move soluble salts through the growing medium. The leach fraction, or the percentage of water that is moved out of the container with each watering, will determine whether the nutrients stay in the pot and accumulate or move out of the pot, maintaining or decreasing the levels in the root zone. With significant leaching, either by quantity or frequency, the nutrient levels are not allowed to build up in the root zone and often fall below the levels needed by the plant, potentially decreasing plant quality.

The quality, or the chemical properties, of the irrigation water greatly affects, either positively or negatively, the growth of a plant. Certain elements such as bicarbonates, chloride, and sodium have been shown to cause less than desirable plant growth. The pH and alkalinity (a measure of how much lime is in the water) of the irrigation supply can greatly affect the availability of nutrients and plant development. There are numerous irrigation sources with various quality characteristics. Some of the traditional sources of water include ponds, wells, rivers, rainfall, and municipalities. Water quality was discussed previously in this chapter.

Growers use various materials combined to make up their growing media. The materials used and the quantities in which they are used can greatly influence the mixes' bulk density, pore space, and water holding capacity. Many perennial mixes consist of bark, peat moss, topsoil, sand, or other substrates. The composition of each growing medium is often dependent on the cost and/or regional availability of the individual components. It is important to communicate to your CRF representative the current composition of the growing mixes and keep them apprised of any changes that you make. This communication will help to ensure the success of the CRF program and crop quality.

Each plant species requires differing amounts of nutrients and has various sensitivities to salts. The labels of the CRF list low, medium, and high rates on the bags to provide growers an indication of which rates work best with the nutrient requirements and salt sensitivities of various crops. Perennials with low nutritional requirements or high sensitivity to salts are usually fertilized using the low recommended rates. Plants with high nutritional requirements or low salt sensitivities are usually fertilized with the medium or high recommended rates. Providing rates above or below the nutritional needs, or tolerances, of the crop will often lead to less than desirable plant growth and could prove detrimental to the crop. Perennial growers commonly use the low or medium recommended rates of CRFs.

Growers use various container sizes when producing perennials, ranging in size from small plug trays up to 2 gal. (8 L) or larger. The size of the container affects both the volume of the growing medium needed and the distribution of the fertilizer granules. When the fertilizer is incorporated using low rates into small container, there is a high likelihood that each container could end up with different fertility rates because the CRF particle distribution is low. To improve the particle distribution in small containers, it is recommended to use CRF products with a size guide number (SGN) of 150 or less. These products contain more particles per pound than the average CRF, which has SGNs of 240–300. Low SGN products contain over twice as many particles per pound than high SGN products, allowing for better, more uniform distribution in the small container.

CRF rates

There are several good preformulated CRFs on the market, and many companies are willing to custom blend fertilizers to meet the needs of the individual grower. Let's face it: We all do things differently. From the quality of the water, to the components of the media, to the quantity and frequency of the irrigation, to the temperatures we grow at—all of these variables make it difficult for growers to use standardized formulations. Custom blending CRFs is often the best option to properly address the variables specific to each grower.

With so many formulations available, it is too difficult to provide rate recommendations based on specific quantities of CRF to apply. For example, one grower may use a three-month formulation of 12-6-6 and another grower uses a three-month formulation of 19-6-12; applying the same amount of these products will provide very different release patterns and growth responses. Using this example, each grower incorporates 6 lb. (2.7 kg) of product into every yard of growing medium. Looking at only the nitrogen component, incorporating 6 lb. (2.7 kg) of product of the 12-6-6 formulation would provide 0.72 lb. (327 g) of nitrogen over the three-month period, while the 19-6-12 formulation would provide 1.14 lb. (517 g) of nitrogen. These growers would observe very different growth responses from the release of the nutrients from these different formulations.

The longevity of the fertilizer is also an important consideration. A formulation such as 15-9-12 may be available with differing lengths of release; a three-, six-, or even nine-month product might be available. Incorporating 8 lb. (3.6 kg) of each of these formulations will provide very different results. The same amount, 1.2 lb. (544 g) of nitrogen, will be released over drastically different time periods. Assuming a grower is producing a crop that can finish in three months, there may be sufficient nutrients available using the three-month formulation, but 1.2 lb. (544 g) of nitrogen released over a six- or nine-month period of time could lead to a shortage of nutrients during the first three months at the rate incorporated.

Fertilizer companies determine the release characteristics of a given fertilizer based on average soil temperatures of 70° F (21° C). Soil temperatures above 70° F will accelerate the release of nutrients, decreasing the product's longevity. Conversely, medium temperatures less than 70° F decrease the release rate of nutrients and increase the product's overall longevity. With polymer-coated products, the release of nutrients completely stops at temperatures less than 40° F (4° C). Growers using polymer CRFs no longer have to worry about the undesirable release of nutrients during the winter months.

The longevity of the fertilizer used most commonly corresponds to the length of time the grower needs to produce the crop. In most cases, growers will build in a little extra time, or fertilizer, to allow for unusual circumstances or to ensure the perennial will remain healthy until purchased by the consumer. For a crop scheduled for eight weeks of production time, a grower might decide to use a three-month formulation, allowing for some but not a lot of fertilizer to be left at the end of the production cycle. If the market window is missed and the crop is going to be carried over to the next production cycle or overwintered, additional fertilizer will need to be applied.

For crops being overwintered, growers will either provide enough fertilizer to get to the overwintering period and then reapply fertilizer prior to the plants breaking dormancy, or they will use long-term fertilizer that will last the entire crop cycle. The difficulty of using long-term fertilizers for overwintering is that the temperatures necessary to release adequate amounts of nutrients need to be warmer than many growers experience in early spring. Perennials often begin actively growing before there are sufficient nutrients available to support the growth. This shortfall of nutrients often results in nutrient deficiencies.

Many trade publications, researchers, and fertilizer companies tend to give CRF recommendations on a pounds of elemental nitrogen per cubic yard (0.76 m^3) basis (or per pot basis). I have provided CRF recommendations in this manner for this book (see table 3.8). Nitrogen is the most important nutrient and can be easily calculated using the formulation of the fertilizer.

To determine how many pounds of product to incorporate per yard, a grower should first determine how much nitrogen is needed. Then, using the medium rate recom-

Table 3.8. Determining How Much Nitrogen to Incorporate with Various Controlled Release Fertilizers

% Nitrogen in Formulation	Pounds Nitrogen per Yard	Pounds of Product Needed per Yard
12	0.50	4.17
12	0.75	6.25
12	1.00	8.33
12	1.25	10.42
12	1.50	12.50
12	1.75	14.58
12	2.00	16.67
15	0.50	3.33
15	0.75	5.00
15	1.00	6.67
15	1.25	8.33
15	1.50	10.00
15	1.75	11.67
15	2.00	13.33
17	0.50	2.94
17	0.75	4.41
17	1.00	5.88
17	1.25	7.35
17	1.50	8.82
17	1.75	10.29
17	2.00	11.76
19	0.50	2.63
19	0.75	3.95
19	1.00	5.26
19	1.25	6.58
19	1.50	7.89
19	1.75	9.21
19	2.00	10.53
21	0.50	2.38
21	0.75	3.57
21	1.00	4.76
21	1.25	5.95
21	1.50	7.14
21	1.75	8.33
21	2.00	9.52

mendation for incorporation on the products label, multiply the pounds listed by the percent nitrogen of the formulation. For example, a 14-9-15 formulation might recommend 8 lb. per cubic yard (3.6 kg/0.76 m^3) be incorporated as the medium rate recommendation. The grower would determine the pounds of nitrogen per yard by multiplying 8 lb. (3.6 kg) of product by 14% (.14), resulting in 1.12 lb. (508 g) nitrogen per cubic yard. If the grower wanted to incorporate 1.0 lb. nitrogen per cubic yard (454 g/0.76 m^3), he would need to reduce the amount of product to incorporate slightly and recalculate the results. Conversely, if the goal was to incorporate 1.25 lb. of nitrogen per cubic yard (567 g/0.76 m^3), the grower would increase the pounds per cubic yard of product and recalculate until the desired amount is achieved.

In some cases, recommendations are given on a grams per pot basis. How do growers determine how many grams of CRF is applied per pot? For topdressing, the CRF is often applied using grams per pot, but when incorporating CRF into the mix, the grams per pot must be calculated. The grower must determine how many pots of any given size can be filled with 1 cu. yd. (0.76 m^3) of growing medium. Let's assume they are filling 1 gal. (3.8 L) containers and have determined that 250 pots can be filled with 1 cu. yd. (0.76 m^3) of growing mix. Using the above example, 8 lb. (3.6 kg) of 14-9-15 is incorporated into every cubic yard (0.76 m^3) of medium. With 454 g equivalent to 1 lb., there are 3,632 g of 14-9-15 being incorporated with every cubic yard of growing mix. Dividing 3,632 g by 250 pots, the grower can calculate that it will take 14.53 g of 14-9-15 for each container.

Handling and storing CRFs

Care should be taken when handling time-release fertilizers. These products are encased in coatings of various materials to provide extended release of the nutrients. The coatings used are not indestructible and can easily be damaged with frequent or improper handling. Any damage to the coating can negatively impact the release characteristics of CRFs. To reduce damage, it is recommended to only handle and move them when necessary.

When applying products containing broken prills, the fertilizer from the damaged particles is immediately available. When only a small amount of damage is present, there are very few, if any, adverse effects. In fact, growers often experience a benefit from having some nutrients available immediately and often provide some nutrients to jump-start crop production. Growers should inspect these products prior to application to predetermine if any adverse effects are likely to occur. If in doubt, contact your fertilizer representative for a professional opinion.

Controlled-release fertilizers should be stored in cool, dry, shaded areas. They should not be exposed to high humidity environments such as greenhouses, as water vapor starts the release of nutrients. Exposure to damp or humid conditions may trigger the release process, causing nutrients to prematurely release during storage. The release can and does occur before these products are applied to the crops.

Combining CRFs with WSFs

Growers who have learned to use controlled-release fertilizers in conjunction with water-soluble fertilizers receive the benefits from each source. There are numerous approaches to addressing the nutritional needs of the crop; it seems there are almost as many fertilization regimes as there are growers. The choice of fertility programs is based on many factors, as discussed in this chapter, and must be customized to fit each crop, production facility, and grower.

Growers producing short-term crops (eight weeks or less) often only use water-soluble fertilizers. In some cases, growers producing short-term crops will topdress the crops they consider to be heavy feeders and still follow the liquid feed program. Many growers routinely use CRFs incorporated at the low rate to provide a nutritional base and follow a liquid fertility program using half of the normally recommended rate. Other growers use CRFs incorporated into the media of all crops at the recommended rates and only use WSFs when the temperatures are not conducive to releasing the fertilizer or when excessive rainfall has depleted the nutrient charge in the containers.

References

Behe, Bridget, Diane Brown-Rytlewski, et. al. *Management Practices for Michigan Wholesale Nurseries*. East Lansing, Mich.: Michigan State University. 2004.

Fernandez, Tom. "Selecting Controlled Release Fertilizers." *The Michigan Landscape*. May/June 2002.

Johnson, John. "Controlling Control Release: Management Decisions Will Affect Your Fertility Program." *NMPro*. March 2004.

Nelson, Paul V. *Greenhouse Operation and Management*, fourth edition. Englewood Cliffs, N.J.: Prentice-Hall. 1991.

Pilon, Paul. "CRFs: Releasing Quality Crops." *Greenhouse Grower*. October 2001.

Robbins, Jim. "Fertilizer Facts: Know Your Options Before Picking Plant-Nutrition Products." *NMPro*. October 2003.

Van Iersel, Marc, Bodie Pennisi, and Paul Thomas. "Figuring Out Fertilization." *Greenhouse Product News*. September 2001.

4

Integrated Pest Management

Integrated pest management (IPM) is a pest control philosophy that utilizes all available control strategies to maintain pest populations of insects and diseases below an economic and/or aesthetic injury level. IPM involves early detection of cultural problems, proper insect and disease identification, assessing damage levels, and selecting optimum control strategies. Cultural problems are controlled using an integration of biological, cultural, and chemical management strategies. IPM programs help to maintain low populations of insects and diseases throughout the production area, while using control strategies that preserve the beneficial organisms that are working in the greenhouse or nursery.

Over the years, there has been much confusion as to what an IPM program really entails. IPM is a decision-making process that utilizes the information gathered through routine monitoring of crops for overall plant health and the presence of pests to determine if, when, and what control strategies should be implemented to keep the pest problems from causing aesthetic and/or economic injury. When determined necessary, a control strategy (i.e., biological, cultural, physical, mechanical, or chemical) or combination of strategies suitable for the current situation is implemented. Most pests can be controlled using biological strategies, cultural techniques, physical strategies, or by applying pesticides. Others problems, such as viral diseases, cannot be controlled with any strategy, and the infected plants should be discarded or destroyed.

The primary goals of IPM programs are to identify pest problems early, isolate or destroy infested plant materials, develop treatment thresholds, use beneficial organisms where practical, and spot-treat using methods that have the least impact on the beneficial organisms and the environment. These programs rely heavily on a scout's data to determine whether a pest population has reached an economically damaging level, or threshold, and when certain control strategies must be implemented. They are relatively simple to implement in operations growing a single crop at one time, especially when that crop has relatively few major pests. IPM becomes more complex for crops with multiple pests or when multiple crops are grown within the same environment.

It is very common for growers to use the term *pests* to describe all of the insects, mites, and diseases that are specific for each crop. Perennial growers should know what the key pests are for each of their major crops. Key pests are those insects or diseases that appear frequently and have a high potential for causing significant aesthetic or economic injury. For example, key pests for hostas may include aphids, black vine weevils, foliar nematodes, leaf spot diseases, root knot nematodes, slugs, and snails. Each plant will have its own collection of cultural woes. A pest on one plant species might not bother another; there are even pest differences among cultivars of the same plant species.

Traditional chemical applications should be used responsibly to reduce the potentially adverse side effects on beneficial organisms and our environment. When chemicals are used, it is recommended to choose pesticides that will affect the beneficial organisms the least. Treat only the infested plants, or a small area around them, to minimize the exposure of the beneficial organisms to these often-lethal insecticides. When applying chemicals, it is important that they are effective on the life stages of insects or diseases present. Application of pesticides to sensitive life stages will increase their effectiveness and provide the greatest results.

Better Pest Management

Over the years, there have been numerous definitions and variations of IPM. Integrated pest management has evolved, and its true meaning is becoming blurred. When looking at pest management strategies used today by greenhouse and nursery operations, each facility practices its own variation of IPM. Some operations use stricter disciplines or may be more devoted to respecting pest thresholds or environmental safety. Others follow many of the guidelines IPM has defined but take more drastic steps to control pests and insure crop quality.

All greenhouses and nurseries practice IPM at some level. Many of the tools and methods of strict IPM programs are similar to those used by the commercial perennial industry. Regardless of what level of IPM is practiced, we all must be responsible: to our environment, our customers, and our employees. Instead of figuring out what the true definition of IPM is or where IPM fits in our operations, we should all commit ourselves to practicing better pest management. A term I use, *better pest management,* is a concept that focuses on managing pests by combining biological, chemical, cultural, and physical strategies that together have a minimal impact on the environment.

With better pest management, there is usually more emphasis on preventative strategies, or avoiding the occurrence of pest problems, versus reactive strategies or trying to solve pest problems when they get out of hand. A reactive approach will usually cost a grower more to control and often causes significant injury to crops, reducing their quality and, in many cases, rendering them unsalable. Better pest management involves detecting pest problems early through scouting programs and minimizing them by maintaining sound cultural practices.

The goals of integrated pest management and better pest management are similar and include protecting crop quality, safeguarding public health, protecting the environment, using pesticides wisely, and reducing pest management costs.

While rating the effectiveness of any insect and disease control program, it is important to not only take into consideration the level of pest or pathogen suppression, but also the practicality and economics of these programs.

Control Strategies

In many cases, there are multiple control options available for growers to use for any given insect or disease problem they might incur. These control methods fall into six basic groups: cultural practices, physical methods, mechanical methods, environmental strategies, biological controls, and chemical controls.

Cultural

Cultural controls are horticultural practices that disrupt or reduce inherent insect or disease populations. Commonly followed cultural practices include sanitation, choosing resistant plant varieties, weed elimination, and irrigation management. Controlling the environment around the plant is also considered a form of cultural control.

Physical

Physical control strategies are those involving some type of physical activity useful to reduce insect and disease pressures. Moving infected or newly acquired plant materials into a quarantine area, eliminating weeds from the production site, pruning infested plant parts, and rouging out infected plants are all examples of viable physical control strategies.

Mechanical

Mechanical control strategies use some physical means to control pest populations. Some examples of mechanical methods include air blasts, screens, pruning, rouging, and vacuuming.

The most commonly used mechanical method is screening, which excludes pests from entering into the production site through the installation of small screens on the vents and openings of production structures. Screens are most commonly installed onto ventilation systems but are also commonly used to separate the production area into smaller zones, keeping insects from spreading throughout the production area. When insects are confined to isolated areas, they are generally easier to control.

Environmental

Environmental control strategies involve the use of environmental elements to reduce pests or encourage beneficial organisms. In many situations, such as outdoor production sites, growers do not have strict control over climatic conditions such as temperature or precipitation. Growers should understand the relationship between environmental factors and the development of pest infestations on their crops. They can then anticipate certain problems or to make decisions regarding control strategies based on the occurrence of certain environmental factors. For instance, certain foliar diseases are promoted by wet weather, so minimal disease infection is likely during a dry spring.

Growers producing perennials under greenhouse conditions have the ability to carefully control the temperature, irrigation type and frequency, and the relative humidity. The ability to control these factors can greatly reduce the occurrence of many common plant diseases and allows growers to follow more consistent procedures.

Biological

Growers using biological control strategies use natural enemies (pathogens, parasites, and predators) of insects and mites, to keep crop pests from becoming too damaging to the crops being grown. These living organisms often occur naturally within the production facility at low population levels, making their benefits not readily visible, though they usually need to be introduced or supplemented into the growing area. The purpose of biological control is to keep insect and mite populations at levels too low to cause any significant crop damage, but just high enough to sustain the population of the natural enemies.

Routine use of pesticides often kills the natural enemies present, by contact or through residual activity. When natural enemies are being utilized, it is beneficial to practice spot treatments of the troubled areas and to avoid sprays that would cover the entire crop area.

Pathogens

Disease-causing organisms, referred to as pathogens, such as bacteria and fungi, are effective for controlling certain pest problems. Bt (*Bacillus thuringiensis*), marketed as Gnatrol, is a familiar example of a pathogen used for controlling a variety of leaf-chewing caterpillars and fungus gnat larvae. Another example is the fungus *Verticillium lecanii*, sold under the trade names Mycotel and Vetalec, which growers use to combat aphid and whitefly populations.

Parasites

Parasitic insects lay their eggs on or in the bodies of a host species. After the eggs hatch, the larval stages of the parasite develop and feed inside the body of the host, eventually killing it. They are usually host specific, which means they only attack one species of pest. Greenhouse and perennial producers often release various small parasitic wasps into the production area for controlling aphids, leaf miners, scales, or whiteflies.

Predators

To control certain insects or mites, some growers introduce various predators into the production area. Predators are organisms that capture, kill, and consume their prey. Whether they are winged or wingless, they are very active and quick movers, allowing them to search for and capture their prey. In most cases, they are larger than the insects or mites they consume and may also have more than one food preference. Perhaps the most familiar example of a predator is the ladybird beetle, which can consume up to fifty aphids per day.

Lacewings (*Chrysoperla carnea*) are predatory insects commonly used to control aphids, mealybugs, mites, scales, and thrips. The larvae of syrphid flies (commonly called flower flies) feed on aphids and small ants. Growers release predatory mites to control two-spotted spider mite populations. Another commonly used predator is the predatory nematode *Steinernema carpocapsae,* marketed as BioSafe or Vector, which can be applied over a crop to control the larvae of fungus gnats.

Chemical

Using chemical controls, such as insecticides and fungicides, is perhaps the most commonly used method of controlling insects, mites, and diseases. All chemicals are generally referred to as pesticides. Pesticides are divided into various classes, depending on what type of pest they are effective at controlling. Insecticides are those chemicals used to control various insect pests. Miticides are effective at reducing mite populations. Growers apply fungicides to reduce or eliminate the presence of fungal plant pathogens. Herbicides are chemicals applied to control any unwanted weeds and plant growth.

Chemicals are generally most effective when applied to a particular life cycle, or stage of development, of an insect or disease. Applications made during nonresponsive life stages will result in less overall control. Growers should first pick chemicals that are appropriate for controlling the desired pest, and apply them to the life stages that will yield the greatest results. Failure to do so will result in inadequate levels of control, wasted time, money, and pesticide.

When using chemicals, care should be taken for plant, environmental, and human safety. Misuse of pesticides can lead to resistant plant pests and/or injury to crops. Each chemical comes with a label, which outlines precisely how the chemical should be used, at what rates, and application frequency. The label is a legal document and is the *law.* Please follow it. Using chemicals is a responsibility that must be taken seriously.

To obtain the highest level of pest control, it is essential that pesticides are properly delivered to the targeted area. Most chemicals are applied as foliar applications, using equipment that creates small droplets with a uniform particle size distribution, allowing for good canopy penetration. It is important to apply pesticides using the right rates, application volumes, and ensuring they are providing adequate coverage over the crop area.

Pesticides are generally categorized into three classes: contact pesticides, systemic pesticides, and insect growth regulators. Contact pesticides control the pest when the pest comes into contact with the active ingredient. Chemicals that must be consumed or absorbed to achieve control are systemic pesticides. Insect growth regulators prevent the juvenile phases of insects from developing into the next life stage.

Many pesticides have some systemic properties that will somewhat compensate for inadequate coverage. Systemic chemicals are those that have the ability to move through the plant and be distributed to other areas where the initial application did not occur. Some chemicals have translaminar properties, which means that systemic activity occurs when chemicals applied to the upper leaf surface move through the leaf to the underside, where they can reach and control various pests located there. Some pesticides, such as the greenhouse insecticide imidacloprid (Marathon), are taken up by the roots and transported through the vascular system throughout the plant, providing control of the targeted pest throughout the entire plant. The effectiveness of systemic chemicals varies with plant age, rate of growth, and the amount of irrigation applied.

Mode of action

The mode of action is the method a pesticide uses against the living systems of the pest. Generally, the mode of action of pesticides within the same chemical class is similar. If the desired level of control is not achieved using one product, understanding which chemical class, or family, it is from will help growers to pick an alternate chemical that belongs to a different class of chemicals, has a different mode of action, and has a greater likelihood of providing an improved level of control. Failure to have this understanding causes growers to pick chemicals with similar modes of action, or chemicals from the same family, which usually will provide no additional level of control. For example, a grower observing insufficient levels of control using the synthetic pyrethroid cyfluthrin (Decathlon) would also receive inadequate control using another chemical in the same class of chemistry such as bifenthrin (Talstar). Both Decathlon and Talstar interfere with the transmission of nerve impulses between nerve cells. Choosing a chemical that does not have the same mode of action (nerve impulse transmission in the above example) may provide better results.

Besides controlling pests in a similar manner, chemicals with the same mode of action also have several other characteristics in common. They are often available in similar formulations and applied in a similar manner as their sister products. Chemicals within the same class tend to break down similarly, or have similar persistence levels, compared to one another. Generally, they will also have similar toxicities and risks to the applicators and the environment.

Tank mixes

Growers often apply combinations of multiple chemicals within the same tank, referred to as a tank mixture, to obtain differing modes of action for controlling the same pest or to manage multiple pests simultaneously with a single application. Tank mix applications used for multiple pests greatly reduce the time and labor usually associated with controlling more than one pest.

Many chemicals list on their label whether they are suitable as tank mix partners. The label may provide an indication of which formulations are compatible and the order they must be added when mixing. Any tank mixes must be prepared consistent to the instructions provided in the label.

Some chemical combinations are not compatible and should not be mixed. The most obvious sign a mixture is not compatible is the formation of a precipitate, or a residue that settles at the bottom of the solution. When two of more ingredients of a solution are incompatible, growers applying these mixtures are likely to observe clogging of equipment, loss of effectiveness against the target species, and increased damage to both target and nontarget plants. If the individual chemical label does not specifically restrict using it as a component of a tank mixture, growers should determine the compatibility by conducting a jar test.

While conducting a jar test, the chemical handler should wear all of the necessary protective equipment as specified on the chemical labels of the individual products. Place one pint (0.5 L) of water in a clean 1 qt. (approx. 1 L) jar. For each chemical, calculate the desired amount to add into the jar. The quantities will be quite small and may require converting them into milliliters or grams for ease of measuring. I often conduct compatibility tests using a 1 gal. (3.8 L) jug so that the measuring becomes more practical.

Add each of the products into the jar (or jug) of water in the following order. Each component should be thoroughly mixed before the addition of a new ingredient.

1. Compatibility agents, activators, or surfactants
2. Wettable powders or dry flowable formulations
3. Water-soluble concentrates or solutions
4. Emulsifiable concentrates or flowables
5. Soluble powder formulations
6. Any additional spreaders or stickers

I prefer to measure the exact amounts necessary for this compatibility test, which is reflective of how the chemicals would interact when a larger quantity, such as 100 gal. (379 L), is mixed. Many growers simply add 0.5 tsp.

(2.5 ml) of all the necessary ingredients, except where wettable powders or dry formulations are necessary, and then they use 0.5 tbsp. (7.5 g). Regardless of what quantities you mix, calculated or ballpark, follow the mixing order listed above.

Close the lid on the container tightly and invert ten times. Observe the solution for uniform mixing. If no clumps, sludge, or non-dispersing oils are formed, wait about thirty minutes and check again. A minor amount of separation that remixes easily with agitation is acceptable. The presence of clumps, gums, or sludge indicates a chemical incompatibility, and this particular combination should not be applied to your crops. When finished, dispose of the compatibility test according to the label directions of the individual products.

Compatibility testing only demonstrates the stability of the mixture; it does not demon strate whether the combination has any pest control activity. Another concern with tank mixes is the potential for increased risk of phytotoxicity, or chemical burn, on target and nontarget crops. Compatible tank mixes should be tested on a small scale, making an application to only a few plants of each desirable species. Observe the test blocks several days after application for any signs of injury symptoms. When no phytotoxicity is observed, it is usually safe to make larger scale applications of the new tank combination.

Pesticide resistance

Perennial growers need to be concerned about insects building up resistance to the chemicals being applied to control them. There are a number of factors involved that have led to the current resistance issues of today. Pest biology, the intensity of past and present insecticide applications, the dynamics of our production facilities, and commercial production practices are all contributing factors leading to resistant insect pests.

The first step to reducing potential insecticide resistance is to reduce the reliance on synthetic insecticides. When and where possible, growers should try to use nonchemical methods of controlling insects, such as implementing cultural, mechanical, physical and biological management options.

When chemicals are being used, growers should avoid persistent applications. For managing resistance, the best chemicals are those that work quickly after application and quickly disappear from the crop area, reducing the length of time pests are exposed to these products. Insecticides that are long lasting or degrade slowly over time often only kill the most susceptible portion of the pest population, leaving behind the resistant individuals, which reproduce, building even more resistant pest populations.

Using tank mixes containing insecticides of two different chemical classes often provides growers with greater short-term control than when either chemical is applied individually. Repetitive use of tank mixtures still has the potential for insects to build up resistance to the active ingredients of the mixture, and a greater risk is that insects may build up resistance to both chemistries simultaneously, dramatically reducing the effectiveness of the two chemicals rather than building resistance to a single active ingredient over time. In general, though, tank mixes lengthen the duration of time it takes for insects to build up resistance.

One way growers combat resistance is by rotating chemicals. There are several approaches growers use when determining when to rotate the chemicals being applied. One approach entails using long-term rotations, which often span at least the duration of one pest generation before rotating. With overlapping insect populations containing all life stages present, many growers use the same insecticide for a least two generations prior to rotating. Other growers practice rotating chemicals with each application. When pesticides are rotated, it is imperative to rotate to chemicals found in different chemical classes, or having different modes of action, than the one previously applied. If the chemicals being rotated are of the same chemical class, the rotation will be ineffective and pests will build resistance.

Introduction to Scouting

Knowledge and early identification are keys to successful insect and disease programs. Scouts need to have a good understanding of plant biology, pest biology, pest life cycles, host plants, beneficial insects, injury symptoms, environmental risk factors, and control strategies.

Scouting entails monitoring the crops, detecting the presence of any potential insect and disease problems, and determining when control strategies must be taken. Carefully monitoring crops for insects and diseases is the backbone of integrated pest management. Whether strictly using an IPM approach or simply managing perennials for current cultural problems, scouting is an essential tool for growers to implement into their production practices. Regular monitoring helps growers identify and prevent potential outbreaks from occurring, which reduces crop damage and plant losses.

Scouting provides the information needed for making good pest management decisions. Regular crop monitoring provides up-to-date information regarding the presence of insects and diseases, while also providing an indication of the effectiveness of previous pest management strategies. Frequent observation of your perennial crops is your best insurance that you will recognize plant problems early, before they get out of hand or cause economic losses. Routine monitoring allows growers the ability to detect when insect or disease problems first appear. Scouting should occur at least once per week, from the beginning of the crop cycle until it is sold. Scouting twice a week is a better practice, but is not very practical for most growers. If detected early, growers can watch closely new insect or disease pressures and will have adequate time to take the necessary actions before little problems explode into big ones.

Scouting can help growers detect and isolate any "hot spots" within the production area. Treating these hot spots is a responsible and economical approach to controlling insects and diseases. Spot treatments decrease the likelihood of misapplication and overapplication and uses less chemical than blanket sprays. Limiting pesticide applications to these hot spots helps to decrease the potential buildup of chemical resistance by the targeted pest and preserves the beneficial organisms throughout the crop.

Monitoring and controlling certain insect pests may prevent other problems from occurring, such as viruses or plant pathogens. These are transmitted and spread to other plants by insects such as thrips or whiteflies. There is some difficulty determining the direct relationship between the pest counts found and the actual population of the pest due to variables such as movement or distribution through the production area and the life stage of the insect detected versus the other life stages that may be present. For example, the number of adult thrips found on sticky cards does not necessarily provide an accurate picture of the actual thrips population. There may be a high number of young, wingless stages feeding in the flowers that may go undetected if only sticky traps are used to monitor populations.

Identifying and tracking the life stage of an insect pest is another useful application growers can receive from scouting. For example, if you know when an insect population has reached a particular life stage that is most susceptible to a particular chemical treatment, you can make your control efforts more effective.

Scouting not only provides an indication of the insect or disease pressure on any given crop before control measures have been taken, it also gives growers an indication of whether their previous treatments have been effective so they can repeat their successes or fine-tune their methods. Routine monitoring also allows growers to observe the overall health of the crop, its growth stage, and evaluate the overall quality of the products they are growing.

Properly identifying insects and diseases is the first step in any monitoring program. Growers and scouts must first determine whether a problem even exists. Some plant cultivars may naturally exhibit peculiar characteristics, such as puckered leaves or unusual colorations that may, in fact, be normal for that plant. If an abnormality does exist, it is important to try to identify its cause (environment, disease, insect, nutrition, etc.). Without the correct identification, a grower may implement a control strategy that is both ineffective and perhaps costly.

A basic understanding of insect biology is also necessary for today's greenhouse scouts. Understanding the pests' life cycles is an essential aspect to effectively control them. For example, controlling fungus gnat adults in the greenhouse requires a different strategy than controlling the larval stages, which are found only within the growing medium. Crop monitoring becomes more efficient when the scout can determine when particular pest problems are likely to occur, what life stages to observe, and what types of crop injury exists. Understanding how to identify pests and their life cycles will help growers to determine if control actions are feasible, what actions are appropriate, and when control measures should be implemented.

Knowing the life cycle will help growers to properly time the application of chemicals or the release of beneficial insects to coincide with the life stages these treatments are most effective on. For example, the insecticides used to control scale insects are primarily only effective during the crawler stage, before these insects develop a protective waxy layer. Therefore, the timing of insecticide applications to scale insects should correspond to the presence of the crawler stage. Certain insecticides are only effective against select stages of insect development; applications applied to the improper life stages are likely to be ineffective.

Growers can increase the level of control they achieve with their treatments when they target the life stage of the pest that is the most numerous. If the scouting results reveal that more larvae are present than adults, treatments controlling the larval stage should be applied. Conversely, if adults predominate, treatments controlling the adult population should be used. In most cases, there are overlapping life stages present and control methods often require controlling multiple life stages simultaneously, or repeating applications until the population is greatly reduced.

Who Should Scout?

The first approach to scouting crops is to devote an employee specifically for this purpose. (For large greenhouses, it may be necessary to divide the production areas into zones with individuals assigned to each zone.) A scout's primary responsibility should be to routinely check the crop areas for potential problems. An in-house employee, often a grower or water technician, has the advantage of being familiar with the crops and the growing systems within the operation. The abilities to properly identify pests that are most common and to know where to find them are very important to fulfill this responsibility. Scouts need to collect information regarding the presence and populations of various pests and to determine when control strategies are necessary.

It is also very important that people with scouting responsibilities learn to ask for help when identifying unknown problems. Many times growers treat for problems that do not exist or use inappropriate or noneffective control strategies. Sometimes treating an improperly diagnosed problem will work, but in many instances control strategies can mask or even make cultural problems more complex. Guessing is not good enough; the crops we grow are too valuable to make inaccurate guesses or decisions. If you are unsure of a particular problem, seek help by contacting a peer in the industry, your county extension agent, or by sending samples to a diagnostic clinic.

Many times, in-house employees lack sufficient training regarding the proper identification of insects and diseases and how to properly control these types of problems. Another disadvantage of having company employees perform the weekly scouting is they are often pulled away from monitoring the crops to perform other activities, such as shipping plants. Many operations do not have the discipline, or devotion, to consistently, without hesitation, recognize the importance of scouting crops on a weekly basis. If you pursue this option, commit this employee's time and training to the task.

The second option available to growers is to hire private crop consultants to perform the weekly scouting activities. An advantage of hiring a professional scout is that regular scouting responsibilities are their primary responsibility and they are less likely to get distracted or moved to other tasks while in the middle of monitoring the crops. Professional scouts, with their experience and resources, can quickly and efficiently monitor various production areas. They have access to the most up-to-date information regarding pest management materials and practices. Crop consultants may also be alert to new or developing problems at other operations in the surrounding area. Research has shown that a good professional scout can save money in both labor and chemicals.

Regardless of who does the actual scouting, it is helpful to train your workers in insect and disease identification. Many times, new pests can enter the production area or existing pest populations can get out of control between the scouting intervals. Having other workers, such as water technicians or shipping crews trained to identify potential pest outbreaks can be very valuable and will compliment any production system.

Scouting Tools for Perennial Growers

In order to maintain an efficient scouting program, it is important that growers and scouts have the proper tools readily available and at their disposal. In most cases, monitoring should be effective and efficient so a large quantity of information can be collected in the shortest period of time. The tools necessary will vary, depending on the types of crops produced and their production environments. For example, perennial producers growing plants directly in the field may require slightly different scouting tools than a plug or a finished container grower. Here is a listing of some of the scouting tools available that I would recommend including in your scouting kits.

Carrying case

A small carrying case such as a gym bag (duffle bag), backpack, plastic toolbox, or even a clean plastic pail is useful to store and carry your scouting tools.

Hand lens

A good quality hand lens, often called a loupe, is an indispensable, must-have scouting tool for any operation. All growers should have and be using one, whether or not their primary responsibilities include scouting. Using a hand lens will allow you to see more clearly some of the details that you cannot see with the naked eye. Hand lenses are usually available with a 10X to 15X magnification factor; I would recommend using the highest magnification factor available. It is helpful to attach a shoestring that goes around your neck to the hand lens, making it easy to carry and readily available at all times.

To use a hand lens, simply place the hand lens in front of your eye and bring the object you wish to view closer to the hand lens until it comes into focus. You may find it awkward to use at first, but with a little practice you will be glad to have one.

Another variation of the hand lens is called an OptiVISOR. The OptiVISOR is a binocular that is attached to a headband or can be worn over prescription glasses and can be

Figure 4.1. Examining for pests with a hand lens.

Shelton Singletary.

flipped out of the way when you are not using it. The biggest advantage of using this type of magnifying lens is that you can use it hands free, meaning both of your hands are available to examine the specimens.

Horticulture knife

A sharp knife is useful for cutting into stems or root tissues when examining for insects or diseases. Keep in mind that sanitation is very important and the knife should be cleaned between uses to prevent the spread of plant diseases. Carry a small container with a lid containing a disinfecting agent such as 70% alcohol, 10% bleach, or quaternary ammonium compounds such as Greenshield or Triathlon to disinfect the knife while scouting.

Plastic bags

Use small zipper bags to collect various samples for observing closer or for sending to a diagnostic clinic. Also bring a waterproof marker for individually marking each bag with the sample enclosed and any relevant information, such as where the sample came from, the plant variety, and the date it was collected. Use small plastic prescription bottles or film containers with lids when collecting insects to prevent them from getting smashed during transit.

Diagnostic clinic forms

Bring a copy of the form required by your diagnostic clinic. It is not necessary to always fill out the form on site while collecting the samples, but in many instances these forms include checklists that help them to narrow down the potential problem. Having a copy of this form with you will help to ensure that you are collecting and recording the appropriate information.

Scouting forms

Scouting forms are very important and should be used religiously by scouts and growers. Using a scouting form will help you to accurately track the exact location of the cultural problem—insect, disease, or otherwise. They can be useful for collecting information about the current conditions or practices that may have contributed to the problems. Or they may be used to help track the effectiveness of previous treatments. These records provide the basis of future scouting activities, identify the current cultural concerns of the crops, and help to determine when action must be taken.

While scouting, it is important to write down your observations. It is often difficult to remember the details later when troubleshooting. If developing your own scouting form, it is important to include such items as: the date, the scout's name, the location, the plant's name, the insect or disease detected, the level of infestation, and any previous actions taken. Be sure to include adequate space for additional comments or observations.

With today's technology, it is fairly easy to develop and modify an in-house scouting sheet specific to your operation and the perennials you are growing. Some growers are using handheld computers to record their scouting information, downloading it into a spreadsheet, and making management decisions or spray lists from there. Regardless of how you collect the information, it is important that you do.

Clipboard

Carrying a clipboard or a notebook will give you a hard surface to write on while performing your scouting activities. Many growers have outdoor sites that can become rather windy, and a clipboard or notebook will help to keep your notes and observations together at all times. Taping a sheet of white paper to the back of the clipboard can be useful when sampling for small insects or mites. Tap the plant stem over the surface, knocking the insects from the stem onto the white background, where they can be easily detected.

Maps

Maps of the property or specific locations, such as greenhouses, are useful references for scouts. With maps, they can record more precisely the location of any of their concerns. The maps can be attached to the scouting

Figure 4.2. **Sample scouting form**

Scouting Form

Date: ______

Scout'sName: ______

Location	Crop	Container Size	Insects Observed	Diseases Observed	Comments

sheets, kept in plastic sheet protectors, or even laminated to prolong their usefulness.

Digital cameras

Digital cameras are very useful for collecting images containing information that might not be obvious with plant samples. Pictures of the production site, where the problem is occurring, and the surrounding areas may be beneficial to make the correct diagnosis. Pictures of the symptoms from a distance and close up can easily be taken, allowing for a more accurate representation of the current situation.

Reference library

Growers should collect a library of resources to use when diagnosing cultural problems and managing them. These resources could include books, trade magazines, trade journals, handouts from seminars, and past scouting notes. This library does not need to be terribly extensive but should at least contain relevant information pertaining to the crops you are growing. References such as these will help aid the pest identification process and management practices, and can provide some general pesticide recommendations.

These tools are intended to make your time scouting more effective and efficient. Generally speaking, they are inexpensive, will help you make the most of your time, and will lead to quicker identification and control strategies, reducing the potential injury to your perennials.

Scouting Specifics

The scouting area

The area to be scouted should be well defined and may consist of the entire greenhouse, specific bays or benches, specific crops, a certain amount of production area, or any other logical layout. A diagram or sketch of the scouting area is helpful in identifying the location of the sampling, the layout of sticky cards, or for denoting the hot spots within the crop.

Sampling units

A sampling unit consists of any plant part where insects or diseases are typically found, such as leaves, stems, or roots. The sampling unit should be consistent with the detectable stages of the insects or diseases being monitored. For example, when monitoring for *Pythium* root rot, a scout would be sampling the root system as opposed to examining the leaves. Similarly, aphids are most likely to appear near the terminal growth and upper stems; sampling from the lower leaves would not provide an accurate indication of their presence.

Sampling techniques

The sampling technique is the method used for collecting information from the sampling location or unit. Active scouting methods involve the scout searching, or examining, plant parts and the surrounding areas for the presence of insects or diseases. Passive scouting methods involve capturing the potential insect pests as they move throughout the crop. There are several techniques available. Regardless of the method used, it is critical to use a standardized procedure throughout the crop cycle, allowing the results to be properly compared.

It is important to know which insects or diseases a crop is most susceptible to, or most likely to have, in order to implement the best scouting techniques into practice. Each pest enters a crop differently, which means there are different methods scouts must use to detect their presence.

Distribution

Pests that are distributed randomly throughout the production area are usually winged insects, such as fungus gnats and thrips. Winged insects are often present throughout the entire production area. Their presence becomes widespread quickly, often shortly after their entry into the site and before their presence is detected. Pests present in isolated areas, often referred to as hot spots, are most commonly non-flying insects, such as aphids or spider mites. Hot spots can, over a short period, encompass the entire production area if they go unchecked. Some insects are known to have an erratic distribution pattern. These insects lay eggs in one area shortly after entering the greenhouse. The result will be patchy infestation in one or more locations. Whiteflies and moths have been known to have patchy distribution patterns. Marginal infestations are those that enter the crop from the outer edge through an opening to a greenhouse or near a pathway and gradually move into the crop.

Counting insects on plant parts

Removing the sample unit, such as a leaf from the plant, and counting the number of insects present, either with the naked eye or using a hand lens, is one type of sampling technique. This method is effective for counting non-flying insects such as aphids and spider mites. Commonly, between five and twenty-five samples are removed for counting, depending on the size of the plant species, the number of plants to be scouted, and the size of the production area. High-value crops and crops with high sensitivity to certain insects or diseases, should be sampled more aggressively. Removing plant parts for sampling purposes is not advisable for plants that are highly valued for their aesthetic quality. An example of this could be perennials being marketed as houseplants.

Timed counts

Timed counts involve a visual inspection of the stems and leaves of an entire plant. It is often recommended to inspect each plant for thirty seconds, using a timer to help keep track of time, as it is difficult to count insects and keep track of time simultaneously. Timed counts are not practical when the pest populations have reached excessive levels because you cannot count the pests fast enough. Timed counts do not damage plant parts or alter the aesthetic quality. This type of sampling technique works well for relatively stationary insects such as aphids, leaf miners, spider mites, and juvenile whiteflies.

Beat samples

This technique involves beating, shaking, or agitating plant branches a standard number of times (usually three to five) above a white surface, such as a tray or heavy sheet of paper. As the plant is shaken, several insects are dislodged and fall onto the tray or paper, where they can be accurately counted. This monitoring strategy is especially effective at dislodging and detecting hard-to-see pests, such as spider mites and thrips. It may be necessary to shake the flowers separately because some insects, particularly thrips, prefer to feed inside flowers (when available to them) as opposed to leaf surfaces. Blowing air onto the flower may also help to bring the thrips out of their hiding places for counting.

Traps

Trapping devices contain synthetic or natural attractants, called pheromones, ultraviolet (UV) light, or attractive colors to physically trap the insects for monitoring purposes. UV lights and pheromone lures are commonly used to detect the presence of moths. Pheromones are the chemicals emitted by female insects for attracting the male of the species for reproductive purposes. Traps using attractants are not an effective tool for detecting the population, nor do they help growers determine when to implement control measures. Traps essentially attract adults, usually males, which reduces the number of males available for reproduction. They do nothing to monitor females or immature stages. Once the males have been attracted to the inside of the trap by the pheromone, they are trapped on one or more sticky surfaces. These traps are useful for early detection of adult pests or to provide an indication of a sudden increase of the population between scouting intervals.

Sticky cards

Colored sticky cards are the most widely used scouting trap chosen by commercial growers. Sticky cards detect the presence of flying insects in the production area. These monitoring devices are made of a stiff paper card, usually bright yellow in color, covered with a nondrying, sticky substance that traps the insects as they land on the card.

Sticky cards should only be used as a method of detecting the presence of and estimating the relative population of winged insects in the production area, not as a means for controlling them. Yellow sticky cards are most commonly used to detect the presence of fungus gnats, leaf miner adults, shore flies, thrips, and whiteflies.

Place the colored sticky cards vertically just above (1–2 in. [2.5–5 cm]) the plant canopy. Much of the insect flight occurs in this area. As the crop grows, it will be necessary to adjust the placement of the sticky cards above the crop. Many growers attach the sticky card to a bamboo stake using a clothespin. This makes it easy to raise the cards as the crop grows. Some growers glue two clothespins back-to-back, one of the clothespins attaches to the stake and can be adjusted as the plant increases in height, and the other clothespin is used to hold the sticky card.

When the detection of thrips is a primary concern, scouts often use blue sticky traps, which attract these insects slightly better than yellow ones do. Blue cards may attract more thrips to them, but it is slightly more difficult to see and count the thrips, compared to using yellow cards.

It is recommended to have one sticky trap per every 1,000 sq. ft. (93 m^2) of production space. Many growers feel they can properly monitor their crops by placing one card every 3,000 sq. ft. (279 m^2). Sticky traps should be examined at least weekly. Additional cards should be placed near vents or doorways, where the introduction of new insects is most common. Be sure to place some cards at ground level (i.e., under production benches), especially in structures with soil, sand, or gravel floors. Many insects such as fungus gnats, shore flies, and thrips thrive in these areas and soon could move up into the crops grown above.

Figure 4.3. **Yellow sticky card positioned just over the plant canopy.**

For improved detection of fungus gnat adults, place the cards horizontally between the potting medium and the lowest leaves. Sticky cards placed in this manner will also help the early detection of newly emerged thrips. For fungus gnat and shore fly control, some growers place sticky tape horizontally above the crop to trap these hard-to-control insect pests. Sticky tape is used as a control strategy rather than a method of monitoring insect populations.

It may be necessary to use a 10x or 15x hand lens while examining the cards to help distinguish between insect pests and natural enemies. Magnification may also be necessary when detecting the presence of thrips, since it is difficult to distinguish a thrips from a small particle of growing medium with the naked eye.

If you have caught an insect on a sticky card that you cannot identify and are not sure if you should become concerned, wrap the card in plastic wrap and send it to an entomologist for identification. Be sure to circle the insect you would like them to identify and attach any information you feel would be helpful in the identification process.

Monitoring insects using sticky cards will help identify population trends of the insects you are monitoring. Sticky cards will provide a quick indication of whether a certain population is remaining stable, increasing, or decreasing. They are useful for determining whether control strategies have been effective or if additional controls are necessary. Many growers make graphs, giving a visual representation of the various insect populations.

It can be very time consuming to obtain accurate insect counts from sticky cards. It is not necessary to obtain exact counts, only counts that reflect how the insect populations are changing. Relative numbers can be obtained by scanning the card quickly, forming an estimate of what insects are present. Usually, when making estimates, I would obtain an actual count of any insects present whose numbers are less than ten. When high populations exist, I would round to the nearest ten the number of insects I felt were present. For example, three whiteflies, ten shore flies, twenty fungus gnats, and forty thrips were estimated on a single sticky card. In reality, there were indeed three whiteflies, but there may actually have been twelve shore flies, twenty-four fungus gnats, and thirty-six thrips if I had taken the time to do an exact count.

Another strategy would be to divide the card into smaller segments such as quadrants, count the insects present in the smaller area, and calculate the numbers that would be present on the entire sticky card. Some manufactures actually have quadrants clearly marked on the sticky traps for growers to use this estimation strategy.

The estimation process will provide fairly accurate information and save at least half of the time necessary to obtain the actual numbers. It is not important to know the actual number of insects at any given time, only the relative numbers over time. What is the trend—up, down, or remaining level?

Be sure to record the number of insects caught on the cards each week, replacing them when necessary. It is important to keep the cards in the same location to help keep track of the insect activity or where the hot spots are developing. To help provide more information, it is important to number the cards, or maintain a map with the card location on it, to show exactly where the insect populations are changing. It is also helpful to write the date the sticky card was placed in the production area on the card itself. The pest population and their locations should be recorded on pest scouting forms or on the maps mentioned above.

Figure 4.4. **This tomato plant is used as an indicator plant for *Lavandula angustifolia* 'Jean Davis'.**

Indicator plants

Growers often use indicator plants, sometimes referred to as sentinel plants, as an early warning sign of the presence of certain insect pests. Indicator plants can provide growers with an indication of the developmental progress of various pests or beneficial insects in the greenhouse. Sentinel plants can help growers determine when to implement a control strategy, such as the release of natural enemies. A good example of an indicator plant is the petunia, whose leaves show clear symptoms of thrips feeding injury or viruses such as impatiens necrotic spot virus (INSV) and tomato spotted wilt virus (TSWV). Some growers utilize insect-infested plants as indicator plants to determine the effectiveness of their control strategies. Indicator plants should be clearly marked and checked every time the crops are monitored. Growers place indicator plants, using approximately the same guidelines they use for sticky cards, one plant for every 1,000 sq. ft. (93 m^2) of production space.

Potato disks

Monitoring the adult population on sticky cards alone does not give an accurate picture of the true population of fungus gnats or of the next generation. Using chemicals that provide "knock down" of adults does not necessarily result in a decrease in the overall fungus gnat population. Detecting the number of larvae present will provide growers a clearer image of the insect pressures to come.

To monitor the population of fungus gnat larvae, many growers insert potato disks into the growing medium. Simply place slices of raw potato on the surface of the growing medium. The larvae will make their way up to the soil surface to feed on the potato. Some growers cut potatoes into 0.25 in. (6 mm) wide wedges and lightly press them into the soil. After three to seven days, examine the bottom of the potato discs and count the number of larvae present. For monitoring the larval stages of fungus gnats, this is currently the only reliable method that has been developed. This technique should always be used in conjunction with monitoring the adult population on sticky traps.

Visual inspections

Conducting visual inspections is a nondestructive method of monitoring the distribution pattern of both flying and non-flying pests, as well as for observing cultural problems such as diseases and plant nutrition. Visual observation is time consuming, but often provides a better image of the true pest populations and their locations than when only using sticky cards or other monitoring methods. Enough observations should be made within each block or variety of plant to determine the extent of a problem and if the incidence is high enough to warrant intervention. These inspections should occur at least once per week.

Randomly select five to ten plants per bench or greenhouse bay for inspection. The plants to be inspected should range in location from the edge to the middle of the production area. Plants near the outside edge or near pathways provide an indication of a pest just entering into the production area either by wind from the vents or workers' clothing as they walk by.

Examine all plant parts including the leaves, stems, and roots. Inspect the upper and lower leaf surfaces of both older and younger leaves. Aphids, mealybugs, scale, spider mites, and whiteflies are most commonly found on the undersides of leaves or on plant stems. Thrips and aphids are commonly found feeding in or on flower buds. It is also important to check the root systems of the crops during the scouting routine. The roots can provide a quick indication to the overall health of the crop. If the root zone is not healthy, pests and diseases are more likely to attack the crop. Check the roots for healthy growth and any abnormalities, such as browning or deterioration, which may indicate a root rot is present.

The production area should be divided into equal areas; the visual inspection should contain plants in each of these areas. The inspection should include areas near doorways, bench ends, pathways, vents, and by cooling pads. Be sure to include the corners of greenhouses or other areas that workers do not regularly visit, where insect or disease problems often go undetected. These areas will become more obvious to the scout after a season or two of experience.

Each crop generally has at least one or two key insect pests and one or two key diseases that commonly attack it. This information should be used to determine the type of sampling technique or scouting procedures necessary when producing these crops. The key insect or disease pests are those that are either the most common or the most difficult to control, both of which could alter the quality of the finished product. While looking for these key pests, be observant of other pests that may be present. Any insects or diseases found should be evaluated, recorded, and, if necessary, controlled.

It is also beneficial to check production areas before a crop is brought in. This pre-crop inspection should include looking for weeds, holes in the structure or plastic, or at other items such as "pet" plants or dead plant materials that might be harboring insects or diseases.

Sampling pattern

Collecting samples randomly throughout the entire production area is the most accurate way of monitoring for insects and diseases. Growers often collect their samples along the perimeter of the crops because the samples are easier to obtain. Collecting samples only from the edges and near walkways will not provide growers with the entire picture regarding insect or disease pressures. The pathways are where there is the most foot traffic in a greenhouse, and there is a greater likelihood for pests to use people as vehicles to move around the greenhouse. Samples collected near the pathways, which are important areas to monitor, should be only a small portion of the samples being collected.

Most greenhouse pests and diseases are not distributed evenly throughout the production area. The entire greenhouse area needs to be monitored in a consistent manner when scouting. Inspections should include plants on the ground, benches, and hanging from the structure, where applicable. The amount of production area, the size of the plants, the number of plants, and the location of the benches or pathways will influence the scouting pattern and the amount of time necessary for the monitoring activities.

It is best to collect samples randomly throughout the entire crop. These can be taken from predetermined areas, coordinates generated from a random number generator such as a calculator, or they can be selected spontaneously as the scout is monitoring the crops. The location of the samples should be documented and easy to relocate for future monitoring and to implement control measures, if necessary. At a minimum, five to ten samples should be taken from each 4,000 sq. ft. (372 m^2) area.

The scouting should begin at the major entrance into the production area. Look more closely at plants around any openings to the greenhouse or plants found on the outside rows of benches. It is also important to check other areas, such as the middle of the production bench, where the plants may have received less spray coverage. It is best to walk through every pathway while scouting. Plants should be selected randomly from every bench or production area. The more plants inspected, the more accurate your scouting results will be. Besides observing the crop, always look around the production site for weeds or items out of the ordinary. For example, is there condensation on the structure that is dripping onto the crops, or are the circulation fans working? The more observant the scout can be, the sooner problems can be addressed before serious cultural problems arise.

Many growers follow a standard M, or zigzag, scouting pattern as they maneuver through the production area. As scouts follow the zigzag pattern, they select plants randomly, choosing plants from every bench, consisting of plants from the outside rows, as well as from the

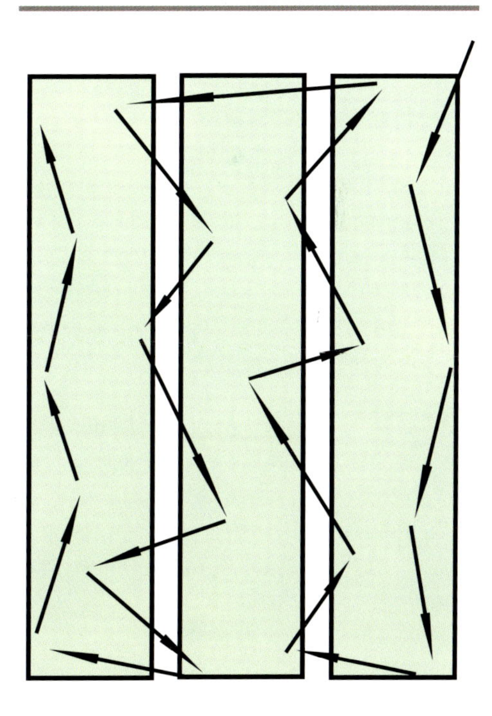

Figure 4.5. **Example of a zigzag scouting pattern.**

middle of the bench. Each plant is monitored from the soil surface up to the top of the foliage, and scouts should also inspect the root systems.

Sample timing

Scouting activities and sampling should occur at least weekly. In this manner, the samples being collected occur at such frequency that an insect species cannot pass through a developmental stage without the grower having adequate time to implement a control strategy, if necessary, and to evaluate its results. The exact interval may vary with the crops being produced, the insect or disease being evaluated, or with production temperatures. Many operations have extended the interval between scouting to every two weeks during periods of lower disease and insect pressure. Evaluate the frequency between scouting activities and increase the scouting frequency back to weekly once cultural problems begin to reoccur.

The time of the day the scouting occurs may have an affect on the results. Some insects may be more active in the morning than they are in the afternoon, which could make detection more difficult and alter the quantities found on the samples collected. Conversely, other insects are more active in the afternoon than during the cool morning hours. Knowing the behavior patterns of common pests for each crop will help you schedule scouting activities.

Blooming crops may attract insects, such as thrips, away from traps and nearby non-flowering plants and may need to be scouted differently once blooming begins.

Number of samples

The number of samples a scout collects will vary with each operation. Many greenhouses have very little tolerance for insects and diseases and may scout more thoroughly, collecting more samples than the average producer. The key is to keep the number of samples realistic, collecting as much as possible in a reasonable amount of time and to ensure that the samples provide a proper representation to the real insect and disease pressures in the production area.

The number of samples should remain constant over the entire growing cycle. At times, growers may find it necessary to collect additional samples, but under no circumstances should the number of samples collected decrease. Although decreasing the number of samples being collected may save time, it also reduces the likelihood of some pests going undetected until significant crop injury has occurred. Since insect and disease problems do not occur uniformly across the crop, taking more samples will provide more accurate population estimates.

There are various recommendations from universities and plant crop consultants as to the number of samples growers should collect. I have seen recommendations ranging from examining twenty plants per 100 sq. ft. (9 m^2) to twenty plants per 5,000 sq. ft. (465 m^2). The most important aspect is to collect enough samples to provide an accurate representation of the true insect or disease populations. It is easy to collect too many samples and even easier to collect not enough. The main point here is to be consistent.

Recording scouting results

Keeping accurate records is essential to any scouting program. Documenting and recording your scouting and sampling results will help growers to spot and track potential problems early enough, allowing them to implement control measures before small problems become big ones. The records do not need to be terribly extensive, but should include the following: the crop name and location, symptoms observed, the pest name, the population present, control method, the date and time of the control strategy, the rates and amounts applied, and the method of application.

It is also helpful to have maps of the production area that show the location of the benches, sticky traps, indicator plants, or any other useful information for the scout or pesticide technician. These maps can be used to help growers detect population trends and to decide where and when management strategies should be implemented.

Records from previous crop cycles can be used to provide an indication of potential problems on particular crops, to determine the sampling techniques and number of samples to be collected, and may suggest that preventative control strategies are necessary. Previous records are useful tools to not only tell what cultural problems are likely to occur, but they may also provide an indication as to when problems are likely to arise.

Comparing insect populations or disease levels, both before and after control strategies are implemented, helps to determine how much control was achieved. Keep in mind that some control measures may provide significant results within a couple of days, while others require more time and may take up to two weeks to provide adequate control. Treatment thresholds can be developed by comparing insect counts to corresponding crop injury or reduction of crop quality.

Control strategies should be documented, recording the treatment used, the application technique, timing of treatments, the number of treatments, and the results. All of this information will be beneficial to modify future scouting and sampling regimes. This information can be assembled with the scouting costs and pest control costs to make comparisons to areas where no scouting occurred. The cost savings achieved and a reduction in overall control strategies, along with improved crop quality, may justify implementing further scouting practices.

Many growers keep track of cultural records such as average temperatures, irrigation applications, and fertility applications in addition to the scouting observations and chemical applications. These records can be very useful to help determine more precisely when cultural problems first began and may give better insights as to the cause and how to manage existing situations. Detailed scouting reports can provide answers to questions about cultural practices that impact plant health quality characteristics.

Simplify your scouting

It is beneficial to develop a written form for recording the relevant information the scout is collecting. These forms, such as the one on page 98, provide a standard method of collecting the right information consistently over time and provide a permanent record of your pest problems. All of the production areas should be checked weekly and recorded on these scouting forms.

Scouting forms should contain the date of the scouting, the crop, the location, any insects and diseases found, observations of items or areas to check in the future, suggested control strategies, and any changes since previous control methods were implemented. Other information, such as recent day/night temperatures, relative humidity, and light levels, may be useful to help diagnose cultural problems or to explain why they have occurred. It provides an organized and consistent method of managing the information the scout is collecting.

To reduce the amount of writing the scout has to do, it may be beneficial to develop a form that contains the plant inventory and items to check or circle as the scouting is conducted. Figure 4.6 is an example of such a form. Notes should be taken on anything that appears to be abnormal. Collectively, scouting forms help the scout record the relevant information, make decisions on the current situations, observe trends over time, predict future pest levels, and to develop data that will assist in future decision-making.

When plant insects or diseases are first observed, it may be helpful to flag the plant with the newly found pest. The flagged plant allows you to relocate the new problem and monitor its development. The flagged plant also serves as an indicator plant, which can be used to evaluate the effectiveness of any control strategies that were implemented.

Diagnostic Clinics

It is nearly impossible to properly identify every cultural problem that may arise. Many perennials exhibit nonspecific symptoms,

Figure 4.6. **Sample scouting form with inventory**

Scouting Form with Inventory

Date: ______
Scout's Name: ______

Location	Crop	Container Size	Quantity	Insects	Disease	PGR	Comments
Greenhouse 1	*Dicentra spectabilis*	1 Gal.	1,077				
Greenhouse 1	*Dicentra* x *hybrida* 'King of Hearts'	1 Gal.	278				
Greenhouse 1	*Pulmonaria* x *hybrida* 'Apple Frost'	1 Gal.	330				
Greenhouse 1	*Scabiosa columbaria* 'Butterfly Blue'	1 Gal.	457				
Greenhouse 1	*Scabiosa columbaria* 'Pink Mist'	1 Gal.	735				
Greenhouse 2	*Aquilegia* x *hybrida* 'Songbird Mixed'	1 Gal.	176				
Greenhouse 2	*Doronicum orientale* 'Magnificum'	1 Gal.	150				
Greenhouse 2	*Heuchera* x *hybrida* 'Mint Frost'	1 Gal.	503				
Greenhouse 2	*Lamium maculatum* 'Orchid Frost'	1 Gal.	64				
Greenhouse 3	*Miscanthus sinensis* 'Strictus'	1 Gal.	23				
Greenhouse 3	*Pennisetum alopecuroides* 'Hameln'	1 Gal.	135				
Greenhouse 3	*Pennisetum alopecuroides* 'Little Bunny'	1 Gal.	191				
Greenhouse 3	*Pennisetum alopecuroides* 'Little Honey'	1 Gal.	290				
Greenhouse 4	*Hosta fortunei* 'Francee'	2 Gal.	250				
Greenhouse 4	*Hosta* 'Gold Standard'	1 Gal.	117				
Greenhouse 4	*Hosta* x *tardiana* 'June'	1 Gal.	80				
Greenhouse 4	*Hosta* 'Twilight'	1 Gal.	165				
Greenhouse 4	*Hosta undulata* 'Albomarginata'	1 Gal.	1,129				
Greenhouse 4	*Hosta ventricosa* 'Wide Brim'	1 Gal.	592				
Greenhouse 5	*Hemerocallis* 'Happy Returns'	1 Gal.	170				
Greenhouse 5	*Hemerocallis* 'Stella de Oro'	1 Gal.	4,265				
Greenhouse 6	*Alcea rosea* 'Chaters Double Mixed'	1 Gal.	250				
Greenhouse 6	*Astilbe* x *arendsii* 'Fanal'	2 Gal.	82				
Greenhouse 6	*Astilbe chinensis* 'Vision'	1 Gal.	104				

Figure 4.7. **Scouting Form with Actions**

Scouting Form

Date: ______________________

Scout's Name: ______________________

Location	Crop	Container Size	Pest Observed	Level of Infestation or Disease	Action Taken	Action Date	Treatment Results

making it difficult to detect exactly what is causing the problem. It can be very helpful to send effected plant material and a description of the problem to a diagnostic clinic. In many cases, the technicians from these clinics can add to your knowledge and experience level. It is not necessary to submit samples of every cultural problem, but the clinics can be very valuable when the problems are too complex for identification without their expertise.

Diagnostic clinics have sophisticated tools available to them, such as high-powered microscopes and identification keys, that are not available to growers. With fungal or bacterial diseases, many labs have the ability to culture pieces of the plant to see if pathogens are present. Some laboratories are using more specialized techniques, such as DNA probes, which are highly accurate but expensive. In general, as the complexity and cost of the testing procedure increases, their usefulness and accuracy for you, the grower, also increases.

Some plant clinics specialize in collecting nutritional information by examining the nutrient concentrations within the growing medium, plant tissues, and water samples. Plant pathology labs do not analyze chemical properties but specialize in looking at plant pathogens. It may be necessary to utilize both types of laboratories to help determine if both chemistry (noninfectious) problems and/or biotic (infectious) problems, are present.

In addition to submitting a sample, it is important to provide the clinic with as much background information as possible. This information, submitted with the sample, is important for them to come up with a proper diagnosis and recommendation. Most clinics have forms to submit with the samples that ask for any relevant background information. If you do not have a submission form from the laboratory, be sure to include the following information: the plant's scientific name, common name, variety, its age, the growing conditions, any fertilizers and chemicals applied to the crop, when the problem first occurred, how quickly it developed, and any relevant observations depicting why you are submitting the sample. It is also beneficial to provide the clinic with photos taken in the greenhouse. Be sure to include the appropriate contact and billing information with the sample.

Most growers usually submit only symptomatic plant tissues as their samples to plant clinics. These samples may not actually harbor the plant pathogens causing the plant to undergo the stress creating the visible symptoms. For example, many times root rots will cause the leaves to turn yellow. Sending only a sample of the yellow leaves to the laboratory will lead to an improper diagnosis. If there is any doubt about what should be included with the sample, send the entire plant, including the roots and the growing medium. If the entire plant is too large to send as a single sample, send representative samples from all of the plant parts and a sample of the growing medium.

Keep in mind that quality samples are the difference between good and bad diagnoses. If an inadequate sample is taken, many plant clinics often will be unable to correctly diagnose the problem and may not be able to make a diagnosis at all. They may request another sample be taken and resubmitted. This will not only waste your time, but you will have given the disease or insect more time to multiply and spread, causing more damage to your crops. The longer it takes, the more difficult it becomes to control the problem, and you could lose the crop altogether.

Do not submit dead plant tissue to the laboratory. It contains a plethora of saprophytic pathogens (bacteria and fungi that live on dead materials), which will likely obscure the pathogen that originally caused the plant tissue to die. It is nearly impossible for a diagnostician to determine the original cause of plant death at this time.

Once the laboratory receives the samples, depending on the problem and the testing procedures involved, it is likely to take one to three days to reach the proper diagnosis. Due to the nature of our business, it is recommended to send any samples via an overnight shipper to the diagnostic clinic. The quicker the problem is properly diagnosed, the quicker a control strategy can be implemented and the economic damage can be kept at a minimum. When possible, submit diagnostic samples to the laboratories early in the week to ensure they will arrive at the diagnostic clinic prior to the weekend.

With good sampling and background information, a proper diagnosis can generally be given. However, in some instances, it may be necessary to resubmit the samples on several occasions until a resolution can been achieved. Some diseases have several causal agents, making them difficult to properly diagnosis at first glance.

The purpose of diagnostic laboratories is to help identify problems associated with plant symptoms, not to determine if plant samples are pathogen free. There are laboratories that provide indexing services, which determine if healthy-looking plants are actually pathogen free. Don't confuse the purpose of diagnostic labs with indexing laboratories; they reach their goals using very different tools and techniques. Be sure you are familiar with what you are trying to accomplish and submit your samples to the appropriate clinic.

It is also beneficial to establish good communication between you, the grower, and the diagnostic clinic. Most clinics are willing to fax you the results or willing to share the information over the phone. Many laboratories have even agreed to report lab results through e-mail. All of these methods allow growers to acquire the results as quickly as possible, without waiting several days for results to be delivered by mail. In any case, it is important to establish that the results of your tests are to be kept confidential and not to be shared with anyone without your approval. If there are going to be any delays for any reason, it is important to work with a lab that will call you immediately. If a lab is not willing to communicate with you and maintain your confidentiality, you should seriously consider changing diagnostic clinics.

Good labs should give rapid and responsive service, provide accurate and reliable results, be willing to communicate delays and complications, are operated by experienced professionals, provide confidentiality, and are cost effective. Be careful when interpreting any cultural recommendations given by a clinic; they specialize in diagnosing plant problems, not producing plants or managing cultural situations. Always formulate your own solutions based on what is practical, effective, and economical.

As stated previously, submit the entire plant, whenever possible, to the diagnostic clinic. If several plants within the population are symptomatic, submit more than one sample. Include plants showing various degrees of symptoms, perhaps one just beginning to show signs of the problem and one with symptoms that are obvious and clearly seen. The larger the sample size or number of samples submitted, the more likely a correct diagnosis will occur. Don't overwhelm the plant clinic with too many samples showing the same symptoms; keep the number of samples submitted realistic and representative. When sending multiple samples, make sure each sample is clearly labeled and identified.

When the samples are expected to be delivered to the laboratory within one or two days, place each sample in a separate plastic bag containing a dry paper towel. The paper towel helps to absorb the moisture inside the bag that accumulates as the plant transpires. Do not wet the sample or the paper towel; wetting these materials could lead to other disease problems during shipping and may render the sample useless. It is also recommended to store the samples in a cooler or refrigerator to keep them fresh until they are sent out or hand delivered.

If delivery of the samples is likely to take three or more days, shake the excess growing medium from the roots, wrap them in dry paper towel or newspaper, and pack them in a box. Do not use plastic to wrap the samples in, as this will trap moisture from transpiration and cause other disease problems, ruining the integrity of your sample. If you are not submitting the sample with the container it was grown in, shake the excess soil from the roots.

It may not always be practical to submit the entire plant to a diagnostic clinic. In such cases, sending samples of the stems or leaves may be appropriate. When submitting stem samples, it is important to cut the stem below the point of injury to include some of the healthy stem tissue. Place the symptomatic stems wrapped in a dry paper towel into a plastic bag. With stem-related problems, it is also beneficial to submit samples of the roots as well. Many root problems will exhibit symptoms on the stems or growing points above the soil line. Root samples should include enough soil to keep the roots moist during transit. Place the root samples in a separate bag from the stem samples.

When submitting leaf samples, it is important to submit enough leaves to demonstrate the various ranges of disease symptoms, from early stages to older more advanced damage. Leaf samples often require several tests and should include enough leaves for the technician to run these multiple tests. Try to submit at least ten leaves with the characteristic symptoms or leaf spots.

Many growers submit insects or mites to diagnostic clinics to be identified. Insect- or mite-infested plant materials should be loosely wrapped in paper towels and placed into a plastic bag that has small holes punched into it. Soft-bodied insects can be submitted in a non-breakable, watertight container containing 70% isopropyl alcohol. Caterpillars, grubs, or worms should be placed in boiling water for thirty seconds prior to placing into the container containing alcohol. Some growers collect pest samples stored in vials of rubbing alcohol for identification and training of new scouts.

Hard-bodied insects should be killed prior to shipment by placing them in a freezer for one or two days prior to submitting them to a laboratory for identification. After they are frozen, gently wrap them in tissue paper and place them securely in a box. When submitting plant materials containing live insects, it is important to get them to the lab quickly, preferably overnight.

Preventing Problems

Sanitation

One of the most beneficial practices growers can do is to ensure that their production areas start clean and sanitized and remain clean throughout the production cycle. Good sanitation is the first line of defense against potential insect and disease infestations entering the production areas. When greenhouses are not emptied between crop cycles, it may be necessary to treat the existing plant materials to control a plant pest that is highly likely to feed on the next crop coming into the facility. Sanitation is the simplest and most cost effective component of a pest control program. Any steps that can be taken to reduce the introduction of insects or diseases into the production area should be considered.

Sanitation does not only refer to starting with a clean crop, but maintaining high levels of cleanliness during the production cycle. Reducing the number of plant pests that enter into the growing area, or decreasing the number spreading from plant to plant, is an important aspect of maintaining sound sanitation practices.

Some general practices include: washing hands between crop activities, wearing clean clothing, and not allowing tools and equipment to make direct contact with floors or other contaminated surfaces. Many growers provide foot mats and hand-washing stations containing disinfectants at the entry points of the greenhouse to prevent the introduction of new plant pathogens into the production site. Disinfecting floors, bench surfaces, plastic pots, tools, and equipment between uses is also critical. Many growers use a 10% solution of household bleach to disinfect propagation/pruning tools and the production area. Several commercial disinfectants are available to growers, including quaternary ammonium products (Greenshield, Physan 20, Triathlon) and hydrogen dioxide (Zerotol).

Figure 4.8. **Weeds such as this annual bluegrass and bittercress should be removed to prevent future cultural problems.**

Weeds

Eliminating weeds and alternate hosts of plant pests before introducing new plant materials into the greenhouse will greatly help to reduce the potential of the new crops from becoming infested with pests that have been harbored on nearby plant materials. Weeds in or near the production area can harbor many insect pests—such as aphids, leaf miners, spider mites, thrips, and whiteflies—that may move onto the crops being produced.

In addition to providing a home for various insects, broadleaf and grassy weeds also serve as host plants for several plant pathogens, including viruses such as impatiens necrotic spot virus (INSV) and tomato spotted wilt virus (TSWV). Some weeds known to host plant viruses include bindweed, bittercress, chickweed, dandelion, jewelweed, lamb's-quarter, nightshades, oxalis, pigweed, and shepherd's purse.

A weed-free zone should be maintained with at least a 10 ft. (3 m) perimeter around the production area. Maintaining a weed-free area will not only reduce pest problems by removing the host, but will also decrease subsequent weed problems by reducing the source of weed seeds. A good weed management program will reduce the resurgence and severity of many pest problems caused from diseases, insects, and viruses. Controlling weeds will be discussed further in chapter 7.

Plant debris

Carefully removing diseased plants from the growing area will help to prevent the spread of fungal spores or bacterial pathogens. Keeping the production areas, such as under benches and walkways, free of decaying plant debris will greatly reduce the presence and spread of fungal spores into the crops. *Botrytis* is a disease that attacks dead or senescent foliage. Once it has infected a plant, it has the ability to attack healthy plant parts, which may cause significant damage to the crops. Flowers that are spent or finished blooming also provide an excellent starting point for *Botrytis* to take hold and move to other living plant parts.

Growing medium left on benches and floors provides an ideal reproductive environment for such insects as fungus gnats, shore flies, and western flower thrips. Any plant debris or compost piles should be located at least 500 ft. (152 m), preferably downwind, from the production area.

Plant shipments

Most operations receive plant materials from outside sources, potentially allowing the entry of insects or diseases into their production areas. All new plant materials—such as unrooted cuttings, plugs, liners, and transplants—entering your facility should be free of any insects or diseases. Bringing in contaminated plant materials from outside sources, unfortunately, is not that uncommon but should by no means become an acceptable practice.

At the very least, new plant materials should be inspected for potential problems upon their arrival. Symptoms to look for include brown, unhealthy root systems (root rot), swellings on the roots (root knot nematode), vein-bounded foliar discoloration (foliar nematodes), mosaic ring spots, dark lines and rings (viruses), and chlorotic leaves (nutritional disorders). These symptoms or the presence of insects, such as aphids, whiteflies, and spider mites, on incoming materials are all indications that a problem is entering your facility. Certain symptoms such as those caused from nematodes and viruses cannot be cured with any control strategy. Infested plants with these symptoms should be rejected and sent back to the supplier or destroyed to eliminate the spread of these problems into your crops.

Many operations have adopted the practice of placing purchased items in a quarantine area, isolating them from the main production facility until they have passed all inspection criteria. The typical isolation period is two weeks. During this period, the plants are observed and carefully inspected for pest and disease symptoms. In-house virus tests are often conducted during this period to verify their absence before allowing the plants enter the production facility.

Cultural practices

There are several things growers can do to reduce the likelihood a crop will become infected with an insect or disease. Growers should be informed about which cultural problems they are likely to experience before production even begins. Having this type of knowledge will greatly aid in your ability to anticipate and minimize many of the cultural problems you are likely to experience.

With this susceptibility information, growers can implement preventative programs, use long-lasting systemic chemistries, or develop stricter scouting regimes. Many perennials under the right conditions are ideal hosts for certain insect pests or plant pathogens. When the presence of a cultural problem is next to inevitable, a preventative program will be beneficial to implement. These programs can consist of multiple applications of various chemistries at a certain frequency or an application of systemic chemicals that will protect plants for several weeks, while the pest pressure is likely to be high. Designing the scouting program to specifically target the most common cultural adversities will allow growers the flexibility to use chemicals in a responsible manner, while not allowing certain insects or diseases to run havoc with their crops.

Crop resistance

It is often possible to produce varieties that are less likely to exhibit certain cultural adversities than are other cultivars. For example, *Phlox paniculata* has the reputation for being plagued with powdery mildew at some point during the production cycle. Plant breeders have brought new varieties, such as the Perennial Plant Association's 1999 Perennial of the year 'David' into the marketplace, which is highly resistant to powdery mildew.

Many plant varieties may exhibit partial or complete resistance to an insect pest or plant pathogen. Partial pest resistance is when the population of a pest is greatly reduced, but its occurrence is not eliminated. Complete resistance occurs when there are no pests found on the plants. Using plant varieties with some level of resistance will often reduce the pest population enough that no control strategies are necessary. Crop resistance often allows any necessary control strategies, biological or chemical, to be more effective, reducing the number of applications or releases, not to mention the costs normally associated with the pest problem at hand.

Fallow periods

Greenhouse growers often experience insects and mites on an ongoing basis, due to the continuous turns of production space or the constant supply of plant materials available to these pests as a food source. Many growers often have extended periods where some of their production sites remain empty, causing these pest populations to become greatly reduced as they die of starvation. During non-crop periods, all plant debris, leftover potting medium, and any weeds must be removed from the production area.

Where the weather conditions allow, greater results can be achieved by closing the doors and vents for long periods, particularly just before crop production. During the summer months, the temperatures within the production structures should be allowed to rise significantly, to 100–120° F (38–49° C) to prevent the pests from going into hibernation. During the winter months, allowing the greenhouse temperatures to reach the freezing point can eliminate many pest populations. Fallowing production sites may not be an option for every operation, as each grower follows different production schedules, but when the opportunity presents itself, fallowing is an easy and effective method of reducing future pest populations.

Water management

Water plays a critical role with the overall health and appearance of any commercially grown crop. Perennials consist of 80–90% water. Water is essential for cell division, plant growth, and for keeping the stems and leaves turgid, or wilt-free. The movement of water from the roots to the leaves is necessary for transporting nutrients, cooling the plant, and removing waste products.

When water is available to the plant at the right quantities and distributed properly to them, everything functions as a complete system, with no major hang up or production problems along the way. When plants are lacking or receiving water in excessive amounts, it can be very hazardous to plant health. Overwatering can lead to increased disease pressures, reduced soil aeration, poor root growth, production of algae and fungi on the surface of the growing medium, and increased populations of fungus gnats and shore flies. Drought-like conditions cause plants to wilt, often resulting in permanent damage and, under extreme conditions, plant death.

Many growers only consider the root zone when thinking about water management. Another aspect of water management is reducing the length of time leaf or flower surfaces are wet. Many foliar diseases, such as *Didymellina macrospora* leaf spots on *Iris* or downy mildew on *Lamium,* require wet leaf surfaces for the disease infections to occur. Irrigating with overhead sprinklers early in the day will allow plant surfaces to dry quicker than irrigations occurring during the late afternoon or early evening hours. Many growers have implemented drip irrigation or subirrigation methods to reduce the wetting of the leaves, as well as reduce the total water needed to grow their crops. Providing adequate spacing and good air circulation will also encourage quicker drying of the leaf surfaces.

Diagnosing Problems

Plant diagnostics is the process growers undergo to come up with the best explanation of why a healthy plant has become unhealthy. Diagnosing plant problems is a complicated aspect of producing plants. There are many tools and methods growers use to quickly identify their cultural concerns.

When cultural problems arise, the cause can be simple and straightforward, but more often than not the cause is complicated and hard to explain. By the time growers recognize there is a problem, what appears to be a single problem has been compounded by other secondary ones. In many cases, growers misdiagnose cultural problems because they often don't look close enough or they make a management decision based on the first symptoms they observe. It is critical to take the time to properly diagnose plant problems before they reach uncontrollable levels or improper, ineffective treatments are implemented. And it is important to begin diagnosing plant problems as soon as plant stress is observed—catching symptoms early could keep the problem from becoming widespread and costly.

Troubleshooting plant cultural problems can be fun if you enjoy solving mysteries, but they are also frustrating and challenging since the possibilities seem endless. Diagnosing plant problems can be complicated and involves a certain level of skill and knowledge that is derived mostly through experience.

Good diagnosticians use a common-sense, systematic approach to problem solving. They keep their eyes and ears open to new information and are not afraid to seek outside help from professionals, often utilizing laboratories and plant clinics. Most growers already have the skills necessary to properly identify most cultural problems; they just have to learn how to use them more effectively. The diagnostic process involves learning and improving the problem solving process.

The first questions a grower should ask are: What is the plant? What does the plant normally look like? How does it grow under normal circumstances? Proper identification of the plant in question is crucial to correctly diagnose any cultural problems that may arise; a normal plant characteristic for one species may be a sign of a serious problem for another, and vice versa. For example, many novice growers often confuse variegated or yellow varieties with viruses or nutrient deficiencies. Try to understand the growing requirements of each plant in production, its normal growth habits, and any seasonal attributes it may exhibit.

The next logical approach should be to identify what cultural problems are common for each plant species. Each perennial has its own set of diseases, insects, and cultural dilemmas. Some perennials appear to be rather problem-free, but most of the commercially produced perennials have several commonly observed cultural woes. Understanding which cultural problems are most likely to occur for each variety can dramatically reduce the time necessary to properly diagnose them. This helps scouts to understand what symptoms they are looking for and what tools or strategies they should be using.

For each perennial variety being produced, a listing of any known cultural problems could be made before the crop is started. This listing of potential crop problems can help a grower decide what types of control strategies may be necessary (preventative versus curative), how detail oriented the scouting might have to be, and even if the production of a certain crop is likely to be profitable.

Whenever detecting crop problems, it is important to clearly consider and list the signs and symptoms that are present, indicating that a problem is at hand. This is where scouts must ask themselves a lot of questions, trying to make accurate observations and narrow down the potential possibilities. Each question asked should be followed by another question. For example, the first question might be: Are there signs of insect or mite feeding? If so, does the injury appear to be from insects with chewing or sucking mouthparts? There are numerous questions that need to be asked and some of them are listed below. Remember to keep an open mind, as many cultural problems are not very obvious and may actually be composed of two or more factors.

Diagnostic questions

I could have listed many pages of questions to ask while scouting for cultural problems. Here are several that should give you an idea of the types of things to be looking for. If the answer to one of these points to a problem, you will need to ask follow-up questions to narrow down the cause.

- What is the affected plant, and how does it normally look?
- Is only one type of plant affected, or are several species or genera?
- What are the key pest problems for the affected plant?
- When did the symptoms first appear?
- Is the problem progressing rapidly in the plant or to other plants nearby?
- Within the production site, where are the affected plants located—near entryways, bench edges, isolated areas, or widespread?
- Does anything appear to be abnormal in the production area? Holes in materials, dripping water, etc.?
- Do the leaves appear healthy? If not, what is abnormal?
- Do the stems appear to be normal and healthy?
- Are there signs of insect or mite feeding or any insects or mites present?
- Are there signs of fungal diseases? If so, where are they located?
- Are leaf spots or cut stems oozing a slimy material, possibly indicating a bacterial infection?
- Are there abnormal growths such as galls on the stems or near the crown of the plant?
- Do the roots appear to be healthy?
- Are effected plants overly wet or too dry?
- Have the leaves stayed wet at night recently?
- When growing on flood floors, does the injury occur where the floor drains most slowly?
- Is the irrigation water (or timing) or potting medium the potential cause of the problem?
- Has the weather changed recently, going from sunny to cloudy or vice versa?
- Are plants spaced to closely, enabling disease or insect spread?
- Is the area maintained and sanitized? If not, what can be cleaned or removed?
- Can prevailing winds or sprinkler head locations be factors?
- Do your propagation techniques predispose the plants to problems later in the production cycle?
- Have there been any changes in chemical/fertilizer applications?
- How much damage has occurred? Are these levels acceptable?
- Are you using scouting data to your best advantage?

After you've made your diagnosis and have begun treatments, the following are a few of the questions to ask:

- Are treatments being applied in a timely manner and properly?
- Are pest populations reduced to acceptable levels?
- Is the current chemical rotation of pesticides effective, or should it be altered?
- What impact are specific treatments having on natural beneficial insect populations?
- Are you able to maintain your standards of plant health and appearance while reducing pesticide use and/or choosing control methods that minimally impact the environment?
- Is your choice of pesticides cost effective when all factors are considered?

Questions such as these will often lead to the discovering of other overlooked symptoms, triggering more questions to be asked and answered. Asking and answering questions will not only help to narrow the potential possibilities but will also help growers determine why the problem occurred in the first place and how best to correct it.

Making determinations

After systematically looking at the symptoms, you will see the problem in its entirety. Once you can see the whole picture, you can more precisely determine the causes of the problems and treat them accordingly. Let us now look at a few of the symptoms.

Typically, when the symptoms appear to be uniform, or affect the entire crop, the problems are usually associated with cultural factors, such as fertility or water quality. If symptoms do not appear to be uniform, or are distributed unevenly within a crop, the problem is often associated with contagious diseases and insects, such as aphids, powdery mildew, or root rots. Contagious diseases often start from a point source and take time to spread throughout the crop and often only will spread when the environmental conditions are conducive. The pattern of symptomatic plants in the production area can be useful to determine whether the problem is associated with environmental/cultural causes or if contagious diseases or insects are responsible.

Cultural problems are, in many cases, brought on by cause and effect. For example, the initial injury to plant roots was caused by high salts, then *Pythium* root rot came in and attacked the already injured root system. *Pythium* root rot was the proper diagnosis, but it probably never would have occurred if the plant had a healthy root system. The *Pythium* needed an entry point or an injured root system in order to attack the plant. In this case, monitoring and keeping the salt levels within an acceptable range during production could have prevented the pathogen.

The list of potential questions to ask is never ending. Each situation will require its own set of questions and becomes more complex when multiple problems are present. It is important to not only closely examine the plant, but to step back and look at it from a distance. As you work your way through the diagnostic process, take notice of any relevant item that may lead to the cause of the initial problem. Once you think you have identified the problem, don't act too quickly, thinking you are finished. Determine if there are more questions that need to be asked.

In the *Pythium* root rot example above, treating for the root rot alone would not be sufficient; the grower needs to understand that the root rot was the effect and not the cause. Treating only for the effect will not make the problem go away. It is important to identify and eliminate the primary problem (cause), which will greatly reduce the occurrence of secondary issues (effects).

References

Behe, Bridget, Diane Brown-Rytlewski, et. al. *Management Practices for Michigan Wholesale Nurseries*. East Lansing, Mich.: Michigan State University. 2004.

Burns, Karol A., Kevin M. Heinz, and Bastian M. Drees. "Scouting Basics." *GrowerTalks*. October 1999.

Chatfield, Jim, Joe Boggs, and Draper Erik. "The 20 Questions of Plant Problem Diagnostics—Part I." *American Nurseryman*. June 1, 2002.

———. "The 20 Questions of Plant Problem Diagnostics—Part II." *American Nurseryman*. June 15, 2002.

Cloyd, Raymond A., and Clifford S. Sadof. "Scouting for Pests." *Greenhouse Grower*. February 1997.

Dutky, Ethel, and Stanton Gill. "Train Your Scouts." *NMPro*. May 2002.

Miller, Fredric. "Be a Better Diagnostician." *NMPro*. July 2000.

Oetting, Ron. "Using Active Scouting to Demystify IPM." *GrowerTalks*. January 1999.

Powell, Charles C. "Correct Solutions Begin with Correct Diagnoses." *NMPro* July 2001.

———. "Integrated Diagnostics: The Basis of Plant Health Management." *Ohio Florist Association Bulletin 869*. May 2002.

———. "Working with Diagnostic Laboratories." *GrowerTalks*. May 1996.

5

Diseases of Perennials

For many growers, detecting and controlling plant diseases are among the most difficult aspects of growing perennials. Unlike other cultural problems, plant diseases are often difficult to detect until significant crop injury is evident. For example, symptoms of nutrient disorders can often be observed early, or the nutrient status can be measured using various testing methods before irreversible injury to the crop occurs. Similarly, insect poulations can be easily monitored and successfujlly controlled prior to causing significant crop damage. Diseases are often more challenging for growers since they cannot easily be detected until some crop injury has occurred. This chapter will familiarize growers with many of the conditions conducive to disease development, how they can better anticipate plant pathodgens, and if necessary, how to control them.

Fungal Diseases

Plant pathogens can provide a number of different signs and symptoms. Quite often fungal diseases may cause the leaf to have a dry texture or to form an area with concentric rings where the disease has attacked the leaf. It is common for many leaf diseases to have definite margins with the center of the spot appearing to be dead, or necrotic. The margins of many leaf spots appear purple or yellow and stand out from the surrounding tissues. These types of changes in the leaf's appearance are considered symptoms of the disease.

With many diseases, such as powdery mildew, mycelium and fruiting bodies of the pathogen are visible. Some fruiting bodies are quite small and may require magnification with a hand lens or microscope in order to see them. These are considered signs of the disease present and can be very useful to properly diagnose the disease.

With fungal diseases, it is helpful to keep track of the weather, as many fungi require a specific range of humidity, moisture, and temperatures for them to develop. If the proper environmental conditions do not exist, the disease cannot develop. But under the right conditions, plant pathogens can develop into a disease outbreak that can be quite extensive.

For any disease to occur, all of the criteria for the disease must be present simultaneously. The criteria for most plant diseases are: a susceptible host, suitable environment, and a causal agent (pathogen). This is often referred to as the disease triangle. (See figure 5.1.) If any of these criteria are missing, a fungal pathogen cannot develop into a problem.

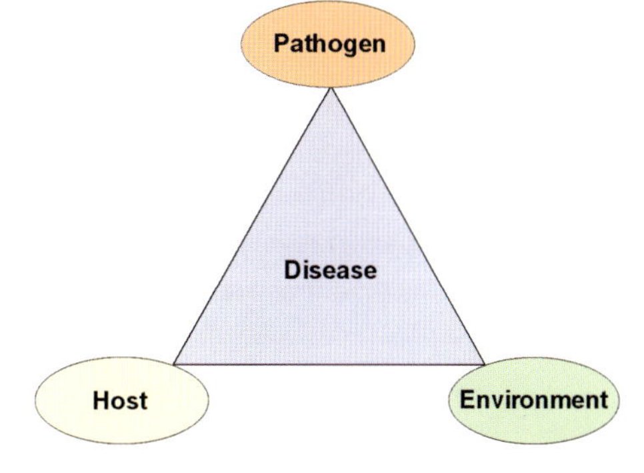

Figure 5.1. **Diseases can only form when interactions between the host plant, plant pathogen, and environment are favorable.**

Since diseases can only develop when they are exposed to a suitable environment, detecting suitable environments for plant diseases is better than detecting the presence of the actual disease. Pathogens can be brought into the production site through air currents, the irrigation water, contaminated growing medium, infected plant materials, infested tools, or by insect vectors such as aphids, thrips, or whiteflies. Stress on the plants from such factors as high salts, poor water management, and improper heating or ventilation can cause a healthy crop to become susceptible to plant diseases.

The most successful disease control programs are those that are planned before a crop is even started. It pays to be familiar with the most common diseases each of your perennial crops is likely to get and how to properly anticipate their presence. It is not uncommon for growers to implement preventative programs for certain diseases, such as powdery mildew, which for some crops (such as *Phlox paniculata*) is nearly inevitable.

Disease management programs begin with a clean start. The production area should be as free of plant pathogens as possible. Weeds and plant debris should be removed from the growing area at all times, from the beginning and throughout the crop cycle. All incoming plant materials should be inspected for disease symptoms before being placed into the production area. Infected plants should be refused, or at the very least placed into an isolated quarantine area for disease control before they are introduced into the crop areas.

Many unrooted cutting suppliers offer perennial cuttings to the industry that are culture indexed. Culture indexing is a process that measures the presence of specific plant diseases. Indexing involves taking thin slices from the base of the cuttings, placing them in a nutrient medium that promotes the growth of certain bacterial or fungal pathogens, if any are present. If fungal or bacterial growth occurs, the cuttings are discarded and not distributed to the perennial propagator. Although indexing prevents diseases from entering the propagation site from the supplier, it does not guarantee that the cuttings will never develop disease or bacterial problems once they are in propagation or during the future stages of crop production.

Botrytis

Most growers are familiar with the disease *Botrytis* (*B. cinerea*), which is commonly referred to as gray mold. *Botrytis* is probably the most common disease observed by perennial growers. It requires dead or damaged plant parts in order for the disease to prosper. Once it begins to grow on wounded or senescent tissues, it can rapidly spread to healthy leaves and stems.

Figure 5.2. ***Botrytis* infects various perennials, including *Phlox subulata*, pictured.**

Gray mold is most commonly identified by the fuzzy gray or brown spore masses that develop on infected plant parts. The damaged tissue first appears as tan to brown water-soaked areas that become gray as they dry out. Infected flowers usually show small water-soaked areas, which enlarge rapidly and turn to a brown or black coloration.

This disease is nearly always present in greenhouses. New infections usually only occur when the conditions are optimal for spreading and germinating the spores. The spores (conidia) are capable of overwintering in growing media and plant debris. The spread of spores throughout the production area occurs primarily with air movement and splashing water. Many normal production activities (irrigation, spraying, harvesting cuttings, etc.) are responsible for the dissemination of *Botrytis* spores. Natural release of conidia occurs midmorning to mid-afternoon, coinciding with a rapid decrease in relative humidity.

Botrytis can rapidly establish at nearly any production temperature (55–84° F [13–29° C]) when plants have wet leaves and the humidity is high. Temperatures of 60–70° F (16–21° C) are most favorable for this disease. Water remaining on plant surfaces for at least eight hours allows *Botrytis* spores to germinate. These conditions are often created from watering late in the day, allowing plant leaves to remain wet into the night, or irrigating during cool, cloudy conditions.

Condensation on plant leaves often leads to *Botrytis*. Water naturally forms on the leaves when the dew point is reached. This moisture is often referred to as free moisture. Free moisture forms when the temperature of the plant drops and the relative humidity is high. When the dew point is reached, water begins to condense on the surface of the plant. This condition often occurs just after sunset, when the plant temperatures drop more rapidly than do air temperatures, or in the early morning as the air temperature rises quicker than the plant temperature does.

After *Botrytis* spores germinate, a germ tube must penetrate the plant; an infection will occur only when there is a source of nutrients for the fungus to uptake following germination. The germ tube penetrates damaged plant tissues, such as senescent flowers, damaged leaves, and cutting wounds. In many cases, infections occur when threads (hyphae) of the fungus grow from infected plant parts into healthy plant parts.

Plants in the propagation phase are prime candidates for gray mold infestations, as they are young, tender plants at high plant densities and are often grown at high humidity levels. Crops at high plant densities with inadequate air circulation should be monitored closely for *Botrytis*, especially when the weather conditions (moist/cloudy) become optimal for this disease.

Preventing free water from staying on plant surfaces for extended amounts of time is the most important aspect of *Botrytis* management. The occurrence of *Botrytis* can be greatly reduced by ensuring adequate air circulation through the growing area. The relative humidity should be kept below 85% by heating or venting the moist air. Reduce any water that may rest on plant leaves, such as dripping from roof condensation or sprinklers. Minimize overhead irrigation late in the day and during cloudy weather, where the leaves are likely to remain wet for long periods of time. Botrytis will not be able to germinate, infect plant tissues, or reproduce without extended periods of wet and humid conditions.

Perennial crops should not be overcrowded and should have an adequate amount of space between each plant. High-density crop spacing limits light and air penetration, promoting senescence of the lower leaves. When wet conditions occur, *Botrytis* readily infects these senescent leaves and produces spores that can spread to nearby healthy plants.

Growers commonly remove (rogue) infected plants from the production site to reduce the spread of *Botrytis* outbreaks. Great care should be taken when rouging out diseased plants, as spores can easily be dispersed, contaminating healthy plants. Simply carrying infected plants out of the production area can release hundreds of spores in the crop area. Bring plastic bags or covered containers to the infection site, place the diseased plants in them, seal the bags or containers, and remove them from the crop area. Removing dead plant materials, such as senescent leaves and flowers, from the crop can also greatly reduce the incidence of *Botrytis* infestations once optimal conditions arise.

Where optimal conditions for *Botrytis* exist, growers often implement preventative spray programs using chemicals such as fenhexamid (Decree) or chlorothalonil (Daconil). These are considered the most effective protectant fungicides. Some *Botrytis* stains have become resistant to certain systemic fungicides, such as thiophanate-methyl (Cleary's 3336) and iprodione (Chipco 26019). Rotate systemic fungicides with protectant fungicides to prevent resistant strains of *Botrytis* from developing.

Downy mildew

The occurrence of downy mildew (*Peronospora* sp.) has become more prevalent in recent years. Downy mildew is closely related to the water mold diseases *Pythium* and *Phytophthora.* Unlike these water molds, which attack the root systems, downy mildew diseases attack the aboveground parts of the plant. Downy mildew diseases are specialized, each one attacking a range of closely related host plants and usually not affecting more than one genus of plants. The downy mildew species that attacks *Rudbeckia* is not identical to the mildew that attacks *Lamium.* Perennials such as *Lamium, Potentilla,* and *Rudbeckia* regularly experience infestations from downy mildew.

During periods of cool temperatures, moist conditions, and high humidity, the fungus begins to infect the plant. It actually grows inside the vascular system of the plant, down through the water vessels. Once inside, the disease becomes systemic and can be very destructive. Reddish purple splotches on the upper surface of the leaf are often a symptom that downy mildew is present. Depending on the perennial and the level of infection, the symptoms can also appear as irregularly shaped dead areas, vein-bound leaf lesions of various colorations (yellow, red, purple, or brown), chlorotic foliage, downward leaf curling, stunted plants, flower buds failing to form, or can resemble chemical or foliar nematode injury. The reproductive structures (sporangia) of downy mildew are the most distinguishing characteristic and almost always appear on the underside of the leaves as white, grayish fuzz. In most cases, the sporangia are directly opposite the discolored patch found on the upper leaf surface. The mauve-gray to brown felt-like mat begins to appear when the humidity levels are high. The spores spread from the plant with the air current. The fungal spores can live for a long time on dead plant materials or in the growing medium. It can take downy mildew a few days or weeks to become established. Downy mildew can be difficult to detect while the infestation is small. Once the environment turns favorable for the disease, an epidemic can occur very quickly. Most growers report, "It appeared overnight."

Table 5.1. Perennial Hosts for Downy Mildew

Abronia	*Duchesnea*	*Hydrophyllum*	*Ranunculus*
Aconitum	*Echinacea*	*Hymenopappus*	*Ratibida*
Agastache	*Epilobium*	*Iberis*	*Rudbeckia*
Agrimonia	*Erigeron*	*Lamium*	*Salvia*
Alyssum	*Erodium*	*Lathyrus*	*Scabiosa*
Ampelopsis	*Erysimum*	*Linaria*	*Senecio*
Androsace	*Fragaria*	*Lupinus*	*Seriphidium*
Anemone	*Galium*	*Matthiola*	*Silene*
Antirrhinum	*Gaura*	*Meconopsis*	*Silphium*
Artemisia	*Geranium*	*Mertensia*	*Solidago*
Aster	*Geum*	*Mirabilis*	*Stellaria*
Aurinia	*Gnaphalium*	*Myosotis*	*Stokesia*
Barbarea	*Hebe*	*Nasturtium*	*Teucrium*
Claytonia	*Helianthus*	*Oenothera*	*Thalictrum*
Coreopsis	*Helichrysum*	*Panax*	*Trautvetteria*
Corydalis	*Hepatica*	*Papaver*	*Verbascum*
Cynoglossum	*Hesperis*	*Phlox*	*Verbena*
Dicentra	*Hieracium*	*Physostegia*	*Verbesina*
Digitalis	*Houstonia*	*Potentilla*	*Vernonia*
Dipsacus	*Humulus*	*Prenanthes*	*Viola*

Figure 5.3. **This *Buddliea* leaf shows downy mildew symptoms as vein-bound leaf lesions.**

Downy mildew is easiest to control on a preventative basis. Scouting should occur weekly, if not more often, to detect this disease, and fungicide applications should begin before an outbreak occurs.

Controlling the humidity (maintaining below 85%) and preventing condensation on the crop are recommended. Downy mildew is a water mold and requires water on the leaves for the spores to germinate and enter into the plant. A film of water on the leaves for more than six hours allows the pathogen to enter the plant. Temperatures of 50–75° F (10–21° C) combined with moisture provide ideal conditions for this disease to run havoc with susceptible crops. Try to maintain plants with dry leaves going into the night to minimize conditions for disease development. Where possible, growers need to regulate the environmental factors conducive to this disease by regulating the humidity using heating, ventilation, air circulation, and irrigation timing.

Once plants have symptoms of downy mildew, it is next to impossible to control. Once it is within the plant, at some point in the future when conditions are favorable, the mildew *will* make new spore structures. Growers should error on the side of caution and implement a preventative program for all perennials susceptible to downy mildew. Many chemicals work well on a preventative basis, but none will clean up an established infestation. Preventative fungicide applications act as a barrier, keeping the disease from infecting the plant. The best strategy is to rotate products such as fosetyl-aluminum (Aliette), trifloxystrobin (Compass), azoxystrobin (Heritage), mancozeb (Junction, Protect), dimethomorph (Stature), and mefenoxam (Subdue Maxx), making applications every seven to ten days beginning at the onset of favorable conditions for this disease. Cleaning up existing populations is usually disappointing. If curative measures must be taken, use products such as fosetyl-aluminum (Aliette), azoxystrobin (Heritage), and dimethomorph (Stature).

Any symptomatic plants should be removed from the production area and destroyed immediately. The spores of downy mildew are thick walled, allowing them to survive for years. Do not carry infected plants through the crop area, as spores will be released, potentially infecting healthy plants. Place these plants directly into a garbage bag to prevent the spread of spores to the noninfected plants. It is best to remove them completely from the production site; sending them to a landfill is preferred. Placing diseased plants in an on-site cull pile may lead to future infestations.

Leaf spots

There are numerous leaf spots that perennial growers are likely to observe. In fact, nearly a thousand fungus species are capable of causing leaf spots on plants. The term *leaf spot* is a very ambiguous description of many plant fungi, bacteria, nematodes, and viruses that often attack perennial crops. Fungal leaf spots vary from the size of a pinpoint to lesions that practically encompass the entire leaf. Many leaf spots are tan to dark brown and may be circular, angular, or irregular in shape. Some of the common leaf-spot-causing fungi are *Alternaria, Ascochyta, Cercospora, Colletotrichum, Fusarium, Gloesporium, Helminthosporium, Phyllosticta, Ramularia*, and *Septoria*. Most fungal leaf spots do not kill perennial crops. However, they can severely affect the appearance of the plant, reducing its quality appeal and marketability.

Bacteria or viruses cause many leaf spots, and it is important to correctly identify them since these types are difficult to manage. The

Figure 5.4. **Leaf spots vary widely in appearance, size, and coloration. Pictured are *Cercospora* leaf spots on *Hosta*.**

bacterial diseases most commonly forming leaf spots are *Pseudomonas* and *Xanthomonas*. The spots caused by bacteria are best described as sunken black spots that appear to be water soaked and are usually angular in shape. Several viruses, such as TSWV and INSV, cause brown leaf spots. Properly identifying and managing bacterial and viral infections will be discussed later in this chapter.

Leaf spots, like all diseases, require four factors for the disease to develop. These factors are a susceptible host, a favorable environment, a causal agent, and time. If one of these factors is missing, the disease cannot develop.

Spores (conidia) are generally produced throughout the growing season. They spread to host plants by insects, wind, rain, or splashing water from irrigation. Splashing water is the most common method in which these pathogens are spread from plant to plant. Once on the leaves, the spores can invade leaf tissues and colonize the plant. Spore germination takes place when water remains on the leaf surface for several hours.

Most leaf spot diseases require cool conditions, wet foliage, high humidity, and little air movement. These conditions occur regularly in commercial operations with high plant densities. The spread of leaf spots can be minimized when growers follow a few guidelines. Provide good air circulation when perennials are grown in enclosed structures. Air movement in the crop can also be increased by placing them at wide crop spacing, minimizing contact from plant to plant. Keep the foliage and flowers as dry as possible; water early in the day to allow the foliage to dry quickly and not stay wet during the night. Using disease-free propagation materials or plug liners can also greatly reduce the occurrence of many of these pathogens. Even seeds can be contaminated with *Alternaria* and other fungal pathogens.

Routine scouting will help detect the presence of leaf spots. At first, the spots usually appear as small necrotic areas of various colors (tan, dark brown, yellow, gray, purple, or black). Some spots may appear shiny or raised. As the disease spreads, the spots often merge together, forming large angular or irregular dead spots. Most fungal leaf spot diseases are host specific, meaning they only attack certain species of plants or plants within the same family. Once leaf spots are observed in one plant species, growers can quickly monitor the other related plant species to determine if the leaf spot has occurred in an isolated area or throughout the production facility.

It is often difficult to properly identify leaf spots by the symptoms alone. There are usually distinguishing characteristics for each type of leaf spot, but there are often no hard and fast rules growers can use to diagnose them. Without proper identification, the best control strategies cannot be implemented. The only method to properly choose an appropriate control strategy is to send samples to a plant pathology or diagnostic clinic for an accurate diagnosis. Without this information, growers are combating problems blindly and may be doing more harm than good.

Growers often treat all leaf spot diseases in a similar manner even though they are often caused by various pathogens and require different management strategies to control them. Many times growers simply apply a broad-spectrum fungicide spray to remedy their leaf spot woes. This strategy may work in most cases, but will not control leaf spots caused from bacteria and viruses, which probably will continue to cause significant amounts of crop damage in the meantime.

There are numerous fungicides available for controlling leaf spots. Many of these are considered general broad-spectrum products. Some of the oldest chemistries are still very effective and widely used by perennial growers today. Many of these products contain the active ingredients chlorothalonil, copper, mancozeb, iprodione, thiophanate methyl, or various combinations of them. More recent fungicide introductions include active ingredients such as fludioxinil, propiconazole, myclobutanil, azoxystrobin, and trifloxystrobin. By using both the new and old active ingredients, growers can select products across a range of chemical classes and modes of action to reduce the threat of leaf spot diseases. Using these products in rotations allows growers to manage the potential for pathogens to build resistance to them.

Most chemicals are protectants, not eradicants; they need to be applied before injury to the crop has occurred. Chemicals can stop fungal leaf spots from spreading, but they cannot make any existing leaf spots disappear. The most effective chemical applications are those that are applied just prior to conditions conducive to spore germination. Fungicides are also highly effective when applied at the first sign of the disease, protecting the plant from new infestations.

Before applying chemicals, growers should take into consideration the amount of disease pressure (threshold) present, the application method, and the application cost (chemical plus labor) to determine if the treatments are cost justified. The most cost-effective method of controlling these diseases is through prevention. Always start with disease-free plugs, liners, or other starting materials. Use cultural practices, wherever possible, to reduce the environmental conditions necessary for these pathogens to survive.

Powdery mildew

Powdery mildew forms whitish grey patches (0.25–0.5 in. [6–13 mm]) of fungal growth, usually on the upper leaf surfaces. These patches, often referred to as colonies, may also appear on the lower leaf surfaces, stems, and flowers during severe infestations. The white powder consists of hundreds of spores (conidia) on the fruiting bodies of the fungus (mycelium). The first colonies seem to appear overnight, and with favorable conditions develop and quickly spread over a larger area, often covering the entire leaf. The talcum-like white colonies are often confused with spray residues from chemical applications. Powdery mildew greatly reduces the aesthetic appeal of the plants being grown, but also has an impact on overall plant quality, number of blooms produced, and a reduction of plant vigor. Many perennial crops—including *Aquilegia, Monarda, Phlox,* and *Veronica* cultivars—are very susceptible to this plant pathogen.

Figure 5.5. **Powdery mildew initially appears as small, powdery, white patches, such as these found on *Scabiosa columbaria.***

Powdery mildew diseases are caused by several different fungi and are usually host specific, meaning that the powdery mildew that infects *Phlox* is different from the mildew that attacks *Veronica.* However, the same fungus will attack plants within the same family. For example, cultivars of *Aster* and *Solidago,* which belong to the Asteraceae family, are both commonly infected by powdery mildew from the genus *Erysiphe.* The primary mildew genera commonly found on perennial crops are *Erysiphe* and *Podosphaera*; these genera each consist of several species. Other common mildews observed on ornamental plants include *Leveillula, Microsphaera,* and *Spaerotheca.*

Table 5.2. Perennial Hosts for Powdery Mildew Diseases

Achillea	*Delphinium*	*Inula*	*Salvia*
Aconitum	*Dendranthema*	*Lathyrus*	*Sambucus*
Actinidia	*Dicentra*	*Leucanthemum*	*Sanguisorba*
Adiantum	*Dipsacus*	*Lithospermum*	*Saxifraga*
Agastache	*Doronicum*	*Lupinus*	*Scabiosa*
Agrimonia	*Echinacea*	*Malva*	*Scutellaria*
Althaea	*Epilobium*	*Mentha*	*Sedum*
Ampelopsis	*Erica*	*Mertensia*	*Semiaquilegia*
Anemone	*Erigeron*	*Mitella*	*Senecio*
Anoda	*Erysimum*	*Monarda*	*Silphium*
Antirrhinum	*Eupatorium*	*Myosotis*	*Solidago*
Aquilegia	*Euphorbia*	*Myriophyllum*	*Sphaeralcea*
Arenaria	*Filipendula*	*Nepeta*	*Stachys*
Arnica	*Fragaria*	*Oenothera*	*Stokesia*
Aster	*Gaillardia*	*Paeonia*	*Tanacetum*
Astilbe	*Galega*	*Papaver*	*Tellima*
Balsamorhiza	*Galium*	*Pedicularis*	*Teucrium*
Baptisia	*Gaura*	*Penstemon*	*Thalictrum*
Boltonia	*Geranium*	*Phlox*	*Thermopsis*
Caltha	*Geum*	*Polemonium*	*Tiarella*
Campanula	*Hedysarum*	*Potentilla*	*Tolmiea*
Castilleja	*Helenium*	*Prenanthes*	*Triosteum*
Centaurea	*Helianthus*	*Primula*	*Trollius*
Chelone	*Heliopsis*	*Prunella*	*Valeriana*
Cirsium	*Heterotheca*	*Pulmonaria*	*Verbascum*
Clematis	*Heuchera*	*Ranunculus*	*Verbena*
Coreopsis	*Hibiscus*	*Ratibida*	*Verbesina*
Cosmos	*Hieracium*	*Rosa*	*Vernonia*
Crepis	*Humulus*	*Rosmarinus*	*Veronicastrum*
Cynara	*Hydrophyllum*	*Rubus*	*Viola*
Cynoglossum	*Hypericum*	*Rudbeckia*	
Dahlia	*Iberis*	*Rumex*	

Spores of this fungus are spread by air movement, such as wind or fans from heating and cooling systems. The occurrence of powdery mildew is prevalent during warm, dry spells or during cool to warm, humid periods. Once spores land on plants, an infection will develop and form visible colonies within three to seven days, provided favorable conditions are present. The spores release, germinate, and cause an infection without a film of water on the plant surface. A fairly high relative humidity (greater than 85% RH, although infections can occur at lower levels) and moderate temperatures (68–86° F [20–30° C]) will promote the initial infection. Once the pathogen begins to infect the host plant, the relative humidity is no longer a factor, as it can prosper regardless of the humidity. The fungi grow into small colonies on the leaf surface, obtaining nutrients from the plant by penetrating into leaf cells with specialized structures called haustoria.

The presence of powdery mildew can be reduced if optimal plant growing temperatures are maintained and high humidity levels or dramatic swings in humidity are avoided. The relative humidity should be kept below 85%. In many cases, mildews occur when condensation forms on the plant leaves at night. Keeping the leaves dry throughout the night will prevent the spores from germinating. It may be necessary to provide ventilation two or three times at night to dehumidify the air. Maintaining sufficient plant spacing and increasing the air circulation are useful to reduce the humidity levels near the plant canopy where infections occur.

It is beneficial for growers to identify and monitor on a regular basis any perennials that are susceptible to powdery mildew. Knowing the type of mildew that attacks various perennial crops allows growers to determine the potential spread of the disease, makes scouting easier, and helps to keep preventative and curative control measures practical.

Routine scouting should occur at least on a weekly basis, more if favorable conditions for this disease have occurred. Scouts should pay particular attention to any areas with susceptible crops, areas with dramatic changes in day and night temperatures, and crops exposed to high relative humidity during the night. Most powdery mildews are observed as white powdery colonies on the upper leaf surfaces. Occasionally, it can be observed on the undersides of leaves, usually directly below a small yellow spot on the upper surface. Severe infestations do not remain as colonies but tend to cover the entire surfaces of leaves, stems, and flowers. Powdery mildews can be distinguished from spray residues since chemical residues do not appear to be fluffy and often have a more droplet-like outline. Under conditions not favorable for spore production, plants infected with mildew do not have the distinguishing white colonies. Instead, the leaves have a purplish or red coloration, often confused with nutritional disorders.

Control strategies should be implemented as soon as colonies are detected. Preventative programs could be implemented, but are not necessary, since powdery mildew can be eradicated using chemicals *if it is detected early*. It is better to apply chemical controls during or just after periods that are optimal for the development of these fungi. It is very difficult to eradicate existing mildew colonies. Some of the most effective chemical controls include: propiconazole (Banner Maxx), trifloxystrobin (Compass), kresoxim-methyl (Cygnus), myclobutanil (Eagle), azoxystrobin (Heritage), piperalin (Pipron), triadimefon (Strike), and triflumizole (Terraguard). If chemical controls are necessary, be sure to rotate between chemical classes to reduce the likelihood of mildews developing resistance to these products.

Several biorational compounds are also effective at controlling powdery mildew diseases. Achieving complete coverage is important, and multiple applications are often necessary when using these products. Some of the most effective biorationals include potassium bicarbonate (First Step, Milstop), neem oil (Triact), paraffinic oil (Ultra-Fine Oil), and hydrogen dioxide (Zerotol).

With the host-specific nature of powdery mildews and several different species of mildews attacking perennials, there is not a single fungicide that works equally well on all mildew species. One fungicide may provide excellent control on one mildew species, but provide little, if any, control on another. Although the powdery colonies look similar between species, they each have different fungicide sensitivities, so you will need to know which pathogen is on your crop.

The dead colonies will remain on the plant surface for several days after treatment. Infected plants should be checked after treatment to observe the efficacy of the treatment, for signs of new growth, or signs of continued mildew outbreaks.

Root rots

Many soilborne fungi cause significant injury to the root systems of perennial crops. The injury caused by these pathogens are commonly referred to as root rots. The term *root rot* is overgeneralized and clumps many different pathogens into one category. In many cases, root rots and crown rots are used interchangeably, as they are caused by similar pathogens. Unfortunately, managing these diseases often requires different control strategies. Using the general term root rot causes many growers to apply the wrong fungicides to their crops, which often leads to more crop injury or losses.

The primary pathogens responsible for injury to root systems of containerized plants are *Pythium, Phytophthora, Rhizoctonia*, and *Thielaviopsis*. Less common pathogens responsible for root rot diseases include *Cylindrocladium, Fusarium, Macrophomina, Myrothecium, Phymatotrichum, Ramularia, Sclerotium*, and *Sclerotinia*. The effects of these diseases can be reduced with the proper identification and an understanding of how they are introduced into the production area, which conditions are conducive for their development, and which chemical and cultural controls are most effective.

Root and crown rots can affect any size or age of perennial. The infection can be light to severe, depending on the environmental conditions, susceptibility of the host plant, and the level of infestation. Soilborne fungi are opportunistic, requiring damaged or stressed tissues to gain entry into the plant. Stresses that damage roots include over- and underwatering, excessively high or low temperatures, high fertility levels, and chemical injury. Root injury from insects, such as feeding from fungus gnat larvae, can provide an entry site for the pathogens through the feeding wounds.

Symptoms of root and crown rot often resemble nutrient deficiencies and include yellow or purple leaves, stunting, wilting, and overall poor plant growth. Many growers fail to check the root system when they see chlorosis; they usually assume the plants are hungry. Fertilizing perennials with root rots will only add fuel to the fire and allow the soil pathogens to become more destructive. Most perennials are susceptible to these pathogens, provided the proper conditions for them exist.

Soilborne fungi are common, naturally occurring organisms that are found practically everywhere. Irrigation water from storage ponds, creeks, or rivers is likely to contain root rot pathogens. Water from deep wells or municipal treatment systems is normally free of these pathogens. Growing mixes containing mineral soils as a component are likely to contain some root rot pathogens. Artificial media components such as perlite, rock wool, and vermiculite start off free of pathogens at

Figure 5.6. **Root rot diseases often affect most plants within a crop, but the level of infestation may vary from pot to pot, such as with this *Agastache*. Several symptoms are present: crop variability, poor plant health, yellowing foliage, and stunting.**

the beginning of the crop cycle but slowly build up low populations of soilborne pathogens during crop production. Naturally occurring media components, such as bark or peat moss, may contain root rot pathogens prior to planting.

Pythium and *Phytophthora*

Pythium root rots can be identified if the roots are generally soft, mushy, and various shades of brown. The outer covering of the root, the cortex, is usually rotted and slides off easily when pulled, leaving the stringlike vascular bundles behind. This is commonly referred to as sloughing of the roots. *Pythium* usually attacks the root tips first and works its way

Figure 5.7. **Aerial *Phytophthora* on lavender is first observed infecting individual stems before spreading over the entire plant.**

upward in the root system. The aboveground symptoms often appear as yellowing, stunting, and wilting. The chlorosis is the result of the plants not being able to transport nutrients up through the damaged root system to the leaves and stems. Most perennials are susceptible to *Pythium* species, namely *P. aphanidermatum* and *P. ultimum.* Plants under stress caused by high soluble salt concentrations, poor drainage of the growing medium, and overwatering are particularly susceptible to *Pythium* infestations.

There are several species of *Phytophthora* that attack perennial crops, the most common being *P. cinnamomi*, *P. citrocola*, and *P syringae.* Plant wilting, chlorosis of the lower leaves, and individual stem death are common symptoms of *Phytophthora* diseases. The roots are typically darkly colored and appear water-soaked. Unlike root rots caused by *Pythium*, the roots do not slough off when they are pulled.

There are some forms of aerial *Phytophthora* that affect ornamental crop production. Aerial *Phytophthora* (*P. parasitica, P. syringae*) is less commonly observed but occasionally can wreck havoc on certain crops. Sudden oak death caused by *P. ramorum* is an example of an aerial *Phytophthora* that has had a major impact in the ornamental industry. Symptoms are often described as sudden flagging or wilting of individual shoots or an entire plant. In some cases, water soaked gray-green lesions at the base of the shoots can be observed, followed by sunken, reddish brown cankers that girdle the affected stems. Plant death often occurs within one to two weeks of the first symptoms. Aerial *Phytophthora* is particularly prevalent when periods of hot, wet weather occur or when plants are grown in greenhouses. High fertility levels and frequent irrigation also contribute to disease development.

Pythium and *Phytophthora* are often referred to as the water molds. Water molds prefer wet conditions for them to infect and spread. The spores (zoospores) move freely in water and are often recovered from recirculated irrigation water and reapplied to crops, further increasing the likelihood of infection.

Rhizoctonia

Rhizoctonia solani is another major disease that can be both a root rot and an aerial disease. It thrives under warm, wet, and humid conditions. *Rhizoctonia* is usually found in the top couple inches (5 cm) of the growing medium and can spread from infected roots and plant debris into the lower leaves. It usually attacks plants at the soil line, causing root loss, girdling and constriction of the stem, and the tops to die back. The infected roots are often reddish brown and appear dry. When the roots are infected, the primary aboveground symptoms are yellowing and stunting. In some cases, the plants have fallen over or broken off at the soil line.

Figure 5.8. ***Rhizoctonia* can occur very quickly, often causing significant injury to the crop, such as on this *Salvia* x *nemerosa.***

Aerial *Rhizoctonia,* often called web or aerial blight, kills the foliage very quickly, causing the leaves to senescence and fall from the plant. In many instances, the leaves are still attached to the stems, dangling, suspended by threadlike hyphae from the pathogen. It is most prevalent when warm and humid conditions are present.

Thielaviopsis

Growers often observe a black root rot caused by the pathogen *Thielaviopsis basicola. Thielaviopsis* is not all that common in containerized perennials but has been observed on *Hibiscus, Phlox,* and *Verbena,* to name a few. *Thielaviopsis* survives in the growing medium or plant debris and is introduced to crops by splashed water or by the roots contacting contaminated surfaces.

The infected roots may have black lesions covering all or just part of the root. Initially, the secondary feeder roots can be seen with bands or dark spots at the tips. These lesions are best observed by rinsing the soil mix off of the roots. As the disease progresses, the entire root system becomes black and water-soaked. The aboveground symptoms, as with the other root rot pathogens, are stunting and yellow leaves. Leaf yellowing is usually observed first, followed by wilting and plant death. This disease prefers high temperatures and wet growing mixes with neutral to alkaline pH levels. Plant stresses caused from high soluble salts and overwatering can contribute to the occurrence of this disease.

Damping-off

Damping-off is the term generally associated with a number of diseases that occur when seeds and seedlings become infected by soil-inhabiting pathogens. Damping-off is used to describe all of the pathogens associated with the rotting of seed and newly emerged seedlings. *Pythium, Phytophthora,* and *Rhizoctonia* are the most common causes of damping-off. Other pathogens such as *Thielaviopsis, Fusarium,* and *Botrytis* are occasionally responsible for the collapse of many young seedlings. Pre-emergence damping-off, stem rot at the soil line, and wire stem are often caused by *Rhizoctonia.* Wire stem is best described as a partial rot near the soil surface; the plants do not fall over but remain stunted and eventually die. The root tips of young seedlings are sometimes invaded by *Pythium,* which usually progresses up the stem causing the young plant to collapse.

Damping-off can often be prevented by following sound cultural practices, using preferably new—or, at least, disinfected—seed trays filled with a porous, well-drained seedling mix. Do not use any growing mixes containing field soils unless they are pasteurized or sterilized. It is also highly advisable *not* to fill the trays with a medium that previous crops have been grown in or to re-sow into plug trays that have had no or poor stands. The seeds should not be sown too densely or too deeply. Provide adequate temperatures for germination, and maintain suitable pH and soluble salt levels for adequate seedling growth.

Water management is very important to reduce the likelihood of damping-off. There is a fine line between maintaining adequate moisture levels for germination and growth of the young seedlings and overirrigating, which creates an optimal environment for many of the soil-inhibiting plant pathogens. Avoid overwatering and excessive overhead misting during the germination process. These pathogens spread rapidly through overcrowded seed flats via plant-to-plant contact and splashing water.

Managing root rots

Root rot diseases are most prone to cause infections under saturated soil conditions. Wet growing mixes often have poor drainage and allow little oxygen to move back into the root zone. Root growth and activity are reduced under these conditions, but root rot pathogens thrive in wet mixes with poor drainage. Maintaining conditions suitable for plant growth will reduce the occurrence and the severity of root rot diseases. General recommendations for controlling root rot pathogens include using a porous, well-drained sterile potting medium, maintaining proper fertility levels, avoiding excess water, controlling fungus gnat and shore fly populations, and thoroughly cleaning and sanitizing the production area between crops.

There is often a fine line between proper and improper irrigation practices. The longer a growing medium remains wet, the greater the chance is for root rots to develop. Plants should be watered only when they need it and when conditions are conducive for it to be applied. It is always best to irrigate early in the day, allowing adequate time for the growing mix to dry. If possible, avoid watering just prior to periods of extended cloudy and cool weather conditions, as water use and loss from evaporation is greatly reduced, creating an optimal environment for many root pathogens.

Always check the root system for root rots whenever there are signs of chlorosis, stunting, or wilting—this step could be the difference between saving the crop or losing it. Growers should remove the plant from the container and look for any signs of decay, brown-mushy roots, lesions, or discoloration. Growers often only observe the aboveground symptoms, thinking nutritional deficiencies are to blame. Without examining the roots, they add fertilizers to the root zone, often causing more root injury and plant stress to occur. Similarly, when root rots occur, plants often appear wilted. The wilted appearance occurs as the fungus has destroyed the plant's ability to move water upward from the roots to the shoots. In the above scenarios, adding fertilizer salts and water to the root zone only creates a more ideal environment for these pathogens. By the time the pathogen is identified, further injury to the crop has occurred, possibly leading to plant death.

I cannot overemphasize the importance of checking the root zone as part of the normal scouting activities and as one of the first places to look when cultural problems arise. In many cases, these diseases can be detected early, and with the appropriate cultural practices and chemical treatments, they can successfully be controlled.

Growers should also consider the effect the production environment may have on plant roots. Many production sites do not have level production areas, and water forms puddles under the pots in the low spots. These puddles create an environment where the root zone remains saturated for hours, perhaps days, after irrigation or heavy rainfalls. Growers should take steps to reduce the puddling of water by crowning the center of the growing areas, allowing the water to naturally move away from the areas where containers are setting.

Sound sanitation practices are important to reducing root rot pathogens. Most root rot pathogens can survive in the soil or in anything containing soil or growing substrates. Any practice or procedure that can move or displace the growing medium can potentially move a root rot pathogen. Prior to planting, all containers, flats, and pots should be new or cleaned and disinfected before reusing. Growing mixes should never be reused, unless they have been properly pasteurized. Avoid setting containers on bare soil or dirty benches and floors.

Reusing or recycling irrigation water can spread the spores of *Pythium* and *Phytophthora* back into the production area. Recycled water should be avoided in any areas where propagation occurs. Growers using recycled water should take measures to disinfect it using sand filtration, chlorine injection, ozone treatment, or UV lighting.

Proper identification is very important to properly manage and control root rot diseases. It is difficult to identify root rot pathogens by only looking at the infected roots. It is recommended to submit infected plants to diagnostic clinics where they can culture the diseases and/or use microscopes to properly identify the pathogen. Implementing control measures without a proper diagnosis can result in further injury to the crop and wasted or costly fungicide applications.

Fungicides are often necessary to prevent root rot diseases from occurring or to control them, provided they have not reached epidemic proportions. Most growers apply fungicides as media drenches where the active ingredients can reach the infected roots. Fungicides are most effective when the pathogen levels are low and conditions are favorable for the disease to develop.

Growers should apply chemical fungicides that are effective at controlling the specific pathogen they have identified. Most fungicides will not control all root rot diseases. There are fungicides more suitable for controlling water mold diseases (*Pythium* and *Phytophthora*), such as fosetyl-aluminum (Aliette), propamocarb (Banol), azoxystrobin (Heritage), dimethomorph (Stature), mefenoxam (Subdue Maxx), and etridiazole (Terrazole, Truban). Other fungicides, such as thiophanate-methyl (Cleary's 3336), etridiazole + thiophanate-methyl (Banrot), iprodione (Chipco 26019), metribuzin (Contrast), fludioxonil (Medallion), PCNB (Terrachlor), and triflumizole (Terraguard), are most effective at controlling *Rhizoctonia* and *Thielaviopsis*. The products controlling *Pythium* and *Phytophthora* usually do not control *Rhizoctonia* and *Thielaviopsis* and vice versa. There are some commercially available products, such as etridiazole + thiophanate-methyl (Banrot) that contain two active ingredients, providing control of all the diseases mentioned above. Growers commonly combine products, such as mefenoxam (Subdue Maxx) and thiophanate-methyl (Cleary's 3336), to achieve control of a broad range of root rot pathogens.

Some root rots can be effectively controlled using biological control products. These products include *Streptomyces griseoviridis* (Mycostop), *Trichoderma harzianum* (PlantShield, RootShield), and *Gliocladium virens* (SoilGard). They are only effective when used on a preventative basis. In most cases, sufficient control is achieved against *Pythium* and *Rhizoctonia*, but only limited control is observed with *Phytophthora* or *Thielaviopsis*. Biological controls are usually incorporated into the growing mix prior to planting or drenched after transplanting, before these diseases become established.

Rust

Rust diseases initially appear as small swellings called pustules, often on the underside of the leaves, but may occur on the upper leaf surface or even the stems of several perennial varieties. Rusts are obligate pathogens and usually require or target a specific host plant or a family of plants, such as those in the Asteraceae

Figure 5.9a - b. **Rust infection on the upper (left) and lower (right) leaves of *Alcea rosea*.**

Table 5.3. Perennial Hosts for Rust Diseases

Abronia	*Canna*	*Galium*	*Monarda*	*Sedum*
Achillea	*Carex*	*Gaura*	*Monardella*	*Semiaquilegia*
Aconitum	*Castilleja*	*Gentiana*	*Moraea*	*Semiarundinaria*
Acorus	*Centaurea*	*Geranium*	*Myosotis*	*Sempervivum*
Actaea	*Chelone*	*Geum*	*Nothoscordum*	*Senecio*
Adoxa	*Chimaphila*	*Gillenia*	*Nymphoides*	*Seriphidium*
Agastache	*Chimonobambusa*	*Gymnocarpium*	*Oreopteris*	*Shibataea*
Agrimonia	*Chrysanthemum*	*Hedysarum*	*Osmunda*	*Sida*
Alcea	*Cimicifuga*	*Helenium*	*Oxalis*	*Sidalcea*
Alchemilla	*Cirsium*	*Helianthus*	*Oxyria*	*Silene*
Allium	*Claytonia*	*Heliopsis*	*Paeonia*	*Silphium*
Althaea	*Clematis*	*Hepatica*	*Pedicularis*	*Sinarundinaria*
Amaranthus	*Clintonia*	*Hesperis*	*Pelargonium*	*Sisyrinchium*
Amsonia	*Convallaria*	*Heterotheca*	*Peltandra*	*Smilacina*
Anaphalis	*Convolvulus*	*Heuchera*	*Penstemon*	*Solidago*
Anchusa	*Coreopsis*	*Hibiscus*	*Petasites*	*Sphaeralcea*
Andropogon	*Corydalis*	*Hieracium*	*Phalaris*	*Stachys*
Androsace	*Crepis*	*Houstonia*	*Phegopteris*	*Tanacetum*
Anemone	*Cryptogramma*	*Humulus*	*Phlox*	*Tellima*
Anemonella	*Cyrtomium*	*Hyacinthus*	*Phyllostachys*	*Teucrium*
Angelica	*Dalea*	*Hydrophyllum*	*Physalis*	*Thalictrum*
Anoda	*Davallia*	*Hymenopappus*	*Podophyllum*	*Thamnocalamus*
Antennaria	*Delphinium*	*Hypericum*	*Polemonium*	*Thelypteris*
Apocynum	*Dendranthema*	*Iberis*	*Polygala*	*Tiarella*
Aquilegia	*Dendrocalamus*	*Inula*	*Polygonatum*	*Tradescantia*
Arabis	*Dianthus*	*Iris*	*Polystichum*	*Tragopogon*
Arachniodes	*Dicentra*	*Lavatera*	*Potentilla*	*Trautvetteria*
Arenaria	*Diplazium*	*Lewisia*	*Prenanthes*	*Trientalis*
Arisaema	*Dodecatheon*	*Lilium*	*Primula*	*Trillium*
Armeria	*Doryopteris*	*Limonium*	*Pseudosasa*	*Triosteum*
Artemisia	*Draba*	*Linaria*	*Pycnanthemum*	*Uvularia*
Arundo	*Dracocephalum*	*Lithophragma*	*Pyrola*	*Valeriana*
Asarum	*Dryopteris*	*Lithospermum*	*Ranunculus*	*Veratrum*
Asclepias	*Duchesnea*	*Ludwigia*	*Ratibida*	*Verbena*
Asplenium	*Epilobium*	*Lunaria*	*Rosa*	*Verbesina*
Aster	*Erica*	*Lupinus*	*Rubus*	*Veronica*
Athyrium	*Erigeron*	*Luzula*	*Rudbeckia*	*Veronicastrum*
Balsamorhiza	*Eriophyllum*	*Lychnis*	*Ruellia*	*Vinca*
Barbarea	*Erysimum*	*Lysimachia*	*Rumex*	*Viola*
Belamcanda	*Erythronium*	*Lythrum*	*Saccharum*	*Waldsteinia*
Blechnum	*Euphorbia*	*Maianthemum*	*Salvia*	*Woodsia*
Boltonia	*Fargesia*	*Matthiola*	*Sambucus*	*Woodwardia*
Calamagrostis	*Filipendula*	*Mentha*	*Saponaria*	*Wyethia*
Callirhoe	*Fritillaria*	*Mertensia*	*Sasa*	*Xerophyllum*
Caltha	*Fuchsia*	*Micromeria*	*Saxifraga*	*Yucca*
Campanula	*Gaillardia*	*Mirabilis*	*Scilla*	*Yushania*
		Mitella		

family. Many rusts have complex life cycles and require two different plant species to use as host plants during various stages of their life cycle. Some rusts can cause systemic infections and plants may not exhibit any visible injury symptoms. Perennials commonly attacked by rust diseases include *Alcea, Aster, Calamagrostis, Iris,* and *Miscanthus.*

Most rust diseases can be first identified by the appearance of pale green or yellow spots on the upper leaf surface. These spots usually develop necrotic or dead areas in their centers. Pustules develop on the undersides of the leaves producing raised blisterlike areas containing white, yellow, orange, black, or brown spores. The spores are spread by air currents from wind or fans from heating and cooling systems. The spores germinate after landing on a host plant when the leaves are wet and the temperatures are optimal. Depending on the rust disease, spores require the leaves to remain wet for three to six hours and temperatures to be 50–75° F (11–24° C). In many cases, spores germinate, infect the plant, and lie dormant in green tissues until conditions favor sporulation.

Most rust diseases overwinter or survive the summers very effectively on infected host plants. Some rust species cannot survive the winters in some parts of the country. The spores of daylily rust (*Puccinia hemerocallidis*), for example, cannot survive the winters in USDA Hardiness Zones 1–6.

Growers can reduce the occurrence of rust diseases by reducing the relative humidity and increasing the air circulation in the production area. The length of time plant leaves remain wet is critical for spore germination and should be reduced to eliminate germination from occurring. Avoid irrigating crops in the evenings or at night; instead, water crops susceptible to rust diseases in the morning, allowing the foliage to dry before nightfall. Using drip irrigation or subirrigation watering systems will avoid getting the foliage wet or splashing water from leaf to leaf. Rust infections also are likely to arise in areas where nightly dew formation occurs (warm days and cool nights) and temperatures remain conducive for spore germination.

Crops should be scouted weekly to detect the presence of these diseases. If control measures are necessary, chemicals should be immediately applied after rust has been detected. Rust diseases can be controlled using such products as propiconazole (Banner Maxx), triadimefon (Bayleton), and mancozeb (Protect). Most chemicals effective at controlling rust act as protectants and do not eradicate rust from an infected plant. Growers should consider preventive programs targeting the crops susceptible to rust diseases or applying chemicals just prior to expecting conditions conducive to this disease.

Wilt diseases

Many perennials may experience wilt symptoms either caused by water stress or plant pathogens. Wilt symptoms occur when the water flow to various plant parts is stopped or slowed. These symptoms can develop slowly or rapidly, may be temporary—such as on hot days—or may become permanent, causing death to the affected plant part. Wilts caused by water stress are usually temporary, unless the water is withheld for extended periods. When the root zone has poor drainage or is excessively moist for an extended period of time, the root zone becomes oxygen deficient, often resulting in a wilt. Other causes of wilts include high soluble salts, insect injury, and bacterial or fungal pathogens.

Table 5.4. Perennial Hosts for *Fusarium* Wilt Diseases

Aster	*Echinacea*
Astilbe	*Freesia*
Centaurea	*Gladiolus*
Chrysanthemum	*Hebe*
Crambe	*Helianthus*
Cyclamen	*Hydrastis*
Dahlia	*Lathyrus*
Dendranthema	*Ligularia*
Dianthus	*Lupinus*
Dicentra	*Nepeta*
Digitalis	*Sedum*

The wilts caused by fungi or bacteria occur when the pathogen and/or its byproducts block the water-conducting vessels within the plant. Symptoms often occur on part of a leaf, an entire leaf, one side of the plant, an individual stem of a plant, or the entire plant. Besides the typical wilt symptom, affected plants may appear yellow, stunted, or have discolored vascular tissues inside of the stem.

Wilts are often caused by various species of the fungi *Verticillium* and by *Fusarium oxysporum*. The two most common wilts affecting perennial crops are *V. dahliae* and *V. alboatrum*. Symptoms will vary with the host plant. Characteristic symptoms of *Verticillium* are wilting and yellowing of the leaf margins progressing upward from the lowest leaves, often appearing one sided, lacking stem and leaf lesions, and maintaining a normal looking root system. *Verticillium* is commonly found in the soil or growing medium, where it can persist. It enters the roots of perfectly healthy plants, where the infection first occurs; fungal wilts do not require injured root tissues to enter the plant. The fungus grows upward through the xylem (water-conducting tissues), eventually causing the wilt symptoms to appear. Since *Verticillium* is located on the inside of the plant, it can be transferred to new production areas if vegetative cuttings are taken from an infected plant (even if the plant is not symptomatic).

Table 5.5. Perennial Hosts for *Verticillium* Diseases

Aconitum	*Helichrysum*
Antirrhinum	*Humulus*
Aster	*Liatris*
Calceolaria	*Ligularia*
Callirhoe	*Mentha*
Campanula	*Paeonia*
Centaurea	*Papaver*
Coreopsis	*Phlox*
Dahlia	*Physalis*
Delphinium	*Reseda*
Dendranthema	*Rudbeckia*
Dicentra	*Salvia*
Digitalis	*Sambucus*
Erigeron	*Senecio*
Fragaria	*Solanum*
Fuchsia	*Vinca*
Helianthus	

Several wilts are also caused by bacteria. The most common are various species of *Erwinia, Pseudomonas,* and *Xanthomonas.* These wilts are most severe with temperatures above 75° F (24° C). Bacteria can enter the plant through its natural openings or through injured plant tissues.

Wilts caused by fungi or bacteria are commonly spread in propagative materials, water movement through the soil, soil movement, contaminated flats or pots, tools such as cutting knives, and splashing water. These pathogens prefer high temperatures combined with high humidity, where they will develop rapidly. Plants under stress are more prone to wilts when the right conditions exist.

Maintaining sound sanitation practices will help reduce the likelihood of wilt diseases. The production site should be free of all old soil and plant debris. The use of pasteurized or sterile growing media is fine, but soilless mixes are preferred. Minimizing plant stresses such as heat stress and avoiding high salt levels will also reduce the occurrence of wilts. Once wilt symptoms are observed, it is difficult, if not impossible, to obtain sufficient control of these fungal and bacterial diseases. Infected plants should be rouged out and destroyed. Preventative control of wilts can be achieved using the appropriate bactericides or fungicides before these diseases are detected.

Bacterial Diseases

Although bacteria are much smaller than fungi, they can cause very serious damage to perennial crops. Bacteria are single-celled, microscopic organisms bounded by a cell wall. The bacteria commonly infecting perennial crops include *Agrobacterium*, *Corynebacterium*, *Erwinia*, *Pseudomonas*, and *Xanthomonas.* Bacterial infections can cause blights, leaf spots, soft rots, and vascular wilts.

Bacteria and related diseases are commonly introduced into the production area by planting infected plant materials. These diseases are spread by water or are carried in infected media, plant debris, and contaminated tools. Simply touching a contaminated plant, followed by touching a noninfested plant, may result in the transmission of bacteria. Dipping cuttings or bare-root materials into fungicide solutions or rooting hormones are great methods of spreading bacteria to large numbers of plants quickly.

The onset of bacterial diseases is most prevalent during warm weather and prolonged wet periods. Bacteria require a wound or a natural opening (stomata) to invade plant tissues. Free moisture on the leaf surfaces is often responsible for moving bacteria to natural openings or wounded areas. Plants that have been recently injured from rough handling or trimming are most susceptible to bacterial infections. Temperature is the other variable determining the rate of disease development. Low temperatures (below 60° F [16° C]) will generally not promote bacterial disease development. As temperatures increase, the rate of disease development also increases, especially if they are stressful to the host plant. Depending on the type of bacteria, there are upper temperature limits (above 85° [30° C]) where the rate of bacterial disease development is actually reduced.

Since bacteria and fungi often occur under the same conditions, it is often difficult to distinguish between fungal leaf spots and bacterial leaf spots. Foliar infections caused by bacteria usually cause leaves and other plant tissues to have a water-soaked appearance accompanied with a rotten odor. The location of the water-soaked spots is usually bounded between the leaf veins; these spots often have angular shapes with straight sides. Holding the infected leaf up to a light source often reveals a definite yellow outline or halo surrounding the leaf spot.

Other symptoms of bacterial infections include wilted foliage accompanied by brown and yellow leaves and dieback of plant parts. The sudden collapse of new leaves or shoots, often with a slimy dark appearance, is referred to as blight. Soft rots cause plant tissues to appear dark and mushy. Sometimes a foul odor accompanies the bacterial soft rot. Soft rots commonly occur near the soil line or the crown of the plant. Bacterial soft rots and wilts often cause plants to collapse when the plants are under water stress.

Some bacterial problems caused by *Agrobacterium tumefaciens* and *Corynebacterium fascians* are swelling of plant tissues (crown gall) or proliferation of tissues at the base of the plant (fasciation). Both crown gall and fasciation

Figure 5.10. ***Corynebacterium fascians* often causes a proliferation of shoots, as shown here on this *Gaura* plug.**

cause a disruption of the normal water and nutrient flow up the stem of the plant.

Diseases caused by bacteria can be devastating and quickly run rampant through the crop. Preventing bacteria outbreaks is possible, but curing infected plants is currently not an option. The first step to controlling bacterial problems is to prevent them from entering the crop area. All incoming plant materials should be inspected for symptoms, such as wilts, black angular spots, galls, or fasciation. Growers with good sanitation practices—keeping the area clean of infected plants, plant debris, and excess growing medium—can greatly reduce the level of bacteria that survive in these materials.

Weekly scouting of the crop can provide early detection of any bacterial problems. All infected plant materials should be removed from the production facility immediately. Growers can manipulate the environment to some extent to reduce the conditions favorable for these diseases. Plants should be grown at an adequate plant spacing (not overly crowded) with plenty of air circulation, and, if possible, overhead irrigation should not occur late in the day. All of these steps are useful methods of reducing the humidity around the plants, which is the most favorable factor for bacterial diseases to become established.

There are only a couple of chemicals growers use as bactericides. These products primarily contain copper. All products claiming to control bacterial diseases should be used only on a preventative basis. They are effective at controlling the bacteria before they enter the plant, but once an infection has occurred, they are largely ineffective and cannot stop the infection. Growers rooting cuttings or producing stock plants have been highly successful implementing preventative bacterial programs to both stock and propagation areas.

Viruses

Plant viruses are caused by very tiny subcellular particles that can only be seen using an electron microscope. These sphere- or rod-shaped particles are composed of small pieces of DNA or RNA encapsulated in a protein coat. Some viruses can completely function using the genetic information of a single particle, while others require the genetic code from multiple particles.

Unlike most plant pathogens, viruses do not divide or produce any type of reproductive structures, such as spores. Instead, they are reproduced inside living cells and can spread into plant tissues; they require the ribosomes and other components of its host cell for multiplication. Virus particles basically take over plant cells, forcing them to replicate more viruses identical to themselves. This causes plant cells to function improperly, and normal plant operations, such as production of chlorophyll or cell division, either do not occur or are seriously affected. In many cases, the cells grow slowly, causing the plant to appear stunted. In other cases, cell division occurs rapidly, causing the plant to grow abnormally or to form galls. Most viruses cause systemic infections, and currently there is no treatment for removing them from infected plants.

There are a number of viruses affecting various perennial crops. Some of the common viruses of perennials include: alfalfa mosaic virus (AMV), arabis mosaic virus (ArMV), cucumber mosaic virus (CMV), impatiens necrotic spot virus (INSV), tobacco etch virus (TEV), tobacco mosaic virus (TMV), tobacco ringspot virus (TRSV), tomato ringspot virus (ToRSV), and tomato spotted wilt virus (TSWV). There is often confusion when a virus such as INSV is diagnosed in hosta leaves; after all, the virus and the hosta seemingly do not belong together. The name of a virus can be misleading; it is usually named after the first plant it was discovered on. Therefore, in most instances, viruses found on perennials are often named after bedding plants or vegetable crops.

Virus symptoms can be difficult to identify, as they often resemble other cultural problems, such as nutrient disorders, herbicide injury, and fungal diseases. Virus infections can remain symptomless or show mild symptoms, or they can be confusing, as multiple viruses could occur simultaneously. Visible symptoms vary with the virus and commonly include abnormal dark green and light green mosaic and mottling of leaves, ring patterns on foliage, necrotic spots, bumps on plant foliage, distortion, stunting, and abnormal formation and discoloration of flowers. In many cases, plants may exhibit multiple symptoms or have more than one virus. Some plants do not express visible symptoms of viruses (latent infections) or do not show them all of the time. With any virus, the visible symptoms relate to how long the plant has been infected, the age of the plant at the time of infection, the type of virus, the environment, and the amount of stress on the plant.

Figure 5.11. **Hosta virus X, as seen on *Hosta* 'Peedee Gold', has become more prevalent in recent years.**

All viral infections start as the virus enters a host plant through some type of wound or injury to a plant cell. These wounds most commonly occur as the result of injury from feeding insects or cultural activities. Viruses can be spread to noninfected plants by numerous means, including fungi, insects, mites, nematodes, parasitic plants, pollen, seed, and vegetative propagation. A *vector* is the carrier of a virus, which transmits it to uninfected plants. For example, certain insects, such as aphids and thrips, are vectors for plant viruses. Other methods in which viruses are transmitted include propagating from virus-infected stock plants or using tools containing plant sap from infected plants.

The two most commonly observed viruses on perennial crops are the cucumber mosaic virus (CMV) and the impatiens necrotic spot virus (INSV). Symptoms of CMV vary but most commonly include distortion, stunting, and a mosaic pattern on leaves and flowers. Aphids are the primary vector of CMV. INSV symptoms are numerous and can include one or more of the following: black, brown, reddish, or yellowish concentric rings; necrotic streaks, rings, or spots on leaves and stems; distorted growth; stunting; and bud drop. Thrips and vegetative propagation are the primary vectors of INSV.

Figure 5.12. **Viruses are most often seen with various symptoms, from rings to mosaics, on plant leaves, as seen on *Ajuga reptans*.**

Tomato spotted wilt virus (TSWV) and INSV are both classified as tospoviruses. Tospoviruses have become widespread in the greenhouse industry and belong to a group of viruses that are primarily vectored by thrips. Only immature thrips are capable of acquiring the virus, but they can transmit it for the remainder of their lives.

Management strategies for plant viruses vary with the specific virus or viruses detected. Some viruses are nondestructive and have very little effect on plant development and performance and require no management techniques. Other viruses are very destructive when they go unmanaged and usually will require some type of strategy to control their presence and risk of spreading to surrounding crops. Unlike animals, plants do not produce antibodies; they cannot become immune to viruses or recover from a viral infection. Once they are infected, they are always infected, even if no visible symptoms are expressed.

Since there are no known controls for infected plant materials, the best management strategies involve controlling the vectors of plant viruses. The first strategy should be to ensure that clean, virus-free materials are entering the production facility. This involves inspecting incoming plants for insect vectors and viral symptoms. Additionally, samples from stock plants should be submitted for virus testing before cuttings are harvested.

Scouting is critical for detecting the vectors and early detection of plants expressing virus symptoms. Scouts usually conduct visual inspections and use sticky traps and indicator plants to determine if viral infections are likely to occur. Using indicator plants and conducting in-house serological tests for tospoviruses, as described below, are affordable options for many perennial growers. Eliminating weeds around the production facility can greatly limit potential sources or viruses that could get transmitted onto the crops. To reduce the likelihood of tobacco mosaic virus (TMV), do not allow any smoking to occur in the production areas. Controlling insect vectors and weed host plants are currently the most effective means of reducing viruses in perennial crops.

When growing perennials susceptible to tospoviruses (INSV and TSWV), growers should place indicator plants within the production area, dispersed within the crop. Indicator plants are plants that rapidly show symptoms of tospoviruses. The most widely used indicator plant for tospoviruses is the fava bean. Fava bean seeds can be obtained through many vegetable seed companies. Indicator plants should be planted in the same size container as the perennial crops being produced. When tospoviruses are suspected and confirmed in the indicator plants, growers should concentrate their efforts on reducing the thrips population and removing the infected indicator plants from the production area. Watch the perennial crop closely for any symptomatic plants, removing them immediately when symptoms become apparent. Replace the indicator plants promptly to continue the monitoring process.

Growers cannot rely solely on visual symptoms to properly diagnose plant viruses. Symptomatic plant tissues should be sent to virus testing laboratories for more accurate testing and identification procedures. These laboratories conduct virus testing using bioassay, serology, nuclelic acid analysis, DNA probing, and electron microscopy identification procedures.

Serological tests, often referred to as immunostrip tests, have been developed to help growers detect the presence of viruses, such INSV and TMV, using on-site testing. These tests are based on two methods, enzyme-linked immunosorbent assay methodologies (ELISA) and genetic probe techniques such as polymerase chain reaction (PCR). The grower should take a small sample of symptomatic tissue, grind it up, place it in the testing receptacle, and find out if the specific pathogen is present and active. These tests are very sensitive but only give results from the tissue being tested. The virus could be highly localized in the plant and may not be able to be detected in all plant tissues.

A negative result means either the virus was not present in the plant or you failed to detect it in the plant tissue that was used. If you feel the symptoms present must be from a virus, then sending a sample to a virus testing lab would be appropriate. Along with the sample, provide detailed information regarding the background of the crop being tested. Samples could be sent to the same lab, a different virus testing facility, or both labs simultaneously.

A positive result from a virus test provides information about the virus detected, the type of damage that may be caused by the virus, either at the present time or in the future, and how to properly manage it. A positive result will also provide an indication of the host range of the specific virus, how the virus is spread, and which vectors are likely to carry the virus from plant to plant. Knowing the specific virus will allow growers to determine which environments tend to intensify the symptoms exhibited by the crops and how destructive the damage is likely to be.

Most unrooted cutting suppliers have implemented virus indexing programs. Virus indexing is used to eliminate viral pathogens from propagation materials. These companies take numerous precautions and test the stock plants diligently to ensure that virus-free materials reach the end consumer. Growers should know that the unrooted cuttings are clean (virus free) initially but could become infected over time if a vector has introduced viral particles into the crop.

Any suspicious plant materials should be tested for viruses immediately. These plants should be placed into a quarantine area or isolated from the main production facility while waiting for the diagnostic results. Once plants are diagnosed with one or more viruses, infected plants should be removed promptly from the growing area to reduce the spread of the virus(es) to uninfected plants. Destroy any plants diagnosed with viruses. Never use virus-infected stock plants for propagation, as many viruses can be spread through vegetative propagation. Again, the only control of infected plants is to destroy them.

With no known control methods, prevention is the only method of managing plant viruses. Growers should take steps to ensure incoming plant materials are virus free and are not harboring any insect vectors. When possible, select virus-resistant plant varieties. Controlling insect vectors, such as aphids and thrips, will help to keep viruses from getting introduced to uninfected plant materials. Taking steps to reduce the mechanical transmission of viruses is also very important. Properly disinfecting the tools used to make cuttings, harvest flowers, or to divide perennials will reduce the potential for viruses to spread from plant to plant. It is also important to properly disinfect pots and the potting surface and to have the employees wash their hands before working in the crop areas. Any symptomatic plants should be quarantined and tested immediately. If they are diagnosed with a virus, they should be removed from the production area and discarded immediately.

Other Disorders

There are several disorders growers are likely to experience that are not caused by insects, diseases, plant nutrition, or viruses. The precise causes of these disorders are often very difficult to isolate.

Pollution is one factor that is often responsible for plant disorders. Pollution can occur in the air, soil, or water. Symptoms of air pollution often resemble other plant pathogens and can easily be misdiagnosed. Injury to plants from air pollutants can occur at very low levels, below the levels where human health would be threatened. Some of the common air pollutants affecting plant growth are carbon monoxide, ethylene, ozone, and sulfur dioxide.

Pollution from exposure to ozone is a rare occurrence. In greenhouses, growers occasionally use metal halide and VHO florescent lights, which can produce ozone. Under certain circumstances where the growing structure has a low air exchange rate and these lights are placed too close to the foliage, brown flecking of the foliage may occur as a result of exposure to ozone.

Carbon monoxide (CO) is a toxic gas that is made as a result of improperly installed oil or gas heaters. Carbon monoxide is an odorless and colorless gas that is a severe health risk to humans, animals, and plants.

Exposure to ethylene gas can have some adverse effects on plant health and quality. Ethylene damage is most commonly associated with incomplete combustion of propane, natural gas heaters, or engines. Symptoms of ethylene injury vary and include twisted foliage, deformed (or "blind") growth, yellowing and dropping of the leaves, burning of flowers, abnormal flower development, or premature bud drop.

Fumes from cleaning agents are occasionally to blame for plant injury. Fumes from commercial strength cleaners such as ammonia can cause leaves to turn black, curl, or fall off. Severe exposure to these elements may lead to plant death for plants that are particularly sensitive.

Mechanical injury is often the culprit of plant injury. Mishandling during transplanting could cause injury to young roots and stems. Improper irrigation practices, such as too much pressure or volume of water applied, often causes injury to some plants. Circulation fans could cause damage if they cause a lot of whipping or rubbing of the foliage. There are numerous ways plants can become injured, and reducing mechanically injury will reduce the incidence of several plant pathogens.

Pesticide injury often occurs after chemical applications have been applied to crops. Under the right circumstances, almost any chemical applied to a crop could cause injury. Generally, they are formulated to be safe when applied at the labeled rates to target crops. Some common causes are overapplication of a chemical in a single application, applying a chemical too frequently, applications to sensitive crops, applications made when plants were under stress, and chemicals applied when the temperatures exceed 80° F (27° C). Spray injury takes on many forms, including leaf and flower spots, distorted growth, and burning of the leaf margins or the entire leaf.

High humidity levels are likely to cause guttation or edema on some crops. Guttation occurs as the seepage of cellular fluids from the margins or edges of leaves. Edema is a disorder that causes cells to rupture due to excessive turgor pressure (internal water pressure), creating callused corky spots on the undersides of leaves and sometimes plant stems.

References

Chase, A. R. "Diagnosis and Control of Downy Mildew Diseases of Ornamentals." *Greenhouse Product News*. November 2000.

————. "Fungicides for Leaf Spot Diseases of Ornamentals." *Greenhouse Product News*. April 2002.

Daughtrey, Margery. "Downy Mildews on Flower Crops." *GrowerTalks*. February 2002.

————. "If It's Powdery, It Must be Mildew." *GrowerTalks*. March 2002.

Hausbeck, Mary. 2004. "Downy Mildew Alert." *GMPro*. April 2004.

————. "Managing Botrytis on Ornamental Crops." *The Michigan Landscape*. July/August 2002.

————. "It's Back: Downy Mildew on Impatiens in the United Kingdom . . . a Problem for the United States. *Greenhouse Product News*. March 2004.

Miller, Fredric. "Foil the Fungi." *NMPro*. May 2002.

Nelson, Paul V. *Greenhouse Operation and Management*, fourth edition. Englewood Cliffs, N.J.: Prentice-Hall. 1991.

Powell, Charles C., and Richard K. Lindquist. *Ball Pest & Disease Manual*, second edition. Batavia, Ill.: Ball Publishing. 1997.

Pundt, Leanne. 2001. "Powdery Mildew." *GrowerTalks*. March 2001.

Storey, Roger, and Charles C. Powell. "Controlling Powdery Mildews with Eradicant and Preventative Fungicides." *Greenhouse Business*. January 1995.

Williams-Woodward, Jean L. "How to Prevent Root Rot: Take Steps to Stop Theses Common Diseases." *NMPro*. September 2004.

6

Insects, Mites, and Other Pests of Perennials

The damage caused by insects, mites, nematodes, snails, and slugs can often be used to identify them. Some insects have chewing mouthparts, which cause the leaf tissue to be destroyed, consumed, or mined. Some of the insects with chewing mouthparts that commonly cause injury to perennials include caterpillars, black vine weevils, Japanese beetles, and leaf miners. Other pests have piercing-sucking mouthparts, which cause the leaves, after feeding, to appear stippled, spotted, or distorted. Insects with sucking mouthparts that create havoc for commercial growers include aphids, leafhoppers, mites, thrips, and whiteflies.

Feeding by insects and other pests is considered a sign (not a symptom) due to the physical evidence the injury leaves behind. Caterpillars and cankerworms often consume the entire leaf. Leaves with only the margins consumed are likely to have been eaten by black vine weevils or leaf cutter bees. Japanese beetles often cause the leaves to appear skeletonized. Leaf rollers cause the leaves to become rolled. Some insects, like leaf miners, mine between the upper and lower leaf surfaces, leaving a visible path, known as a mine. Many insects such as Japanese beetle larvae or chafers prefer to prune or girdle plant roots. Borers and bark beetles make tunnels under the bark of woody plants. Leafhoppers, plant bugs, and mites cause the leaves to appear spotted or mottled with small dots. Aphids cause the leaves to become distorted. Other insects such as wasps and midges cause small galls, or swellings on the stems or leaves. By observing the type of feeding injury, growers will be able to diagnose the insect causing the injury more quickly.

It is helpful to know which host plants certain insects are likely to feed on. The most common insect pests happen to feed on the widest range of plant species, though many insects are more particular about the source of their food. Growers should also understand the life cycles of insects to provide an indication of when plant injury is likely to occur and what life stages cause damage to the crops.

Growers need to understand that some insects are capable of vectoring particular diseases or viruses. In many cases, they may transfer pathogens from plant to plant, or they may cause feeding injury that provides an entry point for diseases into the plant. Managing insects in your perennials may not only reduce injury to your crops, but it will also greatly reduce the potential for plant diseases.

Aphids

It seems that aphids feed on most every perennial produced. The list of perennial host plants is long, and probably would encompass 80% of the commercially produced varieties. Some of their favorites include *Alcea, Bellis, Dianthus, Heuchera, Hibiscus, Myosotis, Monarda, Papaver, Primula, Salvia, Sedum,* and *Veronica*. If you grow it, chances are aphids feed on it.

The most common aphids on perennials are the green peach aphid (*Myzus persicae*), foxglove aphid (*Aulacorthum solani*), and the melon/cotton aphid (*Aphis gossypii*). These species all have wide host plant ranges and occur throughout the world. There are several additional species that tend to feed on perennial crops as well. Generally, the biology and the methods of controlling aphids is the same for all species. The presence of aphids on perennials reduces a plant's marketability, not to mention the potential plant injury, such as stunting and deformities. Aphids become problematic to growers quickly, as they are able to produce a new generation each week. Unlike other insects, aphids are capable of reproducing at cool temperatures (less than 50° F [10° C]), which can cause problems on cool-season crops and overwintered perennial plants.

Aphids are small, soft-bodied, slow-moving insects measuring 1–3 mm in length. They have piercing-sucking mouthparts they use to remove plant fluids from the phloem. Most aphids can be identified by the two tubes, called cornicles, projecting from their abdomen; these tubes resemble small tailpipes. Aphids are the only insects that have these tubes on their abdomens. The color and size of the aphids vary with species, environmental conditions, and the host plant. For example, melon aphids often range from light to dark green in color, and green peach aphids range from light green to a nearly pink coloration.

Figure 6.1. **Aphids are usually observed in groups feeding on young stems, as on this *Leucanthemum.***

Aphids are usually all females that give live birth to young nymphs. Each female is capable of producing fifty to two hundred nymphs in her lifespan of about one month. These nymphs become mature within seven to ten days and begin giving birth to their own offspring. The ability to reproduce and have offspring so quickly is one reason aphids are a major pest for growers.

Most adult aphids are wingless, but often will become winged when the population of aphids reaches high levels and/or the health of the host plant cannot sustain the continued growth of the aphid population. These winged aphids are capable of moving throughout the production area or into the site from outlying areas to find new host plants to establish and sustain a new population. The flight of aphids is not directed; they usually fly with the direction of the wind. Winged aphids usually are colored differently than un-winged adults of the same species.

Scouting for aphids should occur weekly to find localized aphid infestations. They are usually spotted on the young succulent stems of flowering or non-flowering plants, but are also commonly present on nearly every above ground plant part. They are gregarious and quickly form large colonies on plant stems. Aphids produce a digestive byproduct, honeydew, that contains sugars and other chemicals. Under severe infestations, the honeydew can cover leaves and flowers; under high humidity levels, a black 'fungus called black sooty mold often grows on the honeydew. Mealybugs, soft scales, and whiteflies are insects that are also capable of producing honeydew.

One signal that an aphid infestation is occurring is the presence of small white skins left behind during the molting or growth process. These skins are sometimes easier to see than the live aphids themselves. The presence of aphids can also be signaled by the presence of ladybird beetles, parasitic wasps, and lacewing larvae, all of which utilize aphids as a food source. Ants are commonly observed feeding on the honeydew of aphids. If you detect the presence of ants on your plants, observe them more closely for the presence of aphids.

Figure 6.2. **This picture shows many life stages of aphids, from nymphs to young winged adults. The white exoskeletons are left behind as they molt.**
Syngenta Crop Protection

Aphids occasionally transmit viruses to ornamental plants. The potential for viruses to be vectored by aphids is low, due to the slow speed aphids move from plant to plant. Aphid-borne viral diseases include cucumber mosaic virus (CMV), dasheen mosaic virus (DsMV), and tobacco ring spot virus (TRSV). Some perennials susceptible to CMV include *Ajuga, Campanula, Hydrangea,* and *Phlox.* Aphids acquire the viruses through normal feeding on infected plants or weeds. Viruses get transmitted through the aphids' piercing-sucking mouthparts. Each virus is specific to the species of aphid, and not all aphids can be vectors to all viruses.

Aphid control

Cultural control

One strategy is to reduce the likelihood aphids enter the production area by maintaining a weed free environment in and around the production site. Aphids will congregate on weeds and can easily move onto the crops. Good sanitation and weed management will reduce the potential for aphid infestations. Incoming plant materials should be inspected before they are moved into the production facility.

Biological control

There are several biological options available to growers, which when used meticulously do provide adequate control of aphids. Releasing ladybird beetles, parasitic wasps, and lacewing larvae into the production area is how many growers utilize parasites and predators to control this insect pest.

Parasitic wasps lay their eggs inside of the aphid's body. The immature wasp feeds inside the aphid, eventually killing it. The dead aphid appears slightly puffy and is often described as mummified. Close examination of the aphid mummy will reveal a small round hole where the adult parasite chewed its way out. It is important that growers match the proper aphid species up with the correct parasitoid, or little to no control will be achieved. Growers looking to use parasitic wasps to control the green peach aphid or the potato aphid should use *Aphidius matricariae*. When melon aphids are the primary target, *Aphidius colemani* is the recommended wasp species for growers to release.

Using horticultural oil helps to control aphids without taking out the population of natural enemies you are trying to establish. There are several biological-based products available for growers, including azadirachtin (Azatin, Ornazin), *Beauveria bassiana* (BotaniGard, Naturalis T&O), and neem oil (Triact 70), that growers can apply, allowing them to achieve effective control of aphids while preserving the biological system. To achieve the best control using the entomopathogenic fungus *Beauveria bassiana,* I recommend using repeated applications at three- to five-day intervals.

The predatory aphid midge *Aphidoletes aphidimyza* provides excellent control of over sixty species of aphids. The aphid midge is particularly effective at controlling the green peach aphid. Commercially, aphid midges are shipped as larvae and released into the production area. The recommendations are to place one to two larvae on each potted plant or three to five larvae per square foot. Midge releases should occur every two weeks, until the aphid population is under control. When aphid midges emerge as adults, they mate and then lay their eggs near an existing aphid population. Each midge larva will consume fifteen to twenty aphids during its lifetime. Growers who normally battle with melon aphids will get better control if they use the midge species *Aphidoletes colemani*. Midges perform best when released into environments with high light intensities. When released during the shorter days of winter, it is recommended to provide supplemental lighting to keep the midges active.

Chemical control

Detecting aphid infestations early and implementing control strategies before the population gets out of hand is the best method to control these pests. Once the population becomes established and the aphids are deep within the plant canopy or within the flowers, they are increasingly difficult to deal with. Most growers resort to using chemicals as the most effective and quickest method of reducing or eliminating the existing aphid population on their perennials. The industry standard for controlling aphids is the application of imidacloprid (Marathon 60 WP as a drench or Marathon II as a spray). The spray formulation will provide about one month of aphid control, whereas the drench application, when applied properly, provides growers with up to twelve weeks of control.

Several products have come on the marketplace that have systemic or translaminar activity to provide growers with at least one month of aphid control; these include pymetrozine (Endeavor), thiamethoxam (Flagship), acephate (Orthene), and acetamiprid (TriStar). Products that provide satisfactory "knock down" of the aphid population include abamectin (Avid), cyfluthrin (Decathlon), insecticidal soap (M-Pede), bifenthrin (Talstar), and neem oil (Triact 70).

Certain insect growth regulators (IGRs) have proven effective at controlling aphid populations. These products are best when used when aphids are first detected. Pyriproxyfen (Distance) is an IGR that has provided good long-term control.

Root aphids

Most growers are only familiar with the aphids that attack aboveground plant parts. There are aphids that only feed on the root systems of perennial crops. Many perennials in the Asteraceae family are susceptible to root aphids. Controlling root aphids can be difficult because they often go undetected. They are often found in colonies and can be seen feeding around the perimeter of the root zone. When root aphids are present, there are usually areas that appear to resemble a saprophytic mold. These mold patches are indeed mold growing on the honeydew excreted from the root aphid. The best chemical option I have experienced is to drench infested plants with chlorpyrifos (Dursban).

Caterpillars

There are a number of insects, namely butterflies and moths, whose immature or larval forms, often referred to as caterpillars, feed on perennial crops and often cause significant amounts of plant damage. These caterpillars of various sizes include a diverse group of larvae including armyworms, cutworms, leaf rollers, and loopers. Some of the most common caterpillars include beet armyworm (*Spodoptera exigua*), cabbage looper (*Trichoplusia ni*), imported cabbageworm (*Artogeia rapae*), diamondback moth (*Plutella xylostella*), and the European corn borer (*Ostrinia nubilalis*).

Caterpillars only cause injury to plants while they are in the larval stage. Once they become adults, they either do not eat or only feed on nectar. Caterpillars have strong jaws, allowing them to consume large amounts of foliage, tender stems, and even flowers. Many times, their feeding causes serious plant injury, leaving behind ragged, unmarketable plants. Caterpillars are usually host specific, feeding on certain types of plants or plants within a particular family. The damage left behind varies with the species present and includes consumption of part or entire leaves, rolling leaves, or tunneling through stems.

Due to the large number of caterpillar species that feed on perennials, it is difficult to make generalizations about their life cycles.

Figure 6.3. **Looper injury on *Alcea rosea* leaves.**

Many species lay eggs in the soil, while others deposit their eggs on plant parts. There are several species that hide during the day and emerge at night to feed, and others feed only during the day. Pupation is another way many species differ; some species pupate in the soil and others on the plant. Generally, females lay twenty to one hundred eggs that hatch into larvae or small caterpillars, which go through a series of instars (life stages). Depending on the species, three to five instars occur, where there is an increase in size from one instar to the next. The larval stage lasts from seven to ten days before the caterpillar pupates to become an adult. Pupation lasts about one week, and then adults emerge from the pupae. Depending on the species and temperatures, the life cycle from egg to adult takes three to four weeks.

Routine scouting programs can provide early detection for most caterpillar infestations.

Besides the damage to plants, the occurrence of fecal deposits (frass) on plant leaves is a good confirmation that caterpillars are present. If the populations are high and they go unnoticed, significant injury to the crop can occur, resulting in reduced plant quality and possibly crop losses. Growers can use sticky cards to detect the presence of adult butterflies and moths in the crop, though sticky traps are not as effective a monitoring method as they are for other insects, such as thrips.

Depending on the population, control methods can range from simply removing these pests by hand to controlling them using various biological or chemical strategies. Growers can reduce the occurrence of caterpillars by reducing weeds in and near the production facility. Adult moths and butterflies often lay eggs in certain weed species or have pupae overwintering in the plant debris. Using lights at night near the production site often attracts adult moths into the facility, where they may lay eggs on host plants. Growers should reduce the use of lights to avoid luring moths into the production area. The adults can also be monitored using pheromone or black light traps.

The most common biological method is using the bacterium *Bacillus thuringiensis* var. *kurstaki* (Bt). Bt, marketed as Dipel, is very effective at controlling caterpillars and is practically nontoxic to insects, animals, or humans. The active ingredient must be consumed by the foraging larvae and needs to be applied when they are first observed. Multiple applications of Bt may be necessary to increase the residual activity.

Figure 6.4. **Regardless of their size, many caterpillars are difficult to detect because they blend in with the surrounding foliage, as with this white-lined sphinx caterpillar.**

Table 6.1. Perennial Hosts for Caterpillars

Achillea	*Eupatorium*	*Passiflora*
Anaphalis	*Euphorbia*	*Pellaea*
Antirrhinum	*Gaillardia*	*Pennisetum*
Aquilegia	*Geranium*	*Penstemon*
Arachniodes	*Gladiolus*	*Persicaria*
Asclepias	*Gymnocarpium*	*Phegopteris*
Asplenium	*Helianthus*	*Phlox*
Aster	*Hemerocallis*	*Polypodium*
Astilbe	*Hibiscus*	*Polystichum*
Athyrium	*Hosta*	*Pteridium*
Blechnum	*Hymenocallis*	*Pulmonaria*
Campanula	*Hypericum*	*Reseda*
Canna	*Iberis*	*Rudbeckia*
Cirsium	*Iris*	*Salvia*
Coreopsis	*Knautia*	*Scabiosa*
Cryptogramma	*Lamium*	*Sedum*
Cyrtomium	*Lavandula*	*Semiaquilegia*
Davallia	*Leucanthemum*	*Senecio*
Dennstaedtia	*Ligularia*	*Solidago*
Dianthus	*Matteuccia*	*Stachys*
Dicentra	*Matthiola*	*Stokesia*
Digitalis	*Monarda*	*Thelypteris*
Diplazium	*Myosotis*	*Thymus*
Doryopteris	*Nasturtium*	*Tradescantia*
Dryopteris	*Onoclea*	*Verbena*
Echinacea	*Opuntia*	*Vernonia*
Epimedium	*Osmunda*	*Woodsia*

There are a number of parasitoids and predators that can control caterpillar populations. These natural enemies are more prevalent and effective in outdoor production sites, where they are naturally occurring. Greenhouse growers have been successful releasing the parasitic wasps *Trichogramma minutum* and *T. pretiosum* into the production area. The wasps control the caterpillars by parasitizing their eggs, eliminating the larvae before they can develop. The release of *Trichogramma* should coincide with the presence of egg-laying adult moths or butterflies. For information regarding the parasitoids and predators for a specific type of caterpillar, consult a biological control supplier.

Many growers obtain sufficient control by using various insecticide applications. Several effective caterpillar controls are considered to be biorational products, including azadirachtin (Azatin, Ornazin), spinosad (Conserve), tebufenozide (Confirm), and *Beauveria bassiana* (Naturalis). Traditional contact insecticides, namely in the pyrethroid chemical class, are effective and labeled for caterpillar control. These products include permethrin (Astro), cyfluthrin (Decathlon), bifenthrin (Talstar), and fenpropathrin (Tame). Contact insecticides have proven to be more effective than the use of systemic chemistries.

Fungus Gnats and Shore Flies

Most growers are familiar with the presence of the many small black flies around the production area. Most growers consider these pests only to be a nuisance and not a threat to crop production. Both of these insects can cause injury to crops either directly or indirectly. Fungus gnat larvae cause injury to the root systems through their feeding activity. Otherwise, all life stages of fungus gnats and shore flies indirectly cause injury to plants by their transmission of diseases, either on their bodies or fecal deposits. Both of these pests thrive in moist conditions, especially during propagation or before plants establish well developed root systems.

Fungus gnats

In the past, growers thought these annoying insects were only a nuisance, flying around homes and retail shops, but research has shown us that fungus gnats do pose a threat to crop production. The fungus gnats found in greenhouse and nursery production are most commonly *Bradysia coprophilia* and *Bradysia impatiens.*

Adult fungus gnats are small ⅛ in. (3 mm) long, slender, dark brown to black flies. The legs are long and dangling, resembling those of a mosquito. The two wings are clear and have a distinctive Y-shaped vein in each wing. The antennae have many segments and

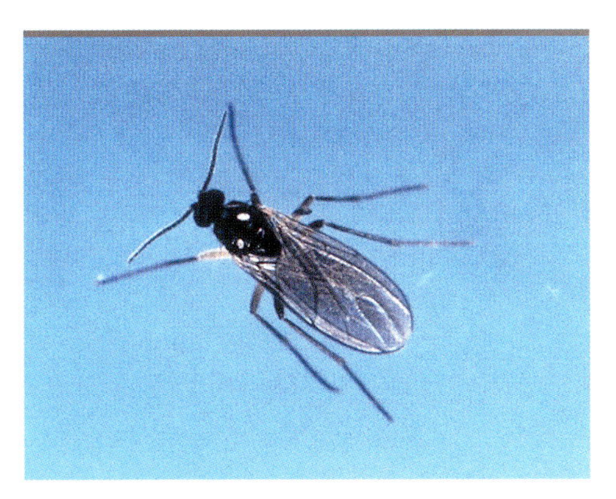

Figure 6.5. **Fungus gnat adult**
Richard Lindquist

are longer than the head. They are weak fliers and are usually observed flying at or near the surface of the growing medium. Often, rather than flying, they can be observed running over the surface of the soil or leaves. Despite being considered weak flyers, they frequently are observed away from the growing medium. The larvae measure ½ in. (13 mm) long or less, are whitish clear, with a black shiny head capsule. The digestive tract is often visible through its translucent body.

Fungus gnats can cause injury to plants directly by larval feeding and indirectly by transmitting plant diseases. Larvae cause the most extensive injury to seedlings or young plants during the propagation phase, consuming their small roots or feeding on the developing calluses. Damage to mature, fully rooted plants is generally less severe unless there are an excessive amount of larvae feeding on them. With young plants, fungus gnat larvae often consume the entire root system faster than it can grow. With larger plants, the larvae often tunnel into the succulent roots or stems at or below the growing medium surface. This tunneling often causes less vigorous growth, wilting, stem collapse, and may even cause plant death.

Adult fungus gnats have been known to carry viable spores of *Botrytis cinerea, Fusarium oxysporum, Pythium, Phytophthora, Thielaviopsis basicola,* and *Verticillium albo-atrum*. The adults can pick up spores from the growing medium when they emerge as adults or simply from moving around on the surface of the growing medium. If the spores fall off the adult's body and the environmental conditions are favorable for spore germination, then the disease may develop.

The female adults mate soon after emergence from the soil and can lay eggs within two days of mating. Female fungus gnats are highly attracted to growing mixes containing peat moss and pine barks, where they lay one hundred to two hundred small (less than 1/32 in. [0.8 mm], oval, creamy white eggs. The eggs are laid singly or in clusters in protected areas, such as the cracks and crevices of the growing medium surface within five days of adulthood.

The eggs mature and hatch within three to six days into small white, slightly translucent, wormlike larvae with shiny black head capsules. There are four larval stages, which are usually found within the top inch or two (2.5 –5 cm) of the growing medium. They can also be observed feeding along the inside edge and bottom of the pots. When the larvae are fully grown they are about ½ in. (13 mm) long. The larvae have mandibles for gnawing and tunneling into plant roots. They consume soil fungus, decaying organic matter (such as softened roots), young plant roots, and stems.

Figure 6.6. **Fungus gnat larvae**
Raymond Cloyd

Depending on temperature, the larvae take approximately two weeks to develop into pupae just beneath the growing medium surface. Once they reach the pupal stage, they take four to seven days to emerge from the pupal skin as adults. The completion of the life cycle is very dependent upon temperature; the development time decreases as the temperature increases. From egg to adult, it typically takes three to four weeks to complete the entire life cycle. There are usually multiple life stages present at any given time. Adult fungus gnats generally live for about ten days.

Fungus gnat larvae may transmit pathogens indirectly by feeding on the roots, injuring them, which creates entry points for several secondary diseases such as *Pythium* spp. and bacterial soft rots. The larvae may also deliver spores from *Pythium, Fusarium, Thielaviopsis basicola,* and *Cylindrocladium* spp. on their bodies directly to the feeding wounds. The larvae are also capable of carrying viable spores and oospores (the sexual stage of water mold fungi) inside their guts and may be excreted at or near the feeding injury, which increases the plant's susceptibility to these pathogens.

Shore flies

Shore flies (*Scatella stagnalis*) are often confused with fungus gnats, but there are several distinct differences that allow growers to properly identify them. Adult shore flies resemble miniature houseflies, measuring ⅛ in. (3 mm) long. They have black bodies, reddish eyes, and dark wings with five or more light-colored spots on each wing. Unlike fungus gnats, the legs of shore flies are small and their antennae are shorter than the head. They are also stronger and faster fliers than fungus gnat adults.

Figure 6.7. **Shore fly adult**

Richard Linquist

Female shore flies lay eggs in areas such as growing medium, benches, and on floors, where algae growth accumulates. Each female can lay three hundred to five hundred eggs, which hatch in two days into maggot-like larvae that are an opaque yellowish brown with no head capsule. The larvae, which measure ⅜ in. (10 mm) long, are most often found just under the surface of the growing medium. The body is usually located in the growing medium and the head in the algae on the medium surface, which is also their food source. It takes seven to ten days for the larvae to form pupae in the potting substrate. It then takes four to five days for the adults to emerge; the adults live for three to four weeks.

The larvae and adults feed on algae and fungi and do not cause injury to plant roots. Shore fly larvae and adults are capable of disseminating the pathogen *Thielaviopsis basicola* in their feces. The adults have also been known to transmit the fungal pathogen *Pythium aphanidermatum* and bacterial pathogens *Erwinia carotovova* and *Pseudomonas cichorii*. While feeding, shore fly larvae ingest spores of plant pathogens and can retain them through pupation and into adulthood. The fecal deposits, containing pathogens, left on the lower leaves and stems of susceptible plants may lead to infected plants if the proper environmental conditions are present for the pathogens to thrive.

Controlling Fungus Gnats and Shore Flies

The first management strategy growers should implement is to examine any incoming plant materials for flying adults. If it is possible, place the incoming plants into a holding area, also referred to as a quarantine area, with sticky cards randomly placed among the plants. The plant materials should be held here for a minimum of five days, with the sticky cards being inspected prior to the

Figure 6.8. **Shore fly larvae**
Richard Linquist

end of this period. If many fungus gnat or shore fly adults are present on the sticky cards, growers should apply an insecticide.

To minimize the favorable environment often created within the production area, discard all dead or diseased plant materials, debris, and any old growing medium or empty pots in the production area. Weeds in and around the production area should be eliminated, as they create a moist environment suitable for shore fly and fungus gnat development. The weed-free environment should include a 20 ft. (6 m) weed-free barrier around the production site's perimeter to prevent weed seeds from blowing into the crop area. All areas within the production facility, including in the crop, behind vents, and underneath benches, if applicable, should also be weed free.

Eliminating algae and freestanding water is the next step in controlling fungus gnats and shore flies. There should be no freestanding water in and around the production area—this includes on benches, pathways, under benches, and in drain troughs. If cooling pads are present, be sure that the distribution tubes are not leaking. Avoid applying excessive amounts of water to the crops, as it is important to allow the top layers of growing medium to become dry between irrigations. With controlling the amount of water in and around the crops, growers can greatly reduce the presence of algae, which provides both a food source and a breeding ground for these insect pests. Benches, areas under benches, walls, and pathways can be treated with disinfectants such as chlorine bleach, quaternary ammonium salt compounds (Greenshield, Physan 20, Triathlon), and hydrogen dioxide (Zerotol) to control algae growth around the production area.

The most common method of detecting the presence of these insects is to monitor using yellow sticky cards. Place the sticky cards 1–2 in. (2.5–5 cm) above the crop canopy and the growing medium. These cards should be inspected weekly using a 10x hand lens, and the adults of each insect should be recorded. It is not terribly important to take a physical count of the adults trapped on each card but to determine each week if the number of adults is increasing or decreasing. Many growers only expose one side of the sticky card each week to save time and labor when counting. Growers specifically interested in the fungus gnat population have found it beneficial to place the sticky cards horizontally on the growing medium surface or on the edge of flats or pots. Monitoring the adult population will provide an indication of whether an infestation is about to occur or if the current levels are nonthreatening to the crop.

Fungus gnat larvae can also be easily detected by placing potato slices or discs measuring 1 in. (2.5 cm) diameter by 0.5 in. (1.3 cm) thick on the surface of the growing medium. Fungus gnat larvae are attracted to the potato slices and will use them as a food source. After two to four days, examine the bottom of these potato discs for the presence of larvae. Some growers slice potatoes into a French fry shape, insert it into the soil to a depth of 2–3 in. (5–8 cm), and examine in a similar manner as the potato disks. Do not leave the potatoes in the growing medium longer than your initial inspection; bacteria often turn the potato pieces into gooey messes. A general recommendation is to use one potato disc/wedge for every 100 sq. ft. (9.3 m^2) of growing area.

To some extent, the type of growing medium used to produce a crop will have an effect on the potential fungus gnat population. No potting mix is exempt from fungus gnats. Potting mixes containing bark seem to be more attractive to fungus gnats and generally produce higher populations of them. Mixes containing coir or coconut fiber may have fewer tendencies for fungus gnat populations to rise; the quick drying of the medium's surface is probably less attractive to the egg-laying adults. Some growers have tried mixes containing diatomaceous earth to reduce the incidence of fungus gnats. Both of these mixes (coir and diatomaceous earth) seem to repel the egg-laying adults but have little effect on the larvae once inside of the mix.

Biological control

Many growers have used beneficial nematodes to successfully control fungus gnats. Beneficial nematodes provide a great deal of control, cause no injury to plants, are safe to workers, and do not promote insecticide resistance. They do not kill the larvae directly, but enter the body of the larvae through their natural openings and release bacteria, which multiplies rapidly and kills them. The most effective strain of nematode commercially available for controlling fungus gnats is *Steinernema feltiae*. This strain of nematode is currently available under the trade names Entonem, Nemasys, and Scanmask. I have had success using nematodes for fungus gnat control, but the nematodes did very little for controlling the shore fly populations. The nematodes should be applied as soon as fungus gnat adults are detected. Applying nematodes regularly during crop production will keep the adult population of fungus gnats relatively low.

Another biological option commercially used to control fungus gnats is to release the predatory mite *Hypoaspis miles*. This mite inhabits the top layers of the growing medium, where it feeds on fungus gnat larvae. The best results are achieved when this mite is released early, such as at planting or shortly afterward.

The last commercial method of controlling fungus gnats by biological means is the application of the bacterium *Bacillus thuringiensis* var. *israelensis*. Currently, this microbial insecticide is available under the trade name Gnatrol. Gnatrol is very effective on fungus gnat larvae when applied as a drench at the recommended intervals. The shortfall occurs when growers do not apply it properly or miss one of the applications. When this occurs, it is very difficult to play catch-up, and other methods of control may need to be implemented.

Controlling the shore fly population using biologicals is a lot more challenging and perhaps not even feasible. The use of beneficial nematodes and *Bacillus thuringiensis* has been very ineffective for growers, providing them with little to no control. Growers have had limited success using *Hypoaspis miles*, which is the same predatory mite growers use to control fungus gnats. Using this mite is only effective in areas where there is no freestanding water. Shore fly larvae can survive in standing water, such as puddles on flood floors, but the predatory mite cannot and are completely ineffective in these areas. They should be released into the production area after planting or at the earliest detection of shore fly adults.

Chemical control

The best chemical strategy for controlling fungus gnats and shore flies involves controlling them while they are in the larval stages. Fungus gnat larvae cause the most damage during rooting or shortly after potting. To prevent damage, it is often beneficial to make an application targeting the larvae within ten days of potting. If an adult population is already present, it might be wise to control their population as well.

There are a number of commercially available chemicals to control larvae. Most of these chemicals are insect growth regulators (IGRs), which often mimic the juvenile growth hormone, causing them to molt prematurely or preventing them from entering their next stage of development. Some of the commercially available IGRs effective at controlling fungus gnat and shore fly larvae are diflubenzuron (Adept), azadirachtin (Azatin, Ornazin), cyromazine (Citation), and pyriproxyfen (Distance).

Chemicals used to control larvae are applied as drenches or sprenches over the crops and usually provide little to no control for the adult population. Drench applications involve distributing the active ingredient uniformly throughout the growing medium without any loss to leaching. Since drench volumes vary with the pot size and the medium type, growers usually apply, as a rule of thumb, 0.5–1.0 fl. oz./1 in. (14–28 ml/2.5 cm) of pot diameter. For example, a grower producing perennials in 6 in. (15 cm) pots would apply 3–6 oz. (85–142 ml) of chemical solution on the surface of the medium of each pot. Sprench applications are often referred to as heavy sprays, which are applied to the surface of the growing medium, followed by irrigating or watering after application. It is recommended for both drenches and sprenches that the medium should be moderately moist (not saturated or dry) prior to application.

Management strategies focused primarily at controlling the larval stages may take some time to reduce the adult population. There are a number of effective chemicals that provide "knock down" control of the adult population. Most of the chemicals to control fungus gnat and shore fly adults are pyrethroids. The effective chemicals include permethrin (Astro), cyfluthrin (Decathlon), pyrethrins (Pyreth-It), lambda-cyhalothrin (Scimitar), and bifenthrin (Talstar). These insecticides do not control the eggs or pupal life stages and may require multiple applications to provide sufficient levels of control.

Leaf Miners

Leafminers are small flies belonging to the family Agromyzidae that commonly attack several perennial crops. The most common species of leaf miner (*Liriomyza trifolii*) are tiny, yellow-and-black flies measuring only 2 mm in length. *Liriomyza sativae* and *L. huidobrensis* have also been observed in various perennial crops. They can cycle through several generations in a year and may reproduce year-round in greenhouses.

Female leaf miners make small punctures on the upper leaf surfaces with their ovipositor (egg-laying organ). These punctures have two purposes; the males and females feed on plant sap that exudes from them, and the females will lay a single egg in some of them. The punctures turn white and take on a speckled appearance over time. Depending on temperature, the eggs hatch in four to five days. The small larvae slash open surrounding plant cells, using their sickle-like mouthparts. As the plant cells rupture, the larvae move through them, destroying more cells as they move forward. As they travel between the upper and lower leaf surfaces, they weave winding trails through the leaf. These trails are commonly referred to as mines and greatly reduce the aesthetic appearance of the plant.

The larvae feed within the leaf for four to six days, where they molt two times. The third-instar larvae chew small slits in the lower leaf surface, allowing them to drop to the lower leaves or soil surface to pupate. The pupal stage lasts from nine to thirty-five days at temperatures of 80° F (27° C) to 58° F (14° C), respectively. Depending on temperature, the complete life cycle, egg to adult, requires approximately fifteen to forty days.

Figure 6.9. **Adult leaf miners resemble small yellow-and-black bees.**
Richard Lindquist

Growers should inspect any incoming plant materials, such as plugs, for any leaf stipples and active mines. Any suspicious materials should be held for several days to observe whether any mines develop from the stipples. The adults can be caught on yellow

Figure 6.10. **Leaf miner larvae on this *Aquilegia* caused these tunnels, or mines, between the leaf surfaces.**

sticky cards to detect their presence and to monitor their population. Sticky cards should be placed one card for every 3,000–5,000 sq. ft. (279–465 m^2) of growing area. The traps should be placed a couple inches (5 cm) above the crop canopy to catch the adults as they fly by and monitored weekly to detect the presence of leaf miner adults and to determine if control measures are necessary.

When the adults are present and control measures have been determined to be necessary, contact sprays of insecticides are effective at controlling the adult population. Sprays should be repeated at four-day intervals for the next ten to fourteen days to control any adults emerging from the pupal stage. There are several products available for controlling leaf miner adults: permethrin (Astro), abamectin (Avid), spinosad (Conserve), chlorpyrifos (DuraGuard), acephate (Orthene), bifenthrin (Talstar), and fenpropathrin (Tame). Insect growth regulators have provided some control of the larvae and are best applied early in the crop cycle when the foliage is less dense. Commercially available insect growth regulators labeled for leaf miner control include azadirachtin (Azatin, Ornazin), cyromazine (Citation), and novaluron (Pedestal). There are some systemic insecticides (imidacloprid [Marathon] and thiamethoxam [Flagship]) that have been effective, when applied as drenches, at controlling the larvae that feed inside the leaves. Drench applications should be made on a preventative basis rather than applying once an established population is present. Biological control from parasitic wasps *Dacnusa sibirica* and *D. isaea* has provided successful control of leaf miner larvae under certain circumstances.

Growers should be aware that leaf miners are notorious for their ability to establish resistance to pesticides with similar modes of action. Perennial producers should practice resistance avoidance through exclusion of adults, spot treatments, and rotating chemical classes. Follow the directions on the label and rotate chemicals accordingly.

Table 6.2. Perennial Hosts of Leaf Miners

Althaea	*Lamium*
Aquilegia	*Leucanthemum*
Asclepias	*Monarda*
Buddleia	*Platycodon*
Coreopsis	*Polemonium*
Delphinium	*Semiaquilegia*
Eupatorium	*Senecio*
Gaillardia	*Tropaeolum*
Geum	*Verbena*
Helenium	*Veronica*
Heliopsis	

Mealybugs

Mealybugs are similar to scales (discussed later in this chapter), except their small, oval, soft bodies (1–8 mm long) are not covered by a hardened cover or shield, but with a layer of white, cottony wax. They can be observed feeding on all plant parts, including the root system. Some species produce short spine-like filaments along the margins of their bodies and have long posterior filaments. They are often observed in groups in leaf axils, undersides of leaves, and near the growing points. Growers commonly observe citrus mealybug (*Planococcus citri*), longtailed mealybug (*Pseudococcus longispinus*), and *Phenacoccus madeirensis* (no common name, unofficially called false Mexican mealybug) feeding on ornamental crops. The citrus mealybug is the most common species found feeding on ornamental plants in greenhouses and nurseries throughout the United States.

Mealybugs have piercing-sucking mouthparts that allow them to feed on plant fluids in the vascular tissues. Feeding injury from most mealybugs causes leaf distortion, particularly on the newest growth. Some species actually inject a toxin, which produces necrotic spots, a general yellowing, or even causes the leaves to drop. They produce clear, sticky honeydew similar to that of aphids and scales, which often promotes the growth of black sooty mold. As with other honeydew-producing insects, the honeydew secretions from mealybugs often attract ants, which can be used as an indication that mealybugs could be present.

Each species of mealybug has a slightly different life cycle, but the following generalization is representative. The females produce three hundred to six hundred eggs in a white cottony covering, called an ovisac, underneath their bodies. The female mealybug dies after laying eggs, but the eggs remain protected under her body. Some species lay eggs in loose masses of cottony wax or in felt-like cocoons. Several species, including long-tailed mealybug, are capable of giving live birth. It takes about two weeks for the eggs to mature and hatch in the ovisac. They emerge as crawlers. As with scale insects, the crawler stage is most sensitive to insecticides. All subsequent life stages are mobile and not stationary, as with scale insects. They move rather slowly, but are capable of moving from leaf to leaf and from plant to plant, spreading the infestation slowly through the production area. Although they are capable of moving, they typically find a place to settle for feeding and stay there for the duration of their life, going through several instars (growth stages). The males are tiny, winged insects that do not feed, live only a short time, and die after mating. The female life cycle takes thirty to

Table 6.3. Perennial Hosts for Mealybug

Arachniodes	*Osmunda*
Asplenium	*Papaver*
Athyrium	*Passiflora*
Blechnum	*Pellaea*
Crassula	*Phegopteris*
Cryptogramma	*Polypodium*
Cyrtomium	*Polystichum*
Davallia	*Primula*
Digitalis	*Pteridium*
Diplazium	*Rudbeckia*
Doryopteris	*Sambucus*
Echinacea	*Sedum*
Euphorbia	*Senecio*
Fuchsia	*Tellima*
Gladiolus	*Thelypteris*
Helianthus	*Thymus*
Hemerocallis	*Tolmiea*
Heuchera	*Tradescantia*
Hibiscus	*Viola*
Hosta	*Woodsia*
Matteuccia	*Woodwardia*
Miscanthus	*Yucca*
Onoclea	*Zantedeschia*
Opuntia	

Figure 6.11. **Citrus mealybug (*Planococcus citri*)**

Raymond Cloyd

seventy days (sixty days average), depending on the mealybug species and temperatures.

Visual inspections are the only method by which mealybugs can be detected. They often appear as a small cottony mass on the leaf or in leaf axils. They are usually clustered together, but can be observed individually. The immature instars resemble miniature adults. The male mealybug looks similar to the female until the last instars, where he wraps himself in an elongate filamous wax cocoon until he emerges as a small fly. The presence of honeydew, causing shiny and sticky leaves, is another indication mealybugs might be present. Growers should inspect all incoming plant materials for their presence and not allow mealybugs to enter the production site.

Once mealybugs are present, they can be difficult to control with conventional pest control products. Due to the waxy covering on the adults, most chemicals cannot reach the bodies of the pest to be effective. Additionally, most control products do not have activity on the eggs. The crawler stage, when no waxy covering is present, is the most susceptible life stage to chemical controls.

Contact insecticides are often applied multiple times to control the crawler stages. Chemicals currently available for controlling mealybugs are foliar sprays such as thiamethoxam (Flagship), imidacloprid (Marathon II), buprofezin (Talus), and acetamiprid (Tristar). These products all provide good to excellent control and offer differing lengths of response time and residual activity. Achieving thorough coverage is essential, as most mealybugs are located in hard to reach areas, such as leaf axils, undersides of leaves, and in leaf sheaths. It is beneficial to add a spreader-sticker to the contact insecticide to improve spray coverage and penetration. Insecticidal soaps, such as M-Pede, contain ethyl alcohol, which can penetrate the waxy covering of mealybugs, allowing for the active ingredient, potassium salts of fatty acids, to contact and kill the mealybug. Other products providing reasonable control include Beauveria bassiana (BotaniGard), cyfluthrin (Decathlon), pyriproxyfen (Distance), chlorpyrifos (DuraGuard), paraffinic oil (1% Ultra-Fine Oil), azadirachtin (Ornazin), acephate (Orthene), pyrethrins (Pyreth-It), and bifenthrin (Talstar).

Systemic insecticides, such as imidacloprid (Marathon) and thiamethoxam (Flagship), have some effectiveness on the feeding stages of mealybug. Systemic products should be applied while plants are actively growing and before the mealybug populations build up to high levels. It generally takes a combination of chemical controls (contact and systemic) to reduce the pest population.

Some growers have found some success using natural enemies, such as the predacious ladybird beetle (*Cryptolaemus montrouzieri*), also known as the mealybug destroyer, and the parasitoid *Leptomastix dactylopii*. Both of these natural enemies have been proven effective at controlling the citrus mealybug and can be used simultaneously. For information regarding the parasitoids and predators for a specific type of mealybug, consult a biological control supplier.

Mites

Two-spotted spider mites

The two-spotted spider mite (*Tetranychus urticae*) is the most common mite encountered by perennial growers and has been a nuisance for greenhouses and nurseries for many years. One reason they are so troublesome for growers is that nearly every commercially grown plant can be an acceptable food source for this mite. Some perennials spider mites often feed on include *Buddleia, Filipendula, Hemerocallis, Lamium, Lavatera, Monarda, Potentilla, Primula, Scabiosa, Verbena*, and *Viola*. Spider mites thrive during hot, dry conditions but are a threat to crops throughout the year.

Mites are actually not insects; they have eight legs, two body regions, lack antennae, and are related to spiders and ticks. The management of mites is very similar to other insects, so growers tend to group them into their insect discussions and control programs. All life stages have eight legs except for the larval stage, which has six. The adult females and most immature stages are oval shaped, usually light yellow to green in color, and have the two large dark green spots on either side of their body, for which they are named.

Females do not require mating and begin laying eggs within three days of becoming an adult. When mating does not occur, they lay eggs that produce all males. A female spider

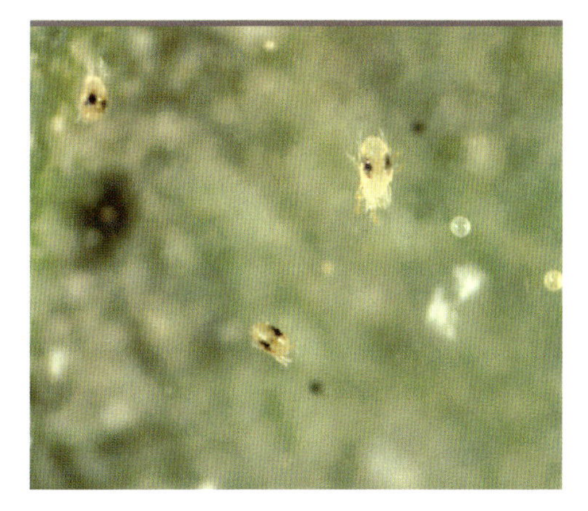

Figure 6.12. **Two-spotted spider mite adult**
Raymond Cloyd

mite lays up to twelve tiny, round, white eggs per day on the undersides of the leaves and is capable of laying one hundred to two hundred eggs in her lifetime. Depending on temperature, these eggs hatch into tiny six-legged larvae in as little as three days. After as few as five days, the larvae change into nymphs. There are several eight-legged nymphal stages before becoming adults. All developmental life stages occur on the plant.

Environmental factors such as temperature, humidity, and host plant all influence the reproduction, developmental time, and survival of the spider mite. Of these factors, temperature is the most important. For example, the development from egg to adult takes about twenty days at 64° F (18° C), but only takes seven days at 81° F (27° C). The typical lifecycle from egg to adult takes seven to fourteen days, but can vary

Table 6.4. Perennial Hosts for Spider Mite

Acaena	*Dicentra*	*Papaver*
Achillea	*Digitalis*	*Pennisetum*
Aconitum	*Dracocephalum*	*Penstemon*
Aegopodium	*Erianthus*	*Perovskia*
Ajuga	*Euphorbia*	*Phyllostachys*
Alcea	*Fargesia*	*Phlox*
Alchemilla	*Filipendula*	*Physostegia*
Alstroemeria	*Fragaria*	*Platycodon*
Althaea	*Fuchsia*	*Pleioblastus*
Ampelopsis	*Gaillardia*	*Polygonum*
Anemone	*Gentiana*	*Potentilla*
Antirrhinum	*Geum*	*Primula*
Aquilegia	*Hedera*	*Pseudosasa*
Aristolochia	*Hemerocallis*	*Reseda*
Artemisia	*Heuchera*	*Rosa*
Arthemis	*Heucherella*	*Rudbeckia*
Aruncus	*Hibanobambusa*	*Sagina*
Aster	*Hibiscus*	*Salvia*
Asteromoea	*Hosta*	*Sasa*
Astilbe	*Humulus*	*Scabiosa*
Ballota	*Imperata*	*Semiaquilegia*
Baptisia	*Iris*	*Semiarundinaria*
Bergenia	*Knautia*	*Senecio*
Boltonia	*Kniphofia*	*Shibataea*
Brunnera	*Lamiastrum*	*Sinarundinaria*
Buddleia	*Lamium*	*Solanum*
Calceolaria	*Lavandula*	*Solidago*
Campanula	*Leucanthemum*	*Teucrium*
Caryopteris	*Liatris*	*Thamnocalamus*
Centaurea	*Lobelia*	*Tolmiea*
Clematis	*Lysimachia*	*Tradescantia*
Coreopsis	*Malva*	*Tricyrtis*
Cosmos	*Miscanthus*	*Tropaeolum*
Cyclamen	*Monarda*	*Verbascum*
Dahlia	*Nasturtium*	*Verbena*
Delphinium	*Nepeta*	*Vinca*
Dendranthema	*Opuntia*	*Viola*
Dendrocalamus	*Pachysandra*	*Yushania*
Dianthus	*Panicum*	

considerable with temperature. As the temperature increases, the rate of development increases, causing a rapid increase in the population as the season changes from spring to more summery conditions. They develop faster on water-stressed plants and prefer hot, dry conditions.

They are hard to find because these sap-feeding insects are so small and feed on the undersides of plant leaves. The adult female is less than a millimeter in size. When scouting for this pest, observe the upper leaf surface for any abnormalities, then turn the leaf over and check the underside of the leaf for the presence of spider mites. It may require the use of a 10X hand lends to be able to see these mites. When high populations of spider mites are present, they produce fine silk webbing that becomes apparent and can eventually completely cover the leaves, flowers, or the entire plant.

They feed on the undersides of plant leaves by piercing and rasping the leaf tissue and injecting a toxin into the plant, which causes leaf drop in some species. The most common injury symptom is pinpoint-sized spots, often described as a mottled or speckled appearance on the upper leaf surface. Usually there will be a patch of these small spots on one area of the leaf, but when spider mites have gone undetected, the entire leaf surface may appear mottled with these spots. Under high mite populations, they may cause severe chlorosis by "stabbing" plant cells with their piercing mouthparts and sucking out the chlorophyll from the plant cells.

When examining plants for spider mites, it is useful to turn over the older leaves first, looking for eggs, immatures, and adult mites. Then work your way up the stem, paying particular attention to buds and bloom for the presence of mites. The eggs and empty egg cases are usually located along the leaf veins. Under low mite populations, they are most commonly found under the lower surfaces of the leaves. Severe infestations usually have mites on all aboveground plant parts, which are covered with the characteristic webbing of these mites.

Once spider mites are observed, focus on other perennial varieties that are particularly susceptible to mite infestations and in production areas that have a history of mite problems. Plants with mites can be flagged so that you can find them the next time you are scouting. Use them as indicator plants to demonstrate the mite's development or the effectiveness of the control strategies.

Generally, spider mites occur in isolated areas of the greenhouse, where workers may have moved them from one area to the next on their clothing or by moving plant materials. If spider mites go undetected, they will spread throughout the crop. It is difficult for growers to control them because they are found almost exclusively on the undersides of the leaves where it is hard for spray applications to make good contact with them. Early detection of two-spotted spider mites is essential and will allow growers to treat them in these isolated areas before they spread across the entire crop. Spider mites are not known to vector any viruses or plant pathogens, but often do cause significant injury to perennial crops.

Cultural control

There are a couple of methods growers can use to control mites culturally, although these methods might not be possible for some growers and only used on a limited basis for others. During certain times of the year, some growers can control greenhouse temperatures and humidity levels to produce a less favorable environment for spider mite development. When possible, growers should avoid sustained periods of high temperatures and low humidity. Even with sophisticated environmental controls, this task is difficult to accomplish during certain times of the year. Frequent misting and overhead irrigation will also help to discourage the establishment of mite populations.

Biological control

There are several commercially available predatory mites available for growers that have been effective at controlling spider mites. The predatory mite species available include: *Amblyseius californicus, A. cucumeris,*

A. fallacies, and *Phytoseiulus persimilis. Phytoseiulus persimilis* is the most widely used predatory mite for spider mite control and has been used in commercial greenhouse production for over thirty years.

If a heavy spider mite population exists, it is usually recommended to apply a miticide before releasing predators into the production area. It is usually necessary to release predatory mites multiple times throughout the production cycle. When done properly, controlling spider mites biologically can be just as effective as using chemical strategies.

Chemical control

There are numerous chemicals available for controlling spider mites. As with all insects, but especially so with spider mites, it is important to rotate chemical families of the miticides being applied to prevent their resistance to these chemistries. There are two good ovicides available to commercial growers, hexythiazox (Hexygon) and clofentezine (Ovation), which primarily act on mite eggs and very young larvae. Ovicides are best used when mite populations are low. A handful of chemicals control various life stages of the spider mite, including fenpyroximate (Akari), abamectin (Avid), bifenazate (Floramite), chlorfenapyr (Pylon), and etoxazole (TetraSan). Many growers have found it beneficial to tank mix various miticides to either gain better control or to control more life stages with a single application. For example, mixing an ovicide such as hexythiazox (Hexygon) with bifenazate (Floramite) allows growers to control nearly every life stage that is present. Many miticides do not control all life stages, particularly mite eggs. For these reasons, it may be necessary to repeat miticide applications after five to seven days.

Regardless of the chemicals being applied, it is very important to ensure good coverage of the upper and lower leaf surfaces for effective mite control. It is difficult to obtain contact with the mite population on the undersides of plant leaves using both low-volume and high-volume applications. I would recommend using products such as abamectin (Avid), chlorfenapyr (Pylon), or etoxazole (TetraSan), which provide translaminar activity. Chemicals with translaminar activity, when applied to the upper leaf surface, move through the leaf from the upper surface to the lower surface, where the mite population feeds. Including these types of chemicals into your mite program can greatly improve the effectiveness of your applications.

To prevent future pesticide resistance, growers should always use at least three different modes of action. There are a number of labeled products for spider mite control that are in differing chemical classes, but some of them have the same mode of action and should not be used in sequential applications. There is a lot more language on the chemical labels of mite control products regarding resistance management than in the past. Usually, each chemical is restricted to a certain number of applications per crop or per year. Always follow the labeled recommendations for the application rates, resistance management, rotation guidelines, tank mixing, and application frequency.

Tarsonemid mites

There are several mites in the Tarsonemidae family that often feed on perennial crops. The broad mite (*Polyphagotarsonemus latus*) and the cyclamen mite (*Phytonemus pallidus*) are members of this family of small mites that have been observed feeding on perennials. Some of the broad mite's perennial preferences are *Antirrhinum, Delphinium,* and *Stachys.*

These mites are very tiny, much smaller than the two-spotted spider mite, making them very difficult to detect, especially when their populations are low. To see them, growers and scouts must used a 12x or 16x hand lens. It may even be necessary to use a dissecting microscope to observe their presence. Look very closely at the leaves near the growing point for mites and eggs. They prefer to hide in protected locations on the host plants, usually near the buds, flowers, or the growing points of the plant. When exposed to light, the mites move away rapidly.

Cyclamen mites feed by piercing plant cells, possibly injecting a toxin into the plant as they feed. The injury symptoms they leave behind include stunted or severely distorted terminal shoots. The expanding leaves from infested growing points often become so curled and distorted that the plants often become unmarketable. The new growth often appears hardened, as if it was sprayed with a high rate of plant growth regulator. Tarsonemid mite injury often confuses growers because it often resembles thrips feeding injury, chemical phytotoxicity, or physiological disorders.

Female cyclamen mites typically lay one to three eggs each day and a total of twelve to sixteen during their lifetime. Like spider mites, mating is not required for egg production. Unfertilized eggs develop into males, and fertilized eggs produce female mites. At 70° F (21° C) the eggs will hatch in four days. Depending on temperature, the life cycle is usually completed in one to three weeks.

Table 6.5. Perennial Hosts for Tarsonemid Mites

Aconitum	*Cyclamen*
Antirrhinum	*Delphinium*
Chrysanthemum	*Stachys*
Clematis	*Verbena*
Cosmos	*Vinca*
Crassula	*Viola*

Broad mites are commonly spread by air currents, direct contact between plants, or by workers who handle or come into contact with infested plants. When controlling these small mites, it is important to use a fine mist spray to make the proper contact. Chemicals such as fenpyroximate (Akari), abamectin (Avid), chlorfenapyr (Pylon), and pyridaben (Sanmite) have been effective at controlling them. It usually requires two or three applications of miticides to control these small mites. Biological control can also be achieved using certain species of predatory mites.

Satisfactory control of tarsonemid mites can be achieved by immersing infested plants in water, which is heated to 110° F (43° C) for thirty minutes. This practice is not commercially feasible or practical but may be useful to treat a small number of infested plants.

Bulb mites

Many perennial growers produce crops from bulbs, corms, tubers, or rhizomes. Many of these crops can possibly become damaged from bulb mites (*Rhizoglyphus echinopus* and *R. robini*) feeding. These small mites are rarely seen unless a grower is specifically looking for them. They live in the potting substrate or within the scales of bulb crops. When present, they are generally observed in small colonies. To find them, examine bulbs, tubers, and corms, by pealing back some of the plant parts and observe using a hand lens. Although they can be found on healthy plant tissues, they are more commonly observed on damaged tissues that have been infected with a fungus.

Table 6.6. Perennial Hosts for Bulb Mites

Crocosmia	*Iris*
Freesia	*Lilium*
Gladiolus	*Narcissus*
Hyacinthus	*Zantedeschia*

Controlling these pests is usually only necessary when their population is high and the injury to the crop is severe. Unfortunately, growers usually do not detect them, applying fungicides to control the fungus or rotting tissues when the mite is the initial cause of the injury. These mites are resistant to several classes of chemicals, which compounds the difficulty of controlling them. One preventative strategy is to soak the starter material in a solution of miticide for thirty minutes prior to planting. Some growers treat for this mite by applying a post-plant drench to the growing medium. Unfortunately, there are not any miticides specifically labeled for controlling bulb mites. Some chemicals that have been used successfully are chlorpyrifos (DuraGuard) and abamectin (Avid).

Nematodes

Nematodes are small microscopic roundworms that can attack plants. Not all nematodes are bad; many nematodes are beneficial, attacking other nematodes or insect pests, such as fungus gnat larvae. Some nematodes (root-knot nematode or cyst nematode) feed on the roots, causing small galls or cysts to form on the roots. A few nematodes known as foliar nematodes may cause damage on plant leaves. The nematodes used to control insects (entomopathogenic nematodes) do not feed on plants. Conversely, nematodes that feed on plants (plant-parasitic nematodes) do not feed on insects. In addition to causing direct injury to plants, some nematodes are vectors for plant viruses, such as the tobacco ring spot (TRSV) and tomato ring spot viruses (ToRSV).

The presence of foliar nematodes (*Aphelenchoides*) can be detected by properly identifying the plant injury they cause. Foliar nematodes often cause deformity of young plant tissue, leaf spots, and stunting of plant growth. Often the symptoms are described to resemble moisture stress and nutrient deficiencies. The most identifiable symptom is the chlorosis or necrosis of leaf areas between the veins, often forming angular lesions (similar to bacterial leaf spots) that are usually first observed on the undersides of the leaves. These lesions usually are small at first, but with favorable moisture and temperature levels may spread over much of the leaf.

Foliar nematodes do not persist in the soil and will cease to exist without a living host. They move from plant to plant with splashing water. Controlling foliar nematodes has become increasingly difficult for growers, as most all of the effective chemicals are no longer available to growers. One of the newest miticides, chlorfenapyr (Pylon), is effective and labeled for controlling foliar nematodes on ornamental crops.

The presence of nematodes affecting the root zone in commercial container production is usually a rare occurrence. They are more prevalent when growers use soil as a component of the growing medium or where bareroot starter plants are used. Due to short crop schedules, there is usually not enough time for the population of nematodes in the root zone to build up to levels where the injury to perennials can be detected.

Root-knot nematodes are perhaps the most widely observed root-effecting nematode. Their presence is most often detected by the occurrence of small galls on the root system. Perennials with galls generally do not perform

Table 6.7. Perennial Hosts for Foliar and Stem Nematodes

Anemone	*Heuchera*
Aquilegia	*Hosta*
Astilbe	*Ligularia*
Athyrium	*Limonium*
Aubrieta	*Lysimachia*
Bambusa	*Malva*
Bergenia	*Matteuccia*
Blechnum	*Osmunda*
Brunnera	*Paeonia*
Calceolaria	*Pelargonium*
Campanula	*Pellaea*
Ceratostigma	*Petroselinum*
Chrysanthemum	*Phegopteris*
Davallia	*Phlox*
Delphinium	*Physalis*
Dendranthema	*Polystichum*
Digitalis	*Potamogeton*
Diplazium	*Semiaquilegia*
Dipsacus	*Tellima*
Doryopteris	*Tricyrtis*
Dryopteris	*Verbena*
Geranium	*Viola*
Hepatica	*Woodsia*

as well as noninfested plants and are usually more susceptible to bacterial and fungal diseases. Plants with root-knot nematodes often appear stunted and easily wilt on warm sunny days. With adequate moisture and fertility levels, the presence of these nematodes is often undetectable.

There have been six identified species of root-knot nematodes that may affect greenhouse/containerized crop production. The northern root-knot nematode, *Meloidogyne hapla*, is the only nematode that survives in outdoor environments and could find its way into your production facility. The occurrence of other species of root-knot nematodes indicates that they were moved into the production area on incoming plant materials.

The average female root-knot nematode lays two hundred to five hundred eggs, although some may lay up to two thousand. The eggs are twice as long as they are wide and are found in a gelatinous mass about the posterior end of the female. The eggs hatch into small, slender worms (larvae), which only measure 1/50 in. (0.5 mm) long. These larvae seek out new roots, where they enter near the tip. Once they enter the root, they become immobile.

The nematode, once inside the root, causes a small gall to form. The gall formation begins with the stimulation of the root cells from nematode saliva to form giant cells, from which the young nematode derives its nourishment. Soon other cells adjacent to the nematode become enlarged, forming the gall. Here the larva will go through three molts before becoming an adult. The soil temperature is critical for the development of the nematode. At 60° F (16° C) it takes fifty-five to sixty days for females to develop from larvae to egg-laying adults, twenty-one to thirty days at 76° F (24° C), and only fifteen to eighteen days at 84° F (29° C). Temperatures below 59° F (15° C) or above 92° F (33° C) are not conducive to the development of mature female adults.

Root-knot nematodes cannot spread from one pot into another. They can migrate within a few feet of their origin when produced in outdoor beds or fields. Once a plant has been

Table 6.8. Perennial Hosts for Root and Bulb Nematodes

Acanthus	*Freesia*	*Narcissus*
Achillea	*Fuchsia*	*Nerine*
Aconitum	*Gaillardia*	*Onoclea*
Ajuga	*Galanthus*	*Opuntia*
Allium	*Geranium*	*Pachysandra*
Aloysia	*Geum*	*Paeonia*
Anemone	*Gladiolus*	*Panax*
Antirrhinum	*Gypsophila*	*Papaver*
Arachniodes	*Hedysarum*	*Passiflora*
Arctotis	*Hemerocallis*	*Polianthes*
Artemisia	*Hibiscus*	*Polypodium*
Asplenium	*Humulus*	*Primula*
Aster	*Hyacinthus*	*Pteridium*
Astilbe	*Hydrastis*	*Pulmonaria*
Bellis	*Hymenocallis*	*Reseda*
Buddleia	*Hypericum*	*Salvia*
Bupleurum	*Hyssopus*	*Saururus*
Calceolaria	*Iberis*	*Sedum*
Campanula	*Iris*	*Senna*
Canna	*Kniphofia*	*Sida*
Carum	*Lamium*	*Sidalcea*
Centaurea	*Lavandula*	*Sisyrinchium*
Cimicifuga	*Leucanthemum*	*Solidago*
Clematis	*Liatris*	*Stachys*
Convallaria	*Ligularia*	*Tanacetum*
Coreopsis	*Lilium*	*Tellima*
Coronilla	*Limonium*	*Teucrium*
Corydalis	*Linaria*	*Thelypteris*
Cryptogramma	*Linum*	*Trachymene*
Cynara	*Lippia*	*Tradescantia*
Cynoglossum	*Lobelia*	*Trillium*
Cyrtomium	*Lupinus*	*Tropaeolum*
Dahlia	*Lycoris*	*Verbascum*
Dianthus	*Lysimachia*	*Vernonia*
Doronicum	*Marrubium*	*Vinca*
Epilobium	*Mentha*	*Woodwardia*
Eriophyllum	*Muscari*	*Yucca*

identified to have root-knot nematodes, it should be discarded immediately. There are no effective treatments to return the plant to normal health. In field soils, the most common control of nematodes is a pre-plant fumigation of the field with methyl bromide. The availability of methyl bromide as a fumigant is diminishing, as the government is restricting its use, and it soon it may no longer be available to growers.

Scales

Scales typically have been a problem for nursery growers producing woody non-herbaceous plant varieties. There are some herbaceous plants, such as ornamental grasses or ferns, on which perennial growers are likely to observe scales feeding. Scales really do not appear to look like insects and are often mistaken for plant parts or bumps on plant leaves and stems. In most cases, scales are covered with a waxy material they secrete over themselves to protect their bodies from predators. This waxy barrier makes it difficult for insecticides to penetrate and reach their bodies.

There are two groups of scales, the armored scale, such as euonymus scale, and the soft scale, such as magnolia scale. These insects most likely get moved into the production areas on infested liners brought into greenhouses and nurseries.

Armored scales

Armored scales are typically smaller than soft scales (1–2 mm); the shape of the body varies between species from a circular to an irregular shape. Their color ranges from shades of white to gray, red, brown, or green and varies with the species, sex, and life stage. They secrete a hard waxy shield over their bodies. Unlike soft scales, the waxy shield may separate from their bodies, and armored scales do not produce honeydew. Some of the common armored scales include: bamboo scale (*Asterolecanium bambusae*), Boisduval's scale (*Diaspis boisduvalii*), euonymus scale (*Unaspis euonymi*), fern scale (*Pinnaspis aspidistrae*), Florida red scale (*Chrysomphalus aonidum*), oleander scale (*Aspidiotus nerii*), purple scale (*Lepidosaphes beckii*), and San Jose scale (*Aspidiotus perniciosus*).

The first indication that armored scales are present is encrustations located on both the leaves and stems. Armored scales cause yellow to brown spots or streaks to form on the leaves. There is also a general yellowing of the foliage along with lackluster growth. Severe infestations can cause dieback of stems and possibly even plant death.

Females produce twenty to four hundred eggs, which are produced next to the female underneath her protective shield. They can reproduce sexually or asexually (no fertilization required), and some species are capable of giving live birth. The eggs hatch into crawlers, which move a short distance from where they hatched. Once they find a suitable place to feed, they settle down and do not move for the rest of their lives. The crawlers are most susceptible to insecticides and have a high mortality rate until they secrete the protective covering over their bodies. The females go through two nymphal stages before becoming adults. The males have two additional short pupal, or resting, stages. The tiny winged males do not have a very long life. After mating, the females produce and lay their eggs. Depending on temperature and scale species, the entire life cycle can take 60 to 129 days to complete.

Figure 6.13. Armored Scales
Richard Lindquist

Several generations may occur throughout the year, and generally all life stages are present at any given time.

Table 6.9. Perennial Hosts for Scale

Ajuga	*Onoclea*
Arachniodes	*Opuntia*
Asclepias	*Opuntia*
Asplenium	*Osmunda*
Astilbe	*Pachysandra*
Bambusa	*Paeonia*
Blechnum	*Passiflora*
Calamagrostis	*Pellaea*
Canna	*Phegopteris*
Coreopsis	*Phlox*
Cryptogramma	*Phyllostachys*
Cyrtomium	*Pleioblastus*
Davallia	*Polypodium*
Dendrocalamus	*Polystichum*
Diplazium	*Pseudosasa*
Doryopteris	*Pteridium*
Dracocephalum	*Salvia*
Dryopteris	*Sambucus*
Euonymus	*Sasa*
Euphorbia	*Semiarundinaria*
Fargesia	*Senna*
Fuchsia	*Shibataea*
Gaillardia	*Sinarundinaria*
Helianthus	*Thamnocalamus*
Hibiscus	*Thelypteris*
Hosta	*Tradescantia*
Hymenocallis	*Verbena*
Iberis	*Vinca*
Iris	*Woodsia*
Lilium	*Woodwardia*
Matteuccia	*Yucca*

Soft scales

Soft scales are usually circular or oval in shape and range in size from 2–5 mm. They are most often shades of gray or brown, but some species have a black coloration. The protective waxy shield is permanent and does not become detached, as does the shield of armored scales. Scales commonly observed by commercial growers include black scale (*Saissetia oleae*), brown soft scale (*Coccus hesperidum*), and hemispherical scale (*Saissetia coffeae*).

Injury from soft scale feeding includes yellow leaves, distorted foliage, and dieback of branches under severe infestations. Scales do not digest the sugars they extract from the plant; these sugars are excreted from their bodies as honeydew. Soft scales produce an incredible amount of honeydew, whereas armored scales do not. This honeydew causes the leaves to appear shiny and sticky, sometimes resulting in the development of the unsightly sooty mold. Honeydew also attracts ants to visit the plant leaves. The presence of ants, honeydew, and sooty mold can all be used by scouts to detect a potential scale infestation.

The life cycles of soft scales are fairly similar to those of the armored scales. They lay eggs or give live births beneath the female's body. The females are capable of producing over a thousand eggs. Crawlers emerge in one to three weeks, moving over plant stems and leaves for a few days until they find a suitable feeding site, where they will remain for the rest of their lives. The crawler stage is the life stage that is most sensitive to insecticides. Prior to adulthood, the females pass through three or four immature stages. The males pass through four immature stages, emerging as tiny winged insects, which live for only a few days. Female soft scales complete their life cycle in forty to eighty days, depending on such factors as host plant, temperature, and species of soft scale. Usually all life stages are present at any given time.

Scales, like aphids, have stylet mouthparts they insert into plant tissues to suck out plant sap, which they feed on. Most scale species are fairly specific to the type of plant they feed on, but several species do have a wide host range. They generally only produce one to two generations per year, depending on the type of scale and the geographic location. The population generally builds up slowly, but once present, it can be a challenge for growers to control them.

Detecting scales

To detect the presence of scales, scouts can look for several signs and symptoms that may occur due to infestations. Some of these symptoms include: premature leaf drop, branch dieback, localized and general chlorosis, discoloration of the foliage, shiny-sticky honeydew (soft scales only), sooty mold build up from honeydew, and plant mortality. Plant inspection is the best method of detecting their presence. All incoming plant materials should be inspected because only a few infected plants can spread to a large infestation over time. The shape of soft scale bodies depends somewhat on where they are feeding; when feeding on leaves their bodies tend to be wide and round, and when they are feeding on narrow branches they take on a narrower, more elliptical shape.

Due to the waxy barrier surrounding their bodies and the fact that certain stages are not susceptible to insecticides, controlling scales can be quite challenging, particularly once a population contains many generations or life stages. In most cases, it requires repeated insecticide sprays at regular intervals to contact all of the susceptible stages in the population. Many growers will discard infested plants instead of spending time and money to control them or risk the further spread of the infestation.

Controlling Scales

Biological control

There are a number of parasitoids and predators available to commercial growers to use to control scale populations. Two species of ladybird beetle, *Lindorus lophanthae* and *Cryptolaemus montrouzieri*, are commercially available predators for controlling both armored and soft scales. Soft scales can be managed by releasing the parasite *Metaphycus helvolus* into the infested crops. The parasite *Aphytis melinus* is used to control some species of armored scale.

Insecticidal control

As mentioned above, scales can be difficult to control with insecticides. Contact insecticides are primarily only effective on the crawler stage and should be applied while this life stage is present (up to a few weeks from emergence). It usually will require repeated sprays (six or more applications may be necessary) to contact this susceptible life stage as it is being produced. Depending on the residual activity of the insecticide, the spray interval is typically every seven to twenty-one days. The addition of a spreader-sticker can help to improve coverage, penetration, and residual activity. Always check the labels of the chemicals being applied for compatibility and potential phytotoxicity that may occur with the addition of spreader-stickers. The insect growth regulators pyriproxyfen (Distance) and s-kinoprene (Enstar II) are effective at controlling the early instars (growth stages) of scales. Thorough spray coverage is critical to successfully control this pest.

Using insecticidal soaps (M-Pede) and horticultural oils (Ultra-Fine Oil) can be effective at controlling more life stages than many contact insecticides, but no residual control occurs, making the application timing and spray interval more critical. It is often beneficial to mix contact insecticides with soaps or oils to increase the level of control and provide some residual activity.

Systemic insecticides such as thiamethoxam (Flagship) and imidacloprid (Marathon) provide some control of the actively feeding stages of scales. The success of systemic chemicals requires that an adequate amount of active ingredient gets translocated to the feeding site. Growers should evaluate the effectiveness of systemic chemicals and consider reapplying once the residual activity is inadequate and no longer provides control of the newly hatched insects.

Slugs and Snails

Slugs and snails are some of the most frustrating pests many greenhouse and nursery operations face. Although they are often considered an insect pest, they are not insects at all. They belong to the phylum Molluska, which contains clams, crustaceans, oysters, and octopi. Even though they are not related to insects in any respect, they are very similar in their biology and behavior. The brown garden snail (*Helix aspera*), the gray garden slug (*Agriolimax reticulatus*), and the greenhouse slug (*Milax gagates*) are some of the most common species threatening perennial crops.

Slugs and snails have bodies that are soft, unsegmented, and slimy. The head has one pair of short tentacles located near the front, used for touching and smelling, and another longer pair with an eye at the distal end of each tentacle located on the top of the head. They move by gliding along on a muscular "foot." This muscle constantly secretes mucous, leaving behind the distinguishing slime trail used for detecting their presence. The biggest distinguishing characteristic between these pests is that snails have an external shell large enough to house their entire body and slugs do not have a visible shell.

Interestingly, slugs and snails are hermaphrodites, possessing both male and female body parts, and all have the potential to lay eggs. Usually they mate with another slug or snail, but self-fertilization can occur. Snails lay numerous spherical, pearly white eggs into a hole in the soil or growing medium, up to six times per year. Slugs lay three to forty clear oval to round eggs under plant debris, in cracks in the growing medium, or in other protected areas. Slugs reach maturity in about three to six months, and snails require about two years before they are mature.

They are active mostly at night, but can frequently be seen moving about on cloudy,

Figure 6.14. Slugs

Richard Lindquist

overcast days. During sunny days they seek refuge from the heat and bright light in plant debris, under pots, or in any other cool dark place. Slugs and snails are typically active from spring till fall in most parts of the country, hibernating during the winter months. In warm climates, they can remain active throughout the year.

Slugs and snails chew irregular holes with smooth edges with their numerous strong small teeth. Feeding injury often resembles the injury from some caterpillars, often causing some confusion for growers. To confirm the injury was caused by slugs or snails, look for the distinguishing silvery mucous trails they leave behind.

Controlling these pests often requires a combination of methods. The first attempt should be to prevent them from entering the production facility. In many cases, they are hitchhikers on plant materials growers either purchase in or move from one location to another. Growers should inspect all plant materials as they arrive to the production facility to detect the presence of slugs and snails. The next measure that can be taken is to reduce the number of hiding places they can hide during the day. Anything sitting on the ground—including boards, stones, boxes, debris, and weeds—can provide adequate shelter for them during the day. Reducing these hiding places will decrease their ability to survive, causing them to look elsewhere for a better habitat. Keeping the environment as dry as possible, reducing the humidity, and preventing surfaces from remaining moist will also deter slugs or snails from staying in the production area.

Small growing facilities often make snail traps and homemade baits to attract and trap these pests so they can be destroyed. One method used to trap slugs and snails involves laying boards with ½ in. (13 mm) runners on the bottom side to allow slugs to enter under them, or placing old pieces of wet carpet on the ground. Growers check under the boards or carpet regularly, removing and destroying any slugs found. Another technique commonly used involves burying traps at ground level with deep vertical sides to keep the snails and slugs from crawling out, and a top to reduce evaporation of an attractant such as beer, a sugar-water yeast mixture, or grape juice. Jars or plastic containers coated with soap or grease can also be sunk into the ground. Once these pests enter the slippery-sided container, they cannot climb back out. All of these trapping methods are only good at trapping slugs or snails that are in the vicinity of the traps but do not work well for trapping pests over large areas.

Commercial baits containing metaldehyde (Deadline), methiocarb (Mesurol), or iron phosphate (Sluggo) are available. Baits work by several modes of action: metaldehyde baits cause overproduction of mucous, which leads to desiccation; methiocarb baits poison the slugs and snails; and iron phosphate baits cause them to stop feeding, leading to starvation. These baits are often broadcast throughout the production site, particularly where the slugs seek refuge during the day. They are best applied under moist conditions, conducive to slug and snail activity. Baits are most effective when they are fresh and more attractive to the slug or snail for consumption. They are less effective during very hot, dry weather conditions or cold times of the year, as these pests are less active during these periods. Baits are valuable tools used to reduce slug and snail populations, but they will not eliminate them altogether.

At times it may be necessary to use chemicals such as methiocarb (Mesurol) applied as foliar sprays to achieve additional control. Growers need to use a combination of control strategies to keep these pests in check. They can be very difficult to manage once they are well established. Growers should try to prevent them from entering the production area, use scouting to detect their presence, and take the appropriate action before their population becomes escalated.

Table 6.10. Perennial Hosts for Snails and Slugs

Acanthus	*Dianthus*	*Lysichiton*
Achillea	*Dicentra*	*Lysimachia*
Aegopodium	*Echinacea*	*Miscanthus*
Agapanthus	*Epimedium*	*Monarda*
Ajuga	*Gaillardia*	*Nepeta*
Alchemilla	*Gentiana*	*Onopordum*
Alstroemeria	*Gladiolus*	*Pellaea*
Aquilegia	*Gypsophila*	*Phalaris*
Arisaema	*Hedera*	*Phlox*
Armeria	*Hemerocallis*	*Polemonium*
Artemisia	*Heuchera*	*Polygonatum*
Asarum	*Hibiscus*	*Primula*
Aster	*Hosta*	*Pulmonaria*
Boltonia	*Houttuynia*	*Rudbeckia*
Calceolaria	*Hydrocharis*	*Rumex*
Campanula	*Iberis*	*Salvia*
Cardiocrinum	*Ipheion*	*Scabiosa*
Clematis	*Iris*	*Sedum*
Coreopsis	*Lamiastrum*	*Sempervivum*
Cosmos	*Lamium*	*Soldanella*
Cyclamen	*Lavandula*	*Stokesia*
Cynara	*Leucanthemum*	*Telekia*
Cypripedium	*Ligularia*	*Tiarella*
Cystopteris	*Liriope*	*Veronica*
Delphinium	*Lobelia*	*Viola*

Thrips

Western flower thrips (*Frankliniella occidentalis*) is the most common type of thrips found feeding on ornamental crops. There are several types of thrips that commercial growers are likely to see, most of which are much more specific to a particular crop than the western flower thrips. The western flower thrips (WFT) is native to the western United States and has moved eastward, covering the entire country within the past twenty years. Thrips can be found year-round inside greenhouses, as long as plants or weeds are available for food and temperatures are favorable.

WFT is the most significant thrips for growers because of the range of plants and the amount of plant injury they can cause, including vectoring plant viruses. They are difficult to control, and their population can increase very quickly. The list of perennial host plants that thrips feed on is too numerous to completely list; some of their most popular perennial food sources include *Alcea, Asclepias, Campanula, Hemerocallis, Iris, Lamium, Lupinus, Malva, Monarda, Penstemon, Platycodon, Polemonium, Primula, Phlox,* and *Tanacetum.*

Thrips eggs are inserted into plant tissues, where they mature and are protected from insecticides. Within two to four days, the eggs hatch into larvae and remain in the terminal or flower buds. There are two larval stages that occur on the plant; the first stage lasts one to two days, and the second stage lasts two to four days. The small larvae are only 0.5–1.0 mm in length. After the second stage, the larvae stop feeding and gradually work their way down to the growing medium, where they undergo two additional transformation stages before emerging as adults.

Once in the soil, they pass through a prepupal stage, which lasts one to two days, and a pupal stage, which lasts one to three days, during which they barely move and do not feed. While in the soil, they are not susceptible to chemicals applied to the foliage. The adults can survive from thirteen to seventy-five days, depending on the temperature and the host plant. Females are

Figure 6.15. **Western flower thrips adults**
Raymond Cloyd

capable of laying 40 to 250 eggs without mating. The life cycle is fairly quick, averaging 7.5 to 13 days from egg to adult, depending on temperature. Development occurs from 50–90° F (10–32° C). At temperatures below 50° F (10° C), thrips can survive, but no development occurs.

The location of thrips feeding changes as the crop develops. On young crops, the immature thrips can often be found feeding on the new leaves. As the plant matures and begins to flower, the thrips move to the buds and open flowers, using them as their primary food source. They can be found feeding on nearly all surfaces of the leaves and flowers.

Thrips are generally distributed evenly throughout a crop. Controlling thrips can be rather challenging, due to their small size and their ability to hide in tight, inaccessible parts of the plant, where direct contact of insecticides can often be impossible. Adding to the difficulty is their high reproductive capacity, rapid life cycle, and resistance to several insecticides. To the untrained eye, the probability of seeing thrips is nearly zero without close inspection and probably requires a hand lens. They are small insects. Adult thrips measure only ⅛ in. long (1–2 mm) and are straw yellow to brown in color. Immature thrips or nymphs look very similar to the winged adults, only they are slightly smaller and do not have wings.

Plant injury from thrips feeding is often the first indication that a thrips population is present. They feed by piercing plant cells with their mouthparts and sucking out the sap, causing the plant cells to collapse, which results in deformed plant growth and flowers. Damaged plant tissue often appears scarred and distorted with silvery white trails, usually visible on the upper leaf surface. The youngest, most tender growth often looks distorted or irregularly shaped. Many growers refer to this injury as stippling of the leaves. These spots are irregular in shape and size. Sometimes this stippling is mistaken for mite injury. Typical mite injury has the appearance of numerous small dots within a small area or over the entire leaf and is often described as russetting. The areas where feeding has occurred are often sunken and

Table 6.11. Perennial Hosts for Thrips*

Achillea	*Dicentra*	*Lupinus*	*Primula*
Alcea	*Digitalis*	*Lychnis*	*Pteridium*
Allium	*Diplazium*	*Lysichiton*	*Reseda*
Alopecurus	*Doronicum*	*Malva*	*Rosa*
Antirrhinum	*Doryopteris*	*Matteuccia*	*Rudbeckia*
Aquilegia	*Dryopteris*	*Melica*	*Sambucus*
Arachniodes	*Erigeron*	*Melissa*	*Scabiosa*
Asarum	*Fallopia*	*Mentha*	*Semiaquilegia*
Asplenium	*Freesia*	*Monarda*	*Solanum*
Aster	*Fuchsia*	*Oenothera*	*Solidago*
Blechnum	*Gaillardia*	*Onoclea*	*Stokesia*
Buddleia	*Geranium*	*Opuntia*	*Tanacetum*
Campanula	*Gladiolus*	*Osmunda*	*Thelypteris*
Centaurea	*Heliopsis*	*Pachysandra*	*Tradescantia*
Chrysanthemum	*Hemerocallis*	*Paeonia*	*Tropaeolum*
Coreopsis	*Heuchera*	*Panax*	*Verbascum*
Corydalis	*Hibiscus*	*Panicum*	*Verbena*
Cosmos	*Hosta*	*Papaver*	*Veronica*
Cryptogramma	*Hymenocallis*	*Penstemon*	*Watsonia*
Cyrtomium	*Iris*	*Persicaria*	*Woodsia*
Dahlia	*Kniphofia*	*Phegopteris*	*Woodwardia*
Davallia	*Lamium*	*Phlox*	*Zantedeschia*
Delphinium	*Leucanthemum*	*Platycodon*	
Dendranthema	*Lilium*	*Polypodium*	
Dianthus	*Lobelia*	*Polystichum*	

* These are the perennial crops thrips have been known to cause significant crop injury on. Thrips may be observed feeding on other perennial crops, especially if they are in bloom.

contain black spots, which are fecal droppings from the young thrips. Under severe infestations, plant growth slows and appears to be stunted. If flowers are present, they often appear deformed, flecked, mottled, or streaked.

Western flower thrips are capable of vectoring tospoviruses, including two destructive viral pathogens to perennial crops: impatiens necrotic spot virus (INSV) and tomato spotted wilt virus (TSWV). Many perennials are susceptible to tospoviruses, including *Aquilegia, Aster, Campanula, Centranthus, Coreopsis, Delphinium, Dianthus, Gaillardia, Lychnis, Lupinus, Monarda, Oenothera, Papaver, Penstemon, Phlox*, and *Primula*. When growing virus-susceptible perennials, a more intensive thrips management program should be in place, and the threshold level, or the number of thrips that can be tolerated, should be lowered.

Thrips initially acquire viruses as larvae when they feed on virus-infected plants and weeds. Once the immature thrips has acquired a virus, it can transmit the virus as soon as it becomes an adult through normal feeding activities. The first-instar larvae (zero to two days old) are better able to acquire a virus and become more efficient vectors as adults, compared with second-instar larvae. The older the thrips is when it acquires a plant virus, the less ability it has to transmit the virus as an adult. Adult thrips are unable to acquire viruses through feeding. Adult thrips that have not acquired a virus as larvae can feed on virus-infected plants without any risk of spreading the virus to other non-infected plants.

Scouting

When scouting, observe the upper leaf surfaces while rotating the plant in your hands, paying particular attention to the terminal growth or flowering parts, if applicable. Look for stippling, feeding scars, and distorted growth. As the plant develops and the leaves harden, thrips injury is more difficult to detect. When flower buds first appear, look for small brown spots on them. Thrips injury on more developed buds will appear as dark spots on the top of the bud and distort the normal opening of the flower. Once the plants flower, both injury to the flower and all life stages of thrips will be present.

Thrips can be detected in flowers using two different methods. The first method involves blowing on the flowers; the carbon dioxide in your breath will cause the thrips to move around and often out of the flower, where they can be seen. The second method is to tap the flower over a white tray or heavy sheet of paper, dislodging them from the flower onto the white surface where they can be observed and counted.

Adult thrips have long, narrow, fringed wings, which allow them to move about the production site or from outside to inside enclosed facilities. They are not strong flyers but can move throughout the production areas or from sprayed areas to unsprayed areas. In addition to flying, thrips commonly move about on wind currents or employee clothing. There is a tendency for thrips to fly less while there is an abundance of flowers present and during the winter months when the temperatures and light levels in the greenhouses are usually lower than during the spring and summer months.

In addition to actively searching for thrips, the adults can easily be caught on yellow or blue sticky traps. Observing the number of thrips on sticky cards can provide an indicator of the number of thrips present in the crop and may be useful when determining if a control strategy is necessary. Place yellow or blue sticky cards just above the crop canopy. Growers typically place one card for every 1,000–3,000 sq. ft. (93–279 m^2) of growing area. At a minimum, growers should use five to ten sticky cards per acre (roughly 4,000 m^2) of growing area. Thrips are more attracted to blue sticky cards than to yellow ones, but most growers still use yellow sticky cards because most other types of winged insects are attracted to yellow and it is slightly more difficult to count thrips against a blue background.

Controlling thrips

Growers are often in situations where they have to manage an established population of thrips. It is generally easier to prevent an infestation than to manage an established one. Prevention should be the first step of any thrips management program.

Thrips can enter your production site through doorways, vents, on wind currents, and on workers. Where possible, use screening (also called microscreening) with 200–400 micron mesh size over the vents to exclude the entry of thrips into the growing area. Due to their small size, it is incredibly difficult to exclude them from greenhouses or other types of enclosed structures. Screening will reduce, but not eliminate, the overall number of thrips that enter the production area. Growers should determine whether thrips are entering the growing area from outside by placing sticky cards in the four primary directions, both inside and outside of the greenhouse, comparing trap counts over time. It may only be necessary to screen certain openings and vents, such as all windward vents, to reduce the entry of thrips into the greenhouse.

One of the best preventative methods to control thrips populations within a production area is to control the weeds in and around the area. Weeds are a refuge for thrips, as well as viruses, which could eventually move into the growing space. Maintain as weed free a production facility as feasibly possible, both before and during the production cycle. Growers should take steps to inspect any incoming plant materials, such as plugs, for the presence of these small pests.

Biological control

Controlling thrips using biological strategies is most effective when the thrips population is relatively low. Growers waiting to apply biological techniques after a thrips population has already exploded will not be able to quickly or easily reduce the population of thrips back down to reasonable or acceptable levels.

The commercial products BotaniGard and Naturalis T&O, which contain the fungus *Beauveria bassiana,* have provided growers who routinely apply these products acceptable control of thrips populations. Researchers are also seeing some success using the fungal pathogen *Metarhizium anisopliae.* One strategy growers have implemented to improve the effectiveness of *Beauveria bassiana* applications is to add 1 lb. (0.45 kg) of brown sugar to every 100 gal. (379 L) being applied. The brown sugar helps to increase the thrips exposure to the spores of the fungus. Similarly, tank mixing azadirachtin (Azatin) and horticultural oil (Ultra-Fine Oil) with *B. bassiana* improves the overall efficacy of the application.

Many growers have found success using the predacious mites *Amblyseius cucumeris* or *Amblyseius barkeri* to control the thrips population within the production site. These mites are generally released every seven days beginning the day the crop enters the production facility at the rate of 4,000 mites/1,000 sq. ft. (93 m^2). Some research indicates that successful levels of control could be achieved by releasing the parasitic wasp *Thripobius semiluteus* or several species of minute pirate bugs in the genus *Orius* into the production area.

Installing microscreening is perhaps the first and most effective method growers can use to keep thrips from migrating into their crops. However, most perennial growers usually produce much of their crops outdoors or in facilities where microscreening is not all that practical, making screening not a viable option. Without screening to keep the thrips from entering the production area, growers usually need to be more aggressive with their biological or chemical applications.

Chemical control

Growers can use numerous chemicals to control thrips. It is not uncommon to apply chemicals repeatedly to achieve satisfactory results due to the thrips' tendency to hide in cracks and crevices of leaves, flowers, and the growing points of the plant. Some of the commercially available insecticides for thrips control include abamectin (Avid), azadirachtin (Azatin, Ornazin), spinosad (Conserve), cyfluthrin (Decathlon), chlorpyrifos (DuraGuard), s-kinoprene (Enstar II), paraffinic oil (Ultra-Fine Oil), insecticidal soap (M-Pede), methiocarb (Mesurol), acephate (Orthene), and novaluron (Pedestal). Spinosad (Conserve) has become the industry standard for controlling thrips.

Thrips have developed resistance to several of the commercially used insecticides, including certain carbamates, organophosphates, and synthetic pyrethroids. Remember to always rotate chemical families to reduce the likelihood of resistance to any one class of chemicals. I recommended using each insecticide for an entire generation of a pest before rotating to another insecticide. For example, given the typical life cycle of a western flower thrips, each insecticide should be used for two to three weeks before switching to another class of insecticide.

None of the commercially available products will control thrips populations with a single application. Several applications should occur at five- to seven-day intervals to significantly reduce the pest population. Making applications at five-day intervals will be more effective than seven-day intervals. Growers will do best to apply insecticides that control thrips with equipment that provides very small spray particles, 100 microns or less. These small particles will penetrate deep into the plant canopy, into the protected areas where thrips are often found, increasing the likelihood of the thrips coming into contact with the insecticide and increasing the overall effectiveness of the application.

To enhance the effectiveness of chemical applications, many growers have adapted the practice of adding sugar to the spray mix. The sugar acts as bait, attracting the thrips out onto the leaf surface, where they will have a greater probability of coming into contact with the insecticides. Typically, 1 lb. (0.45 kg) of sugar is added to every 100 gal. (379 L) of spray solution. Unfortunately, the results have been mixed. Many growers will swear by it, while others seem unimpressed. A potential side effect is the possible growth of unsightly sooty mold on leaf surfaces, similar to what often grows during or following a severe aphid infestation.

Weevil, Black Vine

The first step to identifying weevil infestations is to observe areas within a crop that appear somewhat discolored, stunted, or exhibit a wilted appearance. The foliage of broadleaved plants often has the characteristic C-shaped feeding damage on the leaf margins of the current season's growth. The leaf notching from adult feeding usually does not cause significant injury to the crops, but the larvae are capable of causing significant damage. In severe cases, plants may suddenly wilt and die due to the extensive feeding injury to the roots and crown from the larvae. As few as five larvae can sever the entire root system from the crown, causing the plant top to collapse, often leading to plant death.

Weevils are nocturnal feeders and often seek refuge in plant debris or under the containers during daylight hours, making it difficult for scouts to detect their presence during the day. It is best to look for adult weevils during the evening or early in the morning, before they stop feeding and return to their hiding areas. Black vine weevils (*Otiorhynchus sulcatus*) are beetles, just under 0.5 in. (13 mm) long and are black with tiny tan spots on the back. The adults cannot fly, traveling on foot to find new food sources or places to lay eggs. They also move by hitchhiking on plant materials from nursery to nursery, which is how they have become so widespread.

Similar to aphids, the entire population of black vine weevils consist of females, and they reproduce by parthenogenesis (without fertilization). Two to three weeks after emergence, the females deposit eggs on the soil surface near a host plant. They each lay three hundred to a thousand eggs over several months. The grub-like larvae hatch and feed on the base of the plant and migrate down into the medium, where they feed on the roots. The larvae are white, wrinkled, and legless; they develop through six instars. They feed on the roots from late summer through the fall, until the soil

Table 6.12. Perennial Hosts for Black Vine Weevils

Arachniodes	*Hosta*
Asplenium	*Matteuccia*
Astilbe	*Onoclea*
Athyrium	*Opuntia*
Bergenia	*Osmunda*
Blechnum	*Pellaea*
Cryptogramma	*Phegopteris*
Cyclamen	*Polypodium*
Cyrtomium	*Polystichum*
Davallia	*Pteridium*
Diplazium	*Sedum*
Doryopteris	*Thelypteris*
Dryopteris	*Tiarella*
Epimedium	*Tricyrtis*
Geum	*Woodsia*
Gymnocarpium	*Woodwardia*
Heuchera	

temperatures decrease. Larvae commence feeding in the spring until they pupate during late spring or early summer. The pupation occurs in the soil, lasting three weeks to several months, depending on the soil temperatures. After adults emerge, the life cycle is repeated, resulting in only one generation of black vine weevils per year.

Some growers use "pitfall" traps to detect the first emergence and active populations of black vine weevils. Basically, an 8–12 oz. (0.23–0.34 L) plastic cup is buried at the soil level near the base of a plant, and a smaller plastic cup is inserted into it. Drainage holes are punched into the bottoms of both cups. A funnel-shaped cup with the bottom removed is placed in the larger cup to trap weevils. These pitfall traps should be checked every few days to monitor the presence of adult weevils.

Burlap traps are also commonly used to trap adult weevils. Small burlap squares can be placed around the base of the plant to trap weevils as they emerge or return to the ground after their nightly feeding. These traps should be checked two to three times per week. Burlap traps do not permanently trap the adults or control the population; they simply provide a convenient hiding place where the weevils can be observed and monitored.

Research by Stanton Gill from the University of Maryland has shown the use of entomopathogenic nematodes (*Heterorhabditis bacteriospora*) reduces the black vine weevil population by 90–100% compared to levels found in untreated containers. These microscopic nematodes feed within the weevil larvae, releasing bacteria in their fecal waste, which causes the larvae to become sick and die.

To control adults, the effective chemicals are limited and include permethrin (Astro), acephate (Orthene), lambda-cyhalothrin (Scimitar), fenpropathrin (Tame), and bifenthrin (Talstar). Of these, I have experienced that bifenthrin (Talstar) currently provides the most consistent results. The best results are achieved when these products are applied within the first ten days from the first observation of adults; a second application is usually required seven to ten days following the first. It is more difficult to control black vine weevil larvae. Many of the effective chemicals are no longer available today. The two best options

Figure 6.16. Weevil larvae feeding on the crowns of *Heuchera* often sever the crown, causing plants to collapse.

are incorporating bifenthrin (Talstar) into the growing medium and drenching with imidacloprid (Marathon). Other chemicals labeled for soil application to control weevil grubs include chlorpyrifos (DuraGuard), and acephate (Orthene). To control larvae, it is best to apply the chemical just prior to emergence of the larvae in the late spring or early summer.

Whitefly

The silverleaf whitefly (*Bemisia argentifolii*) and the greenhouse whitefly (*Trialeurodes vaporariorum*) are the two whitefly species growers most often have to contend with. Many growers also observe the presence of the bandedwinged whitefly (*Trialeurodes abutilonea*) on their perennial crops. The adults appear, as their common name suggests, as small white flies. They are sucking insects whose mere presence can reduce the marketability of any given crop.

The greenhouse whitefly is .06 in. (1.5 mm) long, larger than the silverleaf whitefly and holds its wings fairly flat over its abdomen in such a way they appear almost parallel to the leaf surface. The silverleaf whitefly has a slight yellow coloration and holds its wings roof-like over its abdomen, at approximately a 45° angle with the leaf surface. The adults of the bandedwinged whitefly look similar to the greenhouse whitefly except that they have two grayish bands that form a zigzag pattern across each of the front wings. Although it is possible to differentiate between the adults of each of these whiteflies, growers should try to identify the pupal stage as a means of confirmation.

The pupal stages of both whitefly species are most commonly found on the underside of leaves. The greenhouse whitefly has a pupal case with parallel sides that are perpendicular to the leaf surface, giving the pupa a disk-shaped or cake-shaped appearance. There is also a fringe of wax filaments (setae) around the edge of the pupal case of the greenhouse whitefly. The pupa of the bandedwinged whitefly closely resembles that of the greenhouse whitefly except for a dark brown band down the center of the pupal case. The pupal case of the silverleaf whitefly is yellowish and appears more rounded, or dome-shaped, and does not have parallel sides. The silverleaf whitefly does not have a fringe of setae around the edges of its pupal case. Both species have several pairs of filaments arising from the top of the pupa; the filaments on the greenhouse whitefly are typically slightly

Table 6.13. Perennial Hosts for Whitefly

Achillea	*Diplazium*	*Onoclea*
Agastache	*Doryopteris*	*Opuntia*
Ajuga	*Dryopteris*	*Osmunda*
Alcea	*Echinacea*	*Pellaea*
Aquilegia	*Euphorbia*	*Pennisetum*
Artemisia	*Fuchsia*	*Perovskia*
Asclepias	*Geranium*	*Phegopteris*
Asplenium	*Gaura*	*Phlox*
Aster	*Gymnocarpium*	*Physalis*
Astilbe	*Heliopsis*	*Platycodon*
Athyrium	*Heuchera*	*Polypodium*
Bellis	*Hibiscus*	*Polystichum*
Blechnum	*Hosta*	*Primula*
Boltonia	*Houttuynia*	*Pteridium*
Brunnera	*Lamiastrum*	*Pulsatilla*
Buddleia	*Lamium*	*Rosa*
Calceolaria	*Lavandula*	*Rudbeckia*
Campanula	*Liatris*	*Salvia*
Chelone	*Ligularia*	*Sedum*
Clematis	*Lupinus*	*Semiaquilegia*
Coreopsis	*Lychnis*	*Senecio*
Cryptogramma	*Lythrum*	*Thelypteris*
Cyrtomium	*Malva*	*Tiarella*
Dahlia	*Matteuccia*	*Verbascum*
Davallia	*Mentha*	*Veronica*
Dicentra	*Monarda*	*Woodsia*
Digitalis	*Nepeta*	*Woodwardia*

Figure 6.17. **Greenhouse whitefly adult (left) and silverleaf whitefly adult (right).**

Richard Lindquist (left); Raymond Cloyd (right)

larger and more obvious than those on the silverleaf whitefly. These filaments vary in size, depending on the host plant, and are not a reliable trait to properly distinguish between these two types whiteflies.

Female whiteflies can begin laying eggs as early as one to four days after emerging as adults. Mating is not necessary for egg production. They lay 30 to 150 eggs on the undersides of the younger, upper leaves, often in a crescent-shaped pattern. The spindle-shaped eggs are white at first and turn gray with time. At temperatures between 65–75° F (18–24° C), greenhouse whiteflies hatch in about eight to ten days. Silverleaf whiteflies take slightly longer to hatch, at ten to twelve days. Their lifestyle progresses from the first, second, third, and fourth nymphal stages, to the pupal stage, and finally the adult stage, where they may live for one or two months.

The tiny first nymphal stage, or crawler stage, emerges from the egg and moves around the leaf surface, often only a few millimeters, for only a short period of time before stopping at a final location for the rest of their development. The immatures are scale-like, cream to clear colored, and are sessile (immobile) on the underside of leaf surfaces. Since they are immobile and remain close to the leaf surface, they blend in and are not easy to detect unless the population is high and the leaf is speckled with lots of individuals. They molt three times, enter a pupal stage, which lasts four days for both whitefly species, and then emerge into adults. The fourth nymphal stage can be recognized by the red eye spots that are visible through the pupal case. Both the nymphs and adults have piercing-sucking mouthparts and suck plant juices from living plants.

The complete life cycle of the greenhouse whitefly, from egg to adult, takes an average of thirty-two days; the silverleaf whitefly takes up to thirty-nine days to complete its life cycle. Roughly fourteen to sixteen days of both species' life cycles are spent as eggs or pupae, which many of today's chemicals are not very effective at controlling.

Whiteflies do not cause detectable injury symptoms to crops, so scouts must examine the lower leaf surfaces for immatures and adults. Severe infestations could cause some plants to turn chlorotic and mottled from whitefly feeding, but this type of symptom is fairly uncommon. Occasionally, with high populations of whiteflies, honeydew may become visible, causing the leaves to become shiny and sticky. This honeydew is usually not as noticeable as the honeydew produced by aphids and may serve as a food source for a grayish black sooty mold fungus, which interferes with photosynthesis and detracts from the plant's appearance.

Growers should monitor the crop on a weekly basis to determine the presence of whiteflies and to evaluate the effectiveness of their

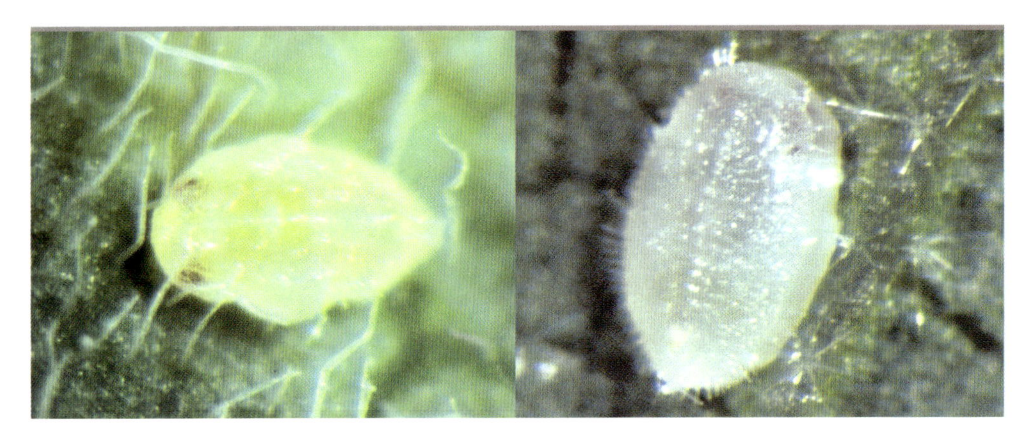

Figure 6.18. **Silverleaf whitefly pupa (left) and greenhouse whitefly pupa (right).**

Raymond Cloyd

management strategies. When monitoring, it is best to use a combination of methods, including sticky cards, random plant inspections, and indicator plants. Quite often when growers walk through a crop, brush up against the leaves, or shake the plant, adult whiteflies will fly from one plant to another. Weekly monitoring will help to determine which life stages (egg, crawler, pupa, or adult) are present and which control strategy will be the most effective at controlling the predominant life stage.

All developmental life stages are normally located on the undersides of the leaf surfaces. For the earliest detection, it is useful to check the lower leaf surfaces as opposed to the newer ones. Because they reside on the undersides of leaf surfaces of older, more mature leaves, it is difficult to reach them with sprays and to control them easily. It often takes multiple applications or the use of systemic insecticides to provide an adequate level of control. Whitefly-infested plants can be marked or flagged to be used as indicator plants, providing an indication of the effectiveness of any control strategies that have been implemented.

Whiteflies are usually found in patches in a production system, unless their population has reached high levels. Yellow sticky cards will provide a good representation to where whiteflies are located while their population is low, but cards are less useful for locating these patches as the population grows. Adult whiteflies are capable of flying at least fifty feet over a twenty-four-hour period, although they generally remain within twenty feet of their emergence site. It is recommended to place one yellow sticky card per every 1,000 sq. ft. (93 m^2) of production area and to place additional cards near doors and vents. Sticky traps are useful to detect the presence of whiteflies and to determine whether the population is increasing or decreasing.

Whiteflies can carry spores of fungal pathogens or bacterial cells that get stuck on their bodies to other plants within the production area. These pathogens get deposited on plant surfaces as the whiteflies feed and may invade the plant cells through the damage caused by the feeding. In vegetable crops, whiteflies specifically can vector a group of viruses known as geminiviruses. Currently, the threat of whiteflies vectoring viruses to perennial crops is low, but the potential for future disease problems is unknown.

Controlling whiteflies

Cultural control

One of the best methods to reduce the incidence of whiteflies in the production area is to start with a clean greenhouse, free of any plant materials including weeds and "pet" plants that may harbor whiteflies. If all the living plant materials can be removed from the production area for at least seven to ten days between crops, any surviving adults will die from starvation. Most growers cannot completely empty their production areas between crops and almost always have plant materials on site. When complete plant removal is not possible, at the very least remove all weeds from the site and any plants that have whitefly immatures or adults on them. Infested plant materials can be moved to another isolated area and treated until the whitefly population has been eliminated. Growers who experience severe infestations may need to consider disposing of heavily infested plants.

Where possible, growers should prevent the entry of whiteflies into the production area. Any incoming plant materials, such as cuttings, plugs, or stock, should be inspected immediately for whitefly immatures and adults before being introduced into the greenhouse. Many growers often only look for the presence of adult whiteflies. Do not assume that plants are uninfested just because adult whiteflies are not present. Be sure to inspect the underside of leaves for the presence of eggs or whitefly immatures.

Perennial producers should implement a scouting program, spot-checking the leaf undersurfaces of several plants throughout the production area on a weekly basis. They should also place yellow sticky cards just above the plant canopy to monitor the whitefly adult populations. The presence of bandedwinged whiteflies on sticky cards or in the production area rarely requires control strategies, as high populations rarely develop. Using screens for whitefly exclusion or to confine these pests has also been a valuable tool for some growers.

Growers should learn which perennials are preferred host plants for whiteflies to feed on. These plants can be grouped together and monitored carefully. If an infestation occurs, it can be easily identified, contained, and controlled.

Biological control

The nymphs and adult whiteflies can be easily controlled using *Beauveria bassiana,* the entomopathogenic fungus found in the commercial biopesticides BotaniGard and Naturalis T&O. The spores of this soilborne fungus infect the whitefly through the cuticle. The whitefly eventually dies as the fungal toxin weakens the immune system. Thorough spray coverage is critical to ensure that the fungal spores contact the insects to begin the infection process. The success of these biopesticides involves repeated application of these products. Growers should not apply *Beauveria bassiana* with any fungicides or use spray equipment containing fungicide residues.

Using predators to control whiteflies can be somewhat challenging. The parasitic wasp *Encarsia formosa* provides some control of the greenhouse whitefly but is not as effective at controlling the silverleaf whitefly. Growers distribute *Encarsia* into the crops by placing small cards face down with parasitized greenhouse whitefly pupa glued to them near the center of a plant. They need to be released as soon as the first whiteflies are detected. The release rate depends on the type of plant and the size and density of the plant canopy. The optimum temperature for their release is 80° F (27° C) with the relative humidity between 50–80%.

The adult *Encarsia* lays its eggs in the third- or fourth-instar whitefly nymphs. The parasitized immatures eventually die and change color; the greenhouse whitefly turns black, and the silverleaf whitefly turns brown. *Encarsia* are very sensitive to pesticides and pesticide residues. It requires some planning to ensure that insecticides with long residuals are not used prior to releasing the wasps. Many of the insect growth regulators, insecticidal soaps, and horticultural oils are compatible with their release and by themselves do not present a threat to the release of *Encarsia.*

Growers have found *Eretmocerus californicus* and other species to be more efficient parasites than *Encarsia formosa,* especially at controlling the silveleaf whitefly. *Eretmocerus* are highly attracted to yellow sticky cards. When releasing this type of wasp, growers should reduce the number of sticky cards they use to monitor their crops. *Delphastus pusillus,* a ladybird beetle, is effective but requires a high prey density to achieve sufficient control and is primarily used to supplement other whitefly biological control strategies. Both immature and adult *Delphastus* beetles are predacious.

Chemical control

There are numerous chemicals available for growers to control whiteflies on perennial crops. Many products, such as pymetrozine (Endeavor), thiamethoxam (Flagship), imidacloprid (Marathon), and dinotefuran (Safari), when applied properly, provide growers with several weeks of whitefly control. These products also protect crops from infestations from other sucking insects, namely aphids, for several weeks. Other commonly used chemicals that have been effective at controlling whiteflies include abamectin (Avid), azadirachtin (Azatin, Ornazin), pyriproxyfen (Distance), and pyridaben (Sanmite).

Many of the above insecticides are only effective at controlling the adult population and are relatively ineffective at killing the eggs or nymphal stages. In many cases, adults will emerge from their immature stages several days after chemicals have been applied but before the next scouting activity, giving the appearance that the chemical treatment was ineffective. It is best to evaluate the effectiveness of chemical treatments after several applications have been made.

The spray intervals will partially depend on the residual activity of each insecticide used, the length of time the whiteflies remain in an insecticide-tolerant stage, and the overall whitefly population. Typically, nonsystemic insecticides are applied at five- to seven-day intervals through the duration of one whitefly generation (twenty-one to thirty days). With whiteflies residing on the undersides of the leaves and most stages being immobile (not crawling over treated leaf surfaces), thorough coverage and good canopy penetration of contact insecticides is critical.

Systemic insecticides are generally more effective at controlling whitefly populations, provided they are applied properly. They are generally most effective when applied early in the crop's development. Some systemic insecticides such as imidacloprid (Marathon 60WP) are applied as drenches and often provide eight to twelve weeks of whitefly control. Other systemic products (imidacloprid [Marathon II], thiamethoxam [Flagship], and pymetrozine [Endeavor]) are available, which are applied as foliar sprays and typically provide up to thirty days of control.

References

Bethke, James A. and Richard A. Redak. "Leafminer Tenacity." *Greenhouse Product News.* September 2002.

Cloyd, Raymond A. "Caterpillar Pests." *Greenhouse Product News.* October 2001.

————. "Watch Out for Mealybugs." GMPro. July 2004.

Flint, M. L. "Snails and Slugs." University of California. UC IPM Online. May 2003. http://www.ipm.ucdavis.edu/PMG/PESTNOTES/pn7427.html. Accessed January 28, 2006.

Gill, Stanton. "Fighting Back: Both Insecticides and Biological Controls Can Help Keep the Black Vine Weevil Out of Your Perennial Crops." *GrowerTalks.* October 2002.

————. "Pests of Potted Plants." *GrowerTalks.* February 2001.

————. "Scale: Pest from Another Planet." *GrowerTalks.* November 2001.

————. "Take Out Mealybugs Before They Take You Out." *Greenhouse Business.* December 2002.

Gilrein, Dan. "Control Black Vine Weevil." *GMPro.* May 2003.

Linquist, Richard. "Tackling a Liriomyza Leafminer Outbreak." *Greenhouse Business.* June 2004.

Miller, Fredric. "Hit the Scales: Controlling These Insects Can be a Challenge for Nursery Growers." *NMPro.* June 2003.

Nelson, Paul V. *Greenhouse Operation and Management,* fourth edition. Englewood Cliffs, N.J.: Prentice-Hall. 1991.

Oetting, Ron. "Slugs and Snails: Constant Nemeses of Ornamentals." *Greenhouse Product News.* January 2003.

Powell, Charles C., and Richard K. Lindquist. *Ball Pest & Disease Manual,* second edition. Batavia, Ill.: Ball Publishing. 1997.

Pundt, Leanne. "Pest Foes of Perennials." *GrowerTalks.* May 2003.

Robb. Karen. "Your Greehouse May Be Just Right for Mealybugs to Destroy Your Crop." *Greenhouse Manager.* August 1993.

7

Controlling Weeds in Perennials

Weeds, and the costs of removing them, are problematic for growers for a number of reasons. They compete with perennial crops for nutrients, water, and light. One weed per container can decrease crop growth and development. Weeds often detract from the perceived quality of perennials grown in containers. In addition to detracting from the aesthetic appearance of our crops, many weeds create persistent problems for growers by harboring insects such as aphids, thrips, and whiteflies, and other pests such as mites, slugs, and snails. Several weeds are host plants to plant viruses such as impatiens necrotic spot virus (INSV) and tobacco spotted wilt virus (TSWV), which can be vectored from infected weeds onto various perennial crops by thrips.

Different weeds bring various levels of concern to commercial perennial growers. Each grower may view the significance of weeds differently, depending on his production systems and customer base. Perennial weeds are typically more difficult to control than many of the annual weeds, causing growers to place more emphasis on controlling them. Certain weeds are considered state or federal noxious weeds and may get shipped into states where they are restricted and may be of concern to several growers. Any weed that is escaping a grower's current weed management program or has few effective control options should be a serious concern to perennial growers.

Controlling weeds in containerized crops is challenging for perennial growers. Many of the registered herbicides for use on ornamentals are fairly safe when applied to woody ornamentals but can cause a great deal of injury when they are applied to herbaceous perennials. Most perennial growers produce a wide array of perennial varieties that have differing levels of tolerance to herbicide applications. Many growers find it necessary to have several different weed management strategies tailored to specific groups of perennial crops. No postemergent herbicide is available for use on perennial crops while they are actively growing, so hand weeding is often the only means of postemergent weed control. For this reason, weed management programs should be preventative in nature.

Figure 7.1. **Weeds can quickly become established in containerized crops, such as this mandevilla vine pictured here.**

Controlling weeds within the crop, the production facility, and the surrounding areas is important to help growers maintain perceived quality characteristics and to reduce potential cultural problems. For perennial producers, a weed management program needs to be comprehensive and include several techniques such as exclusion, sanitation, herbicides, and hand weeding.

Scouting for Weeds

Like scouting for insects and diseases, scouting for weeds forms the foundation of a sound weed management program. Weed scouting is not intended to provide an absolute picture of the population of any given weed species, but to properly identify the weed species that are present, the diversity of weeds, and provide the ability to recognize an infestation before it becomes problematic. Scouting provides an indication to the effectiveness of the current management programs and of which weed species are escaping control measures.

Before scouting activities begin, it is useful to create a map of the production area with each production block clearly labeled. It is also helpful to list the crops present in each block. The initial scouting should be thorough, walking the entire production area to identify the weeds present and to determine their distribution patterns.

After the initial scouting has been done, the scouting pattern varies slightly from traditional IPM scouting. With IPM scouting, growers typically walk a Z pattern, allowing them to walk a crop quickly and efficiently. Weed scouting usually entails a more uniform pattern, walking from the edges of the blocks to the center, as many weeds tend to encroach the crop area from the edges and have a nonuniform distribution pattern.

Similar to IPM scouting, growers should use a scouting form to record where the weeds are located, to identify the weeds present, to determine which weeds are most prevalent, and to identify weeds escaping previous control methods. Growers should notice whether the most prevalent weed is spread evenly throughout the crop or if it is found in isolated areas. Patterns can be used to determine how the weeds have spread, identify recent introductions into the crop, or to indicate nonuniform herbicide applications.

Generally, growers are not concerned with the exact number of weeds present within the production area, but whether the weed population is severe. For example, a rating scale is often used with the following ratings: 0 = weeds are absent, 1 = weeds are rarely observed, 2 = weeds are occasionally seen, 3 = weeds are commonly observed, 4 = weeds are seen throughout the block. The scouting form could have a list of locations down one side and a list of the most common weeds across the top (Fig. 7.2). While scouting, the scout would fill in the appropriate rating for each weed found in the corresponding location. It is important to know which weeds are present or historically have been observed.

Learning to properly identify weeds will be helpful when determining which control strategies will be necessary and most effective. Obtaining a good weed identification guide is helpful. Most guides provide color pictures of various life stages of many weeds commercial growers are likely to observe, as well as some descriptive information. Other useful sources regarding weed identification include extension agents, nursery inspectors, herbicide sales representatives, and the Internet. Some weed identification resources are listed at the end of this chapter.

After properly identifying a weed, it is helpful to know its life cycle to determine the optimal timing for either cultural or chemical control strategies. Life cycle descriptions of various weeds are described in detail in the upcoming section Types of Weeds.

Scouting for weeds should occur prior to potting, closely examining the starting materials. Production areas should be scouted for weeds at least three times a year (spring, early to mid summer, and early to mid fall); a monthly scouting schedule would be even better. Spring scouting helps to identify weeds that have escaped fall herbicide applications and any winter annuals that have germinated (indicating that preemergent herbicide applications have run out). Summer scouting reveals the emergence of summer annuals and persistent spring weeds. Fall scouting can be used to evaluate the overall effectiveness of your weed management program, to find weeds that have escaped your control strategies, and to identify the emergence of winter

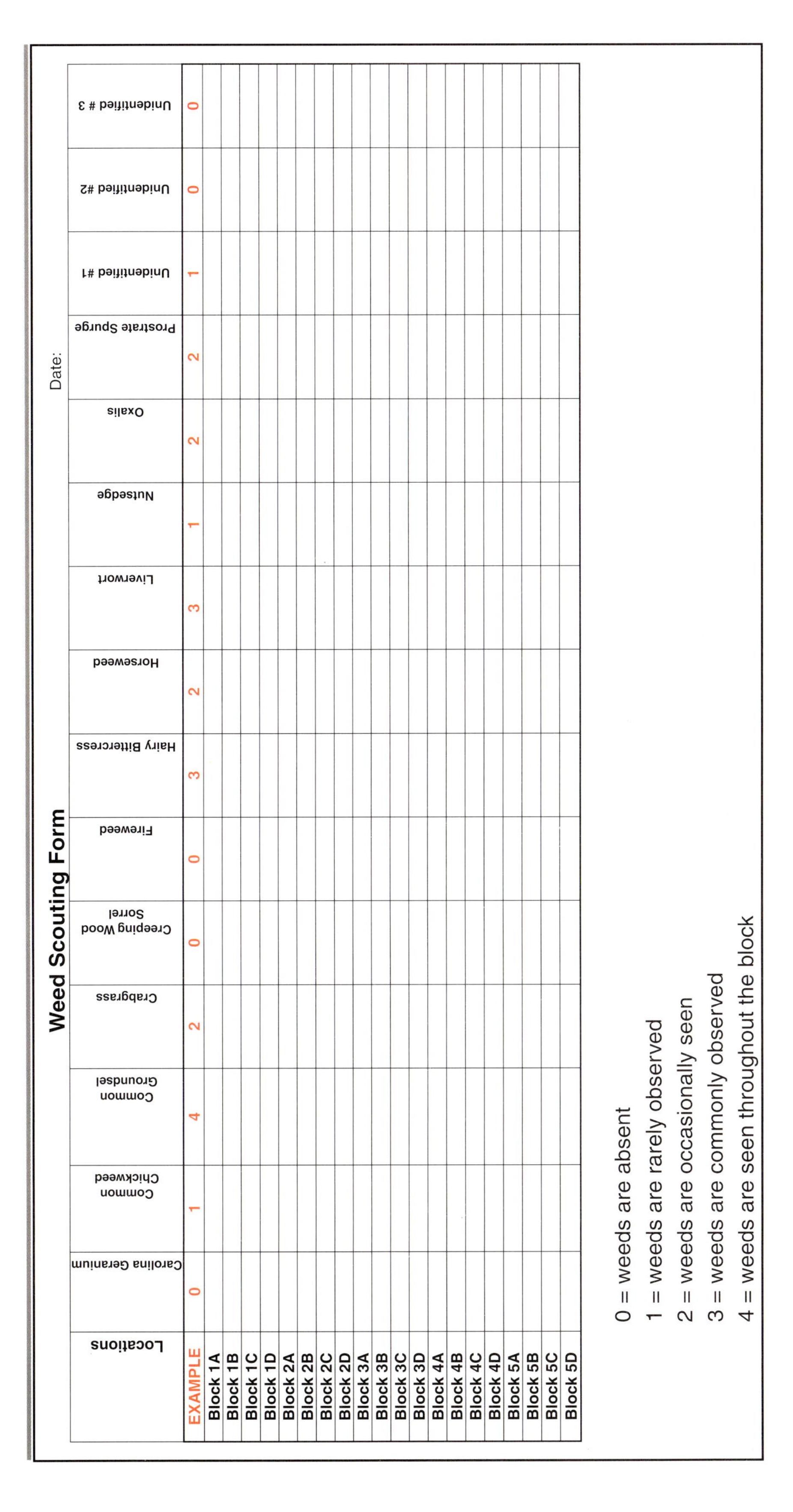

Weed Scouting Form

Date:

Locations	Carolina Geranium	Common Chickweed	Common Groundsel	Crabgrass	Creeping Wood Sorrel	Fireweed	Hairy Bittercress	Horseweed	Liverwort	Nutsedge	Oxalis	Prostrate Spurge	Unidentified #1	Unidentified #2	Unidentified # 3
EXAMPLE	0	1	4	2	0	0	3	2	3	1	2	2	1	0	0
Block 1A															
Block 1B															
Block 1C															
Block 1D															
Block 2A															
Block 2B															
Block 2C															
Block 2D															
Block 3A															
Block 3B															
Block 3C															
Block 3D															
Block 4A															
Block 4B															
Block 4C															
Block 4D															
Block 5A															
Block 5B															
Block 5C															
Block 5D															

0 = weeds are absent
1 = weeds are rarely observed
2 = weeds are occasionally seen
3 = weeds are commonly observed
4 = weeds are seen throughout the block

Figure 7.2. **Weed Scouting Form. Growers fill in the numbers to represent the current weed population. This information can be used for making weed control decisions or evaluating the effectiveness of past control strategies.**

annuals. At a minimum, scouting should occur before preemergent herbicide applications, several weeks after the crops have been weeded, and six to eight weeks after the last herbicide application.

It is important to review the weed scouting reports on an annual basis to provide an indication of the overall effectiveness of the current weed management program. Growers should ask a number of questions when evaluating their current weed control strategies and when modifying existing programs:

- What weeds are not being controlled?
- What new weeds were observed?
- How were they introduced?
- What can be done to reduce weed introduction?
- How are the current herbicides working?
- Are the herbicides controlling the susceptible weeds? If not, why?
- Are there more effective alternatives?
- Does the current management program control them?
- What modifications to the existing program must occur to control future weed problems?

Answering questions such as these will be helpful to determine if your scouting and weed management programs have been effective. It should provide an indication of which areas need to be improved and where to make modifications. Keep in mind that any change in a weed management program will often create an opportunity for new weeds to emerge. Weed scouting and management is an ongoing effort that constantly needs to be evaluated and modified to address the current goals of the operation.

Types of Weeds

To successfully control weeds in container crops, it is beneficial to properly identify the weeds present and determine whether they are annual, biennial, or perennial weeds. It is also helpful to understand the life cycle of the weeds observed; what time of the year do they flower, produce seed, and when do those seeds germinate.

Summer annuals are weeds that emerge in the spring, flower throughout the summer, and set seed before the first frost. Winter annuals are weeds that germinate in the summer or early fall, overwinter, and flower during the spring and summer months. Biennials are weeds that flower and complete their life cycle after two seasons, dying after they flower. Perennial weeds survive more than two growing seasons and are hardest for growers to control. Typically, annual and biennial weeds reproduce from seeds. Perennial weeds often have the ability to reproduce either vegetatively or by seed. Vegetative propagules—such as bulbs, corms, rhizomes, stolons, and tubers—are often resilient to many of the cultural and chemical weed control methods employed by growers.

Some of the most common weeds in containerized perennial production are broadleaf summer annuals (common groundsel, eclipta, prostrate spurge, spotted spurge), broadleaf winter annuals (bittercress, common chickweed, horseweed), perennial broadleaf weeds (creeping wood sorrel, yellow wood sorrel), and annual grasses (crabgrass and annual bluegrass).

Here is a listing of some weeds perennial growers commonly face. This list is not conclusive. For further weed descriptions and identification tools, please refer to the additional references listed at the end of the chapter.

Annual bluegrass (*Poa annua*)

Annual bluegrass is a cool-season grass most commonly observed in cool environments or seasons. It struggles and usually dies during the heat of the summer or under overly dry conditions. When mature, it forms dense, low-growing, spreading tufts up to 8 in. (20 cm) tall. Growers can distinguish annual bluegrass from other types of grasses by the tip of the leaf blade; it is shaped like the bow of a boat and crinkled at the midsection. It has a membranous ligule that is rounded with a slightly pointed tip. Adventitious roots often form at the lower nodes near the medium's surface and are used more for anchorage than for reproductive purposes.

Figure 7.3. **Annual bluegrass discovered in the crowns of dormant daylilies. Most likely the seed was brought in with the bare-root divisions.**

The flowers are produced on terminal panicles that are up to 4 in. (10 cm) in length. Flowering and seed development can occur when plants are only six weeks old. It is a very prolific seed producer—each plant can produce approximately one hundred seeds in less than eight weeks. The seeds are viable in just a few days following pollination.

In container perennial production, the best method of controlling annual bluegrass (and all weeds) is to prevent them from germinating and becoming established. Several preemergent herbicides are effective at controlling germinating seedlings. If established plants exist in the crop, it is usually best to manually remove them, as most postemergent herbicides may cause injury to the desired crops.

Common chickweed (*Stellaria media*)

Common chickweed is an annual broadleaf weed found throughout most of North America. It is classified as both a summer and winter annual. It thrives under the cool, wet conditions often associated with the spring or fall seasons. Chickweed does not tolerate dry conditions very well and often is not a problem during the summer months, except in shady locations.

The cotyledons appear lanceolate as they emerge. Seedlings of chickweed are small, pale green, and only sparsely hairy. The hypocotyl below the seed leaves is slender and often appears reddish. The first true leaves have very pointed tips and petioles. Common chickweed develops tender, prostrate, freely branching stems with opposite smooth, oval leaves with pointed tips. Each plant can spread 12–20 in. (30–51 cm) wide. The stems form adventitious roots at the nodes. The small white flowers have five deeply lobed petals, giving the appearance of ten petals that are located alone or in small clusters at the ends of the stems. It produces small reddish-brown seeds (1.2 mm in diameter) with five to six rows of small bumps that are borne in a capsule. Chickweed flowers throughout the year and is a very prolific seed producer. As mentioned above, it also reproduces by stems rooting at the internodes.

Successful control of common chickweed can be achieved by hand weeding or using pre- or postemergent herbicides. A listing of effective herbicides is provided in tables 7.1 and 7.2.

Figure 7.4. **Common chickweed**

The Scotts Co.

Common groundsel (*Senecio vulgaris*)

Common groundsel is quickly becoming one of the most cumbersome annual weeds in greenhouse and nursery operations. At many operations, it occurs year round. The seed leaves are elongate with a blunt, rounded tip and smooth margins. The first true leaves of common groundsel have shallow teeth, but by the time the plant makes three or four leaves, they are more deeply lobed, alternate, and fleshy. The lower leaves have a short petiole,

Figure 7.5. **Common groundsel seedling emerging between the groundmat and concrete walkway.**

but the upper leaves are attached directly to the stem. It grows 6–16 in. (15–41 cm) tall on succulent, fleshy, and hollow stems. The yellow tubular florets occur in clusters at the ends of stems and branches. The seeds are tipped with a tuft of fine, silky white hairs called pappus (similar to those on dandelion seeds), used to distribute the seeds via wind.

Common groundsel is becoming problematic for growers due to its short life cycle, ability to produce lots of seed, and resistance to many herbicides. The entire life cycle from seed germination to flowering and seed set can occur in as little as six weeks. The seeds can germinate in a few days after landing. Each plant produces thousands of viable seeds.

Growers can reduce the entry of this weed by maintaining a weed-free barrier around the production site. Postemergent products containing glyphosate or glufosinate are effective at controlling established weeds in these areas. Mowing is not a good control method, as this spreads weed seeds and does not kill the roots, allowing them to reflush. Common groundsel has shown resistance to preemergent herbicides in the dinitroaniline and triazine chemical families. These products may still be effective in some areas but may not provide sufficient control in all cases. Commonly used products such as prodiamine (Barricade), pendimethaline (Pendulum), simazine (Princep), and oryzaline (Surflan) are among the preemergent herbicides that may not be very effective at controlling this weed. The new preemergent herbicides Broadstar and SureGuard, containing the active ingredient flumioxazin, are very effective at controlling common groundsel.

Crabgrass (*Digitaria* spp.)

Crabgrass is an annual grass that germinates from late spring and throughout the summer. Once temperatures are warm enough, it sprouts quickly, forming a clump with extensive roots. Once established, it is difficult to remove by hand, as it roots at the nodes (joints) that lie on the soil.

Large crabgrass (*Digitaria sanguinalis*) seedlings are pale green and covered with course hairs. They have membranous, stiff, and papery ligules and no auricles with the young leaves rolled in the bud. As it grows, it forms open clumps reaching two feet tall. The flowering stem has flowers held in three to seven slender fingerlike branches at the end.

Smooth crabgrass (*Digitaria ischaemum*) is a low-growing summer annual grass with 0.25–0.33 in. (6–8 mm) wide pointed leaf blades up to 5 in. (13 cm) in length. The collar has a short membranous ligule, lacking auricles. The upper leaf surface and leaf sheaths are smooth. A few root hairs are often observed on the lower leaf surface. The base of the leaf may have a reddish coloration. Young seedlings of smooth crabgrass can be distinguished from large crabgrass by their shorter, wider leaf and lack of hairs.

Figure 7.6. **Crabgrass**

The Scotts Co.

Figure 7.7. **Bittercress growing around a greenhouse post. A single bittercress weed can produce up to five thousand new seedlings in as little as five weeks.**

Hairy bittercress (*Cardamine hirsuta*)

Hairy bittercress has become a major problem for both greenhouse and nursery operations across the country. Bittercress is commonly observed in propagation houses or overwintering facilities. It is considered a winter annual because it thrives under the cool conditions, in many locations, between the early fall through late spring. With the frequent irrigation practices many perennial growers use, it is able to germinate and grow throughout the year. Besides the most widely used common name, bittercress, *Cardamine hirsuta* is often referred to as pepperweed, shotweed, and snapweed.

Bittercress forms small clumps 6–8 in. (15–20 cm) tall and wide. The leaves are compound, each having four to eight leaflets arranged alternately along the rachis. Flowering occurs on racemes. The white flowers have four petals and measure 3–5 mm in diameter. The seeds are produced in seedpods called siliques. Siliques are dry, two-sided, dehiscent fruit. When the seed is mature, the seedpod splits open into two curling valves, explosively projecting the flattened and finely pebbled seed several feet from the plant. After germination, the seed leaves are round and the first true leaves are often simple and club shaped.

Bittercress completes its life cycle quickly and is known to produce lots of seed in a short period of time. In fact, a single plant can produce up to five thousand offspring in as little as five weeks. Each plant has the ability to forcefully expel seeds up to 42 in. (107 cm) away from the mother plant, potentially carpeting a circle more than 6 ft. (1.8 m) in diameter within approximately a month.

Many growers pull remay fabric over perennial crops to provide protection from the cold during the winter months. This fabric traps solar heat during the day and helps hold heat in during the night. These conditions are very suitable for the growth of bittercress, and quite often large areas under the fabric are completely covered with thriving bittercress plants. Trifluralin + isoxaben (Snapshot), dithiopyrl (Stakeout), and oxyfluorfen + oryzalin (Rout) have all provided an adequate level of control with hairy bittercress. There are some perennials that are sensitive to each of these chemicals and will become injured if these herbicides are applied to them.

Horseweed (*Conyza canadensis*)

Horseweed, also commonly called mare's tail, is considered both a summer and a winter

Figure 7.8. **This horseweed (lighter green shoot) found in a pot of *Buddleia* is hard to detect until it is well established.**

annual. It has a tall, narrow habit, reaching 3–7 ft. (1.2–2.1 m) in height. The cotyledons appear smooth, green, and spatulate. The early leaves are entire, oval, have fine hairs, and form a rosette around the crown. Older leaves are dark green, alternate, without petioles, crowded around the stem, may be entire or toothed, and are often hairy. The leaves may be up to 4 in. (10 cm) in length and become progressively smaller up the stem. Horseweed produces small white or slightly pink ray flowers with yellow discs on at the end of branched stems. The seeds are 1mm long, tapered, achene, with many white bristles useful for wind dispersal. Horseweed is propagated only from seed.

Liverwort (*Marchantia polymorpha*)

Many growers, particularly in the northern United States, experience high populations of liverwort on the surface of their containerized perennials. Eventually the entire surface can become covered with a mat of liverwort, which can restrict the movement of water and nutrients into the root zone and reduce the overall marketability of the crop.

Liverworts are not true plants; they are a more primitive life form, like the mosses. They do not have true leaves; the leaflike structure that covers the surface of the medium is called a thallus (thalli for plural). Another difference between true plants and liverworts is that they do not have root systems; they have rhizoids, which lack xylem and phloem (water-conducting tissues).

Figure 7.9. **Liverwort thalli, which have formed a dense mat over the top of this container.**

Liverworts do not flower or produce seed; they reproduce sexually by spores. They are unisexual, with male and female organs forming on different plants. The male organs (antheridiophores) look like an umbrella while the female organs (achegoniophores) have fingerlike projections. Liverworts can also reproduce asexually by gemmae located in small cuplike structures (cupules). The gemmae are dispersed by splashing water, most commonly from overhead irrigation.

Liverwort thrives in some of the same growing conditions we provide for our container crops—moist environments with high levels of available nitrogen and phosphorus. Liverwort infestations are more evident where crops are under a high-frequency irrigation regime and where nitrogen levels are ambient. Nitrogen applied to the soil surface through liquid irrigation or by topdressing is more likely to promote liverwort growth than when it has been incorporated into the medium.

Hand removal is very difficult and costly for growers. There are very few commercial products that provide long-term control; many times growers are only able to obtain thirty days of control. Unfortunately, many of the remedies growers have found, such as vinegar, cause phytotoxicity to actively growing perennials, limiting the applications to only the dormant period.

Cultural methods can often be used to reduce liverwort establishment. Implementing methods or procedures to dry the medium's surface quickly after each irrigation will greatly reduce the establishment of liverwort. Watering early in the morning allows the surface of the growing medium to dry out by nighttime. Other cultural practices include using a coarse container medium (whose surface dries more quickly), using course mulches on the medium's surface, increasing the air circulation, and incorporating or dibbling fertilizers.

A number of growers have found some success (up to six weeks of control) applying course materials such as hazelnut shells, oyster shells, and copper-treated geotextile discs to the medium's surface. When applying other less-course materials such as coarse sand, perlite, or pumice to the medium's surface, growers have obtained decent control under low-frequency irrigation regimes but not under high-frequency irrigation. Regardless of the mulching material used, it is important to obtain complete coverage of the medium's surface. Any areas where the surface is left uncovered will provide a suitable environment for liverwort growth.

There has been some success controlling liverworts with preemergent herbicides. The most promising herbicides for controlling liverwort are those that contain the active ingredients flumioxazin or oxadiazon. These herbicides have provided good suppression of liverwort for up to twelve weeks. The longevity of the herbicide has a lot to do with the frequency of irrigation: low-frequency irrigation regimes will provide longer control of liverwort, and high-frequency irrigation leads to shorter durations of control. Using herbicides combined with mulches will tend to extend the suppression of liverwort. These herbicides should not be applied to actively growing perennials, as significant injury to the crop is likely to occur.

Nutsedge (*Cyperus* spp.)

Nutsedge is a perennial weed that resembles a grass but it is not a grass species. The stems are three-sided with a triangular cross section (resembles a V), have a distinct ridge, and are without nodes. The leaves are grasslike, smooth, glossy, hairless, and deeply keeled. Umbel-like flower spikes are produced on solitary stems. The three-sided seeds are white to brown in color. Although they produce seeds, most reproduction occurs by small tubers (nutlets) usually present at the ends of the rhizomes. The young plants resemble the older ones, only smaller. With yellow nutsedge (*Cyperus esculentus*), perhaps the most problematic nutsedge for growers, the entire plant appears yellowish to pale green. It is distributed throughout North America.

Figure 7.10. **Nutsedge**

The Scotts Co.

For long-term control, contact herbicides provide initial knock down but do not kill the root system or the small tubers. For complete control, growers should apply postemergent herbicides that can be translocated down to the root system, completely killing the roots and tubers.

Oxalis spp.

There are numerous *Oxalis* species growers observe in containers or the production areas. The two most common species are creeping red sorrel (*Oxalis corniculata*) and yellow wood sorrel (*Oxalis stricta*). *Oxalis corniculata* has purple-green foliage, and the leaves are pubescent (especially along the margins) with a prostrate growing habit, reaching up to 4 in. (10 cm) tall. *Oxalis stricta* has green foliage with much less pubescence and grows in a clump with an

Figure 7.11. **Oxalis**

upright habit, growing up to 8 in. (20 cm) tall. The leaves of both species are compound, with three heart-shaped leaflets, best described as a three-leaf clover. The five-petaled flowers are yellow and produced in clusters (umbels). After flowering, the seeds are produced in long, thin, angular pubescent seedpods. The cotyledons (seed leaves) are small and round. The true leaves are heart shaped, trifoliate, and clover-like.

Both species reproduce from seed. In addition to seed propagation, *Oxalis stricta* spreads by rhizomes and *Oxalis corniculata* spreads with stolons. They produce long taproots, which are difficult to hand-pull from containers.

Spurge (*Euphorbia* spp.)

Perennial growers often have one or more species of spurge growing in the production facility. Prostrate spurge is one of the most common types and contains many different species including: *Euphorbia humistrata, E. maculata*, *E. supina*, and *E. prostrata*. Many of these species are very similar, and it is difficult to distinguish between them. Prostrate spurge is a summer annual as it thrives in hot environments.

The foliage often contains small red spots. The pink flowers are located in the leaf axils and are very small (often requiring magnification to observe). They are prolific seed producers. When germination occurs, there are often countless numbers of young seedlings within a container, making hand-pulling time consuming. They do form taproots but can generally be removed easily. Another distinguishing characteristic of spurge is the milky sap that can be observed when the stems are cut or broken.

Figure 7.12. **Spurge**

The Scotts Co.

Preventing Weeds

The most important aspect of controlling weeds is to control their entry to the production site. Unlike crops grown in the field, where weeds can be controlled efficiently using directed sprays of herbicides, container-grown crops are more complicated. Controlling weeds in containers is complex because many of the crops we grow do not tolerate the application of herbicides. Directed sprays are not feasible, and there are few alternatives to costly hand weeding for removing existing weed populations.

Generally, some form of propagule such as seeds, rhizomes, or tubers from various weeds are introduced into crop areas. They are commonly brought in through the growing medium, the introduction of plant materials from outside sources, splashed into pots by the rain, deposited by birds, or blown in from adjoining areas. If the prevention of weeds does not occur during propagation or early in the crop cycle, weeds are likely to reproduce and become serious problems later.

Weeds are notorious for their ability to produce prolific numbers of seed. Taking measures to control the weeds around the production site will greatly reduce the number of seeds being produced and the number of seeds entering the crop area. Growers should weed frequently to prevent any existing weeds from going to seed. After hand weeding, the weeds should be disposed of, not left on a compost pile in the back corner of the property; this, in many cases, just temporarily moves the weed problem instead of resolving it.

One step growers can take to reduce the number of weeds present is to place the containerized crop on covered ground, such as on gravel, plastic, woven weed fabric (commonly referred to as ground mat), or on floors made of concrete. Growers should eliminate any soil, debris, bark, or growing medium that collects on the surface of these coverings; weeds are often capable of germinating in these areas and could grow, reproduce, and spread to the adjacent crops.

For container production, it is essential to manage the non-crop areas as well as the crop areas. It is recommended to maintain at least 50 ft. (15.2 m) surrounding the container site as a weed-seed-free area. Exclusion of weeds is preferred, but if weeds are present within this barrier, they should not be allowed to go to seed. Growers should use herbicides combined with close mowings to keep the weeds that are in roadways, areas between houses, and the surroundings from producing seeds.

Some weeds, such as bittercress (*Cardanine*) and *Oxalis*, can propel weed seeds several feet from their origin, infesting nearby containers. Other weeds, such as *Eclipta*, can become established in the drainage holes of the containers and have roots that aggressively compete with the crop for nutrients and water.

Growers should clean the beds between crop cycles, removing any plant debris, growing medium, and weeds. Any imperfect areas in the bed, such as torn ground mat, should be fixed if necessary. Many growers apply herbicides directly to the stone or weed fabric covering the bed. Herbicides containing prodiamine (Barricade) work well for these applications, as they have low solubility and last for several months (six to eight months is not uncommon). Many preemergent herbicides are

Figure 7.13. **Besides using herbicides to clean up new production areas, growers can kill existing weeds by burning them.**

suitable for empty bed applications, provided they have low solubility (less than 1 ppm are best). Solubility is explained in more detail later in this chapter (see Herbicide Solubility).

Some growers store media components, such as bark, in outside areas, open to the elements. This can result in potential contamination from wind-blown seeds and vegetative propagules, such as tubers from nutsedge or rhizomes from *Oxalis*. These growers must manage the bark piles continuously to reduce the contamination of various weeds. Covering these piles will help to keep the weed seeds from blowing on them. Some treatments for killing weed seeds and vegetative propagules include composting, fumigation, pasteurization, and solarization. In some cases, nonresidual herbicides such as glufosinate (Finale), diquat (Reward), and pelargonic acid (Scythe) can be effective at controlling weeds on contact. Growers should think carefully before applying herbicides containing glyphosate (Roundup Pro), which is tightly bound and relatively short lived in mineral soils but can be very persistent in peat and other soilless growing mediums and could potentially cause injury to young herbaceous plants. Having sound sanitation practices will reduce the need for these additional treatments.

Used potting medium is often a haven for weed seeds. Growers should only recycle growing medium if they have procedures in place to kill the weed seeds prior to reusing. At the very least, recycled growing medium should only be used in places where a good broad-spectrum preemergent herbicide can be applied.

Another practice that can reduce the entry of weeds into the production area is the use of clean or new pots for propagation and/or potting. Many weed species have small seeds that cling to the sides of the pot with other debris. These seeds often go undetected until after the crop is started, when control measures are more costly. Simply washing the pots with pressurized water will greatly reduce the occurrence of weed seed being transferred into the production site. To determine if weed seeds are coming in with the pots, growers can look for weed seeds germinating near or around the

Figure 7.14. **Weeds in outdoor production sites are particulary cumbersome to growers and can easily make their way into containers.**

edges of the pots. Weeds germinating towards the center of the pot or with a random pattern across the medium's surface are generally not introduced by means of using old pots.

Even with sound weed management programs at your facility, it might be impossible to stop weed seeds from blowing into your property from adjoining nurseries or fields. Using windbreaks, such as fences or hedgerows, should help to reduce the amount of windblown seed entering your property.

Many weeds are imported into our nurseries from outside sources on purchased plugs or liners. Planting weed-infested liners is a very common source of many residual weed problems at greenhouses and nurseries today. It is difficult for plant propagators to control weeds during propagation because most over-the-top herbicides are unsafe to apply to young plant materials.

Growers should inspect all incoming plant materials before potting. If the product is totally infested with weeds or has a weed they are not familiar with, growers should consider refusing the shipment. There should be little tolerance for introducing new weed species into your production system. If weeds do arrive on liners, they should be removed and if necessary, the top 0.5 in. (13 mm) of potting medium should be removed and discarded.

Many weeds, such as liverworts, *Oxalis,* and pearlwort, have extensive root systems (liverworts have rhizoids instead of roots), which can generate new plants when the root systems are not completely removed. A single escaped weed can generate new plants at an alarming rate.

There has been some interesting research in recent years suggesting that the method in which nutrients are delivered to a crop has a great impact on the establishment of weeds. Both topdressing and incorporation of nutrients into the growing medium supply nutrients to the surface of the growing medium, which are readily available for uptake by any weed seeds germinating there. Researchers have found that by dibbling the fertilizer, or placing it directly under the plants roots at potting, nutrients are not readily available at the surface of the medium, often causing germination of weeds to fail and poor growth to occur on those that do sprout.

Most perennials should grow well when dibbling fertilizers. Common groundsel (*Senecio vulgaris*), creeping wood sorrel (*Oxalis corniculata*), and prostrate spurge *(Euphorbia humistrata*) are three weeds that have been shown to reduce their establishment and overall populations with dibbling fertilizers. For many growers, dibbling fertilizers may not be a viable option. Where dibbling is not feasible, incorporating fertilizers reduces weed growth more compared to topdressing them.

Controlling Existing Weeds

It is often necessary to control weeds that have become established in and around the production site. The three most common methods of controlling existing weeds include manual removal, emptying out a greenhouse range, solarizing the weeds that are present, and using postemergent herbicides. Each of these methods only controls the weeds that are present and does not offer any control against weed seeds from reestablishing the weed population. Continuously removing weeds using these methods is both expensive and time consuming.

With perennials, the most common method of controlling weeds is still removing them from containers by hand. On a small scale this practice is acceptable, but as the production scale (size of the operation) increases, removing weeds manually becomes less practical and cost effective. However, hand-pulling weeds should be an important part of any weed management program. Even growers relying on extensive herbicide use should consider hand-weeding the weeds that have escaped the applications or that are not sensitive to the chemicals being applied. Weeds should be pulled before they establish extensive root systems, flower, and go to seed.

Many growers prefer to use nonchemical methods to control weed populations in and around the operation. Growers avoid using herbicides due to safety issues, environmental concerns, or they are worried about causing injury to their crops. Besides manually removing weeds, applying mulches on the surface of the growing medium is a common method of nonchemical weed management.

Mulches are typically applied to containers at, or near, the time of potting to provide a physical barrier to weed emergence for at least one growing season. The thickness of the mulch applied varies with the material being used, but often ranges from 0.5–1 in. (13–25 mm) thick. There are numerous products being used, such as shredded bark, rice hulls, cocoa hulls, pelletized wool sweepings, and pelletized paper. Growers tend to utilize materials that are available regionally.

Depending on the materials being used, the thickness applied, and the uniformity of the application, mulches can provide weed control equal or superior to that of many commercial preemergent herbicides. Mulches are more expensive to apply than herbicide treatments. However, a layer of mulch typically needs to be applied once to provide a whole season of control, whereas preemergent herbicides often need to be applied numerous times.

Herbicide Overview

When determining which herbicides to apply to various weeds, it is necessary to determine the types of weeds (broadleaf, grass, or both) and the stage of plants (seedling versus full-grown plant) to be controlled. Herbicides have either contact or systemic modes of action and are classified as pre- or postemergent. Some herbicides may cause phytotoxicities in many perennial crops. Growers should take great care to read and follow label instructions and trial herbicides on a small scale with each variety before applying them over large crop areas.

All herbicides consist of two key components, an active ingredient and an inert ingredient. The active ingredient is the chemical component of the herbicide that provides the weed control. The inert ingredient portion of the herbicide is a carrier (as with granular herbicides) or a substance that enhances the herbicide's solubility, mixability, or coverage (spray formulations). Sprayable formulations often come as emulsifiable concentrates, dry flowables, water-dispersible granules, or wettable powders and are typically less expensive than granular formulations.

There are only a handful of herbicides labeled for use inside greenhouses, but several are registered for use on containerized ornamentals. There are many herbicide options for woody plants, but the options for application to herbaceous perennials are more limiting. The primary chemicals used to control weeds in perennials are preemergent herbicides such as ammonium sulfate (Corral), isoxaben (Gallery), S-metolachlor (Pennant), napropamide (Devrinol), oryzaline (Surflan), oxadiazon + pendimethalin (Kansel+), pendimethalin (Pendulum), prodiamine (Barricade, Factor), and trifluralin (Treflan). None of these herbicides provide significant control for weeds that have already emerged or established. Growers often apply these products while the perennials are dormant, shortly after potting, or after existing weeds have been removed.

Herbicide selectivity

Herbicides have a property called herbicide selectivity that allows it to kill certain plants while not affecting others. In a general sense, nonselective herbicides control any herbaceous plants they contact, and selective herbicides will control or suppress only certain types of weeds. Both preemergent and postemergent herbicides are often selective. The mechanisms involved in herbicide selectivity include morphological, physiological, and positional. Herbicides are classified into categories such as broadleaf herbicides or grass herbicides largely due to various morphological and physiological selectivities.

There are many morphological characteristics of plants that determine how easily, or quickly, herbicides can get absorbed into them. The smoothness or roughness of the leaf surface, the amount of waxes and/or hairs covering the leaves, and the leaf orientation are all examples of how plant morphology might affect the herbicide's ability to get into a plant.

The specific metabolic plant pathway, or how the herbicide kills the plant (often referred to as the mode of action), is the best way to describe physiological selectivity. If a pathway does not exist for a particular herbicide, it cannot be effective. Some weeds have the ability to compartmentalize, or bind up, herbicides within the plant, causing the herbicide to not have an effect.

Positional selectivity relates to where the herbicide is placed. For example, applying a preemergent herbicide to the soil surface where seeds are germinating will provide greater results than applying them as drenches, which penetrate deep into the root zone, where there are few, if any, weed seeds germinating.

Herbicide tolerance

It is very important to understand not only how herbicides work and what plants they control, but also to have an understanding before any applications are made about what plants or crops can tolerate a particular herbicide application. Each herbicide label includes which ornamental crops the manufacturer has found to have tolerance to the herbicide.

With so many perennials and other ornamentals in commercial production today, it is impossible to test each herbicide on every plant. Keep in mind that not every environmental or cultural situation can be accounted for when testing herbicides for tolerance, and they may perform differently under different conditions.

Before using a new herbicide—or even a familiar herbicide—on a new crop, growers should treat a small number of plants to ensure the plant and the herbicide are compatible under their growing conditions, regardless of whether the plant is listed on the label. Once it is determined that the herbicide did not have any negative effects on the appearance or growth of the plant it was applied to, growers can consider making applications on a larger scale.

Growers can (and should) also test plant tolerances to herbicides on a small scale before applying them to an entire block of plants. One of the best methods is to randomly pull

Figure 7.15. **Conducting trials will help growers recognize and determine the magnitude of injury caused by herbicides. Several perennials, including the *Echinacea* shown here, show sensitivity to the preemergent herbicide flumioxazin (SureGuard).**

ten or more pots away from the main block, clearly mark them, apply the intended herbicide, and return them back to their original locations. Monitor the treated plants throughout the growing season to determine if any injury symptoms have occurred.

Every herbicide has the potential to cause injury to perennial crops. Growers should take the time to learn which perennials a specific herbicide can be applied to. Each herbicide label will provide a listing of the plants to which the herbicide has been applied to safely in experimental tests. These listings are not complete listings of all the plant species the herbicide might be applied to. It would be too costly and time consuming for chemical companies to conduct phytotoxicity trials on every plant species commercially produced. For crops not listed on the label, many herbicide labels allow for growers to conduct plant safety trials for themselves. Always check the herbicide labels and your state laws for this exemption.

Herbicide rotations

Growers who frequently apply herbicides to their production areas should consider rotating the chemical families of the herbicides being applied with each successive application. To prevent herbicide resistance within weed species, herbicides should be rotated as often as possible. When determining a proper rotation for granular products containing multiple active ingredients, at least one of the active ingredients should be replaced with one from a different chemical family.

Constant use of a single herbicide will result in the gradual buildup of a weed's tolerance to that chemical. An example of weed resistance is common groundsel, which has become resistant to dinitroaniline and triazine herbicides where these products have been used exclusively. Resistence is less likely to occur when these herbicides are rotated with other effective chemistries. Constant use of oxadiazon (Ronstar) could result in a gradual resistance in chickweed. Herbicide resistance is not all that common, but it is not impossible. Growers should take measures to ensure they are not creating conditions that over time could lead to a weed species becoming resistant to the herbicides being applied.

The majority of preemergent herbicides used to control weeds in containers contain the active ingredient oxyfluorfen (Goal, OH2, RegalStar, Regal O-O, and Rout). They are becoming more strictly regulated in the United States by the EPA. The restrictions usually include season maximum uses (so many pounds of active ingredient applied per acre per year) and/or a certain amount allowed for each application (so many pounds of active ingredient per acre per application). Rotating herbicides to different chemical classes and modes of action is highly beneficial to reduce the likelihood of weeds developing resistance to these products. The newer preemergent herbicides Broadstar and SureGuard contain the active ingredient flumioxazin and can be a good rotation partner with products containing oxyfluorfen.

Herbicide degradation

Herbicides do not last forever and degrade over time, reducing their ability to control weeds for extended periods. Processes that affect herbicide longevity include chemical degradation, microbial degradation, photodegradation, leaching, and volatilization. Other factors include temperature, light levels, and irrigation practices.

Chemical degradation is a nonbiological reaction that cleaves the active ingredient into inactive secondary molecules. The most common form of chemical degradation is the result of chemical cleavage through a reaction with water.

Microbial degradation occurs as microorganisms use the carbon, hydrogen, nitrogen, and sulfur compounds in the herbicide molecules as a food source.

Photodegradation occurs as ultraviolet light breaks the chemical bonds of the active ingredient, creating secondary molecules, which are less effective at controlling weeds. Photodegradation has the highest likelihood of occurring immediately following application, before the herbicides are bound to the medium components.

Volatilization is the process where the solid form of the herbicide is transformed into a gaseous state. Herbicides that have volatilized no longer provide weed control in the medium and may even cause injury to the foliage of the crop. Trifluralin is a preemergent herbicide that volatilizes easily. Like photodegradation, volatilization is most likely to occur immediately following the application. To reduce the likelihood of volatilization, it is generally recommended to apply irrigation (usually 0.5 in. [13 mm] of water) immediately following the application.

Preemergent Herbicides

Preemergent herbicides control weed seeds as they germinate and up to a few days after germination. They form a chemical barrier (up to 1 in. [25 mm] thick) over the surface of the growing medium after they are applied properly. As weed seeds germinate and grow within the chemical barrier, their growth is either inhibited or the seedlings are killed. Preemergents may have different modes of action, but they all provide control within the chemical barrier. With the chemical barrier containing the herbicides being bound to the top inch of the growing medium, the roots of established perennials are largely unaffected.

The herbicide applied to the soil surface is very concentrated and needs to be incorporated into the top layer of soil or potting mix with irrigation or rainfall following its application. The normal recommendation is to apply approximately 0.5 in. (13 mm) of water shortly after the herbicide application. This is a crucial step in preemergent herbicide applications and often is not administered properly, which decreases the effectiveness and safety of the application.

Some herbicide labels list that it is acceptable to incorporate (water in) herbicides several weeks after they have been applied. Although the herbicide label suggests this is an acceptable practice, it is almost always better to water them in as soon as possible following application. Concentrated herbicide on the soil surface is not as effective at controlling weeds as when a chemical barrier is formed. Leaving the herbicide on the surface also increases the likelihood for it to break down as a result of exposure to ultraviolet light; herbicides properly incorporated offer protection from the sunlight. Watering in preemergent herbicides is important to reduce degradation and volatilization and to increase the overall efficacy of the application.

Growers should water in herbicides using only overhead irrigation or rainfall. Low-volume irrigation sources such as drip irrigation will not properly move the herbicide into the top layer of the medium. It is crucial the herbicide is incorporated uniformly; failure to do so will greatly reduce the level of weed control growers receive.

Growers should consider the crop's tolerance to herbicide, the weed species being controlled, and the herbicide's solubility when selecting which preemergent herbicides to apply. There are a number of preemergent herbicides on the market labeled for application

Figure 7.16. **Improper use and application of herbicides could lead to serious crop injury. This is injury on *Hibiscus moscheutos* due to improper application of oryzalin (Surflan).**

to certain perennials. Unfortunately, they are either safe on a broad variety of perennials and control only a limited number of weed species, or they are safe to only a limited number of perennials and control a large number of weed species. In general, perennials are sensitive to many of the commonly used broad-spectrum preemergent herbicides used to control broadleaf weeds. When considering the application of herbicides to plants not listed on the herbicide label, I recommend checking with the herbicide manufacturer or researchers who have worked with the product to determine if any plant tolerance screens have been conducted since the label was last updated.

Growers have found a strategy to enhance safety to crops with applications of certain foliar-applied preemergent herbicides (isoxaben [Gallery], oryzalin [Surflan], pendimethalin [Pendulum], and prodiamine [Factor]). These herbicides are applied to the crop while the foliage is wet, then irrigated for fifteen to twenty minutes immediately following the application to wash the herbicide off of the leaves of the crop down onto the soil surface.

No herbicide provides control of all weed species. Typically, they are effective at controlling either broadleaf weeds or grasses. There are some herbicides that provide some control of certain broadleaf and grass species. Isoxaben (Gallery), oxyfluorfen (Goal), and simazine (Princep) are effective at controlling emerging broadleaf weeds, but they perform poorly on most grasses. Several preemergent herbicides, including prodiamine (Barricade, Factor, and RegalKade), pendimethalin (Pendulum), and oryzalin (Surflan), control some emerging grasses and many broadleaf weeds. Oxadiazon (Ronstar) controls many species of both broadleaf weeds and grasses.

It is worthwhile to note that young plants are more sensitive to herbicides than are older plants. Plug and liner producers rely more heavily on sanitation and hand weeding, while finished container growers can use preemergent herbicides more widely.

Although each herbicide controls weeds differently, they all essentially control weeds as they emerge through the chemical barrier shortly after germination. If the coverage was inadequate, the chemical barrier will have gaps where weed seeds can germinate successfully. Gaps in the herbicide barrier can be caused by nonuniform herbicide application, containers that have been dropped or blown over, pulling weeds after herbicides have been applied, or holes made in the surface from fingers as workers move or handle the containers.

Some growers apply preemergent herbicides to noncrop areas where existing weeds are present, such as areas between production facilities or outdoor beds. Several products are effective at controlling young weeds that are only a couple inches tall, but in most cases, growers should assume that little or no control of existing weeds will occur. Therefore, it is necessary to either remove the weeds prior to preemergent herbicide application or to mix a postemergent product, such as glyphosate (Roundup Pro), with a preemergent herbicide. If weeds are pulled after a preemergent application has been made, the chemical barrier will be disrupted, leading to reduced effectiveness of the herbicide application.

Many of the weeds often observed in container production can successfully be controlled using one or more of the herbicides listed above. In many cases, it is better to use a combination of preemergent herbicide products, such as one that is especially good at controlling broadleaf weeds and one that is effective against grasses, than using any single product alone.

Herbicide solubility

The solubility of the herbicide is a measure of how readily it goes into solution. Products with high solubility are more likely to go into solution than herbicides with a low solubility. When a herbicide is in solution, it can have a negative effect on plant growth and weed control. Preemergent herbicides tend to have low solubility and tightly bond to the soil particles.

As mentioned earlier, preemergent herbicides applied to the container surface form a chemical barrier. As the weed seeds germinate, the roots and shoots grow through this barrier, where they come into contact with the active ingredient of the herbicide. Generally, perennials grown in containers are irrigated frequently with high volumes of water. When herbicides are applied that have high solubility to containerized crops, they are likely to go into solution and get flushed away from the medium's surface with each irrigation, weakening the chemical barrier and reducing the overall effectiveness and longevity of the herbicide. As the herbicide moves down through the medium, there is also the potential for it to come into contact with the roots of the desired crop, causing injury to the root system, or negatively affecting plant growth.

It is recommended to only use herbicides with a solubility of less than 3 ppm for application to containerized crops. Herbicides with higher solubility can still be applied but should only be used when specific circumstances require it.

It is always recommended to conduct a small trial to ensure plant safety when using herbicides on a new plant species not previously applied at your facility. These trials should be conducted regardless of whether the plant is listed on the label, as there may be conditions specific to your production system that could make the herbicide and plant species incompatible.

Types of preemergent herbicides

Dinitroaniline herbicides, commonly referred to as DNAs, are the most commonly used type of preemergent herbicides. DNA herbicides are highly insoluble and bind tightly to the soil particles, making them difficult to leach. They are absorbed by the root and shoot tips with little to no translocation to other plant parts. Root development is inhibited, and the young seedlings die. Commonly used DNA herbicides include prodiamine (Barricade, Endurance, Factor, RegalKade), pendimethaline (Hurdle, Pendulum), clopyralid (Lontrel), oryzalin (Surflan), trifluralin (Treflan), and benefin + oryzalin (XL2G). DNA herbicides are root-inhibiting chemicals that many herbaceous perennials and ornamental grasses are sensitive to. Most crops are tolerant of DNA herbicides, but some crops have shown sensitivity to them, particularly after potting. Symptoms of sensitivity to DNA herbicides are poor root development, stunting, and lodging. Occasionally, when these herbicides have been applied to young plant materials with poorly developed root systems, growers observe a stunting of the roots. For sensitive crops, use other herbicide types.

Several preemergent herbicides are protox (short for protoporphyrinogen oxidase) inhibitors. These herbicides are held very tightly by the soil, not prone to leaching, and require light to function. The roots, shoots, and stems absorb them. As the seedling passes through the chemical barrier it absorbs the herbicide. Once the seedlings are exposed to light, the reaction with the herbicides form free radicals inside of plant cells, causing mortality to occur. Some of the commercially used protox inhibitors contain the active ingredients oxyflourfen and oxydiazon and include the trade names Goal, OH2, Regal O-O, Ronstar, and Rout.

Other preemergent herbicides are cellulose biosynthesis inhibitors, which, as the name implies, inhibits the biosynthesis of cellulose. They are strongly absorbed by the soil and are not prone to leaching. Cellulose is a component of cell walls that provides strength, support, and stability to the plant. As a result of the herbicide, the cell walls are weakened to the point the plants can no longer grow and develop. Dichlobenil (Casoron), isoxaben (Gallery), and trifluralin + isoxaben (Snapshot) are some of the commercially used cellulose biosynthesis inhibitors.

Formulations

Growers should consider the differences regarding the formulation, granular versus spray, when selecting herbicides. In situations where most of the medium's surface is exposed, herbicides applied as sprays will almost always provide growers with better coverage and a better chemical barrier against weeds. Granular formulations are useful in situations where the crop is actively growing, since the granules can be applied to the medium with minimum risk of burning the foliage of the crop.

Granular formulations are generally safer to apply to containerized perennials than spray formulations are. The degree of safety is achieved because the granules tend to fall to the medium's surface without excessive exposure to the plants leaves. An exception to this is when granular herbicides are applied to a crop that has wet foliage at the time of application. The granules stick to the leaf surfaces and the concentrated active ingredient is released from the granule, causing significant injury to the tender leaves. Granules should only be applied when the foliage is dry, and then the crop should be irrigated to remove the granules from the foliage.

Granular herbicides have the active ingredient formulated onto a dry carrier, or inert ingredient, which makes the application easier. For example, the herbicide Broadstar (flumioxazin) contains 0.25% of the active ingredient flumioxazin and 99.75% inert material. The inert materials used often consist of clay particles, corncobs, granular fertilizers, or a recycled paper product called biodac. They often contain more than one active ingredient and can control a wide number of plant species. For example, Snapshot TG contains one active ingredient to control grasses (trifluralin) and another (isoxaben) to control broadleaf weeds. Compared with spray formulations, granular products are generally more expensive and have additional shipping costs associated with them.

Currently, preemergent herbicides applied as sprays are only available with one active ingredient, not as combination products. However, the spray formulations can often be mixed with one another to form a tank mixture with multiple active ingredients. In most cases, growers observe greater effectiveness when combining two herbicides from different chemical classes than when applying a single active ingredient alone. Most herbicide labels will provide an indication of whether they are compatible with other herbicides, and if so, which ones. With any new combinations they are not familiar with, growers should conduct compatibility tests by mixing the products in a small jar to observe if any reaction occurs or precipitates form.

Spray herbicides usually allow for more uniform applications and create better chemical barriers on the medium's surface than granular products do. However, spray formulations are more prone to cause injury to the existing crop than granular applications are. Some foliar-applied preemergent herbicides have shown some degree of plant safety when applied to certain perennials. Many preemergent herbicides are only safe when they are applied to herbaceous perennials while they are dormant. If preemergent herbicides are applied to the foliage of a perennial crop, wet the foliage prior to spraying the herbicide and rinse off the foliage immediately following the application to reduce the risk of injury to the crop.

With both granular and spray formulations, it is important to water them in with overhead irrigation for fifteen to twenty minutes following the application. This serves two purposes: it washes the herbicide off the foliage and distributes and activates the herbicide in the potting mix (forms the chemical barrier).

Growers should follow each herbicide's labeled recommendations, conduct in-house plant safety screens, and keep good records to determine which crops are tolerant of spray herbicides. Until a grower acquires experience using spray herbicides on container crops, it is generally safer to begin using granular products.

Applying preemergent herbicides

As I have stated earlier, it is very important to apply preemergent herbicides uniformly to create an effective chemical barrier over the container surface. Failure to make uniform applications will often lead to weeds emerging in areas with insufficient amounts of herbicide. Growers should check the calibration of their application equipment and take steps to ensure that the proper rates can be applied. When making the applications, growers should make certain that the rates being applied are accurate. Applying rates that are too low will not provide sufficient amounts of control; conversely, higher than desired rates may lead to crop injury.

Even when growers are using properly calibrated equipment and make great efforts to ensure that the rates delivered are uniform, it is not uncommon for the uniformity of application to vary from half to more than twice the intended rate. One simple method growers can implement to monitor their uniformity is to place collection pans throughout the area being treated. The pans can be used to determine which areas might have inadequate amounts of herbicide applied, leading to a weak chemical barrier, or where excessive quantities may have been applied, possibly leading to crop injury. When using herbicides, it always a good idea to choose rates that will provide sufficient amounts of control when less than desirable amounts are applied yet will be forgiving when slightly higher rates are achieved.

Spray applications of herbicides are generally more uniform than granular applications. In most cases, when the wind is calm, sprays can be applied more accurately, achieving better coverage, than granules applied with a hand-cranked spreader. Applying herbicides by spray offers growers the flexibility to tank mix herbicides—a postemergent with a preemergent or two preemergents together, for example—to achieve differing goals.

It is important to ensure that the spray nozzles are checked and maintained on a regular basis. Spray nozzles and filters routinely get plugged from spray residues. The orifices, or nozzle openings, become worn and are often larger than their original sizes, causing excess chemical to be discharged, increasing the volume of spray being applied. Visually inspect nozzle spray patterns and measure their output by using clear water prior to mixing and applying herbicides to ensure uniform spray output.

To reduce the likelihood of injury to crops, I recommend having sprayers devoted specifically to the application of herbicides and nothing else. These sprayers should be well labeled and only used as intended.

Compared with spray applications, granular herbicides tend to offer growers a higher degree of plant safety but are more difficult to apply uniformly. With typical rotary applicators, trained applicators have a difficult time making uniform applications. Nonuniform applications will almost always result in poor weed control and may result in injury to the crops. Improved uniformity can be achieved using granular herbicides when applied with hand-cranked spreaders by going over the area twice, in opposite directions, to ensure good distribution of granules on the medium's surface.

It is always important to know the square footage of the area being treated and to monitor the output of the calibrated equipment during application. Growers should verify that the proper amount of herbicide was applied following the application. Applying too much or too little herbicide over the intended area can result in poor weed control or injury to the crop.

Timing preemergent applications

It is nearly impossible to determine the best time to apply preemergent herbicides since each nursery has its own geographic location, environment, and cultural practices. Applying herbicides should not be a function of calendar date but determined by certain events such as potting and overwintering. The general practice is to apply preemergents in the spring, just as the overwintering period is coming to an end or soon after potting, and then again prior to overwintering. Some nurseries apply additional applications during the growing season, while others use good sanitation practices and hand weeding between applications.

The most important rule for timing preemergent herbicide applications is to apply the herbicides prior to weed seed germination. Most preemergent herbicides will not control existing weeds at the time of application. A few herbicides, such as flumioxazin (SureGuard) and oxyfluorfen (Goal), will kill many weed species (or the plant parts they contact) that are less than three inches tall. Existing weeds should be removed from the containers prior to applying herbicides. In many cases, even small weed seedlings have a well-developed root systems that have grown beyond the chemical barrier. Failure to remove weeds prior to preemergent applications will result in weeds that continue to grow, flower, and produce seeds, thus perpetuating the weed problem. Applying preemergent herbicides to containers with weeds is a waste of both chemicals and labor.

Preemergent herbicides can be applied to many plant species shortly after potting plugs and liners or shifting plants into larger containers. I recommend irrigating two or three times after potting before applying the herbicide to allow adequate time for the medium to settle. This is often equivalent to 0.5–1 in. (13–25 mm) of rain or precipitation. Some herbicides recommend waiting two to four weeks before applying when bare-root plants have been planted. Remember that preemergent herbicides form a chemical barrier on the medium's surface. If herbicides are applied before the medium has settled, the chemical barrier will shift and not provide the desired level of control. When herbicides are applied to unsettled medium, they could also channel through the medium, making contact with the roots of the containerized plant, potentially injuring or stunting them.

Many growers produce perennials in outdoor settings or in shade houses during the growing season and then cover the houses with plastic or move the plants into an enclosed structure for overwintering. Many weeds, particularly winter annuals such as bittercress, common groundsel, fireweed (*Epilobium angustifolium*), and *Oxalis* thrive during the overwintering period and often require some type of weed control. In many cases, growers like to apply preemergent herbicides prior to overwintering. Since preemergent herbicides are not labeled for applications within enclosed structures, the herbicides should be applied two to four weeks prior to covering. Preemergent herbicides tend to volatilize; if they are applied in an enclosed structure, the risk of injuring plants is greater through the direct exposure to the herbicide vapors or from herbicide condensation on the foliage.

Preemergent herbicides are often applied to containers in early spring as the overwintering protection is removed and before plants begin to flush. Preemergent herbicides cannot be applied inside enclosed structures, such as hoop houses or greenhouses. The plants to be treated either have to be removed from the covered structure or the covering from the structure should be removed prior to herbicide application. The most efficient applications are those that are covering containers that are spaced pot-to-pot. It takes at least three times the herbicide to treat an area where the containers are spaced compared to one where they are pot tight. For example, to treat a crop of 1 gal. (3.8 L) containers spaced 3 in. (8 cm) apart approximately a third of the herbicide will fall into the containers while two-thirds will fall between the containers and onto the ground.

Most preemergent herbicides provide two to three months of control before reapplication is necessary. It is important to reapply them before the level of control subsides, in order to reduce the opportunity of weeds seeds to germinate. The primary factors determining the length a preemergent herbicide will be effective are the chemical used, the rate applied, the uniformity of application, temperature, rainfall/irrigation practices, and herbicide degradation.

Postemergent Herbicides

Postemergent herbicides control weeds at various life stages following germination. Perennial growers typically do not apply postemergent herbicides over their perennial crops; instead, they use them to control weeds in and around the production site. There are two types of postemergent herbicides, contact and systemic.

Contact herbicides kill the parts of the plant the chemical comes into contact with. If a large percentage of the weed is covered with contact herbicides, there is a high probability that it will die. However, they are not very effective at controlling extremely large weeds or weeds that re-grow from underground storage organs.

Systemic herbicides are absorbed by the weed and translocated to other parts of the plant. When enough of the plant is exposed to the herbicide, the weed dies. Systemic herbicides are useful for controlling perennial weeds with underground storage organs. Systemic herbicides tend to be nonselective, controlling a wide range of plant species; therefore, avoid contact with desired plants.

Postemergent herbicides should be applied while weeds are actively growing and before they flower. Several weeds have the ability to develop viable seeds from flowers, even after the rest of the plant has been killed by the herbicide.

Nonselective postemergent herbicides are chemicals that control a broad range of plant species. Growers use nonselective herbicides to clean up fields prior to planting, production sites prior to setting down containers, or to spot-spray weeds that have not been controlled using other methods. Their effects are often very short term or nonresidual, allowing for a crop to be produced in or near the application area shortly after it has been made. Commercial nonselective herbicides include glufosinate (Finale), diquat (Reward), glyphosate (Roundup Pro), and pelargonic acid (Scythe). These products should not be applied over the top of perennial crops, nor should spray drift come into contact with them, since they could become injured or perhaps even killed. Descriptions of the various nonselective herbicides are described in the next section of this chapter.

Selective postemergent herbicides are chemicals that control a limited number of plant species. Most of the herbicides labeled for controlling grasses (clethodim [Envoy], fluazifop-p-butyl [Fusillade, Ornamec], and sethoxydim [Vantage]) are considered selective herbicides. In many cases, selective herbicides can be applied over the top of certain crops. For example, it is acceptable to apply Envoy over the top of actively growing *Heuchera* to control grasses that are growing within them.

Selective weed control products should be applied when the target weeds are actively growing, before they reach maximum size. These products should be applied when the foliage is dry and no irrigation or rainfall is likely to occur for at least an hour, to allow for adequate absorption of the herbicide into the grassy weeds.

To avoid crop injury, apply these products on an area basis, following the labeled recommendations. It is important to achieve thorough, consistent, and uniform coverage. Spray adjuvants may enhance the performance of some of these herbicides. Adjuvants include any substance (surfactants, spreader-stickers, plant penetrants, compatibility agents, pH buffers) added to the spray tank to modify an herbicide's performance, the physical properties of the spray mixture, or both. Always follow the recommendations on the individual chemical labels. Fluazifop-p-butyl (Acclaim, Fusillade II), clethodim (Envoy), and sethoxydim (Vantage) are all selective herbicides that list several perennial varieties on their labels.

During the overwintering period, it is quite common for winter annuals such as hairy bittercress (*Cardamine hirsuta*), Carolina geranium (*Geranium carolinianum*), and horseweed (*Conyza canadensis*) to thrive and for their populations to increase. By the time many of these weeds are observed, it is often too late to apply preemergent herbicides. Most often, growers send crews out to manually weed the winter annuals from the containers, which adds significant labor costs.

In many cases, an option for growers could be to apply postemergent herbicides on top of the containers during the overwintering period. Postemergent herbicides should not be applied to any plants that are not completely dormant, or injury to the crop will occur. Diquat (Reward) and glufosinate (Finale) have shown great results controlling winter annuals without injuring dormant herbaceous perennials. A better strategy for growers would be to tank mix a postemergent herbicide to control existing weeds with a preemergent to provide residual control of germinating seeds.

Since virtually no postemergent herbicide is available for use on perennial crops, or it can only be applied for a limited time (during dormancy), hand weeding is often the only means of postemergent weed control. This is why weed management programs should be preventative in nature.

Postemergent herbicides for weed control in greenhouses

Growers should approach using herbicides in and around greenhouses with extreme care. Herbicide drift (vapors or droplets) through greenhouse vents from outside applications often causes contamination, crop injury, and plant mortality at various operations each year. Occasionally, irrigation sources contaminated with herbicides cause injury to crops inside greenhouses. The most common causes of plant injury inside greenhouses are the improper application of these products or using herbicides not labelled for application inside enclosed structures.

There are only a limited number of herbicides labeled for controlling weeds within greenhouse structures. These herbicides are short lived in the soil and nonvolatile. Great care should be taken when applying herbicides within greenhouses to ensure spray drift does not move over the crops being produced. Growers should apply these herbicides using a course spray at low pressure to avoid the possibility of drift.

Most herbicides have very specific restrictions on their use within enclosed structures, often limited to applications under benches, in walkways, and around the foundation of the greenhouse. Very few products provide labeling for application over crops in containers or crop areas such as ground beds. Deciding to use a herbicide not labeled for use within a greenhouse could prove to be disastrous.

Diquat (Reward)

Diquat (Reward) is a low-cost postemergent contact herbicide used to kill annual weeds under benches, in walkways, and around the foundation of greenhouses. It provides for a rapid kill of young seedling weeds but will only burn, not kill, larger weeds. There is no residual soil activity, and it is not translocated in the plant. Diquat can be applied while a crop is growing within the greenhouse, and if spray drift contacts the crop, only a minimal amount of cosmetic damage is likely to occur. It does not provide good kill to established or perennial weed species. When controlling perennial weeds, it may be necessary to repeat the herbicide applications until control has been achieved. Complete coverage of the target weeds is essential.

Pelargonic acid (Scythe)

Pelargonic acid (Scythe) is another postemergent contact herbicide labeled for use within greenhouse structures. It is similar to diquat in that it burns small weeds on contact. Pelargonic acid will also only burn, not kill, larger weeds. There is no residual soil activity, and it is not translocated in the plant. It is more effective when warm air temperatures (above 80° F [27° C]) are present. This herbicide can be applied when crops are in the greenhouse, but care should be taken to reduce the likelihood of spray drift over the crops being produced. Compared to diquat, pelargonic acid is slightly more expensive, somewhat less effective, and has an offensive odor.

Glyphosate (Roundup Pro)

Glyphosate (Roundup Pro) is a systemic, nonselective, postemergent herbicide that controls both annual and perennial weeds. It is absorbed by green leaf tissues and translocated to the root system of the plant, making it more effective on actively growing weeds. It has very little residual activity, allowing for seeding or transplanting shortly after application. This nonselective herbicide is commonly used to control weeds on greenhouse floors, under benches, and around the perimeter of the facility. It takes five to seven days to see signs that glyphosate applications are effective.

It can be used in greenhouses to control weeds, but great care should be taken to reduce spray drift. Even small amounts of spray drift can cause significant injury to greenhouse crops. The best practice is to only apply glyphosate in empty greenhouses, such as between cropping cycles. Another strategy to reduce spray drift is to shut off the ventilation and circulation fans during and slightly after the herbicide application. If drift occurs within an enclosed structure, it is recommended to wash the benches and sides of the greenhouses within six hours of application to reduce any chance of condensation containing glyphosate from dripping onto the crops below.

Glufosinate-ammonium (Finale)

Glufosinate-ammonium (Finale) is also a systemic, nonselective, postemergent herbicide often used to control weeds within enclosed structures. Glufosinate is translocated down the plant, similar to glyphosate, but not as well, especially in perennial weeds. Since it is not translocated very well, the uniformity of the spray coverage is essential to obtain maximum control. Glufosinate, like glyphosate, has no soil residual activity, allowing for crops to be planted or moved into the production facility shortly after applications have been made.

Growers use glufosinate to control weeds around greenhouse perimeters, under benches, and on floors. Results with Finale can be observed in as little as forty-eight hours. Care should be taken to reduce spray drift and to avoid application to nontarget areas. Some drift can be avoided by using low-pressure equipment with large droplet nozzles.

Controlling weeds outside the greenhouse

Controlling weeds around the production facility will reduce, if not eliminate, a major source of airborne weed seed. Growers often find it beneficial to control weeds in noncrop areas, such as areas between houses, drainage ditches, and roadways.

Weed control around the foundation of the structure will prevent perennial weeds, such as Bermuda grass or bindweed, from growing under it and into the greenhouse. Reducing the weeds around the production site will also help to reduce flying insect pests from entering into the enclosed structure.

Maintain a vegetation-free strip immediately adjacent to the foundation of the greenhouse structure. Weed free noncrop areas are one of the easiest and best methods for reducing weed seeds in container crops. Many growers mow the adjacent areas to keep the weeds from going to seed. Other growers prevent the presence of weeds around the production facility by applying ground mat, gravel, or other inorganic mulches around the production facility.

Some growers apply preemergent and/or postemergent herbicides around the production site to control various weed species. Oryzalin (Surflan) is an example of a preemergent herbicide that is often used for weed control around the production facility. Oryzalin can be applied alone or in combination with various postemergent herbicides such as glufosimate (Finale), glyphosate (Roundup Pro), or diquat (Reward).

Glyphosate applied alone successfully controls weeds outside and around greenhouse or production facilities. Applications of glyphosate do not control the emergence of weed seedlings and would need to be reapplied numerous times a year if no preemergent herbicides are mixed with it. Glyphosate has successfully been mixed with isoxaben (Gallery), oryzalin (Oryzalin, Surflan), simazine (Princep), or flumioxizin (SureGuard).

Herbicides used outside of the production area are most commonly applied early in the spring. Spot treatments are commonly conducted throughout the growing season to control any weeds that have escaped the initial treatment. Many growers apply herbicides as band treatments, along the inside edges of greenhouses or hoop houses, before filling the houses with containerized products.

Growers should not use auxin-type herbicides, such as dicamba + dimethylamine salts (Trimec) and triclopyr + clopyralid (Confront), around the production site. These herbicides are commonly used to control broadleaf weeds in turf. When applied around the greenhouse, they can readily enter the greenhouse through vents and windows. If these types of herbicides are applied, close all vents and windows to prevent spray drift during application (they can be opened shortly after the application is finished). These herbicides volatilize more readily when the temperatures are warm; for the most safety, only apply them during the cool temperatures of spring.

Improving Results with Herbicides

There is not one universal herbicide that will control all weeds under all circumstances. Growers should design comprehensive weed management programs that utilize sanitation, cultural methods, manual weed removal, and herbicides to reduce or eliminate weed populations.

The following are some recommendations for getting the most out of your use of herbicides.

- Identify the weed species to be controlled and verify on the chemical label that it will be effective on these weeds. Applying chemicals that provide little to no weed control is a waste of time and money and could prove to be costly if the weeds get out of hand. Always read the herbicide label to be sure the herbicide will be effective on the target weeds.
- Following the same herbicide program each year could lead to the eradication of the original weed problem while leaving the door open to other weeds not sensitive to the herbicide. A weed scouting program is useful to identify any breakthrough weeds and to determine when a new weed control strategy is necessary.
- Applying herbicides at the right time will greatly improve their effectiveness. Preemergent herbicides work on only one part of the weed's life cycle—the seed germination and/or seedling emergent stage. There are many types of annual weeds (summer and winter annuals) that germinate at various times of the year; herbicide applications should coincide with germination.
- Controlling weeds that are established, whether annual, biennial, or perennial, often requires the use of postemergent systemic herbicides. With most perennials, many postemergent herbicides can only be safely applied while they are in the dormant state.
- Growers can greatly improve their results and reduce injury to crops when they take the time to *read and follow the label* before using any herbicide. Most herbicides are toxic to many of the perennials in commercial production. All chemical labels provide a list of which plant species to which they can be safely applied, and in many cases they also provide a list of which varieties are sensitive to herbicide applications.

Reference Materials

Printed Resources

The Growers Weed Identification Guide. University of California-Davis, 6701 San Pablo Ave., Oakland, CA 94608. 1-800-994-8849. $85.

Weed Control Suggestions for Christmas Trees, Woody Ornamentals, and Flowers. AG-427. North Carolina Cooperative Extension Service, Box 7603, North Carolina State University, Raleigh, NC 27695-7603. $7.50

Weeds of the Northeast. Cornell University Press, P.O. Box 6525, Ithaca, NY 14851-6525. (607) 277-2211. $29.95.

Weeds of Southern Turgrass. Publication Distribution Center, IFAS Building 664, P.O. Box 110011, University of Florida, Gainesville, FL 32611. 1-352-392-1764. $8.

Weeds of the West. University of Wyoming Cooperative Extension Service, Bulletin Room, University of Wyoming, P.O. Box 3313, Laramie, WY 8207-3313. 1-307-766-2115. $24.50.

Web sites

New Jersey Agricultural Weed Gallery, www.rce.rutgers.edu/weeds/index.html
Virginia Tech Weed ID Guide, www.ppws.vt.edu/weedindex.htm
University of Georgia Turfgrass Weed Management, www.griffin.peachnet.edu/cssi/TURF/turf.htm
Weed Science Society of America, www.wssa.net

Extension services

Contact your local Cooperative Extension agent to find the best weed identification guides that are most applicable to your area.

Herbicide labels

Crop Data Management Systems, www.cdms.net
Greenbook, www.greenbook.net

References

Ahrens, John F. "A Weedy Situation." *American Nurseryman*. June 15, 2004.
Atland, James. "Pre-emergent Practices." *NMPro*. January 2004.
———. "Selecting Pre-Emergence Herbicides." *American Nurseryman*. August 15, 2003.
Derr, Jeffrey F. "Comparing Herbicides." *NMPro*. May 2004.
Fernandez, Tom. "Controlling Weeds with Herbicides: General Principles." *The Voice*. November/December 2000.
Mervosh, Todd. L. "Weed Warfare." *American Nurseryman*. August 15, 2001,
Neal, Joseph C. "Managing Weeds in Perennials." *GrowerTalks*. June 2000.
———. "Scout Your Weeds." NMPro. January 2002.
Powell, Chuck C. "Weed Management Basics." *NMPro*. July 2003.
Smith, Tina. "Managing Weeds in Outdoor Cut Flowers." *Greenhouse Product News*. January 2003.

Table 7-1. Preemergent Herbicides Registered for Use on Herbaceous Perennials

Many perennials listed below have numerous species or cultivars that may not have been tested for herbicide tolerance. Always conduct phytotoxicity trials to all perennial varieties before making large-scale applications. Each herbicide listed may contain numerous formulations that may affect the application and plant safety.

X = Registered for some species of this genus. Apply as directed on the label.
* = Postemergent Product

Perennial Genus	Barricade	Broadstar	Corral	Devrinol	Fusilade II *	Gallery	Kansel+	OH2	Pendulum	Pennant	Pre Pair	RegalKade	Ronstar	Rout	Snapshot	Surflan	SureGuard	Treflan, Preen	XL
Achillea	X		X	X	X		X		X	X		X		X	X	X		X	X
Acorus			X						X										
Agapanthus	X		X	X	X		X		X	X	X	X	X			X		X	X
Ageratum				X	X				X	X								X	
Ajuga			X	X	X		X		X	X	X		X			X		X	X
Alcea					X														
Alyssum					X				X	X								X	
Ammophila			X			X			X						X			X	
Anemone									X			X							
Antirrhinum					X				X	X						X		X	X
Aquilegia	X				X				X	X		X							
Arctotheca	X				X				X			X				X		X	X
Arctotis				X														X	
Arenaria					X														
Armeria																		X	
Artemisia	X						X		X	X		X						X	
Arundo			X						X										
Asclepias			X						X	X									
Asparagus			X		X				X							X		X	
Aster	X		X	X			X		X	X		X			X	X		X	
Astilbe		X			X		X		X								X	X	X
Athyrium	X			X								X						X	
Aurinia									X	X								X	
Baptisia							X												
Bergenia							X					X							
Boltonia				X								X							
Buddleia	X	X					X					X					X		
Caladium									X							X			X
Calamagrostis							X												
Calendula					X													X	
Callistephus				X					X							X			
Campanula	X			X	X		X		X	X		X				X		X	X
Canna							X		X	X									
Carex						X			X	X					X			X	
Centaurea					X													X	
Cerastium					X													X	
Ceratostigma												X						X	
Chrysanthemum				X			X		X	X		X	X		X	X		X	X
Convallaria		X															X		
Coreopsis	X	X			X		X		X	X		X	X	X	X	X	X	X	X
Coronilla					X				X									X	
Cortaderia	X		X		X	X		X	X			X			X	X		X	X
Crocosmia									X										
Dahlia				X					X				X					X	
Delosperma				X							X	X			X			X	X
Delphinium	X									X									
Deschampsia			X			X			X						X			X	
Dianthus	X			X	X		X		X	X		X	X	X		X		X	X
Dicentra				X	X				X							X		X	X
Digitalis							X		X							X		X	
Dimorphotheca																		X	X
Doronicum				X			X		X	X			X			X			
Dryopteris				X															
Echinacea	X			X			X		X			X			X	X		X	X
Epimedium							X												
Erianthus															X				

Table 7-1. Preemergent Herbicides Registered for Use on Herbaceous Perennials *(continued)*
X = Registered for some species of this genus. Apply as directed on the label.
* = Postemergent Product

Perennial Genus	Barricade	Broadstar	Corral	Devrinol	Fusilade II *	Gallery	Kansel+	OH2	Pendulum	Pennant	Pre Pair	RegalKade	Ronstar	Rout	Snapshot	Surflan	SureGuard	Treflan, Preen	XL
Erica						X							X		X				
Erysimum				X															
Eschscholzia									X									X	
Eupatorium							X												
Euphorbia					X													X	
Euonymus			X	X	X		X	X	X	X	X	X		X				X	X
Festuca			X			X	X		X						X	X		X	X
Fragaria			X		X				X					X				X	
Freesia									X										
Gaillardia	X			X			X		X	X		X	X		X			X	
Gaura	X						X					X						X	
Gazania			X	X	X		X		X	X			X			X		X	X
Geranium	X			X	X				X			X						X	
Geum									X	X						X		X	X
Gladiolus	X			X	X		X		X	X		X	X			X		X	X
Gypsophila							X		X			X	X	X		X		X	X
Hakonechloa						X									X			X	
Hedera	X	X	X	X	X	X	X		X	X	X	X	X		X	X	X	X	X
Helianthemum	X											X							
Helianthus																		X	
Hemerocallis	X	X	X	X	X	X			X	X		X		X	X	X	X	X	X
Herniaria					X				X									X	
Heuchera							X								X	X		X	
Heucherella	X											X							
Hibiscus				X	X		X		X		X	X		X				X	X
Hosta	X	X	X	X	X	X	X		X	X		X			X	X	X	X	X
Houttuynia	X											X							
Hypericum			X	X		X	X	X	X	X			X	X	X	X		X	X
Iberis					X								X	X		X			
Iris	X				X					X						X		X	X
Jasminium							X	X	X		X			X				X	
Lamiastrum							X												
Lantana				X	X	X			X		X	X	X	X				X	
Lathyrus																		X	
Lavandula	X						X		X			X						X	
Leucanthemum					X				X	X						X		X	X
Liatris					X		X		X				X			X		X	X
Lilium	X	X					X		X	X							X		
Limonium					X				X	X			X			X		X	
Liriope	X		X	X	X	X			X	X		X			X	X		X	X
Lobelia	X		X									X				X		X	
Lupinus									X	X								X	
Lysimachia				X	X				X						X				
Lythrum	X		X						X	X		X							
Miscanthus		X	X			X	X		X			X			X		X	X	
Monarda							X							X				X	
Moraea			X						X	X									
Muehlenbeckia																		X	
Myosotis																		X	
Nepeta							X												
Nephrolepis					X				X										
Oenothera	X				X					X								X	
Ophiopogon	X		X		X	X			X	X		X			X	X		X	X
Opuntia					X														
Ornithogalum									X	X									
Osmunda				X															
Osteospermum	X		X	X	X	X			X		X	X	X			X		X	X
Pachysandra		X	X	X	X	X	X	X	X	X			X		X	X	X	X	
Paeonia	X	X					X		X			X	X				X		
Panicum		X															X		
Papaver																		X	
Pennisetum		X	X		X	X	X		X			X			X	X	X	X	
Penstemon			X	X			X		X									X	
Perovskia	X			X			X		X			X						X	
Phalaris			X	X		X			X	X					X			X	

(continued)

Table 7-1. Preemergent Herbicides Registered for Use on Herbaceous Perennials *(continued)*

X = Registered for some species of this genus. Apply as directed on the label.

* = Postemergent Product

Perennial Genus	Barricade	Broadstar	Corral	Devrinol	Fusilade II *	Gallery	Kansel+	OH2	Pendulum	Pennant	Pre Pair	RegalKade	Ronstar	Rout	Snapshot	Surflan	SureGuard	Treflan, Preen	XL
Phlox							X			X								X	
Physostegia	X								X	X		X							
Platycodon		X							X								X		
Polygonum					X														
Polystichum				X	X													X	
Potentilla			X		X		X	X	X					X				X	
Pulsatilla				X															
Ranunculus																X			X
Rosa	X	X		X	X		X		X	X	X	X	X		X	X	X	X	X
Rosmarinus	X		X		X				X			X			X				X
Rudbeckia	X						X		X			X			X	X		X	X
Rumohra	X				X				X	X									
Salvia					X		X		X					X		X		X	X
Santolina	X				X			X				X							
Saxifraga												X							
Scabiosa	X			X			X							X				X	
Sedum	X			X	X		X		X	X	X	X	X			X		X	X
Sempervivum					X														
Senecio									X	X									
Solanum			X						X										
Stachys										X			X					X	
Stokesia			X						X							X		X	X
Strelitzia				X	X				X							X			X
Tanacetum																		X	
Teucrium												X						X	
Trachelospermum	X						X		X						X			X	X
Tropaeolum																		X	
Verbena			X						X					X		X		X	
Veronica	X			X			X			X		X				X		X	
Vinca	X		X	X	X	X	X	X	X	X		X	X		X	X		X	X
Viola									X	X						X			X
Yucca	X				X				X	X		X				X		X	X
Zantedeschia									X										
Zephyranthes									X										

Table 7.2. Preemergent Herbicides for Controlling Broadleaf Weeds in Container Perennial Production

Herbicides and their use must strictly comply with each product's chemical label. This table is to be used as reference only. Always refer to the label to verify the information and for proper application methods.

X = Effective at or labeled for controlling these weeds.

S = Shows some suppression against these weeds.

Common Name	Scientific Name	Barricade	Broadstar	Corral	Devrinol	Gallery	Kansel+	OH2	Pendulum	Pennant	Pre Pair	RegalKade	Regal OO	RegalStar	Ronstar	Rout	Snapshot	SureGuard	Surflan	Treflan	XL
Ageratum	*Ageratum conyzoides*															X					
Alyssum, hoary	*Berteroa incana*		X															X			
Amaranth, Palmer	*Amaranthus palmeri*		X															X			
Amaranth, spiny	*Amaranthus spinosus*		X				X				X				X			X			
Aster, heath	*Aster ericoides*					X											X				
Aster, slender	*Aster exilis*					X											X				
Beggarweed, Florida	*Desmodium tortuosum*		X															X			
Bindweed, field	*Convolvulus arvensis*					S											S			X	
Bindweed, hedge	*Calystegia sepium*					X											X				

(continued)

Table 7.2. Preemergent Herbicides for Controlling Broadleaf Weeds in Container Perennial Production *(continued)*

X = Effective at or labeled for controlling these weeds.

S = Shows some suppression against these weeds.

Common Name	Scientific Name	Barricade	Broadstar	Corral	Devrinol	Gallery	Kansel +	OH2	Pendulum	Pennant	Pre Pair	RegalKade	Regal OO	RegalStar	Ronstar	Rout	Snapshot	SureGuard	Surflan	Treflan	XL
Bittercress	*Cardamine spp.*					X		X					X		X				X		X
Bittercress, little western	*Cardamine oligosperma*																				X
Bittercress, Pennsylvania	*Cardamine pensylvanica*							X					X								
Bittercress, hairy	*Cardamine hirsuta*	S	X			X	X	X	X	S	X		X		X	X	X	X	X	S	
Bracken fern	*Pteridium aquilinum*												S								
Brassbuttons, southern	*Cotula australis*					X											X				
Bur clover, California	*Medicago hispida*		X			X											X	X			
Bursage, annual	*Ambrosia acanthicarpa*					X											X				
Burweed, lawn	*Soliva pterosperma*					X			X												
Buttonweed	*Borreroa laevis*															X					
Carpetweed	*Mollugo verticillata*	X	X	X	X	X	X	X	X	X	X	X			X	X	X	X	X	X	X
Carrot, wild	*Daucus carota*				S	X											X		S		X
Castor bean	*Ricinus communis*															X					
Catsear, spotted	*Hypochoeris radicata*						X				X				X						
Celery, wild	*Apium leptophyllum*					X											X				
Chamber bitter	*Phyllanthus urinaria*					X											X				
Cheat	*Bromus secalinus*																X				
Cheeseweed	*Malva parviflora*						X				X				X						
Chickweed, common	*Stellaria media*	X	X	X	X	X	X	X	X	X	X	X	X	X		X	X	X	X	X	X
Chickweed, mouse-ear	*Cerastium vulgatum*	X	X		X	X			X			X					X	X	S		
Clover, hop	*Trifolium procumbens*								X												
Clover, white	*Trifolium repens*				S	X			S							X	X		S		
Cocklebur, common	*Xanthium strumarium*				S				S	S										S	
Croton, tropic	*Croton glandulosus*		X															X			
Cudweed, sweet	*Gnaphalium obtusifolium*				X	X		X	X				X				X		S		
Cudweed	*Gnaphalium* spp.			X	X			X	X				X								
Cudweed, purple	*Gnaphalium purpureum*				X	X															
Dandelion	*Taraxacum officinale*		X		S	X		X	S	S			X			X	X	X	X		X
Datura	*Datura* spp.																X				
Deadnettle, purple	*Lamium purpureum*												X								
Dock, curly	*Rumex crispus*					S											S				
Dodder	*Cuscuta* spp.				S												S		S		
Dog fennel	*Eupatorium capillifolium*	S	X			X		X	S						X	X	X	X	S		
Dove weed	*Murdannia nudiflora*		X															X			
Eclipta	*Eclipta prostrata*	S	X			X		X	S						S	X	X	X			
Evening primrose, common	*Oenothera biennis*	S			S	X		X	X	X	X				X	X	X		X		
Evening primrose	*Oenothera* spp.			X					X						S		X				
Evening primrose, cutleaf	*Oenothera laciniata*														S	X					
Fiddleneck	*Amsinckia* spp.			X	X				X						X				X		X
Fiddleneck, common	*Amsinckia intermedia*						X		X						X						
Fiddleneck, coast	*Amsinckia spectabilis*				X	X					X						X				X
Filaree	*Erodium* spp.								X		X										
Filaree, broadleaf	*Erodium botrys*			X					X												
Filaree, redstem	*Erodium cicutarium*		X	X	X	X			X		X				X	X	X	X	X		X
Filaree, whitestem	*Erodium moschatum*			X		X									X		X		X		X
Fireweed	*Epilobium angustifolium*						X	X			X		X		X	X					
Fleabane	*Erigeron strigosus*					X		X									X		X		
Fleabane	*Erigeron* spp.							X					X								
Fleabane, black-leaved	*Conyza bonariensis*					X											X				
Fleabane, dwarf	*Conyza ramosissima*					X															
Galinsoga, hairy	*Galinsoga ciliata*	X	X		X	X				X	X				S		X	X	X		
Garlic, wild	*Allium vineale*												S								
Geranium, Carolina	*Geranium carolinianum*		X		X	X											X	X	X		
Goosefoot	*Chenopodium hybridum*																			X	
Goosefoot, nettleleaf	*Chenopodium murale*					X											X				
Groundcherry, lanceleaf	*Physalis lanceifolia*					X											X				
Groundsel, common	*Senecio vulgaris*	X	X		X	X	X	X		X	X		X		X	X	X	X	X	X	X
Henbit	*Lamium amplexicaule*	X	X	X	X	X			X			X					X	X	X	X	X
Hop clover	*Trifolium agrarium*			X					X												
Horseweed (mare's tail)	*Conyza canadensis*	X	X		X	X		X	S	X			X		X	X	X	X	S		S
Indigo, hairy	*Indigofera hirsuta*		X															X			
Ivy, ground	*Glechoma hederacea*																	X			
Jimsonweed	*Datura stramonium*		X			X		S		S						S	X	X			
Knotweed, prostrate	*Polygonum aviculare*	X		X	X	X	X		X	X	X	X		X	X	X	X		X	X	X
Knotweed, silversheath	*Polygonum argyrocoleon*					X											X				
Kochia	*Kochia scoparia*	X	X	X		X			X			X					X	X		X	
Kyllinga, green	*Keeling brevifolia*																	X			
Ladythumb	*Polygonum persicaria*		X			X											X	X	S		S
Lamb's-quarter, common	*Chenopodium album*	X	X	X	X	X	X		X	X	X	X			X	X	X	X	X	X	X
Lettuce, prickly	*Lactuca serriola*				X	X		X			X						X		S		S
Liverwort	*Marchantia* spp.		X		S	S					X				S			X			
Loosestrife	*Lythrum* spp.					X															
Mallow	*Malva* spp.				X	X					X						X		X	S	
Mallow, common	*Malva neglecta*		X								X				X			X	S		S
Mallow, dwarf	*Malva rotundifolia*					X											X	X			
Mallow, little	*Malva parviflora*		X		X	X					X						X				
Mallow, Venice	*Hibiscus trionum*		X			X											X	X			
Mayweed	*Anthemis cotula*		X			X											X	X			
Medic	*Medicago trunculata*															X					
Medic, black	*Medicago lupulina*					X										X	X				
Milkweed, climbing	*Sarcostemma cynanchoides*																		X		S
Milkweed, honeyvine	*Ampelamus albidus*					S											S				
Morning glory, red/scarlet	*Ipomoea coccinea*																	X			
Morning glory, annual	*Ipomoea* spp.				S	X			X	X					X		X		S	X	S
Morning glory, entireleaf	*Ipomoea hederacea*		X															X			
Morning glory, ivyleaf	*Ipomoea hederacea*		X			X											X	X			

(continued)

Table 7.2. Preemergent Herbicides for Controlling Broadleaf Weeds in Container Perennial Production ***(continued)***

X = Effective at or labeled for controlling these weeds.

S = Shows some suppression against these weeds.

Common Name	Scientific Name	Barricade	Broadstar	Corral	Devrinol	Gallery	Kansel +	OH2	Pendulum	Pennant	Pre Pair	RegalKade	Regal OO	RegalStar	Ronstar	Rout	Snapshot	SureGuard	Surflan	Treflan	XL
Morning glory, smallflower	*Jacquemontia tamnifolia*		X															X			
Morning glory, tall	*Ipomoea purpurea*		X			X											X	X			S
Moss	*Bryum* spp.		X		S				S	X					S	X		X	S		
Moss	*Isopterygium albescens*															X					
Mugwort	*Artemisia vulgaris*	S			S	S													S		
Mullein, common	*Verbascum thapsus*																X				
Mullein, turkey	*Eremocarpus setigerus*					X											X				
Mustard, black	*Brassica nigra*					X											X		S		S
Mustard, Indian	*Brassica juncea*					X											X				
Mustard, tumble	*Sisymbrium altissimum*																	X			
Mustard, wild	*Sinapis arvensis*				X	X				S					X	X	X	X	S		S
Nettle, burning	*Urtica urens*					X											X				
Nettle, stinging	*Urtica dioica*					X											X			X	
Nightshade, black	*Solanum nigrum*		X		S	X			S	X							X	X	S	X	S
Nightshade, eastern black	*Solanum ptycanthum*		X															X			
Nightshade, hairy	*Solanum sarrachiodes*									X								X			
Niruri	*Phyllanthus niruri*						X				X				X						
Oxalis	*Oxalis stricta*	X		X			X	X	X				X			X					
Oxtongue, bristly	*Picris echioides*					X	X				X				X		X				
Parsley piert	*Alchemilla arvensis*		X															X			
Pearlwort, birdseye	*Sagina procumbens*	X	X	X	S	X		X	X	X			X			X		X			
Pennycress, field	*Thlaspi arvense*		X															X			
Pennywort	*Hydrocotyle* spp.					X											X				
Pepperweed	*Lepidium* spp.				X	X		X	X				X		X		X		X		X
Pepperweed, Virginia	*Lepidium virginicum*					X		X					X				X				
Pepperweed, yellowflower	*Lepidium perfoliatum*	X																			
Phyllanthus, long-stalked	*Phyllanthus tenellus*		X			X							X		X		X	X			
Pigweed, prostrate	*Amaranthus blitoides*	X	X						X									X	X	X	X
Pigweed, redroot	*Amaranthus retroflexus*	X	X		X		X	X	X		X		X		X	X		X	X	X	X
Pigweed, smooth	*Amaranthus hybridus*	X	X						X									X	X	X	X
Pigweed	*Amaranthus* spp.	X		X		X				X		X					X			X	X
Pigweed, spiny	*Amaranthus spinosus*	X							X					X						X	X
Pigweed, tumble	*Amaranthus albus*	X	X															X	X	X	X
Pimpernel, scarlet	*Anagallis arvensis*					X									X		X				
Pineapple weed	*Matricaria matricarioides*		X		X	X			S	X	X				X		X	X	X		
Plantain, bracted	*Plantago aristata*					X											X				
Plantain, slender	*Plantago elongata*					X											X				
Plantain	*Plantago* spp.				S	X				X							X		X		
Plantain, broadleaf	*Plantago major*		X			X											X	X			
Plantain, buckhorn	*Plantago lanceolata*		X			X											X	X			
Poinsettia, wild	*Euphorbia heterophylla*																	X			
Pokeweed, common	*Phytolacca americana*	X				X			X	S					S	X	X		X		
Puncturevine	*Tribulus terestris*		X	X					X									X	X	X	X
Purslane, common	*Portulaca oleracea*	X	X	X	X	X	X	X	X	X	X	X			X	X	X	X	X	X	X
Purslane, horse	*Trianthema portulacastrum*				X						X										
Pusley, Florida	*Richardia scabra*	X	X	X	S	X			X	X	X	X					X	X	X	X	X
Radish, wild	*Raphanus raphanistrum*					X											X				
Ragweed, common	*Ambrosia artemisiifolia*	X	X		X	X			X	X	X				X	X	X	X	S		S
Ragweed, giant	*Ambrosia trifida*		X															X			
Ragwort, golden	*Senecio aureus*										X				X						
Redmaids	*Calandrinia ciliata*		X			X											X	X			
Redweed	*Melochia corchorifolia*																	X			
Rocket, London	*Sisymbrium irio*			X		X			X								X		X		X
Rocket, yellow	*Barbarea vulgaris*		X															X			
Rock purslane	*Calandrinia grandiflora*					X											X		X		X
Senna, coffee	*Cassia occidentalis*		X															X			
Sesbania, hemp	*Sesbania exaltata*		X															X			
Shepherd's purse	*Capsella bursa-pastoris*	X	X	X	X	X	X	X	X		X	X	X		X	X	X	X	X		X
Sibara	*Sibara virginica*					X											X				
Sida, prickly	*Sida spinosa*		X		S	X			X	S					X		X	X	S		S
Smartweed, Pennsylvania	*Polygonum pensylvanicum*		X	X	X	X			X	X	X				X		X	X	S	X	S
Sorrel, red	*Rumex acetosella*				S	X											X				
Sowthistle	*Sonchus arvensis*							X					X			X					
Sowthistle, annual	*Sonchus oleraceus*		X		X	X	X	X	X		X				X	X	X	X	S		S
Sowthistle, spiny	*Sonchus asper*					X											X				
Speedwell	*Veronica* spp.				S	X	X		X		X				X		X				
Speedwell, corn	*Veronica arvensis*								X												
Speedwell, Persian	*Veronica persica*	X										X									
Speedwell, purslane	*Veronica peregrina*					X											X				
Speedwell, thymeleaf	*Veronica serpyllifolia*					X											X				
Spurge, petty	*Euphorbia peplus*					X					X				X		X				
Spurge, prostrate	*Euphorbia humistrata*	X	X	X	S	X	X	X	X	X	X	X	X	X	S	X	X	X	X		X
Spurge, annual	*Euphorbia* spp.								X												
Spurge, garden	*Euphorbia hirta*							X			X		X		X	X					
Spurge, hairy	*Euphorbia vermiculata*															X					
Spurge, hyssop	*Euphorbia hyssopifolia*					X											X				
Spurge, roundleaf	*Euphorbia cordifolia*						X														
Spurge, spotted	*Euphorbia maculata*	X	X		S	X		X	X		X		X	X		X	X	X	X		S
Spurry, corn	*Spergula arvensis*				S																
Spurry, sand	*Spergula rubra*	X				X													X		
Starbur, bristly	*Acanthospermum hispidum*		X															X			
Sunflower	*Helianthus* spp.					X											X				
Sweet clover, yellow	*Melilotus officinalis*					X											X				
Swinecress	*Coronopus didymus*					X					X				S		X				
Tansy mustard, green	*Descurainia pinnata*					X											X				
Telegraph plant	*Heterotheca grandiflora*					X											X				

(continued)

Table 7.2. Preemergent Herbicides for Controlling Broadleaf Weeds in Container Perennial Production *(continued)*

X = Effective at or labeled for controlling these weeds.

S = Shows some suppression against these weeds.

Common Name	Scientific Name	Barricade	Broadstar	Corral	Devrinol	Gallery	Kansel +	OH2	Pendulum	Pennant	Pre Pair	RegalKade	Regal OO	RegalStar	Ronstar	Rout	Snapshot	SureGuard	Surflan	Treflan	XL
Thistle, bull	*Cirsium vulgare*					X															
Thistle, Canada	*Cirsium arvense*		X		S													X		X	
Thistle, musk	*Carduus nutans*					X											X				
Thistle, Russian	*Salsola iberica*		X			X											X	X		X	
Velvetleaf	*Abutilon theophrasti*	X		X	S	X			X	X					X		X	X	S		S
Waterhemp, common	*Amaranthus rudis*		X															X			
Waterhemp, tall	*Amaranthus tuberculatus*		X															X			
Willow weed	*Epilobium paniculatum*					X											X				
Wood sorrel, creeping	*Oxalis corniculata*					X											X				
Wood sorrel, Yellow	*Oxalis stricta*	X		X	S	X	X	X	X	S	X	X	X		X	X	X	X	X		X

8

Controlling Plant Height of Perennial Crops

There is a great array of genetic diversity among perennials; literally hundreds of genera, consisting of thousands of varieties are being grown today. Growing such a wide range of genetics doesn't lend itself well to commercial production, where as little as one growing environment may exist. Many perennials are grown in greenhouses with low light levels, warm temperatures, high humidity, and high plant densities, all conditions favorable to promoting stem elongation. With increasing production, more competition, and more emphasis on plant quality, many perennial growers have found it necessary to implement methods of controlling plant height and improving plant quality.

Today's perennial producers are faced with stricter crop specifications from their customers, especially if they are suppliers for garden centers or mass merchandisers. Most crop specifications directly apply to the plant's appearance; plant shape, height, and the total number of flowers are all common specifications. Another challenge many growers face is increasing competition from a number of established greenhouse producers. With increasing competition and an emphasis on plant quality from our customers, we are all expected to deliver the best products we can. The companies that meet these requirements will most likely succeed and pick up market share from companies that cannot.

The height of a plant probably has the most effect on the perceived quality of our products. Each plant will have its own height specifications, or the height at which it looks the most balanced and aesthetically pleasing and is perceived to be the most salable. Besides these quality-related attributes, growers who "grow within the specs" often will experience other benefits such as less shrinkage, increased uniformity, a longer shelf life both in the greenhouse and at the retail site, and the ability to ship more plants per load. To control plant height, it is necessary to "individualize" perennials into specific crops, each with its own best strategy.

Figure 8.1. **PGRs can dramatically improve the look of many perennials. *Delphinium* 'Connecticut Yankee' control on left, one drench application 1ppm on right.**

Non-chemical Methods for Controlling Perennial Height

There are several biological and physical methods growers can use to control the height of their perennial crops. Perennial growers should consider using these methods before using chemical control strategies or in conjunction with using plant growth regulators (PGRs).

Crop spacing

Crop spacing greatly affects the final height of a plant and is one of the most important tools growers can use. Research has shown that most plants will generally grow taller at high-density spacing, as plants must compete with each other for light. Plants at the proper spacing tend to elongate slowly and are well branched, creating a high-quality product. Although providing a wide spacing doesn't always allow growers to maximize productivity and profitability, crop spacing is the best method growers can use to control plant height.

Manipulating photoperiod and light quality

With many perennials, exposure to long-day conditions promotes internode elongation and increases plant height. Many growers force perennials to bloom by specific dates and provide long days in order to achieve flowering. Unfortunately, in many instances the same conditions necessary for flowering (long days) also promote plants to grow taller.

Limiting the exposure of perennials to these photoperiods can be helpful in producing flowering plants that are shorter than they would be if they were produced continuously under long-day conditions. This technique is referred to as limited induction photoperiod. Limited induction photoperiod entails exposing plants to long-day conditions for a period long enough for flower induction (usually two to three weeks), then moving them back to natural day lengths in the winter or under short-day conditions using black cloth when the photoperiods are naturally long (providing nine to ten hours of light). It is usually best to maintain long days until the first flower buds become visible, as the time to flower is slightly increased and the overall flower number is decreased when using limited induction photoperiods. With many perennials, such as *Coreopsis* x *grandiflora* and *Sedum* 'Autumn Joy', this treatment is very effective at reducing plant height while the flowers continue to develop.

When forcing *Echinacea purpurea,* researchers at Michigan State University have found that providing a short night break interval of only fifteen to twenty minutes provides sufficient conditions to induce flowering and develop plants that are about half the size of plants that are forced under the traditional four-hour night interruption method. Currently, there are no additional known long-day perennial varieties that will flower under such short night interruptions. This technique currently may have limited usefulness, but it provides an indication that current production practices may become modified when researchers better understand the mechanisms behind the flowering process, possibly allowing growers the ability to produce shorter perennials in bloom.

Many perennials do not have a juvenility requirement and will bloom readily once the proper photoperiod is provided. Generally, after the correct photoperiod is provided, most perennials still have to produce a certain number of leaves before the flower becomes visible or before the time of bloom. Growers can effectively reduce the final height of certain flowering perennials by simply manipulating the photoperiod while they are relatively young (the plug stage), which initiates flowering earlier than normal and reduces the number of nodes and the overall height of the plant. This method will not work on all perennials, as some varieties would finish too small. Therefore, many perennials must be bulked up before they are exposed to the flower inducing photoperiod. Refer to chapter 10, Forcing Perennials, to determine which varieties should be bulked up prior to manipulating the photoperiod.

The type of lighting used to create long days can affect the height of various perennial crops. Incandescent bulbs, although inexpensive and easy to install, provide a large amount of far-red light. Far-red light is part of the light spectrum that promotes stem elongation. When light sources are compared, the perennials under incandescent lights will often be taller than the same plant produced using a different light source. High-pressure sodium, metal halide, and cool-white fluorescent are the preferred sources of lighting for forcing perennials.

Increasing the quality, or, more specifically, the intensity of light perennials are exposed to can be a tool in reducing plant height. Environments with high light quality tend to produce plants that are naturally shorter. Tight crop spacing, growing plants under hanging baskets, and cloudy weather conditions all create low light levels, causing plants to stretch and become leggy. These conditions should be avoided when possible.

Withholding irrigation

Several perennial growers have withheld irrigation from certain perennials in an attempt to control plant height. Bedding plant growers commonly practice this concept and generally have good success in doing so. Withholding irrigation is most effective during the early spring when the weather conditions are often dark and cloudy. Promoting water stress entails allowing the plants to wilt slightly between waterings but not allowing them to reach the permanent wilting point. During the late spring and summer months, withholding moisture is less effective and creates more risk for growers, as the plants use additional water and dries out more rapidly. The risks of withholding moisture from the root zone include severe damage to the foliage from water stress, delayed flowering, reduction in plant quality, and under extreme conditions even plant death.

Managing fertility

Growers have looked at limiting the nutrients, typically nitrogen and phosphorus, to reduce stem elongation. However, reducing the nitrogen levels does not reduce plant height. Perennials under low nitrogen regimes typically have thin, low-quality growth and usually grow to the same height or are taller than perennials grown at higher fertilizer levels. Withholding phosphorus has also been looked at to reduce the plant height of perennials. This too has been shown to not influence plant height. Limiting fertility levels has little, if any, affect on plant height. Growers run the risk of producing low-quality plants and may experience deficiency symptoms if they withhold too much nutrition. Growers should not provide luxury feed or maintain nutrition levels above and beyond the needs of the plant. Controlling excessive feed levels will have a greater impact on reducing plant height than will withholding nutrients altogether.

Plant genetics

Many perennial species consist of numerous cultivars with differing growth habits. Growers can often choose cultivars that are naturally shorter and less prone to stretching when produced at high plant densities. In many cases, selecting naturally short-growing cultivars greatly reduces or eliminates the need for controlling plant height. Shorter cultivars may not be available for the species you want to grow, which makes cultivar selection a limited height control strategy.

Temperature

Most growers are under the assumption that warmer production temperatures cause perennials to grow taller. Warm temperatures may hasten plant development, causing them to flower earlier and appear taller at the time of flower as compared with the same plant grown at cooler temperatures that has not reached the flowering stage. But the truth is that the final height of a perennial at flower is not increased by warmer production temperatures. Notice the key phrase is *at flower.* They will reach flowering sooner at warm temperatures, causing them to appear taller. But when the plant height at bloom is compared to perennials grown with cooler production temperatures, the warm-temperature plants are usually the same height or shorter than the same plant grown under the cooler conditions.

Many growers use the principle of DIF to control plant height of perennials. DIF is the relationship between day and night temperatures (day temperature – night temperature = DIF). Stem elongation decreases under negative DIF conditions or when the day temperatures are cooler than the night temperatures. Stem elongation increases under the opposite conditions (positive DIF), when the day temperatures are warmer than the night temperatures. DIF is difficult to practice during times of the year when the day temperatures are naturally warmer than the night temperatures. But this is a valuable and effective method to know and implement when the conditions permit.

One method many growers are using is the DIP or DROP method. This variation of DIF provides growers with similar results to using negative DIF conditions. With DIP, the greenhouse temperatures are dropped two hours before sunrise until two hours after sunrise. Most cell elongation occurs at or near sunrise, and providing cooler temperatures during this time will limit how much elongation occurs. An example of a DIP program a perennial grower might follow would be to provide a 68° F (20° C) night temperature, a four- hour drop to 60° F (16° C) beginning two hours before dawn, and a 65° F (18° C) day venting temperature setting.

Root restriction

The size of the container can be used to control the plant height. Generally, plants grown in smaller containers have a restricted root zone and develop less top growth than plants grown in larger sized pots do. Growers can create conditions to cause a restricted root zone by selecting smaller container sizes, planting more plants per pot, and providing optimal environmental conditions (ample light, nutrition, etc.).

Scheduling

Crop timing is a valuable method growers can use to manage plant height. Growers are fairly limited with the number of days or weeks it takes to produce a flowering perennial. However, there is often the need to ship the same variety over an extended period of time. Some growers use this need to their advantage in regard to controlling plant height. For example, instead of producing all of the *Leucanthemum* 'Silver Princess' needed over a two month period at the same time, some growers stagger the finish times by two- or three-week intervals. To stagger the finish date they can either stagger the planting date of the finished container or they can alter the dates the long-day conditions are provided, which triggers the flowering process. Additionally, growers can provide different environments, namely differing temperature settings, if they are producing their entire crop simultaneously. For example, they can produce some at normal temperatures for forcing, others at perhaps 5° F (3° C) less than the normal temperatures, and possibly a third group grown with temperatures at 10° (6° C) less than the ideal settings. Using methods such as these, growers can ensure that a fresh supply of flowering perennials will always be available without the necessity to hold full-grown, flowering plants under conditions that promote undesirable plant stretch and often a reduction of quality.

Mechanical conditioning

There has been some research on vegetables and herbs on the effects of brushing plants as a method of reducing plant height. Brushing entails moving a PVC pipe or wooden dowel over the top third of the plant dozens of times. The foliage should be dry and disease free; otherwise, this method will damage the young, tender leaves and possibly spread disease organisms from plant to plant. Growers with rolling benches have created a method of shaking the plants by rolling (slamming) the benches against each other aggressively, causing the plants to shake. Attaching an agitator (shaker) to the bench and running it for a short duration of time will also produce some results. Plants exposed to the wind or whose leaves move frequently from the air currents made by horizontal airflow fans are typically shorter than the same plants not exposed to these conditions. Mechanical conditioning has limited potential for controlling plant height because it is often difficult and not economical for growers to provide these treatments on a daily basis.

Plant Growth Regulators

One of the best and most underutilized (perhaps misunderstood) tools to control plant height is the use of chemical plant growth regulators (PGRs). Controlling plant height is the primary purpose of using these chemicals, and frequently they are needed in order to maintain a high-quality shippable perennial of desired shape and size.

Using growth regulators is both an art and a science. Because of their perceived difficulty of use, many growers are not comfortable using PGRs. Many growers have either had bad experiences with these chemicals (overcontrol) or have not achieved the results they were looking for (undercontrol). Don't throw in the towel—PGRs are only intimidating if you allow them to be.

Many factors come into play when controlling plant height, making the use of growth regulators more of an art than an exact science. Every PGR application a grower makes yields differing results, even if the application rate never changes. Rest assured that we all have something to learn when implementing these into our production programs. We all can be artists if we understand what it is we are trying to create and what we are using to create it. Follow the ground rules and adjust these guidelines to fit your crops and growing conditions. When all the factors are taken into consideration, the results can be duplicated again and again.

The height of every plant consists of two factors: the number of nodes (leaves) on the stem and the length of stem between these nodes (internode length). The internode length is primarily a function of temperature (more specifically, the difference between the day and night temperatures) as well as other factors, including light quality, light intensity, and humidity. These environmental factors affect the gibberellin biosynthesis within the plant. Gibberellins are plant hormones that stimulate or promote cell elongation. The natural production of gibberellins causes plant cells to become larger, adding to the length of the internodes and ultimately the plant's height. Growth regulators reduce plant height by inhibiting gibberellin biosynthesis, causing the plant to make smaller, more compact plant cells. Each PGR inhibits gibberellin production in a slightly different manner, which allows for greater results when two PGRs are combined, commonly referred to as synergy.

Figure 8.2. **The benefits of PGRs include more compact growth, sturdier stems, and improved leaf coloration, as shown here on *Polemonium* 'Heavenly Blue'. The control plant is on the left, and the right has received three applications of a tank mix of 1,250 ppm B-Nine and 3.75 ppm Sumagic applied at seven-day intervals.**

Besides reducing plant height, PGRs provide many additional benefits to growers who use them. When using PGRs, growers often observe plants have thicker stems that can better tolerate the rigors of shipping and handling. PGRs improve the aesthetic appearance by controlling both plant height and width and, in most cases, improve the color of the leaves, creating a more desirable, healthier looking product. When PGRs are applied, growers often observe increased stress tolerance to dry conditions and to plant diseases. Another benefit is an increased shelf life (up to three weeks) at both the retail and production sites, allowing quality to be maintained during periods of slow sales or adverse weather conditions. When used properly, growth regulators can make a big difference on crop production, plant quality, and your bottom line.

Here are some factors that may influence the use of plant growth regulators on perennial crops.

The crops being grown

Many perennials are naturally tall and may require some form of height control; however, there are often cultivars of various perennial crops that have less aggressive growth habits and can naturally produce shorter finished plants. Some varieties may not be responsive to plant growth regulators and may need a different height management strategy, or conversely may only seemingly respond to chemical PGR applications.

The growing environment

Many perennials are not adaptable to the growing environment many commercial growers provide. For example, many plants tend to stretch when growing inside facilities with a lot of overhead infrastructure (gutters, trusses, bows, and plastic). There are also seasonal differences such as temperature, light levels, and day length that effect how a crop grows. Growers generally observe greater height control from PGRs at cooler production temperatures.

Geographic location

The location of the perennial operation has a great effect on the overall quality of many perennial crops. Factors such as temperature and light have a great impact on crop growth, vigor, and quality growers can achieve. These variables also contribute to the overall effectiveness of various height management strategies, including PGR applications.

Crop density

Most growers try to maximize their production space to maintain a certain level of sales per square foot. In doing so, growers are often faced with high plant densities, which provide optimum conditions for plants to stretch, potentially decreasing crop quality. PGRs may only combat some of this stretch and not be as effective as with lower crop densities. From a quality standpoint, there is often a fine line between realistic crop densities and overcrowding plants. In many cases, the plant density is often the aftermath of a grower's "greed" or the inability to say no to a future sale or opportunity.

Cropping systems

The more perennial varieties an operation needs to grow within a greenhouse, the more difficult it becomes to manage plant height. The more a production environment can become a limited or monocropping system, the easier it is to design height management strategies specifically for that crop. For example, growers can manipulate the environment (providing DIF) and crop culture (controlling irrigation and nutrition) to the particular crop without affecting a wide variety of perennials. When numerous varieties are grown within an environment simultaneously, one of these strategies might provide adequate results with some varieties, have an adverse effect on others, and provide no results to yet others.

Perennial market

The customer has a lot to do with the general level of quality growers must achieve. This is an unfortunate reality; after all, we as an industry should always aim at producing perennials of the highest quality, regardless of the customer. There currently is, and perhaps always will be, a difference in who gets what quality standard. For example, a mass merchandiser may not be willing to pay a premium or fair price for any given perennial, forcing growers to make up for the price difference on volume or by growing plants at higher crop densities, as mentioned above. Unfortunately, the markets served often limit a grower's ability to produce plants of the highest quality and maintain profitability.

Experience level

The amount of experience a growers have using PGRs has a great effect on the response they get from applying them. Without a comfort or confidence level, it is difficult to apply them and achieve consistent results. Growers with PGR experience are more apt to try new application rates and/or techniques when inadequate levels of control have historically been achieved.

The PGRs of Today

Near the time of printing, some new trade names for existing active ingredients came on the market, with more to follow in the coming years. Where there was only one trade formulation available, PGRs are listed under the trade name. Where there were more than one trade product available, PGRs are listed by the active ingredient.

A-Rest

A-Rest (ancymidol) reduces internode elongation as the result of inhibition of gibberellin biosynthesis. When used properly, A-Rest produces no phytotoxic effects (damage to plant tissue). A-rest is mostly applied using sprays or drenches, where it is absorbed by the leaves or roots. It has also been effectively used as a media spray and bulb dip. The label allows for a broad range of chemigation applications (mixing A-Rest with irrigation water), including subirrigation. This chemical is effective on a wide range of plant varieties; its growth regulating activity is greater than daminozide products and Cycocel, but less than paclobutrazol products and Sumagic. Growers commonly use rates ranging from 2–130 ppm. A-Rest is most widely used for plug production and high-value crops, as it is relatively expensive.

Table 8.1. A-Rest (Ancymidol) Spray Dilutions

Desired ppm*	Fluid Ounces per Gallon	Milliliters (cc) per Gallon
1	0.48	14.34
3	1.45	43.02
5	2.42	71.69
10	4.85	143.39
15	7.27	215.08
20	9.70	286.77
25	12.12	358.47
30	14.55	430.16
40	19.39	573.55
50	24.24	716.94
60	29.09	860.32
75	36.36	1,075.40
100	48.49	1,433.87

*Approximate

Cycocel

Cycocel (chlormequat chloride) reduces internode elongation within plants and is a fairly active plant growth regulator but has less persistence and activity than A-Rest, paclobutrazol, and Sumagic. Phytotoxicity from Cycocel shows itself as yellowing (the halo effect) to the actively expanding leaves; little or

Table 8.2. A-Rest (Ancymidol) Drench Dilutions

Drench Solution								
Desired ppm*	Fluid Ounces per Gallon	Milliliters (cc) per Gallon	Drench Volume** (Fluid Ounces) per 4 In. Pot	Milligrams Active Ingredient (a.i.)** per 4 In. Pot	Drench Volume** (Fluid Ounces) per 6 In. Pot	Milligrams Active Ingredient (a.i.)** per 6 In. Pot	Drench Volume** (Fluid Ounces) per 8 In. Pot	Milligrams Active Ingredient (a.i.)** per 8 In. Pot
1	0.48	14.34	2.0	0.059	4.0	0.118	10.0	0.295
2	0.97	28.68	2.0	0.118	4.0	0.236	10.0	0.59
4	1.94	57.35	2.0	0.236	4.0	0.472	10.0	1.18
8	3.88	114.71	2.0	0.472	4.0	0.943	10.0	2.36
12	5.82	172.06	2.0	0.708	4.0	1.415	10.0	3.54
16	7.76	229.42	2.0	0.944	4.0	1.886	10.0	4.72

*Approximate **Labeled rates

Figure 8.3. **Many perennials, including *Dianthus barbatus* (pictured), exhibit phytotoxicity from applications of Cycocel.**

no injury occurs to fully developed leaves. The rates used by growers range from 500–2,500 ppm, although phytotoxicity is commonly observed when rates over 1,500 ppm are applied. This plant growth regulator is effective on a limited number of plant species. Cycocel is labeled for both spray and drench applications. Drench applications are effective but costly—up to twenty times the cost of using other more effective products. Like daminozide, Cycocel is absorbed through the leaves, and it is important to let the sprays dry slowly and to not irrigate overhead for at least six hours after applying. Tank mixing Cycocel with daminozide has been an effective strategy to reduce the phytotoxic effects of the Cycocel and gain more height control than by using either chemical alone.

Table 8.3. Cycocel (Chlormequat Chloride) Spray Dilutions

Desired ppm*	Fluid Ounces per Gallon	Milliliters (cc) per Gallon
250	0.27	8.02
500	0.54	16.04
750	0.81	24.06
1,000	1.08	32.08
1,250	1.36	40.10
1,500	1.63	48.12
2,000	2.17	64.16
2,500	2.71	80.20
3,000	3.25	96.24

*Approximate

Daminozide (B-Nine, Dazide)

Daminozide is the most widely used plant growth regulator in floriculture and controls the height of a wide range of perennial plant species. Daminozide reduces internode length by blocking gibberellin biosynthesis. This chemical is primarily used as a spray and recently has been used as a preplant dip of cuttings. A wide variety of plants are responsive to daminozide, but its activity is less than most other growth regulators. Daminozide is a short-term growth regulator, which provides moderate height control and often requires multiple applications for extended periods of height reduction. It is only absorbed through the leaves of a plant. The plant leaves should be covered with spray solution, often ranging from 500—5,000 ppm, just prior to the point of runoff.

The active ingredient is very soluble in water, which causes it to be absorbed very slowly into the waxy surface of plant leaves. The absorption into the leaf can only occur while the spray solution is still wet on the leaf surface. Once the leaf surface dries, little additional PGR will get into the plant, so this product works best when the spray is allowed to remain on the plant and dry slowly. Apply daminozide either early in the morning or in late evening to allow the leaves to remain wet for a long time (four hours is optimum), ensuring maximum absorption of the active ingredient. Do not irrigate overhead for twenty-four hours after application, or the activity of daminozide will be reduced.

Daminozide generally does not produce any phytotoxic effects; however, it is best to avoid applying during periods of bright sunshine. Growers accustomed to using copper-based fungicides should be aware that the application of daminozide to plants seven days before or after the application of a these fungicides will most likely cause some phytotoxicity.

Table 8.4. Daminozide (B-Nine, Dazide) Spray Solutions

Desired ppm*	Ounces per Gallon	Grams per Gallon
500	0.08	2.23
1,000	0.16	4.46
1,500	0.24	6.69
2,000	0.32	8.92
2,500	0.40	11.15
3,000	0.48	13.38
3,750	0.60	16.73
5,000	0.80	22.30

*Approximate

Florel

Florel (ethephon) is quite effective as a growth regulator for a limited number of perennials. However, it is most often utilized for its ability to stimulate lateral branching and delay of flowering. Unlike most commercially used plant growth regulators, which inhibit gibberellin biosynthesis, Florel controls growth by releasing the gaseous plant hormone ethylene, which reduces cell elongation. Florel is absorbed by the leaves and is only effective as a foliar spray. Like daminozide and Cycocel, this product performs best when the spray solution remains wet on the leaves for a long period of time. Perennial growers commonly use Florel at rates of 250–1,000 ppm. Growers forcing perennials for sale in bloom should not use Florel as a height control tool because it prevents flower initiation and delays flower development. With some perennials, some chlorosis and foliar necrosis may occur. If phytotoxicity is observed, try reducing the rate being applied and altering the frequency of the applications.

Table 8.5. Florel (Ethephon) Spray Dilutions

Desired ppm*	Fluid Ounces per Gallon	Milliliters (cc) per Gallon
100	0.32	9.57
250	0.81	23.93
500	1.62	47.87
750	2.43	71.80
1,000	3.24	95.73

*Approximate

Paclobutrazol (Bonzi, Piccolo, Paczol)

Paclobutrazol reduces internode elongation and is known to be very active and persistent growth regulator. This chemical, when used properly, does not produce any phytotoxic effects, but overapplication can lead to stunting. Paclobutrazol is primarily used as sprays or drenches, consisting of rates of 1–100 ppm, and some growers have also used it as a bulb soak. Some paclobutrazol labels have recently been expanded to include such applications as subirrigation and media sprays. These products have broad-spectrum labels for use on most floricultural crops. Paclobutrazol is primarily absorbed through the stem and the roots. While absorp-

Table 8.6. Paclobutrazol (Bonzi, Piccolo, Paczol) Spray Dilutions

Desired ppm*	Fluid Ounces per Gallon	Milliliters (cc) per Gallon
1	0.032	0.95
3	0.096	2.85
5	0.16	4.75
10	0.32	9.5
15	0.48	14.25
20	0.64	19.0
25	0.8	23.75
30	0.96	28.5
40	1.28	38.0
50	1.6	47.5
60	1.92	57.0
75	2.4	71.25
100	3.2	95.0

*Approximate

Table 8.7. Paclobutrazol (Bonzi, Piccolo, Paczol) Drench Dilutions

Drench Solution								
Desired ppm*	Fluid Ounces per Gallon	Milliliters (cc) per Gallon	Drench Volume** (Fluid Ounces) per 4 In. Pot	Milligrams Active Ingredient (a.i.)** per 4 In. Pot	Drench Volume** (Fluid Ounces) per 6 In. Pot	Milligrams Active Ingredient (a.i.)** per 6 In. Pot	Drench Volume** (Fluid Ounces) per 8 In. Pot	Milligrams Active Ingredient (a.i.)** per 8 In. Pot
1	0.032	0.95	2.0	0.063	4.0	0.125	10.0	0.313
2	0.064	1.9	2.0	0.126	4.0	0.25	10.0	0.626
4	0.128	3.8	2.0	0.252	4.0	0.5	10.0	1.252
8	0.256	7.6	2.0	0.504	4.0	1.0	10.0	2.504
12	0.384	11.4	2.0	0.756	4.0	1.5	10.0	3.756
16	0.512	15.2	2.0	1.008	4.0	2.0	10.0	5.008
20	0.64	19.0	2.0	1.260	4.0	2.5	10.0	6.260
25	0.8	23.75	2.0	1.575	4.0	3.125	10.0	7.825

*Approximate **Labeled rates

tion through the leaves does occur, it has little effect on controlling height since paclobutrazol does not move readily within the plant through the phloem. The spray applications should thoroughly wet plant stems. Once inside plant stems and roots, it moves through the xylem up to the terminals, where it reduces the size of the plant cells being made. The active ingredient has very low water solubility and moves quickly into the plant tissue. These products are absorbed within five to thirty minutes of application, and overhead irrigation can occur sixty minutes after application without reducing its effectiveness. Due to the high level of activity, applying in a uniform manner is critical, or else an uneven level of control will result.

Sumagic

Sumagic (uniconazole-P) is the most active and persistent of all of the growth regulators. Internode elongation is reduced by the inhibition of gibberellin biosynthesis. There are no phytotoxic effects, but overapplication can lead to stunting. Sumagic is most often used as a spray, drench, or bulb dip and is absorbed through the stems and roots. Recently the label has been expanded to include such applications as media sprays and cutting dips. The weakness of Sumagic is the potential for overapplication, but the strength is that Sumagic often shows activity on a wider array of plants when compared with the other plant growth regulators. Due to the high level of activity, application in a uniform manner is critical, or an uneven level of control will result. Like paclobutrazol products, Sumagic is absorbed into the plant within minutes of application, and

Table 8.8. Sumagic (Uniconazole-P) Spray Dilutions

Desired ppm*	Fluid Ounces per Gallon	Milliliters (cc) per Gallon
1	0.26	7.57
3	0.77	22.71
5	1.28	37.85
7.5	1.92	56.78
10	2.56	75.71
15	3.84	113.56
20	5.12	151.41
25	6.40	189.26
30	7.68	227.12

*Approximate

Table 8.9. Sumagic (Uniconazole-P) Drench Dilutions

Drench Solution								
Desired ppm*	Fluid Ounces per Gallon	Milliliters (cc) per Gallon	Drench Volume** (Fluid Ounces) per 4 In. Pot	Milligrams Active Ingredient (a.i.)** per 4 In. Pot	Drench Volume** (Fluid Ounces) per 6 In. Pot	Milligrams Active Ingredient (a.i.)** per 6 In. Pot	Drench Volume** (Fluid Ounces) per 8 In. Pot	Milligrams Active Ingredient (a.i.)** per 8 In. Pot
0.25	0.06	1.89	2.0	0.015	4.0	0.03	10.0	0.075
0.5	0.13	3.79	2.0	0.03	4.0	0.06	10.0	0.15
1	0.26	7.57	2.0	0.06	4.0	0.12	10.0	0.30
1.5	0.38	11.36	2.0	0.09	4.0	0.18	10.0	0.45
2	0.51	15.14	2.0	0.12	4.0	0.24	10.0	0.60
2.5	0.64	18.93	2.0	0.15	4.0	0.30	10.0	0.75
3	0.77	22.71	2.0	0.18	4.0	0.36	10.0	0.90
5	1.28	37.85	2.0	0.30	4.0	0.60	10.0	1.50

*Approximate **Labeled rates

overhead watering can be done after sixty minutes of application without reducing its effectiveness.

Table 8.10. Topflor (F lurprimidol) Spray Dilutions

Desired ppm*	Fluid Ounces per Gallon	Milliliters (cc) per Gallon
1	0.033	0.967
3	0.099	2.901
6	0.198	5.802
10	0.33	9.670
15	0.495	14.505
20	0.66	19.34
25	0.825	24.175
30	0.99	29.01
40	1.32	38.680
50	1.65	48.350
60	1.98	58.02
80	2.640	77.360
100	3.30	96.70

*Approximate

Topflor

Topflor (flurprimidol) is the latest PGR available for ornamental crops. Although it is new to the U.S. market for ornamental crops, flurprimidol has been used in the turfgrass industry under the trade name Cutless for a number of years. Topflor has been labeled and used on a number of greenhouse crops in Europe for years as well. The mode of action is similar to A-Rest, paclobutrazol, and Sumagic; it is absorbed through the leaves, stems, and roots. In fact, it has greater absorption through the stems than does paclobutrazol. Like paclobutrazol and Sumagic, Topflor does not cause phytotoxicity to crops, but it can stunt them when it is overapplied. This product appears to be another valuable tool for growers to manage perennial plant height.

Mixing Plant Growth Regulator Solutions

When preparing to mix PGR spray solutions, thoroughly clean the spray tank and spray hoses to limit contamination from other chemicals. Never tank mix plant growth regulators with fertilizers, fungicides,

Table 8.11 Topflor (Flurprimidol) Drench Dilutions

Drench Solution								
Desired ppm*	Fluid Ounces per Gallon	Milliliters (cc) per Gallon	Drench Volume** (Fluid Ounces) per 4 In. Pot	Milligrams Active Ingredient (a.i.)** per 4 In. Pot	Drench Volume** (Fluid Ounces) per 6 In. Pot	Milligrams Active Ingredient (a.i.)** per 6 In. Pot	Drench Volume** (Fluid Ounces) per 8 In. Pot	Milligrams Active Ingredient (a.i.) per 8 In. Pot
0.5	0.017	0.493	2.0	0.029	4.0	0.059	10.0	0.148
1	0.033	0.987	2.0	0.059	4.0	0.118	10.0	0.295
2	0.067	1.974	2.0	0.118	4.0	0.236	10.0	0.59
3	0.1	2.961	2.0	0.177	4.0	0.354	10.0	0.885
4	0.133	3.948	2.0	0.236	4.0	0.472	10.0	1.18
6	0.198	5.922	2.0	0.354	4.0	0.708	10.0	1.77

*Approximate **Labeled rates

or insecticides, as plant injury may result. Before mixing, growers should determine, to the best of their ability, the volume of spray solution needed. Solutions of PGRs do not store very well; it is often less than a day before they degrade and become less effective. The spray tank should be filled with approximately half the required amount of water. Then after reading the chemical label and calculating the amount of product needed for the required concentration and volume of solution, the proper amount of growth regulator can be measured and added into the tank. After the chemical has been added to the water, the spray tank can be filled with the remaining amount of water

Figure 8.4. **Foliar drench applications of Topflor to *Monarda* 'Jacob Cline'. Left to right: Control, 75 ppm applied twice, 37 ppm applied twice, and 37 ppm applied once.**

Joyce Latimer

needed. Once the solution is mixed, it should be agitated frequently to assure uniform distribution of the product in solution during the application.

In many instances, growers only a mix small amount, perhaps a gallon (3.8 L) or less, of spray solution requiring relatively small amounts of chemical to be measured. It is important to measure PGRs properly; being close enough is not acceptable and can greatly affect the outcome of the application. Growers should obtain various measuring devices that are capable of measuring small quantities accurately. Items such as eye-droppers, graduated cylinders, syringes, teaspoons, tablespoons, gram scales, and precision pipettes can all be used for measuring small amounts of plant growth regulators precisely. You can get these measuring devices from your local pharmacy or a lab supply company.

Growers should not add spreader-stickers to PGR solutions. All of the commonly used PGRs already contain spreader-stickers in them and state on their labels that additional spreader-stickers are not necessary—in some cases, they may burn plant leaves. Adding surfactants to Florel applications is the exception to the above. The surfactant Capsil, in particular, has been shown to improve the effectiveness of Florel applications.

Regardless of where you obtain your PGR information, read and follow the directions found on each individual product label. Care should be taken to apply all PGRs in a manner consistent with the directions outlined on the labels.

PGR Application Techniques

Foliar sprays

The most common method of applying growth regulators is spraying. Sprays are only effective when the chemical comes into contact with the plant tissues that can absorb them. For example, B-Nine and Cycocel are primarily taken up by the leaves. Poor spray coverage limits the amount of leaf surfaces that have contact with the PGR, reducing the uptake of the chemical into the plant and the overall effectiveness of the application. Other PGRs, such as Sumagic, are primarily absorbed through the stems. When the canopy is full, getting the chemical to the stems is difficult because the leaves often block the chemical from reaching them during the application, which reduces the effectiveness of the PGR spray. Spray applications are less labor intensive than applying PGRs using media drenches.

Most growers use high-pressure, high-volume sprayers to apply PGRs, using pressures of 100–250 psi. Adequate results can also be achieved using knapsack (backpack) or hand sprayers using pressures between 20–30 psi, though the most consistent results are achieved with high-volume application equipment. When using plant growth regulators as foliar sprays, it is important to achieve thorough, consistent, and uniform coverage. The objective should be to wet the foliage but not to the point of runoff, where the solution drips down onto the growing medium. To achieve this with most chemicals, it is recommended to apply 2 qt. of spray solution per 100 sq. ft. (1.9 L/9.3 m^2). Situations with well-developed plant canopies may require increasing the spray volume up to 3 qt. per 100 sq. ft. (2.8 L/9.3 m^2). Failure to apply these chemicals properly will lead to inconsistent results.

Depending on the product applied, the rates used, frequency of application, the number of applications, and the crop's stage of development, growers will usually observe a slight delay in flowering time.

Media drenches

Drenching is the second most common method of applying PGRs. Drench applications usually provide longer lasting, more uniform control of plant height than do spray applications. Unlike many PGR sprays, drenching doesn't seem to affect or delay the flowering of perennials. Drenches typically use larger, more diluted volumes of solution when compared with a spray. Applying drenches is more labor intensive than sprays because drenches require a specific volume of solution to be applied to each pot. Recently, some equipment has been developed to help growers deliver exact volumes of solution to each container, but it is still labor intensive, as the solution must be applied pot to pot.

Drenches are applied across the top of the medium of a growing plant. For most chemicals, 2 oz. (59 ml) of final solution is applied to a 4 in. (10 cm) pot, and 4 oz. (118 ml) to a 6 in. (15 cm) pot. More or less drench volume can be used if the ppm is adjusted to keep the amount of active ingredient applied constant. For example, if the desired level of growth control is achieved by applying a 4 ppm solution at a volume of 4 oz./(118 ml) per 6 in. (15 cm) pot, the same level of control can be achieved by applying 8 oz./(237 ml) per 6 in. (15 cm) pot of a 2 ppm solution. The total amount of active ingredient applied is identical in both of the above situations.

The medium of the pots to be drenched must be evenly moist to ensure optimal distribution of the PGRs. A good practice would be to irrigate one day, then drench with growth regulator the next. Growing mixes containing high amounts of organic materials, such as pine barks, are likely to tie up some of the growth regulators and will require higher rates (25–50% higher) when PGR drenches are applied.

Always refer to each product's label to verify the appropriate rates, volumes, and application guidelines for drench applications.

Sprenches

The sprench is a hybrid of the spray and the drench methods. When using this technique, a higher volume of spray is applied using spray or boom watering equipment to achieve more of a drenching effect than that of a spray application. Typically, sprench volumes are two to four times those of sprays. The chemical concentration being applied is usually less than a spray but more than a drench. Sprench applications are used with chemicals that have some persistence in the growing medium and are absorbed through the roots and translocated to the growing points, where they reduce internode elongation.

Bulb dips and seed soaks

Bulb dips or bulb soaks are innovative methods that have been developed to help growers achieve height control of certain bulb crops, such as Oriental lilies. This method involves dipping or soaking the bulbs in a growth regulator solution prior to planting. The concentration of the chemical and the length of time that the bulbs are submerged will vary depending on the species and/or cultivars. Another very similar method growers and researches have been evaluating are seed soaks. Like bulb dips, the results of seed soaks are a function of the chemical concentration and length of time the seeds are left in the solution. Soaking seeds of uniform, high-vigor seed lots can provide predictable, uniform results. Inconsistent seed lots (variable) will provide an exaggerated and undesirable response to seed soaks. At this time, no chemicals are labeled for this procedure on seeds.

Preplant dip of cuttings

Many growers are using a preplant dip of cuttings into a growth regulator solution to obtain height control. This method can be done on either rooted or unrooted cuttings of some plant varieties. To achieve results, the cuttings are placed in the solution only long enough to thoroughly wet all of the leaves and stems; they are then removed and planted.

Subapplication

A fairly new application method, called subapplication, entails applying PGRs through subirrigation systems. Growth regulators that can be absorbed by the roots (A-Rest, Topflor, paclobutrazol, and Sumagic) can be applied in this manner. Delivering PGRs through subirrigation allows the PGRs to get into the root zone faster than traditional drench methods, which must slowly leach down from the surface. Therefore, subapplications provide growers with quicker results. In addition, growth regulators applied from the bottom of a container provide a stronger response than do drench applications. The increased effectiveness occurs because the chemical is delivered to the bottom half of the container, where a majority of the roots are located.

The rates growers apply are typically at least half the rates used for typical drench applications. Similar to drenches, there is a relationship between the volume taken up by the medium and the concentration of the solution. With this knowledge, growers can calculate the uptake of an average pot and determine what the subapplication solution's concentration should be to deliver the desired amount of active ingredient into the root zone. Compared with traditional drenches, plant response is more uniform and less chemical is used, providing growers with more consistent results and reduced cost.

Growers need to ensure that adequate uptake of the PGR will occur while limiting the variability of uptake from pot to pot, which will greatly affect the overall uniformity of the application. To increase the uniformity of uptake and overall response, it is necessary to use the subirrigation method when the growing medium is moderately moist. The crop will most likely need to be irrigated before the treatment to bring the crop to a uniform moisture content. Unlike drenches, irrigating a day before the application may not allow for sufficient uptake of the PGR. Depending on the season and the crop, growers may have to water two or three days before the anticipated PGR application to bring the crop to a moderate moisture level to allow for sufficient uptake into the root zone.

In the past, growers have been concerned about using PGRs in subirrigation systems where the irrigation water is recirculated, fearing the continued use of the PGR solution will cause a cumulative response, adversely affecting the crops. Fortunately, this is not the case. Research at the University of Massachusetts has shown that when stock tanks are refilled by more than 50% of the volume remaining with clear water, the amount of chemical remaining in the diluted solution has little, if any, effect on plant growth.

Media sprays

Media sprays are applied to the growing medium before the crop is sown or transplanted. Some growers have had success applying media sprays after the crop has been started. The purpose is to provide a drenching effect to an early stage of growth. As the crop is watered, the growth regulator is moved down into the root zone, where the plants can take it up. Media sprays are particularly useful for plug producers to provide early yet gentle control, which makes controlling plant height later in the crop cycle even easier. For seedlings, media sprays provide early control of the hypocotyl (the stem between the growing medium and the first leaves), which often stretches as the seeds emerge during dark, cloudy conditions. Media sprays are typically low volume, 2–4 qt. per 100 sq. ft. (1.9–208 L/9.3 m^2), at moderate concentration sprays.

Liner soak/drench

Some growers have adopted the practice of dipping the liner or plug trays of vigorous varieties before transplanting them to their final container. A similar method to the liner soak (plug dip) is a liner drench, also applied before transplanting. These methods allow growers to quickly apply the treatments to a large number of plugs prior to planting, which saves labor and chemical costs if PGR applications are normally necessary shortly after transplanting. Liner soaks or drenches are especially useful for growers producing mixed containers consisting of many different plants and growth habits (drenching the entire mixed container after it is planted would lead to desirable control of some varieties and probable over-control of less-aggressive varieties). These methods provide about two to three weeks of control, which in most cases provide sufficient carryover into the finished container. To reduce potential variability of the height response, bring the plug flats to a uniform moisture content the day before you make the PGR application.

Controlled residue

This innovative approach many growers are trying involves applying the growth regulator directly to the inside of the container prior to filling it with medium. As the roots grow and come into contact with the container, they take up the chemical. This provides control midway through the crop cycle. As the roots reach the plastic cell walls of the flat, the plant begins to respond to the PGR application. Initial research at the University of Florida is indicating that controlled residues are just as effective as drench applications. Controlled residues will probably become more widely used in the future, as growers will have to be accountable for the chemicals leaving their greenhouses through runoff.

PGR Spray Application

Regardless of the PGR, the chemical rates and results will vary with climate, season, genetics, plant stage of development, application method, application volume, and the applicator. Each time a grower uses a growth regulator, the results may vary slightly, even if the application procedures never changed. Individual growers must determine which PGRs, application methods, and rates are most suitable for their crops and growing operations.

Understanding how to use PGRs can be tricky, but following a set of rules will help your applications to become more successful and consistent. The guidelines described below refer specifically to spray applications, but in most cases also can be applied to other application methods.

Choosing a PGR

If height control is necessary, growers must pick a PGR that is effective at controlling the height of each particular plant species. It is a waste of time and money to apply PGRs that provide little to no reduction in cell elongation and will provide disappointing and unacceptable results. When multiple options exist, choose the chemical you have the most experience with and are most comfortable using. Unless a plant is widely grown or has been researched extensively, it is difficult to determine the best chemical to use or where to begin. Often with perennials, a trial-and-error process is used to determine the best growth regulators and the correct application methods and rates.

The Perennial Plant Growth Regulators table (at the end of this chapter) lists over 250 perennial plant species and the chemical plant growth regulators they are responsive to. Within each plant species, there may be several cultivars that are responsive to the PGRs listed. For example, under the perennial species *Alcea rosea* there are several cultivars—'Chater's Double Hybrids', 'Nigra', and 'Powderpuffs Mixture' to name a few—that are all responsive to the listed PGRs. Generally, it is safe to assume that all cultivars within a species will respond similarly to the PGRs listed, unless otherwise noted. *Alcea pallida* is a different species and may or may not be responsive to the same chemicals as *Alcea rosea.* Do not assume what provides height control for one species in a genus will also provide height control for another species.

There are several examples of different cultivars of a plant species that respond differently to specific growth regulators. Most commonly, cultivars within the same species respond to the same growth regulators but may require different rates. For example, *Echinacea purpurea* 'Bravado' is very sensitive to foliar applications of Sumagic, but *E. purpurea* 'Ruby Star' is less sensitive and often requires higher rates in order to achieve the same level of control. In some cases, individual cultivars do not respond to the same plant growth regulators. Perhaps the best example of this is to compare the response of *Gaillardia* x *grandiflora* 'Goblin' with *Gaillardia* x *grandiflora* 'Burgundy'. 'Burgundy' has a moderate response to applications of Sumagic, but 'Goblin' shows no response to this chemical.

It is difficult to properly decipher and communicate these differences to growers. I have seen variability in the results of my trials compared with university research. There are several instances where commercially I achieve satisfactory to excellent results, but information from universities contradicts my observations. I have seen different growers obtain varying results when applying the identical chemical at standard rates. Use the information in the PGR table as a reference, a guideline, to steer you in the right direction. With factors such as plant stage of development, growing environment, and application methods that all affect the response of the PGR, it is extremely difficult to provide absolute information. Use any information provided here, or elsewhere, as a starting point to develop your own PGR programs.

Quite often there are several PGRs listed for a plant species. When deciding which one to use, pick the one you are the most comfortable with. Then determine what rate to use in your region of the country. To optimize

Table 8.12. PGR Rate Recommendation Adjustments Based on Northern Rates Developed in Zone 5

Your Zone	Multiply Recommended Rate By*	Example B-Nine 2,500 ppm		
		Northern Rate	Multiplier	New Rate
Zone 5	1.0	2,500 ppm	1.0	2,500
Zone 6	1.2	2,500 ppm	1.2	3,000
Zone 7	1.4	2,500 ppm	1.4	3,500
Zone 8	1.6	2,500 ppm	1.6	4,000
Zone 9	1.8	2,500 ppm	1.8	4,500
Zone 10	2.0	2,500 ppm	2.0	5,000

*The rate adjustments shown above are to be used only as guidelines.

Table 8.13. PGR Rate Recommendation Adjustments Based on Southern Rates Developed in Zone 10

Your Zone	Multiply Recommended Rate By*	Example B-Nine 5,000 ppm		
		Southern Rate	Multiplier	New Rate
Zone 5	0.5	5,000 ppm	0.5	2,500
Zone 6	0.6	5,000 ppm	0.6	3,000
Zone 7	0.7	5,000 ppm	0.7	3,500
Zone 8	0.8	5,000 ppm	0.8	4,000
Zone 9	0.9	5,000 ppm	0.9	4,500
Zone 10	1.0	5,000 ppm	1.0	5,000

*The rate adjustments shown above are to be used only as guidelines.

height control, it generally requires two to three applications using the appropriate rate at seven- to ten-day intervals. The most benefit is achieved by applying PGRs during active growth and not after the crop is nearly overgrown. Monitor your crop's height control needs on a weekly basis.

Determine a rate

Once a chemical has been chosen, it is important to determine the rate of the chemical to apply. Establishing PGR rate recommendations for perennials is difficult due to cultivar variability, geographic location, and application volumes.

Growth regulator rate recommendations often come from numerous sources, such as trade shows or industry publications. Regardless of the source, it is important to consider the location of the source of these recommendations and make adjustments to fit your growing conditions. For example, the rates applied in Florida are usually at least twice the concentration of the rates commonly applied in Michigan. If growers in Michigan applied the Florida rates, they would almost certainly obtain too much height control, resulting in unsalable plants. Growers in the Northeast and the Northwest often reduce Florida based rates by 25%.

The temperature difference between the North and the South is the most important factor to explain why different rates must be used. Plants grown under warmer temperatures have a faster rate of growth and a greater tendency for stem elongation than do plants produced under cooler conditions. There are currently great debates going on within the industry to decipher how to express PGR rates for growers. Everyone acknowledges the fact that differing rates are necessary, but there is no clear solution for how to interpret, adjust, and apply rate recommendations across a wide geographic area.

A simple method to determine what rates to begin using would be to use a USDA Hardiness Zone Map (see page 2). As mentioned above, rates developed in Florida need to be gradually reduced as production moves north. Using the USDA map, Florida-based rates or recommended rates would apply to Zone 10. Growers in Zones 7 and 8 would have to reduce Florida-based rates by approximately 30%. Producers in Zone 6 may have to decrease their rates by 40%. Perennial growers in Zone 5 should reduce Florida-based rate recommendations by about 50%. See table 8.13.

Conversely, rates developed in the northern United States need to be increased as production shifts to southern locations. If a rate recommendation originated from Zone 5, growers within this zone should not have to adjust their rates from those being recommended. As the location moves south into Zone 6, the rates would need to be increased by approximately 20%. Growers in Zone 8 would have to increase these northern-based rates by 60%. Producers in Zone 10 may have to increase their rates by 100% from those originally recommended from a northern location. See table 8.12. However, using the USDA zones is not a guaranteed method to determine how to prorate chemical rates to apply in varying locations. Growers must conduct trials and base their beginning rates on past experiences using the chemical(s) in question.

The rate ranges listed on the chemical labels, which are often very vague, exist due to the various environmental conditions and geographic locations where the product is distributed. Typically, the low end of the rate applies mostly to northern growers and the high end of the rate is most applicable to southern growers. The vagueness of the chemical labels adds great difficulty in determining what rates to apply. Growers must gain experience to determine the rate that will provide the desired level of control for each crop in their own growing environment.

PGR rates will also need to be adjusted according to the time of the year. For example, in most parts of the country growers must increase the rates of their applications in the summer as compared with applications made during the winter months. There is no proven rule of thumb to help growers with these seasonal rate changes. For a northern grower, it is not uncommon to increase the rates applied in July 50–100% from the rates typically applied in February.

When determining the rate, always start at a low concentration and increase it if the level of control is insufficient. It is easier to reapply at the same or a slightly higher concentration than it is to reverse the effects of a chemical overdose. I find that most perennials require two or three spray applications to provide the best level of control and to build plants of the most desirable architecture or form. For example, better results can be achieved making two 5 ppm applications than with a single 10 ppm concentration. Usually, the applications should be seven to ten days apart, and it may take ten to twenty days to see the full effect of the PGR applications; therefore, be careful to not judge the effectiveness too early.

If you need to adjust the rate for a crop, make those changes on the next crop cycle, after you are sure there have not been satisfactory results from the additional applications. When increasing the concentration, start low and build up. Usually raise the rate by no more than 25% at any time. For example, if unsatisfactory results were achieved using a 100 ppm spray solution, consider increasing the rate for the next crop cycle up to 125 ppm.

Application timing

The best results from growth regulators are achieved when growers use them throughout the growing cycle to regulate plant development, as opposed to an effort to stop rapid elongation just prior to shipping. PGRs work by controlling cell elongation before it occurs and have the greatest effect on *young, actively growing* plant tissues. Many operations decide to use PGRs at the end of a crop cycle just prior to shipping. It is often an application out of panic, afraid that soon their plants are likely to be unsalable. When plants are maturing or becoming reproductive, the application of growth regulators seems to have less of an effect on controlling plant height. The rates and volumes necessary to stop plant growth at this stage are much higher and the results are often disappointing.

It is difficult to describe precisely when PGRs should be applied because there are so many variables that must be taken into consideration. The timing of PGR applications becomes somewhat subjective, as each grower's decision is unique to his/her past experiences and current production scenarios. Experience comes with time. The more often you make PGR applications and evaluate the results, making the proper adjustments, the sooner you will gain confidence in what works for you.

The amount of height control needed also varies for each operation and production scenario. Retail operations may not be as concerned with controlling plant height as would be a wholesale operation that has to ship plants long distances. Some growers may have the luxury of providing adequate space between each plant, while others may have to grow pot tight (which promotes stem elongation) in order to maximize profit margins.

Many growers apply PGRs following the calendar method, making their applications based on the amount of time after planting. For example, it may be normal practice to apply PGRs to *Delphinium* 'Pacific Giant Mix' two weeks after transplanting, and again ten days after the first application. In many cases, these applications are made whether the plants need growth regulator or not. By making calendar applications without first determining the need, growers frequently stunt their plants.

The best method to determine the timing of PGR applications is to evaluate the crops on a weekly basis. The first PGR evaluation should begin about a week after planting or a week after the heat is turned on for overwintered materials. This does not mean that growers should begin applying PGRs at this time; rather, they should begin looking at the crops, observing the rate of growth and development, anticipating how much growth is likely to occur before the next evaluation, and determining if and when the growth regulators should be applied. Every crop needs to be looked at and considered on an individual basis.

Some growers apply growth regulators within the first few weeks, when there are only a couple inches (5 cm) of new growth. In many instances, they reapply PGRs every seven to ten days, if necessary, controlling growth in stages throughout the crop's development. Multiple applications provide growers with the flexibility to make decisions regarding how frequently to make the applications and what rates, based on factors such as plant habit, stage of development, and current weather conditions. Other growers choose to apply PGRs using multiple applications, as mentioned above, but waiting until about a week or so before the plant leaves from surrounding pots begin to touch (1–2 in. [3–5 cm]), before starting applications. This method allows growers to bulk the plants so they reach near close to full size before the height is controlled.

For perennials such as *Echinacea,* it is important to apply growth regulators before or just as the flower spike begins to elongate. Flower spikes tend to elongate rapidly; therefore, any delay in applying growth regulators—even a day or two—can result in several additional inches of plant height and may render the PGR applications ineffective altogether.

As you can see there are a number of methods used to determine the timing of PGR applications. The greatest results are achieved when applications are made before the canopy closes in, or just as the leaves from adjacent plants begin to touch one another. Applying growth regulators at this time allows for good coverage of the spray solution onto the leaves and stems. Once the plant canopy closes in, the coverage of these chemicals is greatly reduced, with penetration of the sprays onto the stems more difficult. It is important to learn each plant's growth curve and apply PGRs during or slightly before its rapid growing phase.

It is important to evaluate on a weekly basis the need for a PGR application from the beginning of the crop until it is sold. Determine the level of control achieved from previous applications and the need for additional applications. Indicators that some control has been achieved include darker green leaves, a more horizontal (or flatter) leaf orientation, and shorter internodes. To help determine whether previous applications are working, growers should always leave some control plants (those with no PGRs applied) to provide a quick comparison of the results being achieved. If additional applications are necessary, allow seven to ten days between applications to avoid overapplication.

I generally apply two or three applications of a PGR to a crop. If you find you are applying four or more applications to any particular crop, consider increasing the concentration, increasing the volume, changing chemicals, or adjusting the crop's scheduling.

It is also very important to time the application of PGRs to maximize their uptake into the plant. PGRs such as B-Nine, Cycocel, and Florel are very soluble in water and are absorbed into the plant very slowly. In most cases, several hours are required to get full absorption of the chemical into the plant. Certain environmental conditions, primarily humidity and temperature, affect the absorption of these products. The chemical can only get into the crop while the leaves are still wet with the spray solution; once they dry, no further uptake occurs. If these products are applied midday and it is sunny outside, the spray applications are going to dry quickly, most likely before the plant has fully absorbed the chemical. It is best apply them while low evaporation conditions are present, such as early mornings, evenings, and cloudy days. When these products are applied and the optimal conditions for application are present, growers will achieve at least 25% more activity from these products than when they are applied under less than ideal conditions. Provided the leaves remain wet from the spray applications, about 50% of the potential activity from the application occurs after about an hour, and over 80% of the potential activity is received when the spray remains on the plant for four hours.

Other products, such as A-Rest, paclobutrazol, and Sumagic, are fairly insoluble in water and are absorbed very quickly into the plant, often within ten minutes. The environmental conditions, such as the weather or the time of day are not so pertinent with these types of products.

Applying plant growth regulators to plants that are under any type of stress will cause the response of the application to be magnified. Water stress, high fertility levels, and root rots are examples of common stresses that can cause the response of plant growth regulators to become intensified. In many instances the reduction of height is intensified so much that the plant becomes stunted and may become unsalable. To eliminate any potential stress caused from drought, thoroughly water prior to the PGR application. If a morning application is going to be made, water the plants the day before; for evening PGR applications, water the plants the morning of the application. Growers should take all the necessary steps to reduce stress on their perennial crops before deciding to apply these height-reducing chemicals.

Know your volume

For foliar applications, the labeled recommendations are to apply 2 qt. of spray solution over 100 sq. ft. (1.9 L/9.3 m^2) of production space. This is equivalent to 5 gal. per 1,000 sq. ft. (18.9 L/93 m^2), or approximately 218 gal. per acre (825 L/0.4 ha).

The volume of spray solution applied over an area has a greater affect on the results than does the concentration of the solution. For example, a grower who applies a 5 ppm solution at a volume of 2 qt. per 100 sq. ft. (1.9 L/9.3 m^2) would achieve nearly the same results from applying a 10 ppm solution at a volume of 1 qt. per 100 sq. ft. (0.9 L/9.3 m^2), providing the grower could achieve adequate coverage. If a grower applied 3 qt. per 100 sq. ft. (2.8 L/9.3 m^2) instead of the intended 2 qt. (1.9 L), 50% more spray solution would be used than is recommend for that area. Thus, an intended spray solution of paclobutrazol at 60 ppm would quickly become 90 ppm. Overapplying PGRs could lead to severe stunting and unsalable product. This demonstrates that the effect of PGRs is both a function of the rate and the spray volume being applied. Of these two factors, volume is the determining factor and will ultimately determine if a grower's application is successful or not.

Proper coverage

The proper volume of spray solution must be applied uniformly over the crop. This is referred to as spray coverage and is also an important aspect of using plant growth regulators. Growers may use the right spray volume, but the application may not be very effective unless uniform coverage is achieved.

Always apply PGRs to crops by applying them from two different directions to ensure that all surfaces of the leaves and stems are covered uniformly with spray solution. Make a first pass over the crop from one side of the bench and a second pass from the other side. If this is not possible, apply by spraying on your way down the aisle from one direction and again on your way back from the opposite direction. This will increase the surface area being treated and increase the activity of the application. Without this bi-directional application, many plants are likely to have one side that does not receive the proper coverage or may not get any of the spray at all, which will reduce the overall effectiveness of the PGR application.

Foliar applications are intended to be applied to the leaves and stems, but in many

Figure 8.5. Always apply PGRs from two directions to get the best coverage.

instances the chemical ends up contacting the soil surface, either directly from the spray application or indirectly through runoff. Some PGRs, such as paclobutrazol and Sumagic, are more active (provide more control) when they are taken up by the roots than when applied to the foliage. In many instances, growers observe results, good or bad, as the result of the PGR being applied to the soil in conjunction with the foliar application. Sprays applied to the growing medium are essentially the same as applying a drench, except growers are likely to obtain twice the response as they would observe using the same concentration applied only to the foliage. This is another reason why certain growth regulators applied to young plant materials tend to be more effective than when applied to older plants—the chemicals are being applied to the foliage and the growing medium simultaneously.

Many times growers receive recommendations or find chemical labels indicating that the growth regulators should be applied as "spray to glisten," not "spray to runoff." For some PGRs such as A-Rest, paclobutrazol, and Sumagic, these recommendations are given to prevent growers from receiving control from the chemical that drips down onto the soil. However, it is often difficult to "spray to glisten" and effectively get the chemical on the stems where they can be absorbed by the plant. Growers should apply PGRs in such a manner that they can achieve the coverage necessary for maximum absorption into the plant.

Growers and spray applicators should also carefully watch the drift of the sprays. Sometimes drift occurs as the result of overspray from one crop to the other. In other cases, drift may be the result of horizontal airflow fans, exhaust fans, or drafts from doors and vents. If drift occurs, there is the potential for unwanted control of other crops. Crops with high sensitivity to certain growth regulators may become stunted.

Since coverage of the proper volume of spray solution is so critical, I recommend growers practice applying clear water over a predetermined area using the same equipment the PGR applications will be made with. Practicing may seem like an unnecessary step, but with the sensitivity of some plants to PGRs and the effect the volume has on the results, practicing is a simple step to ensure growers understand and know exactly what and how much spray solution they are applying.

These applications should be conducted over a known area with a predetermined amount of clear water. For example, a grower might decide to practice applying water over a 1,000 sq. ft. (92.9 m^2) bench, using the recommended application volume of 2 qt. per 100 sq. ft. (1.9 L/9.3 m^2). The grower has calculated that 20 qt. (18.9 L) would need to be sprayed over the bench. The practice applications should be applied at the same pressure as the actual application will be, usually 100–250 psi if a high-pressure sprayer is used.

The applicator should practice by applying the water from two directions, as described earlier. If after the first practice attempt the applicator has not applied the full, predetermined amount of water over the bench, he will need to slightly decrease his walking speed or slightly increase the pressure. Conversely, if too much has been applied, and the bench is not fully covered, the walking pace can be increased or the pressure decreased. The applicator needs to pay attention to how fast he is walking, the appearance of the solution leaving the spray gun, and the amount and size of droplets on the plants. The applicator should keep practicing until the desired coverage is achieved consistently over multiple attempts and the applicator is confident the application of PGRs will be consistent every time. The spray technician needs to observe the spray coverage during an application and make the proper adjustments to ensure the proper volume and coverage is being achieved.

One method growers can use to check the uniformity of their spray coverage is to practice by spraying an area of a paved surface with water in a manner consistent with normal spraying practices. It should be warm and sunny to promote rapid drying of the water from the pavement. Within minutes after

Quick Keys to Successful PGR Applications

1. Choose an effective PGR for the desired perennial. When multiple choices exist, pick the one you are most comfortable using.
2. Determine the application method (sprays, drenches, or other).
3. Calculate the rate. Adjust rate recommendations to fit your growing environment and season. Florida derived rates are typically 2x higher than the rates used in the North.
4. Evaluate the need for PGRs weekly; apply or reapply if necessary.
5. Apply PGRs throughout the production cycle rather than as the crop matures.
6. Applying low rates frequently provides better results than single applications at high rates.
7. Know the volume of PGR applied. The volume applied has a greater affect on the results than does the concentration of the solution.
8. Ensure proper coverage. Apply the proper volume uniformly.
9. Evaluate the effectiveness of your current PGR strategies. Determine the need to modify the current PGR practices for future cropping cycles. Consider changing application method, increasing rates, changing chemicals, or using tank mixes.

applying the spray to the pavement, observe the drying pattern to determine the uniformity of the spray. Fast-drying spots indicate low spray volumes were applied, and slow-drying spots indicate high spray volumes were applied. When the pavement dries uniformly, without streaks or spotting, growers will know their applications are being applied uniformly. This method can be practiced until the uniformity of the application is acceptable.

PGR Tank Mixes

The usual approach to controlling plant height of perennials is to apply one or more applications of a single PGR of a predetermined concentration to a given crop. This approach has provided growers with a wide range of results, from no control to too much. Chemical companies, universities, and growers have recently begun to investigate combining PGRs to enhance the effectiveness of these chemicals on certain perennial varieties. Perennial growers are looking at tank mixing different PGRs together to either provide more control to plants that have poor response to traditional single chemical applications or to decrease the potential stunting that occurs when these products are overapplied.

Tank mixing growth regulators isn't necessarily a new concept, B-Nine and Cycocel have been tank mixed for years. Commercially, growers have only considered using additional tank mix combinations for a few years now. My interest in tank mixes began in 2000, and they currently compromise a major portion of my height control programs. Today, growers are using tank combinations more routinely when controlling height of containerized perennials.

Figure 8.6. *Coreopsis* 'Early Sunrise' responds to several tank mixes of PGRs. Plant on left is the control, and the plant on the right received three applications of 10 ppm A-Rest + 1,875 ppm B-Nine at seven-day intervals.

All of the commercially available PGRs essentially control plant height by interfering with the biosynthesis of gibberellins within the plant. Each plant growth regulator interferes with the biosynthesis process at a different site. Thus, quite often when two different growth regulators are combined, there's a greater response than when a single chemical is used alone. This heightened response to the combination is referred to as synergy. In many cases, as when B-Nine and Cycocel are mixed, the enhanced activity achieved is an additive effect of the two individual products. The level of control observed is dependent on many factors, such as the chemicals being combined, the rates of each chemical used, the volume being applied, the plant they're being applied to, and the age or stage of development of the plant.

Daminozide and Sumagic seem to show the most response across the widest range of perennial species. With the wide range of perennial varieties each of these chemicals control individually, it is no surprise that the combination of these chemicals is providing growers with effective height control. Other encouraging tank mixes for use with perennial crops are daminozide + A-Rest, daminozide + paclobutrazol, and daminozide + Cycocel.

When using tank combinations of chemical growth regulators, growers typically reduce the rate of each chemical 25–50% of the normal rate used individually. Since each growth regulator affects biosynthesis of gibberellins at different sites, tank mixes provide two modes of action, allowing the rates of each component to be reduced while achieving equivalent and often greater results when compared with applying them individually.

Some PGR combinations may be more expensive than traditional growth regulator applications on a per-square-foot basis, but the costs are quite insignificant when amortized over all of the crops being produced. Most

Synergy of Nepeta 'Six Hills Giant' to a B-Nine + Cycocel Tank Mixture

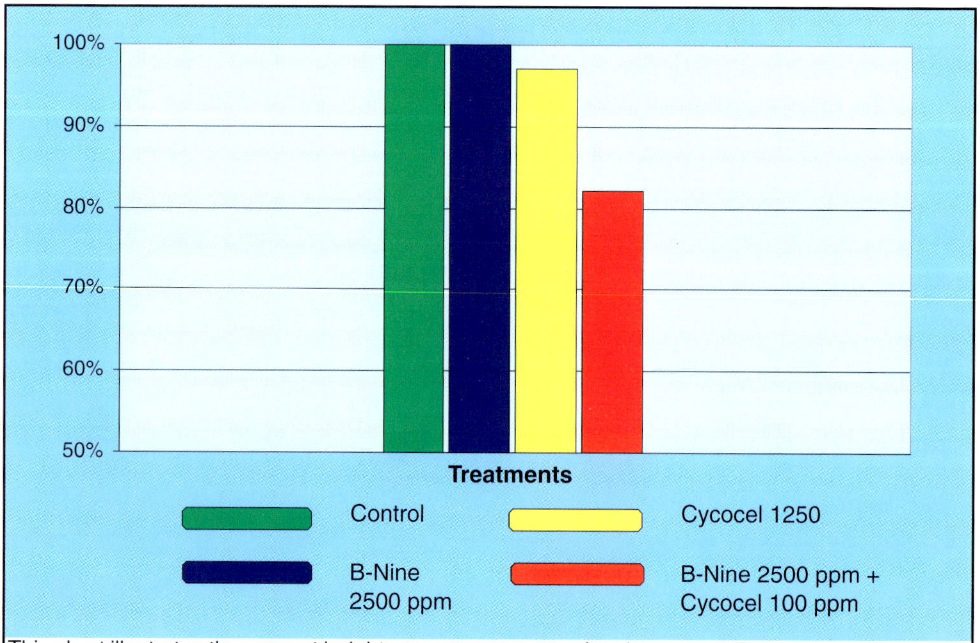

100%
90%
80%
70%
60%
50%
Treatments
Control
B-Nine 2500 ppm
Cycocel 1250
B-Nine 2500 ppm + Cycocel 100 ppm

This chart illustrates the percent height response compared to the control. Three applications were made at seven-day intervals. The plants within the control treatment are the maximum height (100%); any response to a treatment is illustrated as less than 100%. The B-Nine treatments showed no response to the applications. The cycocel treatments showed a slight reponse, 97% the size of the control plants (about a 3% reduction in height). The tank mixture of B-Nine + Cycocel resulted in plants 82% the size of the control plants (18% reduction in height).

Figure 8.7. **Many perennials, such as *Hemerocallis* 'Pardon Me' shown here, are more responsive to tank mixes than to single product applications. Left to right: Control, 5 ppm Sumagic, 2,500 ppm B-Nine, and 1,875 ppm B-Nine + 3 ppm Sumagic. Three applications were made at seven-day intervals.**

growers would rather have high-quality, salable plants as opposed to lanky, low-quality perennials that won't be purchased, even if it costs slightly more to produce them.

Always begin using tank mixes on a small scale to determine their effectiveness before making any major applications. Remember, no one growth regulator will control plant height on all plant varieties; and similarly, no single combination of these products is likely to yield comparable results on all species of plants. Don't make drastic changes to your height control strategies until you have tried tank mixtures on a small scale and are comfortable using them. Although tank mixes have their limitations, they allow growers to streamline their PGR applications into a simple, efficient, cost-effective method of controlling plant height of perennial crops. For growers with relatively little experience using PGRs or new producers of perennials, I would recommend sticking with traditional single product applications or using the relatively safe and widely used tank mix combination of daminozide + Cycocel.

Drenches

Even though spray applications are most growers' preferred method of PGR application, under certain circumstances, applying drenches may be beneficial or sometimes necessary to ensure proper height control. Drenches involve applying a certain volume of chemical solution directly to the top of the medium of an actively growing plant.

Drench applications may become necessary when the spray equipment is either broken, out of service, or being used to apply the necessary insecticides or fungicides to other crops. To successfully control plant height with spray applications, it is often necessary to make multiple applications, which cause several restricted entry intervals (REIs) to be enforced under the Worker Protection Standard. With a single drench application, only one REI has to be enforced, allowing more time for the

Figure 8-8. **Drenches often provide better, longer-lasting control than do spray applications. Left to right: *Hemerocallis* 'Pardon Me' control, three spray applications of 5 ppm Sumagic at seven-day intervals, and a single application of 1 ppm Sumagic drench.**

Table 8.14. Perennial PGR Drenches

The rates listed here are based on Michigan growing conditions. You will need to adjust rates for your own region and growing circumstances. Please refer to page 217 for recommendations on how to do this.

Variety	A-Rest (ppm)	Bonzi (ppm)	Sumagic (ppm)	Topflor (ppm)
Achillea filipendulina		6+		
Agapanthus campanulatus		6+		
Agastache mexicana		4		
Alcea rosea		6	1	6
Argyranthemum frutescens		6		
Artemisia ludoviciana		6+		
Aster x *frikartii*		6+		
Astilbe arendsii		6+		
Baptisia australis		6+		
Bellis perennis		6	1*	6*
Boltonia asteroides		6+		
Buddleia davidii		6		
Calamagrostis acutiflora			1+	6+
Chelone glabra		6+		
Coreopsis grandiflora	5	6	1	6
Coreopsis rosea		6		
Cortaderia selloana	10+	6+	1	
Crocosmia x *crocosmiiflora*		6		
Dahlia x *hybrida*		6		
Delphinium elatum	5	6	1	6
Dianthus barbatus		6	1	
Dianthus deltoides		6		
Dianthus grenadin		6	1	6
Diascia barberae		6		
Digitalis purpurea	5	6	1	6*
Echinacea purpurea		6+	1	
Eupatorium coelestinum		6+		
Fuchsia magellanica		4		
Gaura lindheimeri		6+		
Gazania splendens		4		
Geum chiloense		6	1	6
Heliopsis helianthoides		6+		
Hemerocallis (most varieties)	5+	6+	1	
Heuchera micrantha		6		
Hibiscus moscheutos		5		
Hosta (most varieties)	5+	6	1	
Iris, bearded hybrids		6		
Lantana camara		6+		
Lavandula angustifolia		6		6
Leucanthemum superbum	5	6		6
Lilium (Asiatic lily)		4	0.25–0.5	
Linaria purpurea		4		
Lobelia cardinalis	5	6	1	6
Lysimachia punctata		4		

(continued)

Table 8.14. Perennial PGR Drenches *(continued)*

Variety	A-Rest (ppm)	Bonzi (ppm)	Sumagic (ppm)	Topflor (ppm)
Miscanthus sinensis				6
Monarda citriodora		6		
Nierembergia scoparia		4		
Panicum virgatum				6+
Perovskia atriplicifolia		6+		
Phalaris arundinacea			1+	
Phygelius x *rectus*		4		
Physostegia virginiana		6+		
Platycodon grandiflorus		4		
Plectranthus amboinicus		4		
Polemonium caeruleum				6
Ratibida columnifera		6+		
Rudbeckia fulgida		6+		
Salvia greggii		6+		
Salvia x *sylvestris*		6+		
Sedum spectabile		4		
Sedum spurium		6+		
Solidago sphacelata		6+		
Tanacetum coccineum		6+		
Tricyrtis hirta		4		
Verbena bonariensis		4		
Verbena canadensis		4		
Veronica longifolia		6+		
Viola tricolor	5	6	1	6
Zantedeschia aethiopica		6+		
Zantedeschia hybrida		6+		

*Provides initial control but the effectiveness diminishes quickly.

normal work activities to occur in the greenhouse. With a single application providing control for several weeks, growers do not have to evaluate the effectiveness of the PGR as meticulously as they would need to do with spray applications, allowing more time to conduct other activities.

When application techniques are refined and uniform volumes are delivered, drenches provide more crop uniformity and a higher degree of crop safety (less over-stunting) than when spray applications are applied. There are many variables such as the type and moisture level of growing medium, plant genetics, plant stress, environmental conditions, and the stage of development that can affect the results of PGR drench applications.

It is very important to apply drenches uniformly. As with spray applications, the volume of solution applied to each plant is critical. Differing volumes may lead to different rates being applied to each pot, yielding very different results across a block of plants. An increase in the volume applied will increase the rate and the level of control achieved. Keep in mind that once PGRs are applied, their effects cannot be easily reversed. To achieve the most uniform distribution of PGRs in the root zone, the growing medium must be evenly moist prior to the application of the drench. Growers typically water the perennials well the day before the PGR treatment is to be applied. Growers using bark-based growing media or media containing large amounts of organic matter will have to increase the rate of the

PGR applied (by 25–50%) to compensate for the PGR getting tied up in the potting mix.

Until recently, drench applications were primarily done on a pot-to-pot basis. This method is referred to as the dosage method, where a predetermined volume is delivered to each pot. Delivering plant growth regulators in this manner is very time consuming, and it is often difficult to ensure consistent volumes are applied to each and every pot. A variation of drenching is the watering-in method, also referred to as the feed method. The watering-in method entails applying the PGRs in the irrigation water, or through an injector, usually overhead, at predetermined rates and volumes. Application in this manner allows the application of growth regulators to be applied in a manner more consistent with the normal watering practices of the operation. The rate of the growth regulator is often reduced slightly from the rates using the dosage method of application, but higher volumes of solution are typically applied. For example, the volume typically applied to a 6 in. (15 cm) pot using the dosage method is 4 oz. (118 ml); growers using the watering-in method might apply 8 oz. (237 ml) of solution at half the concentration of chemical on average to each pot. Since it involves overhead irrigation, the watering in-method is a less-labor intensive method of delivering PGR drenches to perennial crops compared with the dosage method.

Before using this technique, growers must calculate the number of pots to be treated and the volume of water to be applied. Like all drenches, the goal is to end up with a predetermined amount of active ingredient in each pot. The volume applied to each pot directly affects how much active ingredient is being delivered. Growers adjust the rate of the solution being applied to compensate for any differences in the volume using the various methods. For example, a grower applying Piccolo using the dosage method would apply 4 oz. (118 ml) of PGR solution at a rate of 4 ppm Piccolo to each 6 in. (15 cm) pot; to achieve similar results using the watering-in method, a grower might adjust the volume applied to 8 oz. (237 ml) and adjust the rate down to 2 ppm. In both instances, the same amount of active ingredient is applied and should reduce plant height very similarly.

Bedding plant and potted plant producers have adjusted their drenches to obtain 10% leachate (run-through). The logic behind applying drenches in this manner is to move the PGR down to where most of the active roots are, typically the bottom half of the pot.

Like spray applications, drenches should not be used with the intention of stopping plant growth but to reduce the rate of elongation during active growth. The effectiveness of PGR drenches should become less with time; plants should grow out of drench applications completely within three to four weeks and resume the normal rate of growth or elongation. If growers determine the rate of elongation was not reduced enough or the plants are growing out of the drench too quickly, they should consider increasing the concentration and/or volume of the drenches being applied. Conversely, if growers determine the rate of elongation was reduced too much or the plants take too long to grow out of the effects of the drench application, they should consider reducing the concentration and/or volume of the drenches.

Growers apply drenches at various times during crop production. Early-season drenches are applied shortly after the crop is started, often after two weeks, and they usually provide growers with two to four weeks of height control. Early drench applications do not delay flower development. Late-season drenches are often applied as a crop reaches a salable size and just beginning to flower. When done properly, late-season drenches do not delay flowering but do effectively control late-season elongation often associated with flowering plants or crops grown at high densities.

To offset the difficulty of delivering drench applications uniformly, I would recommend growers use rates and application volumes that are somewhat forgiving and will not lead to any long term "overcontrolled" situations. It is important to understand that the amount of active ingredient applied is a function of the concentration and volume of solution applied.

Growers need to understand this relationship and make any necessary modifications when applying drenches to ensure the proper or intended amount of active ingredient is delivered to each pot. Until drench rates and practices are refined, growers should view the goal of drenches as a means to provide a base level of height control. Use drenches to achieve up to 80% of the desired level of height reduction, under the presumption that if additional height control is necessary, a single spray application would be sufficient.

In-House Efficacy Trials

It is always a good idea to test a small group of plants for efficacy prior to treating an entire block of perennials. This is particularly useful when growers are producing new plant varieties or when they have no experience applying growth regulators to a particular crop.

The trials should be located in the same environment or receive similar environmental conditions (primarily light levels and temperature) to those the actual crop will be produced in. Growers should determine which treatments to trial. Treatments often consist of an application method, the PGR, and a specific rate. It might be desirable to trial multiple products simultaneously to determine which PGR provides the best control on the variety or varieties being tested. It is very important to always include a control, or some plants with no chemicals applied, to gauge the actual results from each treatment.

Each grower determines the size of the trial. Generally the larger the trial, the more labor intensive and time consuming it will be to collect all of the proper information. Size can be described in two ways, the first being the number of treatments or variables being tested, and the second being the number of plants allocated to each treatment. Trials become increasingly complex with each additional treatment. Sometimes growers make trials so extensive that they are overwhelmed collecting the data and often have difficulty interpreting the results. Each treatment should contain several individual pots, referred to as replications. Typically, growers use three to ten plants in each treatment. Remember, more is better; with more plants in each treatment the results become more accurate (but ten is generally sufficient). For example, a grower is conducting efficacy trials of Sumagic on a new perennial variety and decides to include ten pots per treatment and has four treatments (control, 5 ppm, 10 ppm, and 20 ppm). The trial would, therefore, require forty pots.

When trialing to determine the optimal rates, it is often best to include three rate groupings for each chemical being tested and a control group with no chemicals applied. The rate groupings should include a minimum rate, a mid-range rate, and a maximum rate. For example, a grower trialing Bonzi might use 30 ppm for the minimum rate, 60 ppm for the mid-range rate, and 90 ppm for the maximum rate. It is desirable to observe everything from no control up to too much control, providing an indication of which rate group is more suitable. In most cases, the rate used in production will fall between two rate groupings. Using the example above, the trial might have shown less than desirable control at 30 ppm (the minimum rate) and slightly too much control at 60 ppm (the mid-range rate). The grower might either opt to conduct another trial to narrow down the precise rate, or he might decide to split the difference and apply 45 ppm to the crop.

It is helpful to assign one individual the responsibility of coordinating the trial, applying the PGR treatments, collecting weekly height measurements, and summarizing the results. A single person will provide consistency to the trial. The effects each treatment has on flowering (specifically, any delay of flowering) may be useful for perennials being marketed in bloom by specific dates. When analyzing the data, it is also helpful to calculate the cost of each treatment for cost analysis and justification.

Conducting trials provides valuable information that simply cannot be found on the chemical labels or in trade publications. The time, space, and labor involved to conduct trials can make an incredible impact on future crop production and quality.

Delay of Flowering

Many growers ask about the potential of growth regulators to delay flowering. Their concerns are real, as most growth regulators have the potential to cause delayed flower development. The amount of delay usually depends on how aggressively PGRs are applied. Using high rates tends to delay flowering more than using low rates, even when low rates are applied more frequently. Though there does seem to be a cumulative affect when applying low rates on a regular basis, as each successive application may cause more delay.

How much delay can growers expect? The amount varies based on the perennial variety, the rate applied, and the frequency of application. Growers should be prepared to add three to seven days to the production time to account for any delay of flowering. Unless high rates are used, adding days, not weeks, to the production time is realistic. When growers are very aggressive with chemical rates and application frequencies, it is not uncommon to delay flowering by one to two weeks or more. In these cases, it is not uncommon for growers to observe additional side effects from the PGR applications.

Growth regulators can effect the development of the flower; they often become misshaped, remain small, or abort when high rates of growth regulators have been used. For these reasons, it is important to use plant growth regulators responsibly. Growers should pick the best chemical for each variety and determine which rate and application method will provide the level of control they are looking for. PGR applications should be applied during active growth, with their need evaluated on at least a weekly basis. The closer PGR applications are made to the flowering date, the less effective each application will be and higher rates will be needed, which will delay flowering more.

Application Costs

There is no doubt that applying growth regulators adds cost to the production system. With some crops, the application of PGRs is inevitable; with others, using PGRs is one of several tools available to growers to manage plant height. With the ability to maintain or improve crop quality characteristics such as appearance and shelf appeal, sometimes it is difficult to determine the exact benefit growers receive when using growth regulators on perennials. Is the cost of using these tools justified or profitable? Every perennial grower has to analyze the pros and cons of using PGRs at their operation before they take the cost of applying these chemicals into consideration.

To determine the costs of various PGR applications, it is important to evaluate the cost of the concentrated materials, the concentration of the final spray solution, and the amount of area to be covered. Table 8.15, PGR Costs for Foliar Sprays, demonstrates how the costs of various PGRs and applications of them can be compared and interpreted. The foundation of the table is derived by pricing the individual components. The pricing of PGRs varies widely by the size of the product purchased (quart versus gallon) and the size of the operation (large greenhouses often can obtain chemicals with lower prices due to volume discounts). This table demonstrates pricing skewed to the upper end of the price range, as a small or average-sized operation most likely has to purchase. Most growers can obtain PGRs slightly cheaper than the pricing used in this table.

Table 8.15 illustrates the cost per ounce of concentrated chemicals, the cost of a low rate and a high rate of finished spray, and the application cost to treat one acre of production space for each of the chemical growth regulators. The low rate in the table reflects beginning rates commonly applied to perennials; these are rates that I would use in most circumstances. I have also included the cost associated with various tank mixes commonly used on perennials as well.

Looking at the cost per ounce of each product does not always provide a clear picture to the real costs of applying these products. Each product is applied at different concentrations (ppm), that when diluted in water affects the cost of using the product. For example, when comparing B-Nine and Sumagic, the

Table 8.15. PGR Costs for Foliar Sprays

Chemical Costs				Low Rate					High Rate				
Product	Price	Size	Cost per Ounce Concentrate	Rate	Rate per Gallon Spray (oz.)	Cost per Gallon Final Spray Solution 200 ft²	Cost per Acre When 1 Gal./200 Ft² Is Applied	Cost per Square Foot for 3 Applications	Rate	Rate per Gallon Spray (oz.)	Cost per Gallon Final Spray Solution 200 ft²	Cost per Acre when 1 Gal./200 Ft² Is Applied	Cost per Square Foot for 3 Applications
A-rest	$607.59	2.5 Gal.	$1.90	25 ppm	12.10	$22.97	$5,003.85	$0.34	50 ppm	24.20	$45.95	$10,007.69	$0.69
B-Nine WSG	$418.82	5 Lb.	$5.24	2,500 ppm	0.40	$2.09	$456.09	$0.03	5,000 ppm	0.80	$4.19	$912.19	$0.06
Bonzi	$350.00	1 Gal.	$2.73	30 ppm	1.00	$2.73	$595.55	$0.04	60 ppm	2.00	$5.47	$1,191.09	$0.08
Piccolo	$325.00	1 Gal.	$2.54	30 ppm	1.00	$2.54	$553.01	$0.04	60 ppm	2.00	$5.08	$1,106.02	$0.08
Cycocel	$290.65	1 Gal.	$2.27	750 ppm	0.82	$1.85	$403.07	$0.03	1,500 ppm	1.63	$3.70	$806.13	$0.06
Florel	$45.10	1 Gal.	$0.35	500 ppm	1.60	$0.56	$122.78	$0.01	1,000 ppm	3.20	$1.13	$245.57	$0.02
Sumagic	$338.43	1 Gal.	$2.64	5 ppm	1.28	$3.38	$737.10	$0.05	10 ppm	2.56	$6.77	$1,474.20	$0.10
Topflor*	$105.00	1 L	$3.11	30 ppm	1.00	$3.11	$676.32	$0.05	60 ppm	2.00	$6.21	$1,352.64	$0.09
B-Nine + Cycocel	—	—	$7.51	2,500/750 ppm	0.40/0.82	$3.96	$861.63	$0.06	3,750/1,000 ppm	0.6/1.08	$5.59	$1,218.27	$0.08
A-rest + B-Nine	—	—	$7.13	10/1,875 ppm	4.8/0.3	$10.68	$2,327.07	$0.16	15/2,500 ppm	7.3/0.4	$15.95	$3,474.94	$0.24
B-Nine + Sumagic	—	—	$7.88	1,875/3 ppm	0.3/0.8	$3.69	$802.76	$0.06	2,500/5 ppm	0.4/1.28	$5.48	$1,193.20	$0.08
B-Nine + Bonzi	—	—	$7.97	1,875/15 ppm	0.3/0.5	$2.94	$639.84	$0.04	2,500/30 ppm	0.4/1.0	$4.83	$1,051.64	$0.07

The cost of the chemicals illustrated in this table was an average of three reputable chemical distributors and is reflective of what the average-sized grower would pay for each of these products (2005). The actual prices will vary with suppliers and quantity discounts. *Topflor price is the estimated end-user pricing as determined by the manufacturer.

concentrated chemical is $5.24 per ounce for B-Nine and $2.64 per ounce for Sumagic. B-Nine appears to be the more costly of the two, but when mixed in solution using low rates for each of these products (2,500 ppm B-Nine and 5 ppm Sumagic) B-Nine actually costs 38.2% less for each application.

Growers should not choose to use a particular growth regulator based on the cost of the application alone. Each perennial responds to the individual products and to various rates differently. Therefore, growers need to consider the effectiveness of the various PGRs on the crops they are growing. There are many examples where two products are labeled for the same crop, but the effective rate might be higher with one product, causing the application cost to increase compared to using a lower rate of another product. For example, B-Nine and Sumagic might both be effective at controlling the height of a particular perennial crop. To achieve equal height reduction, the high rate of B-Nine (5,000 ppm) would have to be applied, while the low rate of Sumagic (5 ppm) would be needed. With the pricing shown in the tables, a single application of B-Nine at 5,000 ppm would cost $4.19 per gallon of spray solution, and Sumagic would amount to $3.38 per gallon. In this case, the Sumagic application would be the more affordable choice costing approximately 19.4% less.

Comparing drench and spray costs

When calculating the cost of PGR applications, growers need to look at all the options. For the most part, drenches appear more costly to apply than individual spray applications. However, when growers are faced with applying several foliar sprays—perhaps three to four applications in some instances—to achieve the same efficacy as one drench application, growers must really look at which application method is more costly.

The best method to compare costs is to calculate what the cost would be on an individual container basis. Let's assume that a single drench and multiple spray applications would provide the same amount of height control. To determine the costs of spray applications for each pot, the number of pots per square foot must be calculated.

Let's assume there are two 6 in. (15 cm) pots being grown per square foot. Using table 8.16, a grower would spend approximately $0.013 per pot to apply a Bonzi drench at 5 ppm. If spray applications were to be applied, the grower would apply Bonzi at 30 ppm for each application. Table 8.15 does not clearly show the cost per square foot per application, but it indicates the cost for making three applications ($0.04 per square foot). The cost per application would be $0.04 divided by 3, or $0.0133 per square foot. Using the assumption of two pots per square foot, the cost per pot per application would be $0.00665 ($0.0133 divided by two pots per square foot), or $0.02 per pot ($0.00665 times three applications) when three applications are made. So, a single drench application costs $0.013 per pot, and three spray applications would be $0.02 to apply; the drench application is $0.007 per pot—35%—cheaper to apply.

Let's compare the costs associated with the two application methods using Sumagic. For simplicity, the same assumptions apply; the same level of control is achieved from each method, and there are two 6 in. (15 cm) pots grown in every square foot. According to table 8.16, a drench application of Sumagic using 1 ppm would cost approximately $.021 per pot to apply. To achieve similar results with spray applications, the grower would apply Sumagic using the rate of 5 ppm. At this rate, using table 8.15, three applications would amount to $0.05 per square foot, which amounts to $0.0167 ($0.05 divided by three applications) per square foot per application. At a crop spacing of two pots per square foot, the cost per pot would be $0.00833 per application ($.0167 divided by two pots per square foot), or $0.025 per pot when three applications are made ($.00833 times three applications). Drench applications of Sumagic in this example cost $0.021, and

Table 8.16. PGR Costs for Drench Applications

Product	Cost per Ounce Concentrate	Milligrams of Active Ingredient per 6 In. Pot	Concentration Applied	Fluid Ounces per Gallon of Drench Solution	Fluid Ounces of Chemical Needed per 100 Gallons of Drench Solution	Ounces to Apply per 6 In. Pot	Pots Treated per 100 Gallons	Cost per 6 In. Pot
A-Rest	$1.90	0.125	1.0 ppm	0.5	50	4	3,200	$0.030
		0.25	2.0 ppm	1.0	100	4	3,200	$0.059
		0.375	3.0 ppm	1.5	150	4	3,200	$0.089
		0.5	4.0 ppm	1.9	190	4	3,200	$0.113
		0.625	5.0 ppm	2.4	240	4	3,200	$0.143
		0.75	6.0 ppm	2.9	290	4	3,200	$0.172
Bonzi and Piccolo	$2.64	0.125	1 ppm	0.032	3.2	4	3,200	$0.003
		0.25	2 ppm	0.064	6.4	4	3,200	$0.005
		0.5	4 ppm	0.13	13	4	3,200	$0.011
		0.625	5 ppm	0.16	16	4	3,200	$0.013
		1.25	10 ppm	0.32	32	4	3,200	$0.026
		1.5	15 ppm	0.5	50	4	3,200	$0.041
Sumagic	$2.64	0.06	0.5 ppm	0.13	13	4	3,200	$0.011
		0.12	1.0 ppm	0.26	26	4	3,200	$0.021
		0.15	1.25 ppm	0.325	32.5	4	3,200	$0.027
		0.18	1.5 ppm	0.39	39	4	3,200	$0.032
		0.21	1.75 ppm	0.455	45.5	4	3,200	$0.038
		0.30	2.0 ppm	0.52	52	4	3,200	$0.043
Topflor	$3.11*	0.059	0.5 ppm	0.017	1.7	4	3,200	$0.001
		0.118	1 ppm	0.033	3.3	4	3,200	$0.003
		0.236	2 ppm	0.067	6.7	4	3,200	$0.006
		0.355	3 ppm	0.100	10.0	4	3,200	$0.008
		0.473	4 ppm	0.133	13.3	4	3,200	$0.011
		0.708	6 ppm	0.198	19.8	4	3,200	$0.016

The cost of the chemicals illustrated in this table was an average of three reputable chemical distributors and is reflective of what the average-sized grower would pay for each of these products. The actual prices will vary with suppliers and quantity discounts.
*Topflor price is the estimated end-user pricing as determined by the manufacturer.

three spray applications would amount to $0.025 to apply. The drench application is $.004 per pot—16%—cheaper to apply.

The cost of labor has been omitted here, but it is a real cost that growers must take into consideration. Drenches performed in the traditional manner (pot to pot) can be very labor intensive and time consuming. Many growers are modifying how drenches are applied, shifting to the watering-in method, where PGR rates are modified and they are injected in the irrigation water. This is less labor intensive to apply and can provide effective height control when done properly.

Growers use PGRs to maintain or improve the quality characteristics of a crop and should be familiar with the cost of using chemical plant growth regulators. In many cases, the costs are justifiable without demonstrating the additional value or quality attributes they add to a crop. In other instances, the benefits of using growth regulators are more difficult to justify, causing growers to seek other methods of controlling plant height. Perennial growers should look at all of the variables including the crop, environment, crop density, customer specifications, PGR effectiveness, and cost of the PGR application when determining which PGR strategy to implement. Ultimately the benefits received should outweigh the costs associated with the applications.

PGR calculator info

North Carolina State University has developed software called PGRCALC to help growers determine the cost per application and the cost per treated plant for various plant growth regulators. In addition to calculating PGR costs, the program also assists growers in determining the amount of growth regulator to apply over a given area using the desired rate. When using this calculator, growers can limit the misapplication of PGRs due to incorrectly calculating the rates or the amounts to measure when making the stock solution. The calculator, however, cannot accurately apply growth regulators to the crop; that is ultimately up to the growers. Growers can obtain a free download of PGRCALC at www.floricultureinfo.com. The program is designed in Microsoft Excel 5.0 and can be downloaded for both Windows and Mac systems.

Applying Florel

Whether applying Florel (ethephon) to obtain lateral branching or as a height control tool, it is important to follow a few guidelines. It is generally better to apply low rates (300–500 ppm) more frequently than it is to apply higher rates less often. Using higher rates may cause phytotoxicity with some perennials. The pH of the spray solution after adding Florel should be 5.0 or lower. Growers may need to acidify their water prior to mixing or use a pH adjuster in the spray tank. Using distilled water is another option. Spray solutions at high pH levels cause Florel to turn to a gaseous state while in the

Figure 8.9. **Besides controlling plant height, Florel is used to keep perennials vegetative, which produces more cuttings per stock plant. Pictured at left is *Aster* 'Purple Dome' control and at right with three applications of 500 ppm Florel at twenty-one days apart.**

spray tank, reducing its effectiveness once it is applied to the crop. Spray solutions of Florel should be used within four hours of mixing to ensure maximum effectiveness of the application. Applications made when temperatures are warm will also be less effective.

Any crops Florel will be applied to should not be under any type of stress. Florel is a stress exaggerator, causing the results of the application to be intensified when applied to crops that are currently undergoing stressful conditions.

Florel is absorbed into the leaves slowly and should be sprayed when conditions allow for the spray to remain on the leaves for as long as possible. The leaves should remain wet for at least four hours to allow maximum penetration of the chemical into the plant. When conditions do not allow the leaves to remain wet for this duration, growers can increase the rate of the application or make the applications more frequently. To prevent potential injury to the crop, I prefer to increase the frequency of the applications as opposed to increasing the rate being applied.

Florel applications ideally should occur before the crops have gone to bloom. They can be applied early, even during propagation, provided that roots are present and the plant is stress free and actively growing.

Using PGRs in Plug Production

Many perennial plug growers express the need to use growth regulators during plug production to produce high-quality starter plants. Compact plugs are easier to handle, ship better, and usually produce a more attractive finished product due to the control achieved early in the plants' development. It is very cost effective and efficient to apply growth regulators at this stage since there are so many plugs in a small area. However, applying PGRs to plugs is challenging because plugs are generally more sensitive to these chemicals than larger, more mature plants.

Spray applications are the primary method plug growers use to apply these products to young plants. When applying sprays to plugs, growers must still follow the set of guidelines outlined in this chapter. These rules include choosing an effective growth regulator, determining the optimal rate, applying the proper volume, ensuring uniform coverage, evaluating the results, and reapplying if necessary. Paying particular attention to the spray volume and coverage is important with several PGRs (A-Rest, paclobutrazol, and Sumagic). These products often come into contact with the growing medium (either through direct contact or by runoff from the leaves), where these products provide higher activity than when only absorbed through the leaves. To ensure uniform coverage, the volume of spray to be applied should be the same as volumes used for larger plant materials (2 qt. per 100 sq. ft. [1.9 L/9.3 m^2]). The rates commonly applied to plug trays are typically reduced by 30–50% of the rates recommended for larger plant sizes. Growers generally wait until the first true leaves are present before applying PGRs to plugs.

Plug growers have been applying media sprays onto plug flats either before or after sowing to reduce early stretch of the hypocotyls on some varieties. This method allows growers to control plant height early, before applications are usually made using traditional spray applications. The timing of media spray applications can occur anytime from before the flats are sown up until germination begins. Media sprays are most effective when they can be applied before the

Figure 8.10. *Alcea* 'Chater's Double': Control on left, one foliar application of 1.25 ppm Sumagic on right.

hypocotyls begin to elongate. It is important to provide uniform application of the sprays; inconsistent uniformity will lead to some plug trays receiving more PGR than others, resulting in a variable plug crop. As most growers already know, transplanting variable plugs into larger containers will lead to crops that are not uniform and more difficult to manage. The volume of spray solution applied typically ranges from 2–3 qt. per 100 sq. ft. (1.9–2.8 L/9.3 m^2) The optimal rate to apply varies with the crop. Growers should probably begin media sprays using rates that are equal to, or slightly less than, those applied using traditional spray applications. Remember, as with any PGR application, the results are a function of both the rate and the volume applied. Other factors growers must consider are the light levels, temperature, fertility, spray volume, and the desired amount of height control.

Sprench applications are commonly applied to young seed flats from the time of seeding up until the cotyledons are totally expanded. Sprench applications commonly consist of two to four times the application volumes typically used for spray applications, providing growers with a drenching effect. The primary goal of these early sprenches is similar to the purpose of media sprays: for the roots to uptake the chemical and control early stretch of the hypocotyl.

Regardless of the timing of PGR applications or whether they are intended to be media sprays or foliar applications, the closer the application is to the sow date, the higher the level of control will be. The spray rate and volume applied combined determine the overall effectiveness of the application. Using plant growth regulators during plug production usually has a minimal effect on the time to flower, often causing a delay of three days or less. The delay of flowering largely has to do with the rates being applied and the level of control achieved. Situations where too much control is achieved (overcontrol) usually result in delayed flowering. Growers also need to be aware of seasonal rate differences. The rates being applied may need to be reduced during cooler, darker weather conditions and increased in during the warmer and brighter months.

References

Barrett, Jim. "Bottoms Up with Growth Regulators." *Greenhouse Product News*. September 1999.

————. "Growth Regulators: New Approaches." *Greenhouse Product News*. November 2003.

Carver, Steve, and Peter Konjoian. "Making Sense, or Cents, of PGR Use." *OFA Bulletin* Number 887. November/December 2004.

Faust, James, and Kelly Lewis. "Tank-Mixing PGRs." *Greenhouse Product News*. February 2003.

Heins, Royal D., Erik S. Runkle, et. al. "Forcing Perennials: Follow These Strategies to Regulate Perennial Plant Height." *Greenhouse Grower*. July 1999.

Olrich, Mike, David Joeright, et. al. "Herbaceous Perennials: Plant Growth Retardants." *Greenhouse Grower*. August 2003.

Latimer, Joyce, Holly Scroggins, and Velva Groover. "Asteraceae Response to PGRs." *Greenhouse Product News*. March 2003.

————. "Lamiaceae Response to PGRs." *Greenhouse Product News*. July 2003.

————. "Scrophulariaceae and Verbenaceae Response to PGRs." *Greenhouse Product News*. December 2003.

————. "Using Topflor, Part II: Perennial Plants." *Greenhouse Product News*. February 2003.

Pilon, Paul. "Advanced Height Control of Perennials: Utilizing Tank Mixes Can Help Minimize Excessive Plant Height." *Greenhouse Product News*. April 2002.

————. "Combining PGRs Yields Better Results." *GrowerTalks*. June 2001.

————. "Improving Perennial Crop Production with Growth Regulators." *Greenhouse Product News*. October 2000.

Styer, Roger C. 2003. "Maximizing Chemical Growth Retardants." *Greenhouse Product News*. March 2003.

Warner, Ryan, and John E. Erwin. "A Short Order: Nursery Professionals Can Use Chemical Growth Retardants to Control Perennial Height." *American Nurseryman*. March 15, 2001.

Whipker, Brian E. and Brian Krug. "Methods of Controlling Plant Growth." *OFA Bulletin* 882. January/February 2004.

Whipker, Brian, Ingram McCall, et. al. "Using Topflor, Part 1: Bedding and Potted Plants." *Greenhouse Product News*. January 2003.

Table 8.17. Perennial Plant Growth Regulators

The rates listed here are based on Michigan growing conditions. You will need to adjust rates for your own region and growing circumstances. Please refer to page 218 for recommendations on how to do this.

Variety	A-Rest (Ancymidol) (ppm)	B-Nine or Dazide (Daminozide) (ppm)	Bonzi, Piccolo, or Paczol (Paclobutrazol) (ppm)	Cycocel (Chlormequat Chloride) (ppm)	Florel (Ethephon) (ppm)	Sumagic (Uniconazole-P) (ppm)	Topflor (Flurprimidol) (ppm)	A-Rest (Ancymidol) + B-Nine (Daminozide) (ppm)	B-Nine or Dazide (Daminozide) + Bonzi, Piccolo or Paczol (Paclobutrazol) (ppm)	B-Nine or Dazide (Daminozide) + Cycocel (Chlormequat Chloride) (ppm)	B-Nine or Dazide (Daminozide) + Sumagic (Uniconazole-P) (ppm)
Achillea filipendulina		2,500	30		500	5					2,000/3
Achillea millefolium	25	2,500	30		500	5				2,500/1,000	2,000/3
Achillea x *hybrida*		2,500									2,000/3
Achillea x *hybrida* 'Moonshine'		2,500	30			5					
Aconitum nacelles											2,000/3
Aegopodium podagraria		2,500									
Agapanthus campanulatus			30								
Agastache mexicana			30					10/2,500			
Agastache rugosa		2,500				5				2,500/1,000	2,000/3
Ajania pacifica		2,500									
Ajuga pyramidalis						2.5					
Ajuga reptans						2.5					
Alcea rosea	10	2,500	15	1,250	500	2.5					
Alyssum saxatile		2,500									
Amsonia tabernaemontana			30								
Anchusa azurea		2,500									
Anthemis hybrida		2,500				5				2,500/1,000	2,000/3
Anthemis tinctoria		2,500				5				2,500/1,000	2,000/3
Aquilegia alpina		2,500									2,000/3
Aquilegia caerulea		2,500	30	1,250		5					2,000/3
Aquilegia vulgaris											2,000/3
Aquilegia x *hybrida*	25	2,500	30			5		10/2,000	2,500/15	2,500/1,000	2,000/3
Aquilegia chrysantha		2,500									2,000/3
Arabis blepharophylla		2,500									

(continued)

Table 8.17. Perennial Plant Growth Regulators *(continued)*

Variety	A-Rest (Ancymidol) (ppm)	B-Nine or Dazide (Daminozide) (ppm)	Bonzi, Piccolo, or Paczol (Paclobutrazol) (ppm)	Cycocel (Chlormequat Chloride) (ppm)	Florel (Ethephon) (ppm)	Sumagic (Uniconazole-P) (ppm)	Topflor (Flurprimidol) (ppm)	A-Rest (Ancymidol) + B-Nine (Daminozide) (ppm)	B-Nine or Dazide (Daminozide) + Bonzi, Piccolo or Paczol (Paclobutrazol) (ppm)	B-Nine or Dazide (Daminozide) + Cycocel (Chlormequat Chloride) (ppm)	B-Nine or Dazide (Daminozide) + Sumagic (Uniconazole-P) (ppm)
Arabis caucasica		2,500									
Arenaria montana			30								
Argyranthemum frutescens			30			5					
Artemisia arborescens		2,500				5					2,000/3
Artemisia ludoviciana		2,500									2,000/3
Artemisia schmidtiana		2,500	30			5				2,500/1,000	2,000/3
Artemisia vulgaris						5					2,000/3
Aruncus aethusifolius											2,000/3
Aruncus dioicus											2,500/5
Asclepias tuberosa	25	2,500	30			5					
Asclepias tuberosa 'Royal Red'		2,500	30							2,500/1,000	
Aster alpinus		2,500	30								
Aster dumosus											2,000/3
Aster novae-angliae											2,000/3
Aster novi-belgii		2,500									2,000/3
Aster tongolensis		2,500									2,000/3
Aster x *frikartii*		2,500	30								
Astilbe x *arendsii*	25-50	2,500	30	1,250		5				2,500/1,000	
Astilbe chinensis						5					2,000/3
Astilbe japonica						5					2,000/3
Astilbe taquetii		2,500									
Astilbe thunbergii	25-50	2,500	30	1,250		5					
Astilbe x *rosea*						5					

Aubrietia x *hybrida*		2,500									
Aurinia saxatilis		2,500									
Baptisia australis			30								
Bellis perennis		2,500		1,250		5					2,000/3
Boltonia asteroides			30					10/2,000			2,000/3
Brunnera macrophylla						5					2,000/3
Buddleia davidii	25	2,500	30			5	30			2,500/1,000	
Buddleia x *weyeriana*						5					
Campanula carpatica	25	2,500	15	750		2.5					
Campanula glomerata		2,500	30		500	5					
Campanula medium											2,500/5
Campanula persicifolia	25	2,500	30			5					2,000/3
Campanula punctata	25	2,500		1,250		5					
Campanula rotundifolia	25	2,500				5					
Campsis radicans						5					
Centaurea dealbata						5					
Centaurea macrocephala						5					
Centaurea montana	25	2,500	30			5					2,000/3
Centranthus ruber						5					
Cerastium tomentosum						2.5					
Ceratostigma plumbaginoides		2,500	30	1,250		5					
Cheiranthus cheery						5					
Chelone glabra			30			5					
Chelone lyonii			30			5					
Chrysanthemum coccineum	25	2,500		1,250		5					
Chrysanthemum x *morifolium*		2,500			500	5					
Chrysanthemum parthenium			15	1,250		2.5					

(continued)

Table 8.17. Perennial Plant Growth Regulators *(continued)*

Variety	A-Rest (Ancymidol) (ppm)	B-Nine or Dazide (Daminozide) (ppm)	Bonzi, Piccolo, or Paczol (Paclobutrazol) (ppm)	Cycocel (Chlormequat Chloride) (ppm)	Florel (Ethephon) (ppm)	Sumagic (Uniconazole-P) (ppm)	Topflor (Flurprimidol) (ppm)	A-Rest (Ancymidol) + B-Nine (Daminozide) (ppm)	B-Nine or Dazide (Daminozide) + Bonzi, Piccolo or Paczol (Paclobutrazol) (ppm)	B-Nine or Dazide (Daminozide) + Cycocel (Chlormequat Chloride) (ppm)	B-Nine or Dazide (Daminozide) + Sumagic (Uniconazole-P) (ppm)
Cimicifuga racemosa						5					
Clematis x *hybrida*	25–50		30			5					
Coreopsis auriculata						5					
Coreopsis grandiflora	25–50	2,500	30	1,250		5	45		2,000/15	2,500/1,000	2,000/3
Coreopsis rosea		2,500	30	1,250		5					
Coreopsis verticillata		2,500	30	1,250		5			2,000/15	2,500/1,000	2,000/3
Coreopsis x *hybrida*						5					
Delphinium belladonna						5					2,000/3
Delphinium elatum	25–50	3,750	30			5				2,500/1,250	2,500/5
Delphinium grandiflorum	25–50	2,500	30			5			2,000/30	2,500/1,000	2,000/3
Delphinium 'Magic Fountain' series			30			5			2,500/30	2,500/1,250	2,500/5
Delphinium 'Pacific Giant' series	25-50		30			5			2,500/30		2,500/5
Dendranthema grandiflora		2,500			500	5					
Dendranthema zawadskii		2,500	15			2.5				2,500/1,000	
Dianthus barbatus		3,750	45			10				3,750/1,250	3,750/5
Dianthus caryophyllus	25–50	2,500	30	1,250				10/2,000		2,000/1,000	2,500/5
Dianthus deltoides			30	1,250		5					2,000/3
Dianthus gratianopolitanus											2,000/3
Diascia barberae		2,500			500						
Dicentra eximia		2,500									2,000/3
Dicentra formosa		2,500									2,000/3
Dicentra x *hybrida*		2,500									
Dicentra spectabilis	25–50	2,500	30			5					2,000/3

Digitalis ambigua						5					2,000/3
Digitalis grandiflora						5					2,000/3
Digitalis x *mertonensis*						5					
Digitalis purpurea	25	2,500	30			5					
Doronicum orientale		2,500									2,000/3
Echinacea paradoxa											2,500/5
Echinacea purpurea	25–50	3,750	30	1,250	500	5	30			2,500/1,000	2,500/5
Echinops ritro						5					2,000/3
Erigeron glaucus											2,000/3
Erigeron speciosus						5					2,000/3
Erysimum cheiri						5					
Erysimum linifolium			30			5	30				
Eupatorium coelestinum	25	2,500				5					
Eupatorium maculatum											2,000/3
Eupatorium rugosum	25–50	3,750		1,250		10					2,500/5
Euphorbia hybrids			30			5	45				
Euphorbia polychroma		2,500									2,000/3
Evolvulus nuttallianus			45								
Fallopia japonica						5					
Fuchsia magellanica		2,500	30		500	5	30				
Gaillardia aristata		2,500	45			10			2,500/15		
Gaillardia grandiflora	25–50	3,750	45		500	10				2,500/1,250	2,500/5
Gaillardia grandiflora 'Goblin'		NR	NR			NR				2,500/1,250	2,500/5
Galium odoratum				1,250							
Gaura lindheimeri		2,500	30	1,250		5					
Gaura lindheimeri 'Corrie's Gold'		2,500	30		500	5				2,000/1,000	
Gaura lindheimeri 'Siskiyou Pink'		3,750	30			NR			2,000/30	2,000/1,000	2,500/5

(continued)

Table 8.17. Perennial Plant Growth Regulators *(continued)*

Variety	A-Rest (Ancymidol) (ppm)	B-Nine or Dazide (Daminozide) (ppm)	Bonzi, Piccolo, or Paczol (Paclobutrazol) (ppm)	Cycocel (Chlormequat Chloride) (ppm)	Florel (Ethephon) (ppm)	Sumagic (Uniconazole-P) (ppm)	Topflor (Flurprimidol) (ppm)	A-Rest (Ancymidol) + B-Nine (Daminozide) (ppm)	B-Nine or Dazide (Daminozide) + Bonzi, Piccolo or Paczol (Paclobutrazol) (ppm)	B-Nine or Dazide (Daminozide) + Cycocel (Chlormequat Chloride) (ppm)	B-Nine or Dazide (Daminozide) + Sumagic (Uniconazole-P) (ppm)
Gaura lindheimeri 'Whirling Butterflies'		NR	30			5				2,000/1,000	
Geranium x *cantabrigiense*											2,000/3
Geranium endressii						5					2,000/3
Geranium himalayense		2,500	30	1,250	500	5					2,000/3
Geranium hybrids						5					2,000/3
Geranium macrorrhizum						5					2,000/3
Geranium x *magnificum*						5					2,000/3
Geranium phaeum						5					2,000/3
Geranium sanguineum											2,000/3
Geum chiloense											2,000/3
Geum coccineum	25		30	1,250		5					
Gypsophila elegans						5					2,000/3
Gypsophila paniculata			30	1,250		5					2,000/3
Gypsophila repens						5					
Helenium autumnale		2,500	30								
Heliopsis helianthoides		2,500			500	2.5				2,500/1,000	
Hemerocallis (most varieties)	50	3,750	45			10				3750/1,250	2,500/5
Heuchera micrantha						5					
Heuchera sanguinea			30			5					
Heuchera x *hybrida*			30			5					
Heucherella x *hybrida*						5					
Hibiscus moscheutos	50	3,750	45	1,250		10				2,500/1,000	
Hibiscus x *hybrida*		2,500		1,250						2,500/1,000	

Hosta (most varieties)											2,500/5
Hosta fortunei	25–50					10					2,500/5
Hosta montana											2,500/5
Hosta nigrescens											2,500/5
Hosta plantaginea											2,500/5
Hosta sieboldiana											2,500/5
Hosta x *tardiana*											2,500/5
Hosta tokudama											2,500/5
Hosta undulata						10					2,000/3
Hosta ventricosa											2,500/5
Hypericum calycinum		2,500	30			5				2,500/1,000	
Iberis sempervirens	25		30			5					2,000/3
Iris pallida											2,500/5
Iris siberica											2,500/5
Knautia macedonica											2,000/3
Kniphofia uvaria		3,750	30			5					
Lamiastrum galeobdolon			30			5					
Lamium maculatum		2,500	30	1,250	500	5					
Lantana camara	25–50	3,750	45			10	45			2,500/1,000	
Lathyrus latifolia	25					5					
Lavandula angustifolia	25	2,500	30			5					
Lavandula officinalis						5					
Lavandula x *intermedia*		2,500	30			5				2,500/1,000	
Leucanthemum superbum	25	2,500	30		500	5			2,500/15		2,000/3
Liatris spicata	25–50	3,750									2,500/5
Ligularia dentata						5					
Lilium (Asiatic lily)	25		15			2.5					
Lilium (Oriental lily)			30			5					

(continued)

Table 8.17. Perennial Plant Growth Regulators *(continued)*

Variety	A-Rest (Ancymidol) (ppm)	B-Nine or Dazide (Daminozide) (ppm)	Bonzi, Piccolo, or Paczol (Paclobutrazol) (ppm)	Cycocel (Chlormequat Chloride) (ppm)	Florel (Ethephon) (ppm)	Sumagic (Uniconazole-P) (ppm)	Topflor (Flurprimidol) (ppm)	A-Rest (Ancymidol) + B-Nine (Daminozide) (ppm)	B-Nine or Dazide (Daminozide) + Bonzi, Piccolo or Paczol (Paclobutrazol) (ppm)	B-Nine or Dazide (Daminozide) + Cycocel (Chlormequat Chloride) (ppm)	B-Nine or Dazide (Daminozide) + Sumagic (Uniconazole-P) (ppm)
Limonium tataricum		2,500									
Linum perenne	25		30			5					
Lobelia cardinalis	25	2,500	30	1,250		5	30			2,500/1,000	
Lobelia fulgens	25–50	2,500	30	1,250		5					
Lobelia x *hybrida*	25–50	2,500	30	1,250		5					
Lobelia x *speciosa*	25–50	2,500	30	1,250		5					
Lunaria annua						2.5					
Lupinus x *hybrida*	25	2,500		1,250		5					2,000/3
Lupinus polyphyllus	25	2,500		1,250		5					2,000/3
Lychnis chalcedonica						5					
Lychnis coronaria		2,500	30		500	5					
Lysimachia clethroides						5					
Lysimachia nummularia						5					
Lysimachia punctata						5					
Lythrum salicaria		2,500	30			5				2,500/1,000	
Lythrum virgatum		2,500				5					2,000/3
Malva alcea	25	2,500	15	1,250		2.5					
Malva sylvestris						2.5					
Miniature roses			25								
Monarda citriodora			45								
Monarda didyma	25	2,500	30		500	5	45			2,500/1,000	2,000/3
Monarda x *hybrida*			30			5					2,000/3
Myosotis alpestris		2,500									

Myosotis sylvatica		2,500				5					
Nepeta x *faassenii*	25	2,500	30		500	5					2,000/3
Nepeta mussinii		2,500		1,250		5					2,000/3
Nepeta subsessilis											2,000/3
Oenothera fruticosa		2,500				5					
Oenothera missouriensis						5					
Oenothera speciosa						5					
Oxalis crassipes				1,250		5					
Pachysandra terminalis											2,000/3
Paeonia lactiflora						2.5					
Papaver orientale		2,500									2,000/3
Pennisetum setaceum	50					10					
Penstemon barbatus						5					
Penstemon campanulatus	25	2,500		1,250		5					
Penstemon digitalis	25	2,500	30	1,250		5					2,000/3
Penstemon x *hybrida*		2,500	30								
Penstemon smallii		2,500									
Perovskia atriplicifolia	25–50	2,500	45	1,250	500	10	30			2,500/1,000	2,000/3
Phlox arendsii											2,000/3
Phlox carolina											2,500/5
Phlox maculata											2,500/5
Phlox paniculata		3,750	45	1,250		10	45				2,500/5
Physalis franchetii						2.5					
Physostegia virginiana		2,500		1,250	500	5					2,000/3
Platycodon grandiflorus	25	2,500	30			5					2,000/3
Polemonium caeruleum		2,500	30			5			2,000/15	2,000/1,000	2,000/3
Polemonium yezoense											2,000/3
Polygonum aubertii			30								

(continued)

Table 8.17. Perennial Plant Growth Regulators *(continued)*

Variety	A-Rest (Ancymidol) (ppm)	B-Nine or Dazide (Daminozide) (ppm)	Bonzi, Piccolo, or Paczol (Paclobutrazol) (ppm)	Cycocel (Chlormequat Chloride) (ppm)	Florel (Ethephon) (ppm)	Sumagic (Uniconazole-P) (ppm)	Topflor (Flurprimidol) (ppm)	A-Rest (Ancymidol) + B-Nine (Daminozide) (ppm)	B-Nine or Dazide (Daminozide) + Bonzi, Piccolo or Paczol (Paclobutrazol) (ppm)	B-Nine or Dazide (Daminozide) + Cycocel (Chlormequat Chloride) (ppm)	B-Nine or Dazide (Daminozide) + Sumagic (Uniconazole-P) (ppm)
Primula obconica		2,500									
Primula polyanthus						5					
Pseuderanthemum lactifolia		2,500	30							2,000/1,000	
Ranunculus repens		2,500									2,000/3
Rosmarinus officinalis		2,500	30			5				2,000/1,000	
Rudbeckia fulgida	50	3,750	45	1,250		10				2,500/1,250	2,500/5
Rudbeckia hirta		3,750	45			10					2,500/5
Rudbeckia nitida						10					
Rudbeckia triloba		3,750	45	1,250		10				2,500/1,250	
Rumex sanguineus						2.5					
Salvia farinacea	25		30		500	5					
Salvia greggii		2,500	30			5				2,500/1,000	
Salvia guaranitica		2,500		1,250							
Salvia leucantha		2,500	30	1,250	500	5	30			2,000/1,000	
Salvia lyrata											2,000/3
Salvia nemorosa		2,500	30		500						2,000/3
Salvia officinalis		2,500									
Salvia x *superba*	25–50	2,500	30	1,250		5					2,000/3
Salvia x *sylvestris*		2,500	30							2,000/1,000	2,000/3
Salvia verticillata						5					
Saponaria ocymoides	25–50										
Scabiosa caucasica						5					2,000/3
Scabiosa columbaria		2,500	30		500	5	30				

Sedum spectabile		2,500	30		500	5	45		2,000/15	2,000/1,000	
Sedum spurium			30			5					
Sedum x *hybrida*			30			5					
Solidago canadensis		2,500	30			5					
Solidago luteus		2,500	30						2,000/15	2,000/1,000	
Solidago sphacelata	25–50	2,500	30	1,250		5					2,000/3
X *Solidaster luteus*		2,500	30							2,000/1,000	
Stachys byzantina						5					
Statice tataricum		2,500									
Stokesia laevis	25	2,500	30	1,250	500	5				2,000/1,000	2,000/3
Stokesia cyanea		2,500	30							2,000/1,000	2,000/3
Tanacetum coccineum	25	2,500		1,250		5					
Tanacetum parthenium			15	1,250		2.5					
Thymus x *citriodorus*						5					
Tiarella x *hybrida*						5					
Tradescantia x *andersoniana*						10	45				
Tradescantia virginiana						10	45				
Tricyrtis hirta	25-50	2,500	30			5					
Verbena bonariensis		2,500	30			5	30				
Verbena canadensis			45		500					2,000/1,000	
Veronica alpina		2,500	30			5				2,000/1,000	
Veronica longifolia	25	2,500	30	1,250		5			2,000/15	2,000/1,000	
Veronica peduncularis		2,500	30	1,250					2,000/15	2,000/1,000	
Veronica spicata	25	2,500	30			5				2,000/1,000	
Veronica x *hybrida*	25	2,500	30	1,250		5	30		2,000/15	2,000/1,000	
Viola cornuta						2.5					
Viola tricolor						2.5					

NR = No response The information provided in this chart is a compilation of research primarily from Paul Pilon and Joyce Latimer (Virginia Tech).
The author would also like to acknowledge contributions from Chemtura, Michigan State University, OHP Inc., SePRO Corp., and Valent USA.

9

Perennial Propagation

Propagation is the method used to reproduce plant materials. The two types of propagation are sexual propagation, such as with seeds, and asexual propagation, as with vegetative cuttings. Each type of propagation entails ensuring the cultural requirements for the species being propagated is delivered at the proper time and is consistent with the method of propagation.

Young plant materials are often referred to as plugs or liners. Sometimes these terms are used interchangeably. Plugs are generally items started from seed, and liners are most commonly started vegetatively.

Many growers self-propagate nearly everything they grow, while others purchase propagated materials from growers who specialize in plugs or liners. Some only propagate varieties they have been successful with in the past and then purchase the remainder. "To propagate or not to propagate", is a question each operation must answer for itself.

Propagation from Seed

Seeds are the result of sexual propagation and contain the entire embryonic plant covered with a protective coating. They are living organisms that remain dormant until they are exposed to favorable environmental conditions. These conditions include moisture, temperature, and light. Many growers like propagating perennials from seed, because, in many instances, perennial seed is relatively inexpensive, easy to handle, and grows better than other forms of propagation.

There are literally thousands of perennial varieties that can be propagated by seed, and each has its own set of germination requirements. Many perennial varieties, such as *Astilbe, Asclepias, Baptisia, Bergenia, Campanula, Delphinium, Liatris, Primula,* and *Sempervivum,* are often challenging to germinate, resulting in frustrations for many growers.

Perennials are sown at various times throughout the year, depending on the type of finished product (non-flowering vs. flowering) desired and the growing stage during the winter months (active vs. dormant). Many spring-flowering plants, such as *Aquilegia,* will usually not flower during the same year as the seed was sown. It is best to sow the seeds of spring-flowering plants in the late summer or early fall the year before flowering is desired. Many of the summer-flowering perennials will bloom when started from seed during mid to late winter. Typically, perennials that are going to be overwintered in quart (1 L) sized or larger containers are sown in the late spring to early summer months.

Perennial varieties often have poor germination rates, low vigor, seed dormancies, and they may contain weed seeds and/or may not be cleaned well, containing plant debris. Performance of each seed lot varies from supplier to supplier, and even from lot to lot from the same supplier. Due to these factors, predictably germinating perennial varieties can be somewhat challenging.

It is fairly easy to learn how to successfully propagate a few varieties of perennials from seed, but most growers' product lines consist of dozens, if not hundreds, of species and cultivars. It is beneficial to conduct some research on each variety you are producing, such as each plant's background, origin, and natural habitat, to better understand its germination requirements. With wide product lines, it takes a greater understanding of the germination requirements of each of the varieties, not to mention the ability to provide for each plant's specific needs. This may limit the production of certain varieties some operations can offer. For this reason, many growers purchase perennial plugs from growers who specialize in perennial seed propagation.

Perennial Seed

In recent years, there have been some seed suppliers offering enhanced perennial seed. Enhanced seed has basically been treated in some manner to improve the ease of sowing and/or improve the germination percentages and times. Some of the commercially available methods of enhancing perennial seeds include cleaning, de-tailing, and treating to alleviate seed dormancies.

Refined seed

Seed that is marketed as refined seed has gone through a process separating it by physical characteristics such as color, density, shape, and size. Refined seed consists of uniform characteristics throughout the seed lot, making it more reliable for the grower. Growers using refined seed observe higher germination percentages, quicker germination, and improved uniformity.

De-tailed seed

There are numerous perennial varieties that have hairs, spines, or other structures that, when left in the "raw" form, make them difficult to sow with automated equipment. It is often beneficial to remove (de-tail) these structures before growers sow the seed. Defuzzed and dewinged are other terms meaning the same thing. Removing these unnecessary structures makes handling easier and allows growers to utilize their automated equipment. The procedures used to de-tail seeds often cause injury to the seed coat and may decrease the storage life of the seed. It is best to use these seeds within a short time after receiving them.

Coated seed

Several seed suppliers apply a thin film coating made of clay or other materials to the seed coat. Film coatings are used to smooth out the rough surfaces of many perennial seeds and do not change the overall shape of the seed. Seeds with smooth edges are easier to handle and to sow using automated seeding equipment. Some film-coated seeds have slightly reduced germination rates; however, any reduction in germination caused by film coatings is more than offset by improved handling and sowability.

Pelleted seed

Pelleting is the process of applying a clay coating for the purpose of making small seeds larger. Pelletizing very small seeds improves the ability to handle and properly sow these varieties. The pellets are colored, usually yellow or white, improving their visibility during the sowing process.

Many suppliers offer multipellets, which consist of multiple seeds within the pellet. Typically, three to five seeds are within each pellet, but there are some varieties that contain twenty or more seeds per pellet. Multipellets are a good option for growers who intend to multiple sow and have difficulty handling small seeds.

Primed seed

Many annual seeds are primed, or osmoconditioned, prior to sowing. As the demand for more uniform germination of perennials continues to increase, priming of perennial seed will become more prevalent. Priming entails hydrating the seeds using an osmotic solution to activate the metabolic processes of germination. Before the seeds germinate, the seed is dried down to its original moisture content and handled as raw seed would be handled. After a grower sows the seed, germination occurs quickly since many of the metabolic processes have already been completed. Primed seed allows growers to increase the germination speed and rate, improve the germination percentage, increase seedling vigor, and achieve greater uniformity. Every variety requires a slightly different priming process. All primed seed has a reduced storage life and should be used in a timely manner.

Enhanced seed

Many seed suppliers have developed or are developing their own proprietary methods of treating seed to break seed dormancies and to improve germination. Enhanced seed is designed to overcome some of the grower's problems by providing faster germination, increasing the overall germination rate, and providing more uniform germination, as compared with raw seed. Enhanced seed will not solve all of a grower's problems and by no means substitutes for having sound cultural practices, but it can improve the germination success of several difficult varieties.

There are several treatments used by seed suppliers that involve one or more of the commercially known methods, such as hormone treatments, priming, or scarification. Other treatments being used are proprietary and are highly guarded. Perennial seed from Jelitto's Gold Nugget line is an example of commercially available enhanced seed. Enhanced seed has a limited shelf life and should only be purchased just prior to sowing. Storing treated seed will result in inconsistent germination results.

Seed quality

There are many variables regarding seed quality with perennial crops that are not issues with growers accustomed to producing annuals from seed. Some of the attributes associated with seed quality include germination, vigor, dormancy, cleanliness, genetic purity, and product form. The product form refers to the type of seeds being offered commercially, whether it is seed in the raw form or some type of enhanced product, such as seeds that have been primed, pelleted, pregerminated, detailed, defuzzed, or coated.

There are a number of perennial seed suppliers from whom growers can obtain seed. It is usually advantageous to acquire perennial seeds from large suppliers who have their own testing labs where they check for vigor, viability, and germination rates. I recommend purchasing seed from a reputable company that labels each package with a lot number and germination percentage.

The vigor within a seed lot influences how well seeds germinate. For example, a seed lot with high vigor may germinate successfully at non-optimal temperatures, while a seed lot with poor vigor may not germinate well when temperatures are outside the optimal range for that variety. Some seed companies measure the dry weight of a seed lot to determine its physiological maturity, or the point at which the seed has the greatest potential for maximum germination and vigor. Ball Horticultural Company has developed a patented computer-imaging system, called the Ball Vigor Index, to evaluate seed vigor. Unfortunately, measuring the true vigor of a seed lot is difficult to do, and with perennial crops the seed vigor is rarely provided when the seed has been ordered.

Disinfecting seed

It is possible to receive perennial seed from suppliers that contains plant pathogens on the seed coat. When growers have histories of always having a plant pathogen during the germination process or even after the seeds are germinated, there is a good chance a pathogen is coming in with the seeds. To verify plant pathogens on the seed coats, growers can send samples of seed to a good diagnostic laboratory, which can attempt to culture disease organisms from the seed.

Once pathogens are identified, growers can take steps to reduce the likelihood pathogens will pose any serious threat to the crops. Disinfection of the seed prior to sowing can help to reduce problems associated with infected seed lots. Soaking the seed for five to thirty minutes in a 10% bleach solution can be very helpful for disinfecting the seed coat and reducing plant pathogens. Soaking in a hydrogen dioxide (Zerotol) solution at 1.25 oz. per 1 gal. of water (37 ml/3.8 L) is also an effective method of disinfecting the seed coats.

I have also tried soaking seed in various fungicide solutions, such as copper sulphate pentahydrate (Phyton-27) or mefenoxam (Subdue Maxx). It is difficult to precisely determine the effectiveness of any of these seed soaks unless growers set aside a control (sowing some seed that has not been treated). In most cases, there are no negative or detrimental effects on germination or young seedling growth when seeds are soaked in disinfectants or fungicide solutions. In fact, I have often observed a slight (up to 5%) increase in the overall germination rate when using various seed soaks. But there is the potential for some varieties to have a reduced rate of germination following seed soaks. Test these treatments on a small scale to evaluate the effect on germination before treating large amounts of seed.

Seed storage

It is best to receive seeds just prior to sowing. Since seeds are living organisms, the fresher the seed, the greater their odds are of being viable. If the seeds need to be stored, they should be stored at cool temperatures (40–50° F [4–10° C]) with low relative humidity (20–40%). Providing cool temperatures and low relative humidity conditions helps to reduce the respiration or natural processes of the seed, which increases the life and viability of the seed, allowing it to be used at a future date. Many growers use refrigerators, with or without dehumidification systems, to provide the proper conditions for seed storage. Seeds should also never be allowed to freeze, as freezing can kill them when the cell membranes burst from the rapid expansion of the freezing water in the cells. Even when seeds are placed under ideal storage conditions, their vigor and germination decline over time. Improper storage of perennial seeds will dramatically reduce plant vigor, particularly with poor quality seed lots.

The seeds should be well labeled and placed in sealed containers prior to placement in a cool storage area. Any seeds being stored should be dry and protected from exposure to moisture during the storage period.

When seeds are removed from storage, there is the potential for them to absorb moisture, which can decrease viability. Any changes in temperature should occur gradually to allow the seeds adequate time for acclimation.

Although many varieties of perennial seeds store well, it is difficult to determine how a seed lot will perform after being in storage, regardless of the duration. Many seeds can be successfully stored for over one year with very little reduction of viability and vigor, while other perennial seeds do not fare as well, even under optimal storage conditions. It is impossible to predict how seeds that have been in storage for over six months will perform. If seeds must be stored, a short duration of time, three months or less, is preferred.

Germination testing

There is a quick and easy test growers can conduct on site to determine the germination rate of a particular seed lot. The germination rate is simply the percentage of seeds that sprout.

First, moisten a paper towel, draining off the excess water. Place seeds (ten, twenty, or one hundred are common quantities) on the wet towel, making sure you record how many seeds you are testing. Fold the towel in half over seeds and press, don't squeeze, the paper towel so it is firmly in contact with the seeds. Place the paper towel and seeds in a small zipper plastic bag and seal it. Then place the bag in a warm place.

Each day check the paper towel to ensure it remains moist and record the number of seeds that have germinated. Depending on the type of seed, the germination test should last from three to twenty-one days. Many perennials will finish germination within ten days of starting this test. Count the seeds that have sprouted and remove them from the towel to prevent them from affecting the other seeds in the test. If seeds turn moldy, remove them and consider them as dead. Write down the number of seeds that have sprouted and turned moldy.

Calculate the percentage that sprouted out of the total tested—this is the germination rate. The germination rate is the average number of seeds that germinate over a reasonable amount of time. The mathematical formula is:

Germination % = (Number of Seeds Germinated/Total Seeds) x 100

For example, if eighty-six seeds germinated in a towel of one hundred seeds, you would figure it as follows: Germination % = (86/100) x 100, which equals 86%. If fourteen seeds germinated in a towel containing twenty seeds, the equation would be (14/20) x 100, which comes to 70%.

Seed dormancies

Seed dormancy is a common problem perennial propagators face. Dormancy is a natural protective mechanism that has evolved to help plants survive in the wild. The seeds are unable to germinate due to their anatomy and/or physiology until certain criteria are met. The mechanisms behind these dormancies are not fully understood. When we understand a plant's origin, natural climate, habitat, and bloom time, we will better understand how to commercially germinate its seed.

Commercially there are many techniques, some of them proprietary, that are being used to overcome seed dormancy of perennial seeds, improve germination rates, and increase crop uniformity. Until perfected, any treatments growers implement to overcome seed dormancies should occur over multiple seed lots and should always contain a control for comparison.

Hard seed coats

Some seed dormancies occur when the seed coat is impermeable to water. Perennials with hard seed coats are commonly found in the Convolvulaceae, Fagaceae, Geraniaceae, and Malvaceae families. To overcome this type of dormancy, see the scarification methods described on page 255.

Chemical dormancies

Chemical dormancies occur when various chemicals outside of the embryo inhibit the germination process. Seeds that are in a fruit often contain these germination-inhibiting chemicals. *Iris* is an example of a perennial with a chemical-induced dormancy. To overcome chemical dormancies, leach the seed with copious amounts of water to rinse the chemical inhibitors off the seed coat.

Physiological dormancies

There are physiological dormancies that often occur with perennial seeds. Some perennials, referred to as frost germinators, germinate poorly, or not at all, until they pass through a cold period, or stratification. Stratification entails several months of cool, moist conditions or multiple of cycles of cool, moist conditions followed by warm temperatures. Generally, the stratification period for most perennials with this type of dormancy is twelve weeks or less. Some plants may require up to twenty weeks of stratification before they will germinate properly. Frost germinators are discussed in more detail on page 254.

Another type of physiological dormancy is an absolute light or dark requirement for germination to occur. A lot of perennials have a light requirement; many of these are shown in table 9.2 at the end of this chapter.

Some researchers and growers are fortunate enough to overcome certain physiological dormancies by soaking the seed in potassium nitrate or gibberellic acid (GA) solutions. When using GA as a substitute for stratification, try using a solution with 200–1,000 ppm GA and soaking the seed for twenty-four hours. This will not work on all perennials, and the rates to use are not clearly known at this time, but this should provide a good starting point for growers. Soaking seeds may help to improve the germination but usually does not completely eliminate the dormancy.

Morphological dormancies

Morphological dormancies occur when the embryos are undeveloped, often resulting in a permanent dormancy, which cannot be overcome. There are some instances, when given the proper conditions, the embryo can develop prior to germination. Sometimes the area beneath the seed coat is not large enough to accommodate the embryo. This type of dormancy is impossible to detect and often gets confused with poor germination that often occurs due to poor cultural practices.

Double dormancies

There are also seed dormancies that consist of two factors, often referred to as double dormancy, such as seeds having both underdeveloped embryos and hard seed coats. When overcoming multiple dormancies within the same seed, it is important to overcome them in the proper order. In this case, it is necessary to overcome the morphological dormancy by providing warm conditions to develop the embryo, followed by scarification to allow water to penetrate the seed coat. Certain *Anemone* cultivars experience these double dormancies.

Secondary dormancies

Secondary dormancies often occur when germination fails due to unfavorable environmental conditions. Temperatures either too warm or too cold, inappropriate moisture levels, or some combination of these often send seeds into a secondary dormancy. Seeds can enter these dormancies even after a primary dormancy has been broken.

Once a seed enters a secondary dormancy, it is difficult to remove the dormancy, allowing for normal germination to occur. For example, with seeds having secondary dormancies caused from excessive temperatures, it is difficult to achieve normal germination by simply returning the temperatures back to optimal levels. Commercially, priming is used to overcome secondary dormancies. Priming is a treatment that involves hydrating the seed, starting the germination process, but not allowing the seed to germinate. Priming is effective at overcoming certain secondary dormancies, as well as improving the overall uniformity of germination and the germination rate of several perennial varieties.

Frost germinators

Some perennials, referred to as "frost germinators," require a cold period called stratification before germination can occur. The term *frost germinator* is a little misleading and is used interchangeably with *cold germinator,* as these varieties require a period of low temperatures to break dormancy. Stratification is the process of using a cool, moist period to overcome dormancy. Many perennials can be successfully stratified by placing the moistened seed flats at 40–44° F (4–7° C) for six to twelve weeks following a warm, moist period. Exposure to warm temperatures or dry conditions during the stratification process will often inhibit or delay germination.

Seed sown in the autumn and overwintered in cold frames is a good example of how growers stratify certain frost germinators. Growers commonly sow the seeds in a plug tray, keeping it moist and placing it in a warm environment (65–70° F [18–21° C]) for two to four weeks. For many perennials, such as *Asarum, Corydalis,* and *Helleborus,* the warm period is needed for the embryo to finish developing to a size that is capable of germinating. Following the warm period, the plug trays are moved to a cold environment (25–39° F [-4–4° C]) for four to six weeks before placing them at ambient temperatures for germination. To increase the likelihood of breaking the dormancy, the temperatures during the cold treatment for most perennials should not rise above 41° F (5° C). Many perennials in the *Ranunculus* family have a slightly colder chilling requirement (23° F [-5° C]) to overcome their seed dormancy.

If the temperatures fluctuate above the recommended ranges, some time should be added to the cold period to compensate for the inadequate temperatures. It is very important to moisten the seed during the stratification process, as dry seed exposed to cold conditions will not break dormancy and could be injured when exposed to freezing conditions. Temperatures below 20° F (-7° C) usually do not directly harm the seed itself but affect the germination process and delay the dormancy.

Following the cold treatment, the plug trays should be exposed to ambient temperatures (41–54° F [5–12° C]) to promote germination. Exposing them to more "normal" greenhouse temperatures (above 65° F [18° C]) will often induce a second dormancy, which is more difficult to overcome than the first dormancy.

Inspect the trays for any germinated seedlings, which should be transplanted, and the flat should be returned to the cold environment. Typically, cold germinators emerge over a period of several weeks (sometimes months) rather than germinating simultaneously. Sometimes the warm-cold-warm treatments do not produce satisfactory or commercially acceptable germination rates. If the germination rate is lower than expected, it is usually beneficial to repeat this process again rather than discarding the plug flats.

Certain perennials respond well to the methods described above, while others, such as *Cimicifuga,* take considerably longer (often a year or more). For these, it might be more practical to stratify using a slightly different method. This method involves creating layers of moist sand alternating with a thin layer of seeds, followed by moist sand and so on. Growers have used wood or concrete boxes to hold the layers of sand and seed. These stratification boxes are kept in a shady location with a fine wire mesh covering to protect the seed from rodents and birds. The boxes should be checked frequently during the spring for germination. Once germination has started, the seeds should be sown immediately in plug trays or open flats. One advantage of this method is that it reduces the amount of production space needed and ensures the seeds stay moist.

Scarification

Some perennials species, such as *Baptisia, Malva,* and *Sidalcea,* have seeds with very hard seed coats that prevent the intake of the water necessary to trigger germination. It is often beneficial to scarify the seed coats prior to sowing. Scarification is a process of chemically or physically breaking down the seed coat, allowing water and oxygen to enter more easily and allowing germination to take place.

One method to scarify seed when planting small amounts is to nick or file off a portion of the seed coat. This is an effective but slow method of scarification. Another method for treating small amounts of seed is to scarify the seeds by grinding them in dry, sharp sand or rubbing them with sandpaper. To make this process slightly faster, or to scarify larger amounts of seed, some growers nick the seed coats mechanically by placing them in a rock tumbler.

Commercially, growers chemically scarify seeds using sulfuric acid. They typically soak the seeds in sulfuric acid for five to twenty minutes. The proper precautions should be taken any time growers or workers are using acids. Some seed suppliers scarify seeds before they are delivered to growers. Unfortunately, scarifying seeds can cause injury to the embryos, adversely affecting the germination rate.

Another scarification method involves soaking seed in hot water. The water should be brought to a boil and then allowed to cool to 180° F (82° C). Place the seed in the hot water for several hours or even overnight. If available, maintain the temperature of the water using a Crock-Pot. Do not place the seed directly in the boiling water, which would damage the seed, resulting in poor or even no germination. Also, the seed coats of some varieties can be softened by soaking them for several hours in lukewarm water prior to sowing.

Figure 9.1. **Following a soak in hot water, these *Baptisia* seeds are layered on a paper towel to dry.**

Fresh germinators

There are many perennials, such as *Helleborus,* that are considered to be "fresh germinators," whose seed is short lived and cannot be stored for extended periods, or possibly not at all.

Many of these plants are difficult to germinate commercially because it may take up to two years for the seeds to sprout. When possible, sow fresh germinators immediately after they have been harvested. If these varieties must be stored, place them in moistened peat or growing medium until you are ready to sow them. Some growers have luck providing stratification to varieties that have been stored, although the results have been variable and it may not always be effective.

Many of the fresh germinators, such as *Helleborus,* have fleshy seeds that should not be allowed to dry out. The shelf life of these varieties is often six months or less. For fresh germinators, it may be beneficial for growers to collect their own seed from mother plants grown specifically for seed production and sow the seeds immediately after harvesting.

Plugs

Plugs are the chosen method for producing young plants from seed. Plugs are seedlings that are produced in their own container or cell. Plug trays contain many cells (twenty-one to eight hundred) and are commercially produced by many growers. Large plug sizes, such as twenty-one-cell or thirty-cell trays, are commonly used for supplying starter plant materials to the industry for quicker finishing times.

Figure 9.2. **Growers produce various plug sizes for selling or transplanting into larger containers. Left to right: 220-cell *Echinacea,* 128-cell *Aster,* 72-cell *Artemesia,* and 21-cell *Hosta.***

Some advantages of plugs include faster growing times, less labor inputs, more cost effective than vegetative propagation, less transplant shock, and uniform growth of plants in the final container. There are some disadvantages of producing plugs over the open flat method of geminating seedlings, such as higher costs per seedling, greater space requirements during production, and the need for specialized equipment. Plug production is not suitable for all perennial growers, but most operations utilize plugs either grown themselves or purchased from other growers.

Five stages of plug production

Stage 1

The period of time from sowing until radicle emergence is a critical phase. The conditions for this stage are uniform temperatures, adequate moisture, and, depending on variety, light. After seeding, it is beneficial to place plug flats into a germination chamber. Chambers allow growers to maintain sufficient temperature, humidity, and light levels. Many times this stage is done in the greenhouse with great success; however, using a chamber improves germination and crop uniformity. Once the radicle emerges, move the flats directly to the greenhouse. Germination chambers are discussed in detail on pages 262–263.

Stage 2

The primary purpose of this stage is to develop a root system. Stage 2 usually lasts until the first true leaves develop. The first set of leaves are the cotyledons. They often do not even resemble the appearance of the plant being grown. The next sets of leaves are true leaves and often resemble the variety being grown. It is still important to manage the watering at this stage, until they become established. Providing adequate soil temperatures is also important during this stage. Bottom heating is beneficial to promote rooting but is not necessary. The relative humidity can be reduced to 70–80% during this stage. Fertilizers are usually applied once the first true leaves are present.

Stage 3

This stage entails producing the plugs from the first true leaves until they reach either a shippable or transplantable size. Most of the growth occurs during this stage. Supplemental fertilizer is necessary to promote healthy plant development. Most perennials do not require high fertilizer levels to finish. Fertilize with a balanced fertility program including a minor nutrient package at 50–100 ppm nitrates at nearly every watering. Depending on factors, such as water quality, it may be beneficial to acidify the irrigation water.

Stage 4

This stage involves preparing the seedlings for shipping, transplanting, or holding. Consider this the toning or acclimation phase. The plugs are usually adjusted to growing at lower temperatures and humidity levels (35–40%).

To acclimate plugs for overwintering (described in stage 5), the temperatures should be dropped to 35° F (2° C) gradually over a four- to six-week period. For example, dropping the temperature by approximately 1° F (0.6° C) per day or 5° F (3° C) per week should provide for proper acclimation.

Stage 5

Holding the plugs until they can be transplanted or shipped is the purpose of this stage. Consider it the maintenance stage. For perennials, the length of time involved often varies from stage 5 of annuals. For example, many perennial growers often offer vernalized plugs (vernalization is a cold period often required for flowering) to their customers. Before vernalization begins, many varieties also have a juvenility requirement, meaning they have to be of a certain size or age before entering the cold period.

Vernalization is usually done in cold frames or greenhouses during the winter months but can effectively be achieved using a cooler. The cold requirement for most perennials is at least ten weeks at 41° F (5° C) or less. With smaller plug sizes, greater plant losses occur at temperatures below freezing; therefore, plugs should be overwintered between 35–40° F (2–4° C).

Water management is a major aspect of the holding period, especially during the dormant season. The plugs should contain some moisture, but not be saturated. It is extremely easy to keep plugs too wet, potentially leading to root rots and poor nutritional uptake. With low light intensity, low temperatures, and little water uptake from the plants, it takes a great deal of time to dry dormant plugs out during the fall and winter months. To prevent future losses, make sure that the plugs are not under any nutrient stress or disease pressure before entering the overwintering period.

Sowing

There are a number of seeders available to growers, with a wide range of prices. Although many seeders work well, they generally only work as well as the individuals operating them. Sowing is a detail-orientated task, and the people who are placed in seeding positions should be able to pay attention to details and have good observational skills.

It is important to sow seeds accurately to optimize the germination and overall plant stands. The best results are achieved when the seeds are placed near the center of the plug cell; seedlings tend to root poorly when they germinate near the edges of the cells. Many growers use a dibbler to create a depression in the medium to help guide the seeds toward the center of the cell. With seeds consisting of various sizes and shapes, it is important to make the proper adjustments while sowing each variety, to improve seeding accuracy and overall germination success.

Some perennials, such as *Lobelia,* have small seeds that are extremely difficult to handle and sow properly. It is very easy to sow too many, too few, or to improperly place them in the plug flat or plug cell. To obtain more even distribution of seeds, growers often mix them with a very fine sand or talcum powder and sprinkle this mixture, using a shaker (salt shaker) over the growing medium of a premoistened plug flat or tray. Growers usually do not cover these small-seeded varieties, but they may press the seeds very lightly into the

Figure 9.3. **Sowing equipment, such as drum seeders, allow growers to accurately sow large quantities of plug flats economically.**

growing medium. Great care should be taken to not wash the seeds away when irrigating.

After the seeds are sown, it is recommended to pass the flats through a water tunnel of some sort or to gently water the flats to moisten but not saturate the growing medium. Care should be taken to not overwater and wash or displace the seeds from the individual cells. After sowing, the trays should be handled carefully because the seeds can and do bounce around.

To improve plant stands, most growers sow multiple seeds per cell. The number of seeds to put into each cell depends on the variety being sown, the anticipated germination rate, and the grower's preference. Multiple sowing helps growers to hit their numbers and/or helps to improve the quality of their products. Most commonly, growers sow two to five seeds per cell when multiple-sowing perennials. There are numerous seeders in the industry that have the ability to sow several seeds at a time into individual plug cells. Otherwise, growers could run the plug trays through the seeder more than one time to achieve the desired number of seeds per cell.

The other method, to ensure growers make their planned numbers, is to sow more trays than they actually need (also called over-start). For most perennial varieties, the over-start quantity is 10–25% more than the desired quantity. When the anticipated over-start is greater than 25%, growers will often acquire these varieties from plug suppliers rather than tying up their valuable production space.

It may not be necessary or beneficial to multiple sow all perennial varieties. Growers should first determine if multiple seedlings in each cell are desirable. Several seedlings in a cell might produce fuller containers after transplanting or can gain a perceived quality advantage over the competition. Unfortunately, multiple seedlings per cell sometimes lead to competition between the plants, increased height, softer growth, and the potential for plant diseases to set in.

Price is another consideration. Many perennial varieties' seeds are relatively affordable, and multiple sowing can be easily justified. But some varieties are very expensive and, for some growers, it too costly to multiple sow. The anticipated germination percentage or rate is also beneficial to know when deciding how many seeds to sow into each cell. Cultivars with high germination rates are often sown with one seed per cell.

Growing mixes

The type of growing medium or substrate used for germination is very important to germination, plant development, and overall success.

Germination mixes are usually light and porous, retaining sufficient amounts of moisture while allowing for proper drainage. Commercially available mixes usually contain a blend of peat moss, perlite, vermiculite, and sometimes pine bark or sand. These mixes must be free of weed seeds, insects, and disease organisms. They generally have a low nutrient charge with a pH range of 5.4–6.4.

Unfortunately, there is not one growing medium that is perfect for all perennial growers or situations. The most important factor with a plug medium is consistency of the physical and chemical properties over time. Once a blend has been determined, the components should remain constant without alterations to the ratios used or their particle sizes.

Since most growers are using plug sizes that are relatively small, the size of the bark, perlite, and vermiculite particles should also be small. Avoid mixes that appear chunky unless these mixes are used for very large plug sizes and smaller finished plugs are going to be transplanted into them. Growers should also avoid growing mixes containing excessive amounts of fines that can lead to improper air and water holding. Mixes containing excessive fines often contain sand, soil, or compost.

Figure 9.4. Many seed propagators sow perennials using Gro Mor Vibro Hand seeders. These sowing tools are especially useful for small, hard-to-handle seeds.

The porosity, or the pore spaces available to hold air and water, should also not change over time. With any plug mix, the compaction that often occurs while filling the flats and the initial watering practices during stages 1 and 2 has the greatest affect on the pore spaces and the ability of the mix to hold air and water. Take care while filling flats to reduce the compaction and maintain the optimal physical properties of the growing medium.

The initial chemical properties—such as the pH, soluble salts, and any nutrient charges—should all be within acceptable ranges. Many commercial plug mixes contain a starter charge of nutrients, which can help young seedlings get off to a good start by satisfying their early nutrient needs. Since plugs contain a relatively small volume of growing medium, growers should pay particular attention to their early watering practices to prevent leaching of these nutrients. Growers should regularly monitor the pH and electrical conductivity (EC) of their plug crops. For plugs, the general recommendations for most perennial crops is to maintain the pH of 5.6–6.2 using the 1:1 extraction method and the EC at 0.3–0.7 using the 2:1 extraction method.

The key is to find a mix that is suitable for the sizes you wish to grow and fits your growing style, watering practices, and facilities. Make small modifications over time, not drastic rapid changes. Quite often growers will make a dramatic change in the growing medium to save a few dollars. When this occurs, they often have to change their cultural practices and often experience plant losses. In many cases, it may be more costly than the savings they incurred.

Great care should be taken to properly fill the plug trays with growing medium. Most growers take filling plug trays for granted, perhaps not appreciating the effect it has on germination or other cultural problems. Improperly filled trays and poor watering practices can lead to compaction and poor oxygen levels, particularly with small cell sizes. It is not uncommon for some plug trays to be filled with growing medium that is overly compacted, resulting in reduced germination and increased losses from plant diseases. Many times the individual cells within a flat have differing amounts of growing medium; some are overly full, while others are only partially filled. When the flats are improperly filled, they do not hold moisture uniformly, making it difficult to manage the irrigation, which often leads to erratic germination.

To properly fill plug trays, it is important to use moistened growing medium, moist but not saturated. It should be moist enough to hold together loosely when squeezed, with no water droplets coming out. If the medium is too wet, it is difficult to fill the trays without having unwanted air pockets. Conversely, filling with medium that is too dry causes undesirable settling of the medium, leading to reduced root volumes and aeration. Most commercial mixes do contain some moisture, but it is often insufficient for proper flat filling. Growers should have a water source at the flat filling station. When the flats are filled, the growing medium should be moist enough to just hold a dibble at sowing.

In addition to the moisture content, the flats should be packed uniformly with soil. The smaller the cell size of the flat, the more difficult it is to properly fill them. The goal of flat filling should be to obtain the same volume of medium in every cell in the flat. Differing volumes will lead to uneven drying out, making irrigation management more difficult throughout production. Take some time to examine how each size plug flat is being filled and adjust your methods accordingly.

Growers also commonly use plug flats that have been filled for several days, or even weeks, prior to sowing. When possible, plug trays should be filled no more than one day prior to sowing. Prefilling flats may seem convenient, but it may cause the growing medium to dry out before sowing. If plug flats must be filled in advance, cover them with plastic and place them in a cool area to help reduce the loss of moisture. Avoid stacking filled plug trays directly on top of one another; this leads to compaction and a dramatic reduction in the air porosity.

Moisture

All seeds have a moisture requirement, and without water, germination cannot occur. Moisture management during germination is critical. Improper moisture levels can significantly reduce germination and possibly kill small, newly germinated seedlings. Growers should provide the proper moisture levels with minimal amounts of fluctuation or swings of moisture levels. Many growers commonly overapply irrigation, creating saturated conditions, which reduces the oxygen levels in the growing medium, puts unnecessary stress on germinating seeds, and increases the likelihood of plant pathogens.

Moisture levels necessary for germination vary with each plant species. The optimal moisture levels for most perennials can generally be split into three groups: drier than average, average, and wetter than average. (See table 9.2. at the end of this chapter.) Until the radicle is emerged from the seed, it is recommended to maintain 80–90% relative humidity. The relative humidity can be maintained at this level using a germination chamber with a fogging system or by simply covering the plug flats with plastic. Once germination has occurred, the humidity levels can be dropped to 70 to 80%.

Where seeds are covered, the moisture levels around them remain stable longer, holding onto the optimal level needed for germination. It's particularly challenging for growers to manage the moisture level of perennials with small seeds, such as those of *Heuchera,* during the germination process. Typically, small seeds are not covered with growing medium or vermiculite when they are sown. Avoid heavy overhead watering following sowing, which could dislodge the seeds or wash them out of the plug flat altogether.

In addition to the moisture levels, the temperature of the irrigation water can also play a critical role with growth of perennial plugs. It is not uncommon for the medium temperatures to drop by at least 10° F (6° C) following irrigation, especially in northern climates or during the winter months. This type of temperature drop could delay root development and reduce overall crop quality. Most growers do not take precautionary measures to deliver water with consistent temperatures to their crops. It may be beneficial to consider installing a hot water heater with a valve for regulating water temperatures in your plug production facilities.

Temperature

Providing the proper temperatures during the germination process is critical. The general temperature range for germinating perennial seed is 68–86° F (20–30° C). There are basically three temperature regimes growers use for germination: cool (60–65° F [16–18° C]), average (68–75° F [20–24° C]), and warm (75–85° F [24–30° C]). Most perennials fall in the average temperature category.

Growers should monitor soil temperatures rather than the air temperatures of the greenhouse or germination chamber. During stage 1 (radical emergence), perennials with cool germination temperatures should be started maintaining 60–65° F (16–18° C) average soil temperatures, and those requiring average temperatures should be at soil temperatures of 68–75° F (20–24° C). Methods of maintaining soil temperatures are discussed on page 272.

To provide the desirable temperatures for cool and warm germinating crops, it is often beneficial to utilize germination chambers, where the temperatures can be controlled. Perennials that fall within the average germination temperatures are commonly germinated in chambers or placed directly in the greenhouse for germination.

Light

Light is an important factor, which perennial growers often overlook when propagating perennials by seed. The role of light during germination is variety specific and, depending on the variety, promotes, inhibits, or has no effect on germination. The light level can be the determining factor with germination success.

Many varieties of perennials have an absolute light requirement for germination to occur. The light intensity needed to trigger germination is relatively low—often less than 10 f.c. (108 lux) is sufficient. There are some perennial species that must have complete darkness for the germination process. To satisfy the dark requirement, growers either cover the seed with growing medium or place the trays in a dark germination room. Most perennial cultivars do not have a light requirement and can successfully be germinated with exposure to light or dark conditions. (See table 9.2 at the end of this chapter.)

As mentioned above, during emergence (stage 1) only 5–10 f.c. (54–108 lux) of light are necessary for those light-requiring varieties. It is recommended to limit light levels to 2,000–2,500 f.c. (22–27 klux) for the first few days following germination or after removing from a germination chamber. During stage 2, the light levels can be greatly increased. High light levels will help to produce shorter, tougher plants. The light levels for stages 3 and 4 should be ambient, as much as possible. It is not uncommon for growers to provide some shade during stages 2 through 5 to reduce the overall temperature and to reduce the drying of the plugs.

Recently, there has been some research conducted at the University of Minnesota showing that many perennial varieties benefit from night break lighting during plug production while the natural day lengths are short. A few of the known varieties that respond favorably to night interruption lighting are *Asclepias, Astilbe, Campanula, Echinacea, Oenothera, Rudbeckia, Salvia,* and *Scabiosa.*

Covering seeds

For many perennial varieties to geminate properly, it is often beneficial to cover the seeds at the time of sowing to exclude light and to prevent them from drying out. Covering the seeds also helps to direct the emerging roots down into the growing medium, and with some varieties it helps to reduce the initial stretching of the hypocotyl below the cotyledons. For seeds that require a covering, the general rule of thumb is to cover the seed by a loose layer of propagation substrate as thick as a quarter the length of the seed. This covering commonly consists of medium-grade vermiculite, perlite, the germination mix the flats are filled with, or some combination of these. In many cases, growers use equal amounts of the germination medium and fine vermiculite mixed together. Using straight vermiculite makes it difficult to properly determine the moisture levels around the seed.

Many perennial crops do not require that the seed be covered, while others must have a heavy covering. Perennial seeds that perform better with a heavy cover, which buries the seed completely, include *Althaea, Echinacea, Lupinus,* and *Viola.* Other varieties, such as *Aquilegia, Astilbe, Bergenia, Campanula,* and *Heuchera,* prefer to be covered lightly or have the growing medium just surrounding the seed. There are several varieties that benefit from having a covering not at sowing but following radical emergence.

In addition to covering with some form of substrate, some growers cover the flats with white or black plastic to prevent rapid drying of the germination mix and to keep the moisture levels high and uniform. After the radicals are emerging from the seed, or the seedling protrudes from the substrate surface, the plastic covering is removed.

Germination chambers

Many growers utilize germination chambers to provide the ideal environmental conditions for seed germination. Inside the chambers, they can maintain the optimal temperatures, moisture levels, and light intensities necessary for germination. Although germination chambers are not a necessity for germination, chambers tend to improve the germination percentages by 5–10% and decrease the time needed for complete germination. Growers using chambers can schedule crops more precisely and achieve greater crop uniformity because they can easily control all of the factors necessary for the germination process. Starting seedlings in a greenhouse can be difficult because the amount of light and temperatures varies each day. Germination chambers take away those variables.

Germination chambers vary in size, proportionate to the size of the greenhouse operation. For example, small greenhouses may only have a germination cabinet large enough to hold the flats needed for the peak season, while large operations may have germination rooms that will accommodate numerous (hundreds or thousands) flats at any given time.

Here are some of the considerations growers are faced with when designing germination facilities.

- It may be beneficial to insulate the germination chambers to allow greater ability to maintain uniform temperatures.
- A heat source such as heat cables or a gas furnace may be needed to ensure that minimum temperatures can be provided.
- In most cases where lights are used, the light fixtures generate enough heat that additional heat sources may not be necessary.
- Providing air circulation using one or more small fans will help to keep the temperature more uniform.
- Cooling may be necessary and can be provided using small exhaust fans and air intake louvers. Commercial refrigerators (air conditioners) may also be an option. Humidity decreases the effectiveness and longevity (not to mention increased maintenance) of conventional air-conditioning systems.
- Most growers will need lighting inside of the germination facility. Fluorescent lights are the most common source of lighting used. Mixing cool white and warm white bulbs will give a balanced light spectrum.
- The walls and ceilings inside germination facilities are often white to provide a reflective surface for the light.
- There should be adequate power available

to run the heating, cooling, and lighting systems.

- Maintaining a high relative humidity (90–95%) can be achieved by hand wetting the floors, misting the crops, using a humidifier, or using a fogging system.

Figures 9-5. **There are numerous types and sizes of germination chambers used by growers. Chambers are designed used to provide optimal conditions for seed germination.**

Shelton Singletary

It is more important to maintain the desired temperatures in the growing medium (the temperatures immediately surrounding the seeds) than to maintain air temperatures in these ranges. The best results usually occur where growers are using root zone heating, or bottom heat, to maintain adequate temperatures in the root zone for germination. When using bottom heat, growers often can run the air temperatures 5–10° F (3–6° C) cooler than the desired medium temperature.

Germination chambers are especially useful for plants that have germination requirements that are different from the conditions commonly found in the greenhouse. For example, plants with dark requirements can be easily germinated in chambers where the dark conditions can be maintained. They are also useful for crops that need to be stratified, require cool temperatures, or where germination is difficult in greenhouses in sunny climates.

Watering

Watering is the most important job in the floriculture industry. No matter how carefully other aspects of production are planned and managed, irrigation practices can make or break crop production. Proper watering is important with all container sizes, but with plug production it is exponentially important. Plug cells do not contain much room for error. For example, under normal conditions plug cells contain less than 10% air in the pore spaces; if the plugs are overwatered, the pores become full of water, reducing the air in the cells to virtually nothing. Conditions such as this will dramatically reduce root growth, plant development, and increase the incidence of disease. Any variances in the application of irrigation water, whether it is timing, frequency, or amount applied, can adversely affect plug production.

Watering is a very difficult job to master. It appears to be an easy task, and most operations delegate watering responsibilities to relatively inexperienced individuals. With the importance of proper irrigation practices and the impact water has on our crops, growers should invest a great deal of time learning about the impact poor watering can have on the crops and thoroughly training detail-orientated individuals for this task. It is the hardest job to learn and to teach, but the results received from training and implementing sound water management will be well worth the efforts. Learning to water properly is beyond the scope of this book and involves hands-on training, where the applicators make observations and see the results. Following are some guidelines that can lead to more successful training and irrigation practices.

There are several indicators that can be observed to determine when irrigation is necessary. A slight change in surface color, from a darker coloration to a lighter one, indicates the surface is drying out. For germinating seedlings, it is very important to recognize these changes in surface color. Touching the medium's surface or pushing a finger into the plug will allow growers to feel the moisture content of the plug. Picking up the plug flat allows producers to determine the moisture content by weight. Growers should have a rough idea of the ideal weight of each of the various plug sizes they are producing. Dry flats will weigh significantly less than properly watered ones.

The plants themselves will provide indications that water is needed. Slight changes in the leaf color (taking on a slightly darker coloration), flagging or wilting, different leaf orientation, and leaf rolling all suggest that irrigation might be needed. Sometimes these signs are enough to determine the need for applying water, but confirmation can be made using the indicators mentioned above.

Conversely, there are indicators that plants have been receiving too much water. Tall, soft shoots with thin, big leaves may suggest plants are being grown too wet. Looking further, the root system, particularly poor, weak roots or the presence of root rots, can also indicate the plants are overwatered.

The environmental conditions can also be used to determine when irrigation should be applied. There are great differences between cloudy and sunny days in regard to water uptake and loss. When the light levels and temperatures are high, evaporation and transpiration occur at a faster rate, causing plugs flats to dry out quickly. During dark, cloudy conditions with high humidity and poor ventilation, evaporation and transpiration are greatly reduced, decreasing the need to apply irrigation.

Fertility

It is not necessary to apply fertilizers to germinating seedlings. There is generally enough stored food in the seed to provide sufficient nutrients to last from germination until the first true leaves are formed, or stages 1 and 2. Many commercial growing mixes contain a starter charge that is available for the first couple weeks of production. The starter charges in plug mixes usually have half the nutrients of other mixes used for bedding or potted plants.

After the true leaves are present, a light fertilizer solution can be applied. During the plug stage, it is usually beneficial to utilize continuous liquid fertilizer programs using 50–75 ppm nitrates and a balanced micronutrient package. Growers not fertilizing with each watering will need to apply fertilizers one or two times per week using higher rates of application, typically 100–150 ppm nitrates. When fertilizing, plug growers should supply less than 10 ppm phosphorus, as higher levels are likely to promote undesirable stretching of the plugs. Plug growers use nitrate-based fertilizers to provide for toned growth and ammonium sources for quick growth and cosmetic toning.

For most crops, maintaining a pH of 5.5–6.0 throughout the crop is ideal. Plug mixes containing limestone typically start off with a slightly lower pH (5.3–5.8) but after a week or so of production will often increase. Unless levels are severely off, problems associated with pH do not show up until stages 3, 4 and 5.

The water quality and irrigation practices have a great affect on the nutritional status of the plugs and how growers manage their fertility. Growers should frequently test the water source to determine any changes in the quality of the water. When growers use irrigation sources with high alkalinity levels, each watering has a liming affect, unless acid is injected into the water. When acids are used to control the pH and/or alkalinity levels, nutrients are added to the irrigation water. Depending on the source of the acid, nutrients—such as phosphorous or sulfur—are added and should be considered when designing a fertilizer program for perennial plug production. The alkalinity of the irrigation water should be maintained at 40–100 ppm (40–100 mg/L of calcium carbonate).

Fertility is covered in more detail in chapter 3, Fertility for Perennials.

Insect and disease management

It is important when growing plugs to start off with a clean environment. Begin producing plugs in an environment that has been free of other plant materials for at least ten to fourteen days. Remove all old plant debris and any weeds that are in or around the production facility. When possible, use new plug trays. If old trays must be used, wash them out well and clean with a commercial disinfectant or a 10% bleach solution.

Perennial plugs are prone to the same insects as crops typically grown under greenhouse conditions. The most apparent pests are fungus gnats and shore flies; neither poses a real threat to plug production unless their populations reach epidemic proportions. To control both fungus gnats and shore flies, keep the production area clean and free of algae and debris. Water management, biological controls (such as nematodes and parasitic wasps), and chemicals (such as the insect growth regulators), are all effective at controlling these pests.

Other common insects found in plug production include thrips, aphids, and whiteflies. There are numerous products labeled for control of these insects. Refer to the insect management chapter for descriptions of these insect pests and their controls. It is recommended to not apply insecticides during stages 1 and 2, as the young seedlings are more sensitive to chemicals and susceptible to phytotoxicity.

Diseases are usually less evident than insect outbreaks. The most common diseases are *Phytophthora, Pythium, Rhizoctonia,* and *Thielaviopsis.* All of the conditions that promote the spread of these diseases are similar; too much moisture, high salt levels, reused medium, and improper sanitation. Controlling these factors and the environment will go a long way to prevent these diseases.

In most cases, fungicides should not be applied during stages 1 and 2. They can be applied for later stages of plug production to treat problem spots as necessary. It is most beneficial to approach disease control from the preventative standpoint by controlling the factors that promote these diseases and implementing a monthly broad-spectrum fungicide drench, such as mefenoxam (Subdue Maxx) + fludioxanil (Medallion).

Foliar diseases such as leaf spots, mildews, and *Botrytis* are less common, but monthly preventive sprays can be implemented. Rotating between trifloxystrobin (Compass) and chlorothalonil + thiophanate methyl (Spectro 90) has worked well for me. Botrytis is more problematic in the later stages (3 through 5) of plug production, and preventative sprays of chlorothalonil (Daconil), fenhexamid (Decree), or the chemicals listed above can be applied. Monitor the plug crop daily—any signs of *Botrytis* or dampening-off should be treated immediately.

It is always best to head off diseases by creating an environment where they are unable to prosper. For example, providing adequate air circulation by using horizontal airflow (HAF) fans will decrease the chance of diseases from such pathogens as *Botrytis* and powdery mildew. For disease prevention, there is no substitute for sound cultural and sanitation practices. Refer to the disease management chapter for descriptions of these diseases and their controls.

Plant growth regulators

Controlling plant height is an important aspect of plug production. Initially, when plug production was new and little information about controlling plant height of small seedlings was available, growers would use various cultural methods to reduce the overall size of their plugs—withholding water, decreasing fertility, cutting back the plants several weeks before shipping, or manipulating temperatures (DIF). Today, growers use a combination of cultural and chemical strategies to reduce the height of many perennial varieties.

To maintain a particular size of plant, to improve quality, and to lengthen the holding period, plant growth regulators (PGRs) are often used. For perennials, A-Rest, daminozide, paclobutrazol, and Sumagic are the most commonly used PGRs.

Some growers have success sprenching PGRs onto plug flats prior to or just after sowing. Sprenching is a method of applying PGRs that delivers more volume of solution than a spray application but less than a drench. Sprenches are essentially applied as heavy sprays using about three times the volume of typical spray applications. They are very effective at controlling the initial stretch that often occurs with many perennial varieties during early plug production.

Approach the application of growth regulators on plugs with more caution than you would with finished crops. Generally, the rates used should be decreased slightly because young plugs are often more sensitive (responsive). For example, if you usually apply Sumagic at 5 ppm to quart-size (1 L) perennials, reduce the rate used on plugs to about 3 ppm. Applying PGRs to plugs is an art; it takes a great deal of time and experience to learn what rates to apply and at what frequency. For application guidelines, refer to chapter 8.

Vegetative Propagation

Vegetative propagation is a type of asexual propagation where new plants are developed from plant parts such as leaves, stems, buds, and roots. This type of propagation produces plants that are clones, or genetically identical, to the mother plant. Vegetative cuttings are an increasingly popular and reliable method for growers to obtain plant materials. It gives growers varieties that are true to type and provides for more uniform growth in the propagation tray. Vegetative propagation is advantageous for reproducing varieties that produce seed that is not viable, contains various dormancies, or are not economical to grow.

Some examples of vegetative propagules, or starter plants, used for commercial perennial production are bulbs, corms, rhizomes, stolons, and tubers. A bulb is a specialized organ consisting of compressed leaf tissues that can be sectioned into smaller pieces for propagation, provided a portion of the basal plate is attached. A corm is a specialized organ composed of compressed stem tissues that can be cut into smaller pieces, provided at least one bud is located on it. A rhizome is a thickened, horizontal underground stem used as a propagule when there are several nodes on it. *Bergenia* and *Dicentra* are examples of perennials commonly propagated from rhizomes. A stolon is a horizontal, aboveground stem that can either be rooted quickly or layered to form roots over time. A tuber is a thickened underground storage organ made of compressed stem tissues.

Cuttings

Perhaps the most common type of propagule used for perennial propagation is a cutting. Creating a cutting entails taking a portion of the plant stem, sticking it in a propagation medium, providing the proper conditions, and allowing it to root.

Tip cuttings are the most commonly used type of cuttings for propagation and consist of the growing point of a terminal or lateral branch and one or more sets of nodes. *Phlox subulata* is a good example of a perennial that is propagated by tip cuttings.

Internode cuttings, often called stem cuttings, are similar to tip cuttings; they consist of one or more nodes but do not

Figure 9.6. ***Lavandula intermedia*** **'Grosso' tip cuttings showing the growing point and several nodes.**

contain the growing point. *Ajuga* and *Lamium* are examples of perennials that can be successfully propagated using internode cuttings.

Leaf cuttings, not commonly used for propagation of perennials, contain a leaf with a small section of the stem attached. Although not the preferred method of propagating *Sedum,* it can be propagated using leaf cuttings.

Basal cuttings are useful when propagating certain perennials, such as *Leucanthemum,* and consist of a stem that has part of the crown (the crown is the base of the plant where the stems join with the roots) and sometimes a few roots attached.

With increased interest and production of vegetative cuttings, the industry is much more integrated and now involves plant breeders, stock plant mangers, propagators, brokers, and growers. Breeders develop new plant varieties by conducting massive evaluation trials for selecting desirable genetic material. Once selections have been made, the genetic selections can be replicated through stock plant or tissue culture production. In most cases, stock plants are produced offshore, mainly in Central America, where environmental conditions are ideal and labor costs are low. Cuttings are harvested from these plants and shipped to growers or rooting stations in the U.S. Many growers produce their own stock plants and root the cuttings on site or ship them to clients, either regionally or nationally.

Most of the new genetic material being developed each year is getting patented, so it is protected from unauthorized propagation. Patented varieties cannot be replicated without permission. Growers must pay royalty fees when propagating or selling patented varieties. Due to the large investments firms make for producing clean, disease- and virus-free plant materials, the industry is closely monitored for plant quality, business integrity, and payment of royalty fees.

Culture indexing is one method used to ensure quality plant materials are being produced and distributed into the industry. Scientist basically grow shoot tips from stock plants on sterile medium in a laboratory and test them over several months to ensure the stock is clean. This method allows them to screen out any potential viruses, fungi, and bacteria.

If you are taking your own cuttings, a key to success is to maintain clean and healthy stock plants. Cuttings with various insects, diseases, or viruses most likely will not root well and may perform poorly after transplanting. In many cases, the problem on the stock plants is exaggerated exponentially on the cuttings during propagation.

It is difficult to maintain clean stock, even when the plants are clean to begin with. The stock plants should be grown in a facility that maximizes plant growth and minimizes plant pests. Stock plants should be replaced at least once per year to maintain plant vigor and to keep them from obtaining plant viruses. Stock plants should be tested routinely to ensure new viruses have not been introduced or transmitted into the production site.

If this is too much of a commitment for your operation, you should consider buying in unrooted cuttings. Whether you grow your own or buy them in, you will need to follow the same steps outlined below for propagating cuttings.

Unrooted cuttings

Many perennial growers are using unrooted cuttings as their source of propagation materials. These cuttings are self-produced or acquired from domestic or offshore producers. Unrooted cuttings acquired from offshore sources allow growers to meet their propagation numbers or to increase the number of varieties produced without growing their own stock plants. When taking into consideration the labor and overhead (greenhouse space, heat, labor, and chemicals) associated with maintaining a stock plant program, it can be very economical to purchase unrooted cuttings.

Cutting suppliers

It is important to obtain unrooted cuttings from reputable suppliers with proven track records of supplying perennial cuttings to the industry. They should be committed to supplying clean, healthy cuttings. The major sources of unrooted perennial cuttings originate from Central America, Israel, and Mexico. Many commercial plant brokers can obtain unrooted perennial cuttings from one or more sources.

At times, it is difficult to locate all of the perennial varieties you may need for your perennial program. Many perennials commonly come from other sources, such as from seed or tissue culture, and may not be available from most (if any) unrooted cutting suppliers. Some suppliers may supply the varieties you are looking for but may have difficulties delivering the quantities you are seeking by the date you need them.

Projecting future cutting availabilities is very challenging for cutting suppliers. Many times, cutting availabilities are projected without adequate time for growing the stock plants. Until the stock plant producer has enough time to grow multiple crops of stock for each plant variety, the supplier may make some errors in projecting cutting availability. It is common for growers' orders to get backordered or cancelled due to inadequate cutting availabilities. Try to receive cutting availability listings from multiple suppliers to show what your options might be for each variety desired.

There are a number of costs associated with obtaining unrooted cuttings from offshore suppliers. These costs include the cost of the cuttings, normal shipping and handling charges, and customs fees. Depending on your location and the origin of the cuttings, the costs associated with shipping and customs range from $30–150 per box. There is often a number of volume or early order discounts that could offset the costs of bringing the unrooted cuttings into the country.

There are certain perennial varieties that may not be allowed into this country from offshore suppliers. Some states may even have additional regulatory and inspection requirements for imported perennial cuttings. Check with your state's department of agriculture inspector prior to placing an order for unrooted cuttings to determine which regulations have to be followed.

Building a good relationship with the cutting supplier or the broker that deals with these suppliers is important. These relationships will help them to be more responsive to your needs and to establish the communication channels necessary for order acknowledgement, order changes, backorders, and pricing. Communication will also help growers obtain prompt response to problems and to obtain technical and cultural support.

Receiving cuttings

When purchasing unrooted cuttings from offshore suppliers, it may take two to three days from the time the cuttings are harvested until the grower receives them. The cuttings should be packaged properly to improve the ease of handling once they are received, as well as to endure the shipping process. The cuttings should be arranged neatly, all facing the same direction, in well-labeled bags. Consistency with cutting sizes, quantities, and arrangements in each bag will help growers to better handle, apply rooting compounds, and to propagate (stick) them. The bags of cuttings should be placed in sturdy boxes and packaged in such a manner that the cuttings will not become jarred, jumbled, or crushed during transit. Each box should contain adequate insulation, such as foam inserts, and cool packs to ensure the cuttings remain fresh. Cuttings improperly packaged will most likely become worthless or have lower quality characteristics than those that were packaged and shipped properly.

Figure 9.7. **Most unrooted cutting suppliers clearly label each package of cuttings and orientate them in a similar direction within the bag.**

It is important to act promptly when receiving unrooted cuttings. The first task is to verify the varieties and that the box count matches what is listed on the packing slip. Did you receive the right order? Are there any unexpected surprises? Any missing boxes should be reported immediately to the carrier and the supplier. Open all the boxes and verify the count of the cuttings against the packing slip. Inspect the plant materials for any abnormalities, such as heat or freeze injury, dehydrated cuttings, or any soft stems or rots that may be present. Any type of plant damage or shortages should be recorded and reported immediately after receiving them, not several days later.

Stick the cuttings as soon as possible after you receive them. I don't recommend storing the cuttings prior to sticking them. If cuttings must be stored, keep them out of direct exposure to sunlight, heat, and freezing temperatures. They can be placed in a cooler at 35–45° F (2–7° C) for several days until they can be processed and placed in the propagation facility. Growers trying to hold cuttings at temperatures outside of this range will usually experience some damage to the cuttings. Unrooted cuttings deteriorate quickly when they are not stuck and exposed to temperatures above 45° F (7° C).

Cutting consistency

Regardless the source of unrooted cuttings, consistency of quality is important when propagating perennials. Vegetative cuttings often vary in size, age, and overall quality, which makes it more challenging for growers to propagate them. To obtain uniform cuttings, it is important to set certain specifications, such as the cuttings' length or stem diameter. The more consistent the batch of cuttings, the more uniform will be the rooting times and propagation results.

Every cutting supplier will have quality specifications for each variety they produce. These specifications often vary from supplier to supplier. Currently, there are no universal quality specifications for each perennial cultivar in production. You should look for individual suppliers to have similar cutting quality attributes with each delivery over a period of time. Growers often find that they prefer one supplier's specifications to another's, or they prefer to purchase particular varieties from one supplier while buying other varieties from another. Depending on the relationship you have with your cutting supplier and the volumes purchased, growers can often request the cuttings are custom cut to their specifications.

Sticking the cuttings

As stated earlier, cuttings should be stuck into propagation flats or finished containers immediately. Failure to act quickly will reduce the rooting success and the overall quality of the rooted materials.

Perennial cuttings should be stuck into a premoistened, well-drained growing medium. Depending on plant species, they should be stuck, or placed into the growing medium, just deep enough that they are anchored by the growing mix, usually about 0.25–0.5 in. (6–13 mm) deep. It is important for some varieties to have a node below the surface of the growing medium. Cultivars of *Caryopteris, Monarda,* and *Phlox paniculata* are a few perennials that have been shown to have lower winter survival rates when the cuttings are stuck with the node above the growing medium. Pay attention to the needs of each variety when sticking cuttings.

Rooting hormones

Most growers apply the rooting hormone IBA or IBA-K to the young cuttings, either during or shortly after the sticking process. IBA is an auxin commercially used to promote rooting. The application rates vary from 500–10,000 ppm, depending on the plant species and cultivar. Generally, the rate used for most perennials is 500–1,500 ppm. Most perennials do not require the use of rooting hormones to successfully root, but often the uniformity of rooting is increased and the time for rooting is decreased when they are used.

Figure 9.8. **Many growers dip the basal end of *Perovskia atriplicifolia* cuttings in rooting compounds to hasten rooting and improve the uniformity of the plugs.**

There are liquid and powder forms of IBA available. The powder forms are harder to work with and provide variable results at times, as each cutting may have more or less of the rooting compound (providing a slightly different rate applied and plant response with each cutting).

Many of the commercially available liquid forms of IBA contain alcohol as the carrier. There are several plants that are sensitive to the alcohol and may demonstrate a phytotoxic response after application. This negative response often appears as twisting of the foliage. It is commonly observed when the cuttings are dipped too deeply or if the solution was applied to the growing point of the cutting. IBA-K is a salt that can be diluted in water to provide the IBA safely without using an alcohol solution.

Several growers dip the ends of the cuttings into the rooting hormone solution just prior to sticking. Many growers apply IBA to cuttings within twenty-four hours of sticking as a light uniform spray application. This spray should be applied while the leaves of the small cuttings are dry to ensure that the spray rate does not become diluted. After the IBA spray has dried, the misting regime can resume.

Conditions for rooting

Successful propagation of cuttings entails starting with healthy, high-quality cuttings and providing the proper environment for rooting. Growers need to provide the proper amount of light, temperature, humidity, and moisture throughout the propagation process. The cultural requirements change as a cutting goes from an unrooted cutting to a rooted plant. These changes require that growers make adjustments to their propagation regimes during the propagation phase in order to limit cultural dilemmas and to improve rooting success.

Humidity

It is imperative to reduce the wilting or dehydration of the cuttings during the propagation phase. Maintaining high relative humidity levels during the early stages of propagation will help the newly stuck cuttings to remain turgid and reduce the stresses on them. The proper environment, in addition to the humidity level, should be provided to prevent desiccation of the cuttings until the roots are formed well enough to support a transpiring plant. Growers typically maintain the humidity level using intermittent misting or fogging systems; watering alone is often not sufficient.

The amount of misting necessary varies with the crops being propagated, the stage of propagation, and the time of the year. For the most part, growers should only apply the minimum amount of water necessary for the cuttings to remain turgid. Generally, enough mist should be applied to evenly coat the leaf surfaces but not to drip off of them. The highest quality cuttings generally come from propagation houses where high humidity levels can be maintained, evaporation is limited, and the cuttings are misted relatively infrequently.

More water during propagation is not better. Providing excess water creates saturated conditions, which reduce the oxygen levels in the soil, resulting in slow root development and possibly other cultural problems. It is better to use mist nozzles providing low volumes of water and small droplet sizes rather than using the high-output types. Using less water for misting will also decrease nutritional leaching, decrease rooting times, increase plant health, and decrease plant diseases. Growers should carefully monitor root development and the cuttings' ability to remain turgid as they gradually reduce the amount of mist applied.

There are a number of methods used by growers to apply mist to their young cuttings. These techniques vary in application methods and precision. Many operations use misting nozzles or traveling booms attached to time clocks, turning them on at predetermined intervals. Some growers have fully automated, computerized systems that are very effective and adjust the misting frequency and durations by monitoring the humidity, light levels, and temperatures within the propagation facility. It is not usually necessary to provide misting to the cuttings at night, provided the humidity can be kept high. Some growers find it necessary in their environments to provide misting once or twice at night during the first few days of propagation.

One option to increase or maintain the proper humidity for propagation is to use a fogging system. Growers commonly use high-pressure fog or fan-driven water atomizer fogging systems to meet the humidity needs of the crop during this stage. These systems can be used alone or most commonly in conjunction with misting systems during propagation. It is easier to maintain high humidity during the winter months in double poly houses than in glass or single poly structures, as less water condenses on the lower surface on the glazing material.

Growers have always been taught to provide a lot of air circulation inside the greenhouse in order to provide more uniform air temperatures and reduce plant diseases. I am a strong advocate of providing adequate air circulation during crop production, but during propagation it is best to reduce air movement within the propagation area. Rapid air circulation increases the evaporation rate and reduces the temperatures of both the leaves and the growing medium. Air circulation from horizontal airflow (HAF) fans should be stopped during the early stages of propagation to provide growers with more adequate control of the humidity. As the cuttings root and begin to grow, HAF fans should be turned back on since high humidity levels no longer need to be maintained.

Figure 9.9. **Fogging is an excellent method of maintaining the appropriate humidity levels in a propagation facility.**
Blooms of Bressingham North America

As the cuttings callus and develop root hairs, usually within seven to ten days, the importance of maintaining high humidity levels is reduced, and the intervals between the misting cycles can be increased. As the cuttings continue to root, the misting and fogging should be reduced gradually, acclimating the plants to a lower humidity environment. As the rooting occurs, they can also gradually be exposed to higher light levels. Most perennials will be rooted in fourteen to twenty days. After checking to verify the rooting status of the cuttings, the misting can be completely removed when they appear rooted and remain turgid throughout the day. It takes four to six weeks for a plug flat (128 cell to 72 cell) to become fully rooted and ready for transplanting to larger sized containers.

Temperature

The best conditions for rooting involve maintaining adequate temperatures in the growing medium. Providing an optimal temperature (73–77° F [23–25° C]) will create ideal conditions for cell division and root development, often accelerating root development and reducing the overall rooting time. It is more important to maintain these temperatures in the growing medium than in the air. During rooting, the air temperature has little to do with the rooting process.

During propagation, growers should at least ensure the temperature of the root zone does not fall below 65° F (18° C). Temperatures less than 65° F (18° C) dramatically reduce root development and are often ideal for *Botrytis.* Conversely, as temperatures increase above 75° F (24° C), the cuttings are placed under stressful conditions, which slow the rate of root development.

The best results usually occur where growers are using root zone heating or bottom heat to maintain adequate temperatures in the root zone. Growers commonly use propagation mats, which have electrically heated wires encased in rubber, heating tubes containing warm water on the benches, or steam pipes located under the propagation bench. The bottom heat is typically shut off or reduced once the roots reach the bottom of the cell.

When bottom heat is provided, growers can maintain air temperatures slightly cooler than the root zone (68–73° F [20–23° C]). If growers propagate perennials without bottom heat, they must run the air temperatures warmer (77–80° F [25–27° C]) to maintain optimal soil temperatures. The quality of the cuttings is usually better when growers can provide bottom heat while maintaining cooler air temperatures. This reduces stem elongation during propagation and also reduces the overall misting requirement. Maintaining the optimal propagation temperatures not only decreases the overall rooting time but decreases the likelihood of pathogens, such as *Botrytis,* from attacking the cuttings.

Light

During the propagation phase, light levels have a tremendous effect on the success of rooting cuttings. During high-light times of the year, such as the summer months, growers should provide some shading over the propagation area. Using shade cloth or whitewash on the surface of the structure, aim to reduce the light levels by approximately 50%. Conversely, during the winter months, growers should consider increasing the light levels using supplemental lighting. For most perennials, maintaining light levels from 1,000–2,000 f.c. (11–22 klux) is most beneficial.

The evaporation of water during propagation is primarily a function of light. For example, it is necessary to mist cuttings more frequently on a sunny day than on a cloudy one, due to the rate the water evaporates from the leaf surface. The cuttings are most sensitive to wilting during the first few days of propagation. During the first week of propagation, it is recommended to provide light levels no higher than 1,500 f.c. (16 klux). Higher light levels are likely to cause more rapid drying of the moisture and require more frequent misting. High light levels can also cause more stress on the cuttings, possibly leading to leaf scorch, bleached leaves, or wilt. Initially, limiting the light level and increasing the humidity with little air movement is essential to creating an ideal rooting environment.

It is important to deliver a minimum quantity of light to the propagation area. Although shading is often necessary and beneficial, too little light over the cuttings will delay or prevent rooting. The cuttings obtain their energy from the light to make food. Without this energy, they cannot maintain cell function, cell division, or growth. To promote sufficient rooting and cell functions, growers should ensure that at least 1,000 f.c. (11 klux) are provided over the propagation area during the middle of the day.

Growers reduce the natural light levels at times of the year when the light intensities are high using a number of methods. Some growers provide shade by applying a shade fabric over the structure or by applying whitewash on the surface of the greenhouse. Several growers utilize a retractable shade cloth system, which can be automatically opened or closed, depending on the outside light levels. These automated systems allow better light transmission on cloudy days than the more permanent, fixed shading systems.

As the cuttings become rooted, the light intensities they are grown under can gradually be increased. When they are fully rooted, they should be acclimated to their natural, normal light levels.

Growers who can effectively control the photoperiod during this stage will most likely be able to produce better plug liners than growers who do not. Most perennials are long-day plants, meaning that they will go into a reproductive phase while they are produced under long-day conditions (more than thirteen hours of light). For some perennial cuttings, such as *Sedum,* being in the reproductive stage during propagation does not allow for sufficient rooting or the desired vegetative growth following rooting. The optimal photoperiod for maintaining vegetative growth and propagating most perennials is less than thirteen hours. The majority of perennial propagators do not manipulate the photoperiod during propagation, and with a few exceptions, most perennials will root successfully under longer day lengths.

Growing medium

The growing medium used for vegetative propagation is another important factor that influences the success rate of the propagation process. Propagation mixes should support the cuttings in an upright position and retain moisture, yet provide adequate amounts of oxygen to optimize rooting. Mixes that hold too much moisture often become saturated during the propagation process, have little oxygen available for rooting, and may lead to devastating diseases, such as *Erwinia.*

Most propagators use mixes containing peat, perlite, and vermiculite. There are a number of commercially available mixes that are suitable for vegetative propagation. It is also common for propagators to make their own propagation mix; a commonly used combination contains 35% peat moss and 65% perlite. Growers often loose-fill propagation flats using their own or a customized propagation medium or a commercially available media system. Many growers prefer to use Ellepot or Fertiss sytems, which contain propagation medium bound by a root-permeable sleeve. Several propagators use water-absorbing foam products, such as Smithers-Oasis Root Cubes or Wedges. Most commercial propagation systems are designed for optimal callus and rapid root formation.

Water quality

Where possible, growers should use water sources that are low in salts and bicarbonates during the early stages of propagation. Water with high amounts of bicarbonates or high alkalinity levels will often delay the rooting process and may have a negative effect on overall quality of the finished plug. If necessary, the bicarbonates can be reduced to 60–80 ppm using various sources of acids. Sulfuric and phosphoric acids are both acceptable and commonly used for this purpose. However, they both react with the calcium in the bicarbonates to form a water insoluble byproduct that may leave an undesirable residue on the leaves of the cuttings. Typically, this residue is not significant and is not problematic for most growers.

Nitric acid is a better source of acid for neutralizing bicarbonates during propagation. The reaction that occurs produces nutrients that are beneficial for plant development and does not form a byproduct that leaves a residue.

Fertility

During the first few days of propagation, the unrooted cuttings do not require any nutrition from the growing medium. As the cuttings become callused or start to root, they benefit from having some small amounts of nutrients available to them. Fertilizer applications should begin once root primordia are present. It is a common practice to fertilize one to two times per week using a water-soluble fertilizer at 100 ppm nitrates.

Applying nutrients during the rooting process is important because many of the nutrients become leached from misting or the water applied during the rooting process. When the cuttings appear chlorotic, they should be fertilized with a complete fertilizer with minors immediately to improve plant health, appearance, and quality.

Sanitation and disease management

To prevent diseases, great steps should be taken to ensure that sound sanitation practices are performed throughout the propagation phase. Disinfectants, such as bleach or commercially available products containing quaternary ammonium salts, such as GreenShield or Triathlon, should be used to sanitize benches, flats, pots, knives, scissors, floors, and walls before conducting any propagation activity.

The incidence of disease problems during propagation depends on several factors, including the plant species, the overall health of the cutting, the environment, and the propagation procedures being implemented. There are several species—for example, silver-leaved plants such as *Achillea* 'Moonshine'—that are sensitive to overhead misting during propagation and often experience foliar decay during the rooting process. When cuttings are taken from unhealthy stock plants (nutritionally or diseased), the problems usually run rampant during propagation.

It is important to always start with clean, healthy cuttings. Make sure the cuttings you received were free of any symptoms. If problems are observed during propagation, it is beneficial to rogue out any dead, dying, or diseased foliage on a regular basis.

Maintaining high humidity levels during propagation also creates ideal conditions for fungal and bacterial growth. *Botrytis* is the primary threat to young cuttings, as it thrives in moist, stagnant environments. It is very important to practice sound cultural and sanitation practices during this stage of propagation.

Many growers apply preventative spray applications of broad-spectrum fungicides such as trifloxystrobin (Compass) or azoxystrobin (Heritage) within forty-eight hours of sticking. For the best results, preventative fungicide applications should be made at the end of the day when the misting has been reduced. Until the cuttings are rooted, it is beneficial to make weekly preventative fungicide applications using a rotation of broad-spectrum fungicides such as thiophanate-methyl (Cleary's 3336), fludioxonil (Medallion), and the products mentioned above.

In addition to foliar applications of broad-spectrum fungicides, many growers have been successful applying a fungicide drench within the first week of propagation to reduce the likelihood of any soilborne pathogens developing during the rooting process. Commercial products such as Medallion and Cleary's 3336 are safe and effective when applied as a drench to young cuttings.

Exposure to high temperatures (over 85° F [29° C]) can lead to insufficient rooting and quite possibly bacterial infections. Bacterial problems are more likely to occur when high temperatures are combined with high misting frequencies and when saturated growing medium is present. Many growers implement preventative programs using copper-based fungicides such as Camelot (copper salts of fatty and rosin acids) or copper sulphate pentahydrate (Phyton-27) to reduce the likelihood that bacterial problems will arise.

Insects

The biggest threat from insects during propagation is fungus gnat larvae. In many cases, the larvae are already present in the growing medium before the cuttings are even stuck. Once a population of larvae is present, they are capable of eating new roots faster than the young cuttings can grow them. They may even burrow into the young stems, causing them to become stressed and possibly even killing them. If adult fungus gnats are visible, there are almost always some larvae present in the growing medium.

There are numerous products available for growers to apply to the soil for controlling fungus gnat larvae. Some of the most effective biologicals, insect growth regulators, and chemicals include diflubenzuron (Adept), azadirachtin (Azatin, Ornazin), cyromozine (Citation), pyriproxyfen (Distance), chlorpyrifos (DuraGuard), *Bacillus thuringiensis* (Gnatrol), and *Steinernema feltiae* (Nemasys). It is usually most beneficial to apply some larvae-controlling insecticide early (within the first few days) in the propagation cycle to control any fungus gnat larvae that might be present.

For a more thorough discussion regarding insect control, please refer to chapter 6.

Tracking results

As growers try unrooted cuttings for the first time, it is important for them to track their successes and failures along the way. Unless there are no other options available to you, start off slowly and gradually increase your reliance on unrooted cuttings as you gain a better comfort level with them. When testing what will work at your facility and what will not, order minimum quantities of several varieties. Try to schedule multiple deliveries of the same varieties from various suppliers over a period of several weeks to provide an indication of each supplier's consistency and quality.

Keep good records of all aspects of the propagation phase, taking notes regarding the size of the propagation materials, rooting hormones, misting schedules, temperatures, chemical applications, and rooting times. This will help you refine your practices and increase your quality with each crop.

Tissue culture

Tissue culture, or micropropagation, is another method used to vegetatively propagate perennials. Tissue culture is commonly used to produce large quantities of plants in a relatively short period using a small amount of stock plants. Tissue culture has become a popular method of generating enough plant materials to supply the marketplace when new cultivars are first introduced into the industry. Micropropagation is also useful for propagating certain plant species that are difficult to propagate using other vegetative methods.

Micropropagation requires specialized equipment and facilities that are not affordable to most commercial growers. It takes a great deal of knowledge regarding plant hormones and media cultures, as well as the ability to maintain a sterile environment to be successful with micropropagation.

There are four phases, or stages, involved in the micropropagation process: establishment and stabilization, shoot multiplication, root formation, and acclimatization.

In the first phase, establishment and stabilization, it is critical for everything to become disinfected and sterilized. The explants (tiny plant parts) are established in a medium and stabilized using light and temperature.

The second phase involves forming clusters of microshoots using a medium containing cytokinins. Cytokinins are plant hormones used to promote cell division and tissue differentiation. In this phase, cytokinins are needed to induce shoot development. These clusters are then separated, isolated, and subcultured inside individual test tubes.

The third phase involves forming roots on the shoots by moving them into a different medium that contains auxins. Auxins are plant hormones used to promote cell elongation and root initiation. In this phase, auxins are necessary to initiate roots on the clusters of microshoots.

Figure 9.10. **Stage 3 *Phlox subulata* are ready to be transplanted into a plug flat and moved into a propagation house.**

The last phase of micropropagation is the acclimation phase. It involves removing the plants from the agar medium by washing off of the young plants' roots. The small plants are then transplanted into a plug flat with a commercial plug or growing mix. They gradually need to be exposed to increased light levels while humidity and moisture levels are also slowly reduced. Growers often acclimate plants from tissue culture inside modified greenhouses or germination chambers.

Root cuttings

Taking root cuttings is another method some growers use to propagate certain perennials. Root cuttings entail taking small pieces of root and forming adventitious shoots on them creating new plants. This method of propagation is variety specific and will not work for all perennial varieties or all of the species within a genus. Some commonly root-propagated varieties are *Anemone, Brunnera, Papaver, Phlox paniculata,* and *Stokesia.*

Root cuttings are generally taken from dormant plant materials; many growers harvest root cuttings from existing container crops they wish to sell the following spring. The roots from the bottom third of an individual plant are removed and replaced with new growing medium, and the plant is then returned to the overwintering facility. Removing more than one third of the root system will adversely affect the performance of the stock plant.

The roots measuring 0.125–0.25 in. (3–6 mm) thick are most commonly cut up into 1–4 in. (3–10 cm) long pieces and planted into plug trays, open flats, or scattered across propagation medium. When scattering thin roots on the growing mix, they should be covered with additional growing medium and kept relatively dry until they become rooted. Thicker roots tend to root better than do thin roots. Many perennials root best when the roots are about the diameter of a pencil. When placed in open flats, there is the potential to have over one hundred root cuttings per square foot (0.93 m^2).

With some perennials, such as *Phlox paniculata,* it is important to place the roots in the proper orientation—top of the root facing up and the bottom of the root down. To help identify the root orientation, it is helpful to mark the roots by the way they are cut. For example, cut the top of the roots straight and the bottom of the roots with an angled cut.

The root cuttings are often left at cool temperatures for the duration of the winter. Growers often provide minimum heat during this period, keeping temperatures above freezing, but below 44° F (7° C). Roots should be kept moist, but not wet, during this period, as they rot easily under saturated conditions. After the roots have gone through about six weeks of cold, the temperatures are gradually increased up to about 65° F (18° C) to promote root and shoot development. Following the cold period, it often takes six to eight weeks for rooting to occur. Root cuttings of certain varieties are taken during the growing season and do not require a cold treatment; these varieties often root in four to eight weeks.

Bare-root perennials

Another popular method of propagating perennials is by division. Divisions are plants that have been grown (usually in a field), dug, and divided into smaller sizes. These small plants are referred to as bare-root divisions or bare-root perennials. Divisions are commonly obtained by pulling offsets from crowns with root initials attached, dividing the crown into smaller sections, or cutting woody crowns to smaller sizes.

Despite the popularity of perennial plugs, bare-root perennials are still a popular source of starter materials for many growers. Bare-root perennials can be purchased domestically or imported as washed soil-free crowns or roots. They consist of clumps, crowns, fans, roots, or tillers of the parent plant.

Commercially, bare-root perennials are usually dormant plants with the majority, if not all, of the soil removed from the root system. Many perennials are available by division throughout the year. They are typically

Figure 9.11. **Daylilies and other perennials are often harvested from actively growing plants and may not always be dormant upon their receipt.**

grown in a field for one year prior to harvesting, although some crops, such as *Hosta*, may take two to three years in the field before they are harvested. When the plants are dug or harvested, the tops are usually cut back to about 1 in. (3 cm) above the crown and the roots are trimmed to 3–4 in (8–10 cm). Several varieties, such as *Dianthus* and lavender cultivars, are evergreen perennials and are usually shipped without the tops being removed.

There are four types of root systems, or rooting structures, that are being supplied to the industry: fibrous roots, taproots, rhizomes, and corms. Fibrous roots are root systems with profusely branched roots and significant amounts of side rootlets. Taproots consist of a main descending root of the plant with few adjoining lateral roots. Perennials with rhizome root systems have a swollen stem with branching close to the soil surface. The fourth type of root system used for bare-root production is the corm, which has a bulblike portion of the stem located under the ground.

There are several advantages for using bare-root divisions compared with starting from plugs or liners. Using bare-root divisions is often a cost-effective method to grow a finished crop in less time than starting them from smaller plugs. Some varieties, like many *Hosta* and daylily cultivars, produce a higher quality finished product, often with more shoots and flowers, when they are started from bare-root sources. Bare-root divisions can be stored or frozen, allowing growers to alter their potting and flowering dates, which can be useful to provide flowering plants over an extended period of time. Most domestic and offshore suppliers offer a wide assortment of plant varieties and cultivars in various grades or sizes, making them suitable for various finished container sizes.

Growers have also observed a few disadvantages from using bare-root divisions. It is often difficult to evaluate the quality of the bare-root materials or to determine whether they are healthy and alive or unhealthy or even dead. Planting bare-root materials is often difficult, as it is hard to identify which end is up, determine the proper planting depth, and place the roots properly inside of the container. Once they are received, it is very easy to mishandle or to improperly store the divisions, often causing injury or excessive dessication to occur. As a root system with few distinguishing characteristics, such as stems and leaves, it is nearly impossible to identify one cultivar of a plant species from another. Bare-root divisions are often only available certain times of the year, and seasonal variances in growth have also been observed.

Bare-root plants are available as different grades or sizes of materials. Each grade represents the number of stems or branches per plant and gives an indication of the finished size of the plant. For example, the grade of plant material needed as the pot size is increased from 1 qt. (1.9 L) to 2 gal. (7.6 L) should also be increased with the container size. Smaller grades are more suitable for small container production, and larger grades are best used for large container production. Perennials such as *Astilbe* and *Hosta* are graded by the number of "eyes" per plant. One to two eye divisions are suitable for 1–2 qt. (1–1.9 L) container production, and two to three eye divisions are suitable for 1 gal. (3.8 L) containers. With some plants, such as daylilies, the divisions are graded by the number of "fans" per division. Similar to *Hosta* mentioned above, as the number of fans per plant increases, the pot size should increase correspondingly.

Perennial bare-root materials being imported into this country must be free of soil per regulations of USDA-APHIS. These regulations are to minimize the danger of unwanted plant pests, such as nematodes, from entering the country. The main method of removing soil from the plant materials is to perform a series of moderate to high-pressure sprays (referred to as "washing"). Some facilities perform high temperature dips, with or without fungicides, along with washing procedures.

Bare-root perennials should be inspected upon arrival, prior to planting. The roots should be relatively dry, firm, and appear light brown in color. Occasionally, a light surface mold may be present. This mold is fairly common and caused from the humidity levels necessary for bare-root storage. The mold will disappear quickly, and is not a threat to plant health. Fungicide applications are not necessary for controlling this pathogen.

It is important to handle bare-root materials with care and not subject them to unnecessary stresses. Subjecting them to conditions that could cause them to dry out, such as exposure to direct sun, removing excess materials from the cooler, and exposing them to temperature extremes could all lead to poor development and growth after planting.

Sometimes bare-root plants are frozen when they are received. They can be kept frozen for several weeks if they are not needed immediately. They should be thawed out slowly before handling. It is better to thaw them out gradually in a cool room (40–55° F [4–13° C]), as opposed to placing them at warm temperatures (65 to 75° F [18–24° C]) to thaw them out rapidly. Once thawed, if they cannot be planted immediately, store them loosely wrapped in plastic bags in a cooler at 35–40° F (2–4° C) for no more than three to five days.

Some growers soak the roots in water for approximately an hour to rehydrate them. By rehydrating the roots, growers have found that bare-root materials will establish more quickly and produce more uniform crops. Although soaking does a good job to rehydrate the dry roots, it also increases the potential spread of pathogens from plant to plant within the water tank. Several growers have changed from soaking in clear water solutions to soaking in a solution containing broad-spectrum fungicides. This still does not prevent the spread of bacterial diseases or pathogens not controlled by the fungicides in the dipping solutions. Because of this potential threat, most growers do not soak the roots prior to planting and instead water thoroughly, often with fungicides, after planting.

There are a number of growing mixes available to perennial producers. For bare-root production, it is important to use a potting substrate with a total porosity of 50–60%. Of this porosity, approximately half of it should consist of air. This amounts to an aeration porosity of 20–25% and a water retention porosity of 30–35%, which is optimum for root development.

Most bare-root perennials should be planted with the crown being approximately 0.5–1 in. (13–25 mm) below the soil surface. This is referred to as the deep planting method. To encourage root development, it is often beneficial to fan, or spread out, the roots when they are transplanted. Improper placement of the root system in the soil will lead to crop variability and possibly cultural problems that could lead to plant death. If the root systems on the bare-root materials are too long, they can generally be cut back to fit the size of the container. Because there are so many varieties commercially available, a general rule of thumb on trimming roots cannot be made. Contact your bare-root supplier for the correct trimming procedures.

Research at Cornell University has shown that many perennials perform best when the crown is planted at or above the surface of the growing medium (0.25–0.5 in. [6–13 mm]) than the traditional deep planting method described above. See table 9.1 for a listing of perennials that generally perform best when they are planted slightly above the surface of the growing medium.

Table 9.1. Bare-root Varieties to Plant High

Aconitum	*Hemerocallis*
Astilbe	*Hosta*
Athyrium	*Iris siberica*
Campanula	*Liatris*
Echinops	*Ligularia*
Epimedium	*Lysimachia*
Euphorbia amygdaloides	*Salvia nemorosa*
Filipendula	*Tradescantia*
Geranium	*Trollius*
Geum	*Verbascum*
Helenium	

Source: "Handling Bareroot Perennials" by William B. Miller and Amy Bestic, *GrowerTalks,* May 2004.

Regardless of the planting height, it is important to plant bare-root materials properly. In many instances, the roots are often folded or wadded up in a ball to allow for easy planting. This practice generally leads to poor plant stands and plant mortality. The roots should not be folded to fit the hole made at the time of transplant. They should be placed on the surface of a semi-filled container with the roots spread out, and the remainder of the medium added to fill the pot.

It is best to plant evergreen varieties first because they will dry out and become desiccated faster than will varieties that have had their shoots removed. After planting, water the pots thoroughly to help reduce undesirable air pockets. As mentioned previously, many growers apply a fungicide drench after potting to help reduce the occurrence of root rot problems and to ensure the perennials get off to a good start.

The success growers have using bare-root as starter material varies widely from grower to grower. There are usually slightly more plant losses associated with bare-root plants as compared with using plugs. Proper handling and planting methods will go a long way toward reducing crop losses and improving crop uniformity. Growers should consider the effects of temperature in regard to the rate of growth and success of growing bare-root perennials. It is recommended to keep them from freezing after they have been potted. Several growers plant bare-root as containerized perennials in the early spring directly outside. If this is done too early, there is the risk of experiencing plant losses if severely cold weather returns. I recommend providing temperatures of 50–55° F (10–13° C) for the first ten to fourteen days after transplanting to promote root development, then raising the growing temperatures up to 65° F (18° C), depending on the crop and the anticipated date the product must be finished.

Vernalization of Plugs

It is important for perennial growers who produce finished containers to provide plants to the marketplace that are flowering. In many cases, plugs must receive a vernalization (cold treatment) period to promote flowering once they are transplanted into the final container and receive any additional requirements necessary for blooming. Many growers vernalize their perennials while they are in plug trays in order to promote flowering after they are transplanted.

The majority of the perennial plugs being vernalized receive the cold treatment in structures during the winter months. It is possible

Figure 9.12. **Perennials are often overwintered in Quonset houses, as shown here, where minimum heat can be provided. Larger plug suppliers may instead opt for vernalizing plugs in large, gutter-connect houses.**

to vernalize the plugs in coolers during times of the year when natural vernalization cannot occur or where temperatures do not get cold enough to satisfy the vernalization requirement.

Temperatures less than 41° F (5° C) are usually acceptable to satisfy the cold requirement of most perennial varieties. The plugs are typically grown to full size (at least eight weeks) and fully rooted before growers acclimate them to cold temperatures. Depending on the time of the year, they can be acclimated by exposing them to natural temperatures or they can be gradually acclimated by reducing the average temperatures by 3–5° F (2–3° C) until the average temperatures reach 35–40° F (2–4° C).

The cold requirement varies with each plant species and cultivar. Some perennials have a very short cold requirement (less than four weeks); others require extended durations of cold (up to sixteen weeks); and some perennials do not have a cold requirement. The majority of perennials require six to nine weeks of cold to promote flowering. To ensure complete flowering, growers should deliver the minimum amount of cold required for each variety.

Commercially, most growers try to deliver a minimum of ten weeks of cold temperatures to most perennials. Providing more than the minimum number of weeks of cold is usually advantageous and only rarely becomes problematic. When vernalizing plugs in greenhouses, providing a few extra weeks of cold will offset the fluctuating temperatures often experienced within these structures.

Plants without a cold requirement will perform just fine following exposure to a vernalization period. Many perennials without an absolute cold requirement actually benefit from exposure to cold. With many perennial varieties, the longer the cold period they receive, the shorter the length of forcing time they require to reach bloom. The forcing time of many perennials is often reduced by seven to ten days for every two weeks of cooling they receive. Other benefits of cold include more uniform flowering, increased flower number, and with some perennials, such as *Aquilegia,* the flower stalk rises above the foliage, an improved quality characteristic.

Larger plug sizes (72 cell or larger) usually survive vernalization better than smaller plugs (128 or smaller). With smaller plug sizes it is harder to manage the moisture levels during a cold period, and they do not have a lot of buffering capacity to tolerate cold temperatures. However, there are growers who successfully vernalize 128-cell plugs. Smaller plug sizes may not flower regardless of how much exposure to cold they have received because they are often not mature enough to perceive and respond to the cold treatment.

Small plugs struggle to survive during the vernalization period because there is little soil surrounding the roots. The growing medium and the moisture in it act as insulation and help to protect the root system while the temperatures are cold. Generally, the larger the plug, the colder temperatures it can tolerate. Regardless of size, small plugs or large containers, all plants grown in containers may experience injury to extended exposure to temperatures less than 25° F (-4° C). It is always best to maintain a moist but not saturated root zone just prior to and during the vernalization period.

The smaller the plug size, the more critical it is to run minimum heat during this period. Minimum heat usually entails trying to maintain temperatures of the root zone above freezing. Many growers use heat set points of 35–40° F (2–4° C). The goal is to receive the benefit of the cold while not allowing the moisture in the plug cells to freeze, which, with small cell sizes in particular, could lead to plant death. Light freezes generally cause no harm to perennial plugs, but hard freezes (less than 25° F [-4° C]) may cause significant damage to the root system.

When vernalizing perennial plugs inside coolers, growers should maintain temperatures at 35–38° F (2–3° C). Many varieties can be vernalized in the dark, while others, particularly the semi-evergreen to evergreen varieties, perform best when a small amount of light is provided. Providing 10–15 f.c. (108–161 lux) of light during vernalization will help keep green leaves from getting diseases. It is also beneficial to apply a preventative foliar fungicide application prior to cooling. Providing any plant maintenance, such as irrigating and reducing the incidence of plant diseases, is more difficult in coolers than in greenhouses. To reduce potential disease or regrowth problems, growers should remove plants as soon as possible after the minimum amount of cold has been achieved.

It may be advantageous to either purchase vernalized plug materials (there are a number of good perennial plug growers who supply vernalized materials to the industry) or to provide a cold treatment to perennial plugs in your own facility. Growers with empty greenhouse or cooler space might be more apt to consider in-house vernalization than would be growers with limited space and year-round production schedules. Sometimes vernalized plant materials of all the cultivars growers wish to grow are not available, which means growers have no choice but to provide their own cold treatments.

References

Friel, John. "Handling Unrooted Cuttings." *Greenhouse Product News*. August 2003.

Joeright, David, Dan Tschirhart, et. al. "Herbaceous Perennials: Propagation." *Greenhouse Grower*. April 2001.

Nau, Jim. *Ball Perennial Manual: Propagation and Production*. Batavia, Ill.: Ball Publishing. 1996.

Panter, Karen. "Propagation Station." *American Nurseryman*. September 1, 2004.

Pyle, Allen. "Evaluate Perennial Cutting Suppliers." *GMPro*. December 2002.

———. "Maximize Perennial Germination: How to Successfully Germinate Difficult Species." *GMPro*. December 2003.

———. "Planting the Seeds of Success." *American Nurseryman*. February 1, 1999.

Simeonova, Neda. "Bare Root Perennials." *Greenhouse Product News*. July 2003.

Styer, Roger C., and David S. Koranski. *Plug and Transplant Production: A Grower's Guide*. Batavia, Ill.: Ball Publishing. 1997.

Table 9.2. Germination Requirements of Perennials

Variety	Weeks to Finish 128 Cell	Germination Temperatures	Seeds per Cell	Type of Covering	Moisture Requirement	Light Requirement	Comments
Achillea filipendulina	8	65–72° F (18–22° C)	3	Light cover	Below average	Light beneficial	
Achillea millefolium	7	65–72° F (18–22° C)	3	Light cover	Below average	Light beneficial	
Achillea ptarmica	8	65–72° F (18–22° C)	3	Light cover	Below average	Light beneficial	
Achillea tomentosa	8	65–70° F (18–21° C)	3	Light cover	Below average	Light beneficial	
Agastache cana	8	68–75° F (20–24° C)	2	Light cover	Average	Light required	
Agastache rugosa	7	70–75° F (21–24° C)	2	Light cover	Average	Light required	
Alcea rosea	6	65–72° F (18–22° C)	2	Cover	Below average		
Alchemilla mollis	9	65–72° F (18–22° C)	3	No cover	Above average	Light required	Cold germinator, provide stratification
Alyssum montanum	7	65–72° F (18–22° C)	4	No cover	Above average	Light beneficial	
Alyssum saxatile	8	65–72° F (18–22° C)	4	No cover	Above average	Light beneficial	
Anacyclus depressus	10	60–65° F (16–18° C)	3	Light cover	Average		
Anaphalis margaritacea	9	68–72° F (20–22° C)	3	No cover	Average	Light required	
Anaphalis triplinervis	9	68–75° F (20–24° C)	3	No cover	Average	Light required	
Anemone hupehensis	12	60–65° F (16–18° C)	2	Light cover	Above average	Light beneficial	
Anemone multifida	9	60–65° F (16–18° C)	2	Light cover	Above average	Light beneficial	
Anemone sylvestris	11	64–68° F (18–20° C)	2	Light cover	Above average	Light beneficial	
Anthemis tinctoria	8	68–75° F (20–24° C)	3	Light cover	Average		
Aquilegia alpina	8	70–75° F (21–24° C)	3	Light cover	Above average		Provide moist treatment at 40° F (4° C) for four weeks prior to germination

Aquilegia caerulea	9	70–75° F (21–24° C)	3	Light cover	Above average		Provide moist treatment at 40° F (4° C) for four weeks prior to germination
Aquilegia chrysantha	8	70–75° F (21–24° C)	3	Light cover	Above average		Provide moist treatment at 40° F (4° C) for four weeks prior to germination
Aquilegia flabellata	8	70–75° F (21–24° C)	3	Light cover	Above average		Provide moist treatment at 40° F (4° C) for four weeks prior to germination
Aquilegia hybrida	8	70–75° F (21–24° C)	2	Light cover	Above average		Provide moist treatment at 40° F (4° C) for four weeks prior to germination
Aquilegia vulgaris	7	70–75° F (21–24° C)	3	Light cover	Above average		Provide moist treatment at 40° F (4° C) for four weeks prior to germination
Arabis blepharophylla	9	68–75° F (20–24° C)	3	Light cover	Above average		
Arabis caucasica	8	68–75° F (20–24° C)	4	Light cover	Above average	Light beneficial	
Arenaria montana	8	60–65° F (16–18° C)	5	Light cover	Below average	Light beneficial	Cold germinator, provide stratification; cover lightly with medium after germination
Armeria hybrida	9	64–68° F (18–20° C)	3	Light cover	Average	Light beneficial	Cover lightly with medium after germination
Armeria latifolia	9	64–68° F (18–20° C)	2	Light cover	Average	Light beneficial	Cover lightly with medium after germination
Armeria maritima	10	64–68° F (18–20° C)	3	Light cover	Average	Light beneficial	Cover lightly with medium after germination
Armeria pseudarmeria	9	64–68° F (18–20° C)	2	Light cover	Average	Light beneficial	

(continued)

Table 9.2. Germination Requirements of Perennials *(continued)*

Variety	Weeks to Finish 128 Cell	Germination Temperatures	Seeds per Cell	Type of Covering	Moisture Requirement	Light Requirement	Comments
Asclepias curassavica	9	65–72° F (18–22° C)	3	No cover	Above average	Light required	
Asclepias tuberosa	9	65–72° F (18–22° C)	3	No cover	Above average	Light required	
Aster alpinus	8	65–72° F (18–22° C)	3	No cover	Average	Light required	
Aster amellus	8	65–72° F (18–22° C)	3	No cover	Average	Light required	
Aster novae-angliae	7	65–70° F (18–21° C)	3	No cover	Average	Light required	
Aster novi-belgii	8	65–72° F (18–22° C)	3	No cover	Average	Light required	
Aster tongolensis	9	65–72° F (18–22° C)	3	No cover	Average	Light required	
Astilbe arendsii	12	65–72° F (18–22° C)	5	Light cover	Above average		
Aubretia x *cultorum*	8	68–72° F (20–22° C)	3	No cover	Above average	Light required	
Aurinia saxatilis	9	63–68° F (17–20° C)	4	No cover	Above average	Light required	
Baptisia australis	11	65–72° F (18–22° C)	2	Cover	Below average		Cold germinator, provide stratification; cover lightly with medium after germination
Bellis perennis	7	70–75° F (21–24° C)	4	Light cover	Average	Light required	
Bergenia cordifolia	10	68–75° F (20–24° C)	3	No cover	Above average	Light required	
Campanula carpatica	12	65–72° F (18–22° C)	5	No cover	Above average	Light required	
Campanula cashmeriana	12	65–72° F (18–22° C)	3	No cover	Above average	Light required	
Campanula cochleariifolia	11	65–72° F (18–22° C)	3	No cover	Above average	Light required	
Campanula garganica	10	50–60°F (10–16° C)	3	No cover	Above average	Light required	Cold germinator, provide stratification
Campanula glomerata	11	65–72° F (18–22° C)	4	No cover	Above average	Light required	
Campanula lactiflora	11	65–72° F (18–22° C)	4	No cover	Above average	Light required	Cold germinator, provide stratification

Campanula medium	10	65–72° F (18–22° C)	3	Light cover	Above average	Light required	
Campanula persicifolia	11	65–72° F (18–22° C)	5	No cover	Above average	Light required	
Campanula poscharskyana	11	65–72° F (18–22° C)	5	No cover	Above average	Light required	
Campanula rotundifolia	12	65–72° F (18–22° C)	5	No cover	Above average	Light required	
Carlina acaulis	9	68–75° F (20–24° C)	2	Cover	Average		
Centaurea dealbata	8	70–75° F (21–24° C)	2	Heavy Cover	Below average		
Centaurea macrocephala	8	70–75° F (21–24° C)	2	Heavy Cover	Below average		
Centaurea montana	8	70–75° F (21–24° C)	2	Heavy Cover	Below average		
Centranthus ruber	9	65–72° F (18–22° C)	3	No cover	Average	Light required	
Cerastium tomentosum	8	65–72° F (18–22° C)	4	Light cover	Above average	Light beneficial	
Chasmanthium latifolium	10	68–75° F (20–24° C)	3	Cover	Average		
Cheiranthus cheiri	7	65–72° F (18–22° C)	2	No cover	Average	Light required	
Cheiranthus suffruticosum	7	65–68° F (18–20° C)	2	No cover	Average	Light required	
Coreopsis auriculata	7	68–75° F (20–24° C)	3	Light cover	Below average	Light beneficial	
Coreopsis grandiflora	9	68–75° F (20–24° C)	3	Light cover	Below average	Light beneficial	
Coronilla varia	8	68–75° F (20–24° C)	2	Light cover	Average	Light required	
Cortaderia selloana	10	65–72° F (18–22° C)	3	Cover	Below average		
Delosperma cooperi	10	68–75° F (20–24° C)	3	Light cover	Average		
Delosperma floribundum	10	68–75° F (20–24° C)	3	Light cover	Average		
Delphinium belladonna	11	60–65° F (16–18° C)	3	Light cover	Below average		
Delphinium elatum	11	64–68° F (18–20° C)	3	Light cover	Below average		Cold germinator, provide stratification
Delphinium grandiflorum	9	64–68° F (18–20° C)	3	Light cover	Below average		

(continued)

Table 9.2. Germination Requirements of Perennials *(continued)*

Variety	Weeks to Finish 128 Cell	Germination Temperatures	Seeds per Cell	Type of Covering	Moisture Requirement	Light Requirement	Comments
Delphinium nudicaule	9	64–68° F (18–20° C)	3	Light cover	Below average		
Dianthus 'Allwoodii'	8	68–75° F (20–24° C)	3	No cover	Average	Light required	
Dianthus barbatus	9	65–72° F (18–22° C)	3	Light cover	Average	Light required	
Dianthus caryophyllus	9	68–75° F (20–24° C)	3	Light cover	Average	Light required	
Dianthus deltoides	9	65–72° F (18–22° C)	3	Light cover	Average	Light required	
Dianthus gratianopolitanus	8	68–75° F (20–24° C)	3	Light cover	Average	Light required	
Dianthus grenadin	9	70–75° F (21–24° C)	3	Light cover	Average	Light required	
Dianthus hybridus	8	65–72° F (18–22° C)	3	No cover	Average	Light required	
Dianthus plumarius	7	68–75° F (20–24° C)	3	Light cover	Average	Light required	
Dianthus superbus	8	68–75° F (20–24° C)	3	No cover	Average	Light required	
Dicentra eximia	20	64–68° F (18–20° C)	4	Light cover	Average		Cold germinator, provide stratification
Digitalis grandiflora	8	70–75° F (21–24° C)	4	Light cover	Above average	Light required	
Digitalis mertonensis	8	70–75° F (21–24° C)	4	Light cover	Above average	Light required	
Digitalis purpurea	8	70–75° F (21–24° C)	4	Light cover	Above average	Light required	
Doronicum orientale	9	75–80° F (24–27° C)	3	Light cover	Average		
Draba aizoon	7	65–72° F (18–22° C)	4	Light cover	Average		
Echinacea paradoxa	10	68–75° F (20–24° C)	2	Heavy Cover	Below average		
Echinacea purpurea	10	68–75° F (20–24° C)	2	Heavy Cover	Below average		
Echinacea tennesseensis	10	68–75° F (20–24° C)	3	Heavy Cover	Below average		
Echinops bannaticus	8	70–75° F (21–24° C)	2	Light cover	Below average	Light beneficial	

Echinops ritro	10	70–75° F (21–24° C)	2	Light cover	Below average	Light beneficial	
Erianthus ravennae	9	68–75° F (20–24° C)	3	Light cover	Average		
Erigeron aurantiacus	9	68–75° F (20–24° C)	3	Light cover	Average		
Erigeron glaucus	8	70–75° F (21–24° C)	3	Light cover	Average	Light beneficial	
Erigeron karvinskianus	11	60–68° F (16–20° C)	4	Light cover	Average	Light beneficial	
Erigeron speciosus	9	68–75° F (20–24° C)	4	Light cover	Average	Light beneficial	
Erinus alpinus	8	68–75° F (20–24° C)	5	Light cover	Average		
Eryngium planum	8	60–70° F (16–21° C)	3	No cover	Average	Light beneficial	
Erysimum perofskianum	7	60–70° F (16–21° C)	3	No cover	Average	Light beneficial	
Erysimum suffruticosum	7	60–70° F (16–21° C)	3	No cover	Average	Light beneficial	
Euphorbia myrsinites	7	70–75° F (21–24° C)	3	Cover	Below average		
Euphorbia polychroma	9	68–75° F (20–24° C)	3	Cover	Below average		Cold germinator, provide stratification
Festuca amethystina	9	68–75° F (20–24° C)	3	Light cover	Below average		
Festuca glauca	9	63–68° F (17–20° C)	3	Light cover	Below average		
Festuca valesiaca	9	63–68° F (17–20° C)	3	Light cover	Below average		
Gaillardia aristata	8	65–72° F (18–22° C)	2	Cover	Below average	Light beneficial	
Gaillardia grandiflora	8	70–75° F (21–24° C)	2	Heavy Cover	Below average		
Gentiana septemfida	10	65–72° F (18–22° C)	4	Light cover	Above average		Cold germinator, provide stratification
Geranium bohemicum	9	65–75° F (18–24° C)	3	Light cover	Above average		Cold germinator, provide stratification
Geranium pratense	9	65–75° F (18–24° C)	3	Light cover	Above average		Cold germinator, provide stratification

(continued)

Table 9.2. Germination Requirements of Perennials *(continued)*

Variety	Weeks to Finish 128 Cell	Germination Temperatures	Seeds per Cell	Type of Covering	Moisture Requirement	Light Requirement	Comments
Geranium sanguineum	10	65–75° F (18–24° C)	3	Light cover	Above average		Provide moist treatment at 40° F (4° C) for four weeks prior to germination
Geum chiloense	11	70–75° F (21–24° C)	3	No cover	Average	Light required	Provide moist treatment at 40° F (4° C) for four weeks prior to germination
Geum coccineum	10	70–75° F (21–24° C)	3	No cover	Average	Light required	Provide moist treatment at 40° F (4° C) for four weeks prior to germination
Gypsophila pacifica	8	70–75° F (21–24° C)	4	No cover	Average	Light required	Provide moist treatment at 40° F (4° C) for four weeks prior to germination
Gypsophila paniculata	8	70–75° F (21–24° C)	4	No cover	Average	Light required	Provide moist treatment at 40° F (4° C) for four weeks prior to germination
Gypsophila repens	11	70–75° F (21–24° C)	4	Light cover	Average	Light required	Provide moist treatment at 40° F (4° C) for four weeks prior to germination
Helenium autumnale	9	68–72° F (20–22° C)	2	No cover	Average	Light required	Provide moist treatment at 40° F (4° C) for four weeks prior to germination
Helianthemum nummularium	10	70–75° F (21–24° C)	6	Light cover	Average		Provide moist treatment at 40° F (4° C) for four weeks prior to germination
Heliopsis helianthoides	9	70–75° F (21–24° C)	2	Light cover	Average	Light beneficial	
Heliopsis scabra	9	70–75° F (21–24° C)	2	Light cover	Average	Light beneficial	
Heuchera americana	10	65–72° F (18–22° C)	5	Light cover	Above average		

Heuchera micrantha	11	65–72° F (18–22° C)	5	Light cover	Above average		
Heuchera sanguinea	11	65–72° F (18–22° C)	5	Light cover	Above average		
Hibiscus moscheutos	6	70–75° F (21–24° C)	1	Heavy Cover	Below average		
Hibiscus x *hybrida*	6	70–75° F (21–24° C)	1	Heavy Cover	Below average		
Hieracium villosum	8	68–75° F (20–24° C)	2	Light cover	Average		
Hypericum calycinum	11	65–72° F (18–22° C)	3	No cover	Above average		Cold germinator, provide stratification
Hypericum x *polyphyllum*	10	65–72° F (18–22° C)	3	No cover	Above average	Light beneficial	Cold germinator, provide stratification
Iberis sempervirens	10	60–65° F (16–18° C)	4	Light cover	Below average		
Incarvillea delavayi	7	68–75° F (20–24° C)	3	Light cover	Average		
Inula ensifolia	8	68–75° F (20–24° C)	3	Light cover	Average		
Inula orientalis	8	68–75° F (20–24° C)	3	Light cover	Average		
Jasione laevis	8	65–72° F (18–22° C)	5	Light cover	Above average		
Knautia macedonica	8	72–78° F (22–26° C)	3	Light cover	Average		Provide moist treatment at 40° F (4° C) for four weeks prior to germination
Kniphofia uvaria	11	68–75° F (20–24° C)	2	Cover	Average		
Lathyrus latifolius	10	65–70° F (18–21° C)	2	Cover	Below average		
Lavandula angustifolia	13	60–70° F (16–21° C)	3	Light cover	Below average		Provide moist treatment at 40° F (4° C) for four weeks prior to germination; cover lightly with medium after germination
Lavandula vera	11	65–72° F (18–22° C)	3	Light cover	Below average		

(continued)

Table 9.2. Germination Requirements of Perennials *(continued)*

Variety	Weeks to Finish 128 Cell	Germination Temperatures	Seeds per Cell	Type of Covering	Moisture Requirement	Light Requirement	Comments
Leontopodium alpinum	10	65–72° F (18–22° C)	4	Light cover	Average	Light beneficial	
Leucanthemum superbum	8	68–75° F (20–24° C)	3	Light cover	Average		
Leucanthemum vulgare	8	68–75° F (20–24° C)	3	Light cover	Average		
Lewisia cotyledon	9	60–65° F (16–18° C)	3	Light cover	Below average		Cold germinator, provide stratification
Lewisia longipetala	10	64–68° F (18–20° C)	3	Light cover	Below average		Cold germinator, provide stratification
Liatris spicata	11	65–72° F (18–22° C)	3	Light cover	Average	Light beneficial	
Lilium willmottiae	10	65–72° F (18–22° C)	3	Light cover	Average		
Limonium caspium	10	68–72° F (20–22° C)	3	Light cover	Above average	Light beneficial	
Limonium latifolium	11	68–72° F (20–22° C)	3	Light cover	Above average	Light beneficial	
Limonium perezii	10	68–72° F (20–22° C)	3	Light cover	Above average	Light beneficial	
Limonium tataricum	11	68–72° F (20–22° C)	3	Light cover	Above average	Light beneficial	
Linaria purpurea	8	62–68° F (17–20° C)	3	Light cover	Average	Light beneficial	
Linum flavum	7	64–68° F (18–20° C)	4	Light cover	Average		
Linum perenne	8	68–75° F (20–24° C)	4	Light cover	Average	Light beneficial	
Lobelia cardinalis	12	68–75° F (20–24° C)	5	No cover	Above average	Light required	
Lobelia speciosa	12	68–75° F (20–24° C)	5	No cover	Above average	Light required	
Lobelia x *hybrida*	12	68–75° F (20–24° C)	5	No cover	Above average	Light required	
Lunaria annua	8	70–75° F (21–24° C)	2	Cover	Below average		
Lunaria biennis	8	70–75° F (21–24° C)	2	Cover	Below average		
Lupinus polyphyllus	6	65–72° F (18–22° C)	2	Cover	Below average		

Luzula nivea	8	64–68° F (18–20° C)	3	Light cover	Average		
Lychnis chalcedonica	9	65–72° F (18–22° C)	3	Light cover	Average	Light required	
Malva sylvestris	9	70–75° F (21–24° C)	2	Cover	Below average		Cold germinator, provide stratification
Melica transsilvanica	8	68–72° F (20–22° C)	3	Light cover	Average		
Monarda didyma	7	70–75° F (21–24° C)	3	No cover	Average	Light beneficial	
Myosotis sylvatica	7	68–72° F (20–22° C)	3	Light cover	Above average	Dark beneficial	
Nassella tenuissima	10	68–72° F (20–22° C)	2	Light cover	Average		
Nepeta mussinii	6	70–75° F (21–24° C)	3	Light cover	Average	Light beneficial	
Nepeta x *faassenii*	7	65–72° F (18–22° C)	3	No cover	Average	Light beneficial	
Oenothera missouriensis	9	64–68° F (18–20° C)	3	Light cover	Above average		Cover lightly with medium after germination
Oenothera speciosa	9	60–65° F (16–18° C)	3	Light cover	Above average		Cover lightly with medium after germination
Papaver alpinum	10	64–68° F (18–20° C)	2	Light cover	Above average	Light beneficial	Cover lightly with medium after germination
Papaver miyabeanum	9	64–68° F (18–20° C)	3	Light cover	Above average	Light beneficial	Cover lightly with medium after germination
Papaver nudicaule	9	64–68° F (18–20° C)	3	Light cover	Above average	Light beneficial	Cover lightly with medium after germination
Papaver orientale	10	64–68° F (18–20° C)	3	Light cover	Above average	Light beneficial	Cover lightly with medium after germination
Pennisetum alopecuroides	7	72–78° F (22–26° C)	3	Light cover	Average		
Penstemon barbatus	7	64–68° F (18–20° C)	4	No cover	Average	Light required	Provide moist treatment at 40° F (4° C) for two weeks prior to germination for some cultivars

(continued)

Table 9.2. Germination Requirements of Perennials *(continued)*

Variety	Weeks to Finish 128 Cell	Germination Temperatures	Seeds per Cell	Type of Covering	Moisture Requirement	Light Requirement	Comments
Penstemon digitalis	8	70–75° F (21–24° C)	4	No cover	Average	Light required	Provide moist treatment at 40° F (4° C) for two weeks prior to germination
Penstemon virgatus	10	70–75° F (21–24° C)	4	No cover	Average	Light required	Cold germinator, provide stratification
Physalis alkekengi	8	70–75° F (21–24° C)	2	No cover	Above average	Light required	
Physalis franchetii	8	70–75° F (21–24° C)	2	No cover	Above average	Light required	
Physalis virginiana	7	68–72° F (20–22° C)	2	No cover	Above average	Light required	
Physostegia virginiana	9	70–75° F (21–24° C)	4	No cover	Average	Light required	
Platycodon grandiflorus	10	68–75° F (20–24° C)	3	No cover	Average	Light required	
Polemonium boreale	9	68–72° F (20–22° C)	3	No cover	Above average	Light required	
Polemonium caeruleum	9	68–72° F (20–22° C)	3	No cover	Above average	Light required	
Polemonium pauciflorum	8	68–72° F (20–22° C)	3	No cover	Above average	Light required	
Potentilla nepalensis	8	72–75° F (22–24° C)	2	No cover	Average	Light required	
Primula x *polyantha*	11	62–68° F (17–20° C)	3	Light cover	Above average		Cover lightly with medium after germination
Primula x *pubescens*	10	62–68° F (17–20° C)	3	Light cover	Above average		Cover lightly with medium after germination
Primula rosea	10	62–68° F (17–20° C)	3	Light cover	Above average		Cover lightly with medium after germination
Primula vialii	11	62–68° F (17–20° C)	3	Light cover	Above average		Cover lightly with medium after germination
Pulsatilla vulgaris	11	64–68° F (18–20° C)	2	Cover	Above average		Cold germinator, provide stratification; cover lightly with medium after germination

Ranunculus repens	10	64–68° F (18–20° C)	3	No cover	Average	Light required	Cold germinator, provide stratification
Rosa chinensis	9	65–70° F (18–21° C)	2	Light cover	Average	Light beneficial	
Rudbeckia fulgida	11	72–80° F (22–27° C)	3	Light cover	Below average	Light beneficial	Provide moist treatment at 40° F (4° C) for two weeks prior to germination
Rudbeckia hirta	8	70–75° F (21–24° C)	2	Light cover	Below average	Light beneficial	
Rudbeckia speciosa	9	72–80° F (22–27° C)	3	Light cover	Below average	Light beneficial	
Rumex sanguineus	9	68–75° F (20–24° C)	2	Light cover	Average		
Sagina subulata	8	65–72° F (18–22° C)	5	No cover	Above average	Light required	
Salvia lyrata	8	68–75° F (20–24° C)	3	No cover	Average	Light required	
Salvia officinalis	7	65–70° F (18–21° C)	3	No cover	Average	Light required	
Salvia x *superba*	8	68–75° F (20–24° C)	3	No cover	Average	Light required	
Salvia x *sylvestris*	9	70–75° F (21–24° C)	3	No cover	Average	Light required	
Santolina chamaecyparissus	10	68–72° F (20–22° C)	3	Light cover	Average		
Saponaria ocymoides	9	64–68° F (18–20° C)	4	No cover	Above average	Light required	Cover lightly with medium after germination
Saxifraga x *arendsii*	12	68–72° F (20–22° C)	5	Cover	Above average		Cold germinator, provide stratification; cover lightly with medium after germination
Scabiosa caucasica	10	68–72° F (20–22° C)	2	Cover	Below average		
Scabiosa columbaria	8	70–75° F (21–24° C)	2	Cover	Below average		Cold germinator, provide stratification
Sedum acre	10	65–72° F (18–22° C)	5	No cover	Above average	Light required	
Sedum reflexum	10	65–72° F (18–22° C)	5	No cover	Above average	Light required	

(continued)

Table 9.2. Germination Requirements of Perennials *(continued)*

Variety	Weeks to Finish 128 Cell	Germination Temperatures	Seeds per Cell	Type of Covering	Moisture Requirement	Light Requirement	Comments
Sedum selskianum	8	65–72° F (18–22° C)	5	No cover	Above average	Light required	
Sedum spurium	10	65–72° F (18–22° C)	5	No cover	Above average	Light required	
Sempervivum spp.	13	68–72° F (20–22° C)	4	Light cover	Above average		
Sidalcea x *hybrida*	8	64–68° F (18–20° C)	2	Cover	Below average		Cover lightly with medium after germination
Silene maritima	9	68–75° F (20–24° C)	4	No cover	Average	Light required	
Silene orientalis	8	68–75° F (20–24° C)	4	No cover	Average	Light required	
Silene schafta	8	68–75° F (20–24° C)	3	No cover	Average	Light required	
Solidago canadensis	9	65–72° F (18–22° C)	3	Light cover	Average		
Stachys byzantina	8	68–75° F (20–24° C)	3	No cover	Average	Light required	
Tanacetum coccineum	8	68–72° F (20–22° C)	3	No cover	Average	Light beneficial	
Teucrium chamaedrys	11	65–72° F (18–22° C)	2	Light cover	Below average		
Thalictrum aquilegiifolium	10	68–75° F (20–24° C)	3	Cover	Average		
Thymus serpyllum	8	60–65° F (16–18° C)	4	No cover	Above average	Light required	
Thymus vulgaris	7	60–65° F (16–18° C)	4	No cover	Above average	Light required	
Tiarella wherryi	13	60–65° F (16–18° C)	4	No cover	Above average	Light required	Cold germinator, provide stratification
Trollius chinensis	9	68–72° F (20–22° C)	2	Light cover	Below average		Cold germinator, provide stratification
Trollius x *cultorum*	8	70–75° F (21–24° C)	2	Light cover	Below average		Cold germinator, provide stratification
Verbascum bombyciferum	7	68–72° F (20–22° C)	4	Light cover	Average		
Verbascum phoeniceum	7	68–72° F (20–22° C)	4	Light cover	Average		

Verbascum x *hybrida*	7	70–75° F (21–24° C)	3	Light cover	Average		
Verbena bonariensis	7	65–70° F (18–21° C)	3	Cover	Average		
Verbena rigida	7	65–70° F (18–21° C)	3	Cover	Average		
Veronica incana	7	68–72° F (20–22° C)	3	No cover	Average	Light required	
Veronica longifolia	6	68–72° F (20–22° C)	3	Light cover	Average	Light required	
Veronica repens	8	68–72° F (20–22° C)	3	No cover	Average	Light required	
Veronica spicata	8	68–72° F (20–22° C)	3	Light cover	Average	Light required	
Veronica teucrium	7	68–72° F (20–22° C)	3	No cover	Average	Light required	
Viola cornuta	7	60–65° F (16–18° C)	2	Cover	Average		
Viola labradorica	8	60–65° F (16–18° C)	2	Cover	Average		Cold germinator, provide stratification
Viola mandshurica	7	60–65° F (16–18° C)	2	Cover	Average		
Viola sororaria	7	60–65° F (16–18° C)	2	Cover	Average		Cold germinator, provide stratification
Viola tricolor	7	60–65° F (16–18° C)	3	Cover	Average		
Viscaria alpina	8	65–68° F (18–20° C)	4	Light cover	Average		

10

Forcing Perennials

Selling perennials blooming out of season is the current wave washing across the floriculture marketplace. Perennial production practices are changing due to the needs or demands of the customer. The marketplace is directing the way growers produce perennials, to what degree they are blooming, and when they want them. There are many factors that are influencing these market-driven production changes, but the single most important would be "flower power." Flowering perennials sell themselves.

Mass merchandisers, retailers, garden centers, and consumers are all demanding more color at the point of sale. Due to the undisputed fact that a majority of plant purchases are made on impulse, growers and retailers are trying to capitalize by providing flowering plants, adding color to the perennial displays. In many cases, green, non-flowering plants may perform best for the consumer, but it can be a challenge to sell perennials green. Big colorful care tags may improve the sale of some green perennials, but tags alone do not replace the impact of seeing real flowers for most consumers.

Since 1992, professors at Michigan State University (MSU) have been researching the flowering requirements for a broad range of perennial crops. The main vision entailed allowing growers to produce flowering perennials for any predetermined date. This was highly criticized and controversial at the time (and maybe even so now in some circles). With the success of MSU's research and perennial production on the rise, several universities are adding perennial research to their programs.

Figure 10.1. Retail displays featuring colorful perennials generate impulse buying.

Getting plants to bloom out of season is not a new concept. Growers have forced plants into bloom since the first greenhouse was constructed. Many of the potted crops, such as spring bulbs, pot mums, and calla lilies, to name a few, are crops that are grown and sold in bloom outside their normal flowering seasons. In fact, the production of annual bedding plants is a type of forcing that allows growers to produce flowering plants for the marketplace.

Forcing is the term used most commonly by greenhouse growers to describe the process of producing a plant into bloom out of season. Perennials forced into bloom provide growers with a great marketing tool that has several advantages over selling perennials without bloom. Flowering plants with color create a great amount of impulse buying. What would the average consumer prefer: a blooming *Rudbeckia,* or a green plant with a care tag showing what the flowers will look like in the future? A flowering plant provides instant gratification and assurance that consumers are getting what they paid for. Additionally, the ability to schedule flowering perennials out of season allows growers to extend the shipping window, broaden the product line, utilize greenhouse space, and maximize sales opportunities. With blooming perennials at the retail site, the space can be turned more frequently, increasing the sales per square foot, and overall profitability.

From a marketing perspective, it is hard to argue that growers can increase sales by increasing the percentage of blooming plants at the retail site. From the business point of view, it really makes sense. There is, however, some controversy as to the ethics of supplying blooming plants to the market when they would naturally still be vegetative. Are we deceiving customers? Will they be disappointed next year when they bloom in a different month? What effect will forced plants have in landscape designs when they bloom in subsequent years? Do we have to educate consumers about forced perennials?

When selling perennials that have been forced to bloom out of season, I believe it is necessary to provide color care tags that clearly list what time of the year the perennials usually flower. Educate the employees in the garden center that certain perennials have been forced to bloom out of season and explain how they will perform in the landscape, both this season and next. Encourage them to pass this information on to customers, and to answer any questions that may arise. As long as this information is presented to the consumers, growers and retailers are not deceiving their customers and are operating in an ethical manner.

With the ability to force many perennials into bloom, growers will probably begin to market perennials alongside bedding plant flats, allowing perennials to gain more of the overall bedding/garden plant market share. Many growers can extend their normal sales windows by offering flowering perennials for late summer and early fall sales. Flowering perennials could conceivably be produced year-round, creating many new marketing opportunities. By manipulating flowering, growers can even promote the added value of some perennials as houseplants that when finished flowering can be planted outside in the garden. There are many undiscovered opportunities with perennials, and we are just beginning to take advantage of these new markets.

Growers who attempt to force perennials out of season are faced with some challenges and must understand the general requirements and each variety's specific requirements for flowering. Many plants have a juvenility requirement; they must reach a certain size or have a certain number of leaves before they will flower. Vernalization, or a cold treatment, at a certain temperature and duration is required for many varieties. Most plants also have a photoperiod requirement; they must receive a certain number of hours of light each day to produce flowers.

To achieve flowering, growers must take some steps to ensure that these requirements are met. The starter material used can often be purchased as vernalized plugs, which have the juvenility and cold requirements already achieved. Otherwise, growers can use cold frames or coolers to provide the cold treatment prior to forcing. To provide the proper photoperiod, it is often necessary to use supplemental lighting to extend the day length or supply night interruption for long-day plants. Perennial forcers should anticipate a longer production time and usually a greater space requirement than are needed for growing non-blooming perennials. These extra steps require some extra planning, organization, and expense. If the grower is rewarded with a higher price and the sell through (percent sold) is increased due to impulse buying, the reward can be greater than the extra costs associated with forcing.

Juvenility

The first requirement for perennials to flower is they must be old enough to bloom. The inability to bloom is often called juvenility and refers to the early stages of growth when plants will not flower, even if all of the other criteria necessary for flowering have been provided. Juvenile plants are incapable of perceiving and responding to the environmental stimuli that cause mature plants to flower. The duration of the juvenility varies by species and usually lasts from a few weeks to a few months. Juvenility is usually more prevalent with perennials propagated from seed and occurs less with vegetatively propagated or bare-root materials. Plants that have finished the juvenility phase are considered to be mature and can be induced to flower under the proper growing conditions.

Figure 10.2. ***Campanula carpatica* cultivars have a juvenility requirement and must form nine to eleven leaves before they are mature enough to flower.**

For perennials that have a juvenile phase, it can be difficult to determine the age of the plant simply by looking at the size of the plant or of the starter plug. Growers should not just assume a 50-cell plug is older or more mature than a 72-cell or 128-cell plug. Larger plugs may look older, but size alone does not provide an accurate indication of plant maturity. A plant's actual age, as determined by a calendar, does not reflect the environmental conditions such as temperature that greatly influence the plant's rate of development and maturity. The most practical method to determine plant maturity is to count the number of nodes or leaves the plant has made. Keep in mind though that the juvenility requirements and number of leaves necessary before plants will flower have not yet been determined for all of the perennials being produced. For the perennials with known juvenility requirements and leaf counts, see table 10.1.

Once a certain number of leaves have been made, referred to as the critical leaf number, the juvenile period is over and the plant is allowed to flower, provided all the other requirements for flowering have been met. Growers should use leaf counts as an indication of the average age of a population of plants. It is not necessary to count the leaves on every plug being planted. It is best to count the leaves randomly from plants within a population and take an average to determine what the average age of the group of perennials is. A perennial requiring ten leaves for 100% of the population to flower will often have a percentage of the population bloom even when fewer leaves are formed. As with the example above, once the average leaf count reaches ten leaves, 100% of the plants will bloom, but if the population averages eight leaves, only 70%

Table 10.1. Juvenility Requirements

Generally, individual cultivars within the species listed below will have similar juvenility requirements. Cultivar-specific differences can and do exist. Use the information provided here only as a guideline.

Variety	Cultivar	Juvenility # leaves
Alchemilla mollis	Cultivars	20+
Anemone hupehensis	Cultivars	Until leaves become trilobed
Aquilegia alpina	Cultivars	16+
Aquilegia caerulea	Cultivars	16+
Aquilegia flabellata	'Fairyland'	15
	'Cameo Blue & White'	5+
	'Mini Star'	7–9
Aquilegia x *hybrida*	Cultivars	12–14
	'Crimson Star'	12+
	'Musik White'	14
	'Song Bird' series	12–16
	'Swan' series	12–14
	'McKana Giants'	12
	'Origami' series	7–9
	'Winky' series	9–12
Aquilegia vulgaris	Cultivars	16+
Astilbe x *arendsii*	Cultivars	7
Astilbe chinensis	Cultivars	5
Astilbe japonica	Cultivars	7
Astilbe thunbergii	'Ostrich Plume'	7
Campanula medium	'Champion' series	8–10
Campanula carpatica	'Clips' series	9–11
Campanula isophylla	'Stella'	15
Campanula portenschlagiana	'Birch Hybrid'	4–5
Cineraria hybridus	'Jester'	8–10
Coreopsis grandifolia	Cultivars	16
Digitalis grandiflora	'Carillion'	20
Digitalis obscura	Cultivars	60
Digitalis purpurea	'Foxy'	12–15
Digitalis thapsi	'Spanish Peaks'	35
Euphorbia epithymoides	Cultivars	25+
Gaillardia x *grandiflora*	Cultivars	12–16
Heuchera x *hybrida*	Cultivars	15
Heuchera sanguinea	Cultivars	15–16
Iberis sempervirens	'Snowflake'	40+
Lavandula angustifolia	Cultivars	25+
Leucanthemum x *superbum*	'Snowcap'	12–16
Lobelia x *speciosa*	'Compliment Scarlet'	6–7
Oenothera fruticosa	'Youngii-lapsley'	18–19
Oxalis crassipes	'Rosea'	13–17
Penstemon digitalis	'Huskers Red'	14

Variety	Cultivar	Juvenility # leaves
Physostegia virginiana	Cultivars	20
Platycodon grandiflorus	'Sentimental Blue'	13
Ranunculus repens	Cultivars	9–16
Rudbeckia fulgida	'Goldsturm'	10–15
Salvia x *superba*	'Blue Queen'	4–5
Scabiosa columbaria	'Butterfly Blue'	10–14
Stokesia laevis	'Klaus Jelitto'	10
Tiarella x *hybrida*	Cultivars	10
Veronica longifolia	'Sunny Border Blue'	4–5

of the population may be able to reach flowering. Growers will want to grow the plants until the proper average is reached for uniform flowering and increased marketability.

Many growers sow multiple seeds per cell. When counting the leaves, be sure to only count the leaves on a single stem, not all of the stems within the cell. Leaf counts may be difficult, or impossible, to count once they have gone through the vernalization process because many plants are trimmed back prior to receiving the cold treatment or have lost all of their leaves during the dormancy process.

Bulking

Bulking is a term often used to describe the growth period before a perennial is placed into the other treatments necessary for flowering. Bulking can be used to ensure a plant has passed through the juvenility phase and is mature enough to flower. It can be used to build the root system, allowing the plant to become well established prior to going through a cold treatment. Bulking can also be used to build the size of the plant, regardless of juvenility requirements, to a size more suitable for forcing. Depending on the crop and the availability of space, bulking can occur in the plug flat or in the finished container.

Generally, perennials are bulked under non-flower inductive conditions, keeping them in the vegetative state. The length of the bulking period varies for each variety, but often averages two to four weeks. The duration of this period is variety specific and is often determined by the size of the material being bulked and the size of the finished container it is intended for. For example, a 72-cell plug for a 1 qt. (1 L) container will require less bulking than a 72-cell intended for a 1 gal. (3.8 L) pot.

Consistent Starting Materials

One of the biggest challenges growers face when putting together forcing programs is the variability in the starting materials for various crops from year to year. Growers who purchase a majority of their starting materials are prone to experience variability, which most often gets carried over into the finished crop.

Often, growers piece forcing programs together without an adequate amount of time for planning or crop preparation. In these instances, to obtain a single perennial cultivar, growers are often at the fate of obtaining whatever products are currently available, of various starter sizes, and often from several suppliers across the country. Piecing forcing programs together, with so many exceptions up front, leads growers to produce inconsistent crops even before other variables are factored in. Without being in full control of the starting materials, growers often have no choice but to accept many unknowns regarding their starting materials.

There are a number of factors leading to variable and unpredictable crop production. It is not uncommon for growers to receive plugs that are fully rooted and appear ready for transplanting only to discover (usually too late) they were still in the juvenile stage and not able to

Figure 10.3. **When growers plant plugs that have not completely become mature, the crop will contain both juvenile (non-flowering) and non-juvenile (flowering) plants, as shown with this crop of *Aquilegia* 'Songbird Cardinal'.**

perceive the treatments necessary to achieve flowering. Growers often receive plug trays of perennials that are already blooming, which for many varieties does not allow growers the ability to bulk them up before the cold treatments or proper photoperiods are delivered. Growers might receive plug flats that are different ages or have varying degrees of maturity, which often leads to difficulty forcing a uniform crop of blooming perennials. Bare-root starter materials work well for several varieties of perennials, but in many instances they create a great deal of diversity in the size of the starter materials, uniformity of emergence, and may possibly carry over diseases from the fields they were produced in or from injury they received during harvesting, storage, and shipping.

An extreme example—that very easily could occur—is when a grower orders a hundred flats of 72-cell *Coreopsis* 'Baby Sun' and expects them to be vernalized in the plug flats before they are received for potting in the early spring. The plug supplier offers but *does not guarantee* vernalized plugs. When the grower receives the plugs, he gets seventy flats of vernalized plugs and thirty flats of non-vernalized plugs. If the grower chooses to plant all one hundred flats, he will get two distinctly different results: a large group of flowering plants and a small group of non-flowering plants.

Growers must allow adequate time to preplan forcing programs, reviewing the schedules and the various requirements of the crops to be produced. It is important for growers attempting to force perennials to work closely and to establish relationships with perennial plant suppliers. Plug suppliers are willing to work with individual customers to supply plugs that meet their needs and expectations.

Vernalization

Many perennials require a cold treatment, referred to as vernalization, to induce flowering before they are grown at warmer temperatures. In the landscape, perennials receive the necessary cold during the winter months in many parts of the country. Researchers have found three distinct cold requirement categories that perennials can be placed into: no cold required, obligate cold required, and cold beneficial. The first category represents the perennials that do not require or have any apparent advantage to receiving a cold treatment for flowering. The second category represents the perennials that absolutely must receive a cold treatment for them to flower; they have an obligate cold requirement for flowering. The cold beneficial category contains those perennials for which a cold treatment is beneficial but not required for flowering. Successful vernalization depends on the maturity of the plant, or its ability to perceive cold, the cooling temperature, and the length of the cold treatment.

No cold required

Many perennials do not require any cold exposure in order for them to flower. These perennials may have other requirements for flowering, such as juvenility or a proper photoperiod. Though these perennials do not require cold for flowering, it does not mean that they cannot experience vernalization or that cold is harmful to the flowering process. With most perennials in this category, it is acceptable and often beneficial for

Table 10.2. Perennials with No Cold Requirement

Many perennials do not require a cold treatment in order for them to flower. Many plants in this category are also classified as cold beneficial since providing cold can reduce the time to flower, produce more flowers per plant, and bloom more uniformly following exposure to cold. Providing cold to most non-cold requiring perennials is acceptable and usually has no detrimental effects on plant health and performance.

Generally, individual cultivars within the species listed below will have similar cold requirements. Cultivar-specific differences can and do exist. Use the information provided here only as a guideline.

Achillea millefolium	*Erysimum hybridum*	*Platycodon grandiflorus* *
Agastache foeniculum *	*Erysimum linifolium*	*Polemonium caeruleum* *
Agastache rugosa	*Erysimum suffruticosum*	*Polemonium viscosum*
Armeria x *hybrida* *	*Eupatorium rugosum*	*Primula elatior*
Armeria pseudarmeria	*Gaura lindheimeri*	*Primula japonica*
Asclepias tuberosa	*Hakonechloa macra*	*Primula* x *polyantha* *
Aster dumosus	*Helianthus angustifolius*	*Rosa chinensis*
Bellis perennis *	*Heliopsis scabra*	*Rudbeckia hirta*
Buddleia davidii	*Hibiscus moscheutos*	*Rudbeckia speciosa*
Campanula carpatica	*Hibiscus* x *hybrids*	*Salvia guaranitica*
Campanula longistyla	*Hypericum polyphyllum*	*Salvia hybrid*
Centaurea montana *	*Kniphofia uvaria* *	*Salvia* x *superba* *
Centranthus ruber	*Lamium maculatum*	*Salvia* x *sylvestris* *
Ceratostigma plumbaginoides	*Lavandula stoechas*	*Sedum aizoon*
Cheiranthus suffruticosum	*Leucanthemum superbum* *	*Sedum bithynicum*
Chelone glabra	*Limonium perezii*	*Sedum cauticolum*
Chrysogonum virginianum	*Lupinus* x *hybrida* *	*Sedum maximum*
Clematis x *hybrida* *	*Lysimachia clethroides*	*Sedum spectabile* *
Coreopsis auriculata *	*Lysimachia punctata*	*Sedum telephium*
Coreopsis grandifolia *	*Malva alcea*	*Sedum tetractinum*
Delphinium cultorum	*Myosotis hybrida* *	*Sidalcea hybrida*
Delphinium grandiflorum	*Nepeta faassenii*	*Silene maritima*
Delphinium nudicaule	*Panicum virgatum*	*Sisyrinchium angustifolium*
Delphinium belladonna	*Papaver alpinum*	*Sisyrinchium bellum*
Delphinium elatum	*Papaver miyabeanum*	*Sisyrinchium tinctorium*
Dianthus barbatus *	*Papaver nudicaule* *	*Solidago canadensis*
Dianthus hybridus	*Pennisetum setaceum*	*Thymus serpyllum* *
Dianthus plumarius	*Penstemon barbatus* *	*Tricyrtis hirta*
Dianthus superbus	*Penstemon campanulatus* *	*Verbascum phoeniceum*
Diascia barberae	*Penstemon mexicale*	*Verbascum* x *hybrida*
Erigeron aurantiacus	*Penstemon smallii*	*Verbena bonariensis*
Erigeron karvinskianus	*Persicaria affinis*	*Verbena rigida*
Erinus alpinus	*Phlox paniculata* *	*Veronica spicata* *

*Only select cultivars

Table 10.3. Perennials with Obligate Cold Requirements

Many perennials have an absolute cold requirement that must be met in order for them to flower. Generally, individual cultivars within the species listed below will have similar cold requirements. Cultivar-specific differences can and do exist. Use the information provided here only as a guideline.

Providing more weeks of cold than the recommendations listed here is acceptable and may continue to increase the benefits with many perennial varieties.

The cold duration is the recommended minimum amount of cold to provide for each species.

Variety	Weeks of Cold	Variety	Weeks of Cold	Variety	Weeks of Cold	Variety	Weeks of Cold
Achillea filipendulina	6–9	*Arenaria montana*	6–10	*Campanula persicifolia*	9–12	*Dianthus gratianopolitanus*	6–9
Achillea tomentosa	6–9	*Aster alpinus*	10–12	*Campanula portenschlagiana*	10–15	*Dicentra eximia*	9–12
Aconitum napellus	7–10	*Aster amellus*	10–12	*Campanula poscharskyana*	10–15	*Dicentra spectabilis*	9–12
Ajuga reptans	6–9	*Aster tongolensis*	10–12	*Campanula punctata*	10–15	*Digitalis obscura*	6–9
*Alcea rosea**	6–9	*Astilbe* x *arendsii*	10–12	*Carlina acaulis*	9–12	*Digitalis purpurea**	6–9
Alchemilla mollis	6–9	*Astilbe chinensis*	10–12	*Centaurea macrocephala*	6–9	*Digitalis thapsi*	6–9
Alyssum saxatile	5–8	*Astilbe* x *hybrida*	10–12	*Cerastium tomentosum*	6–9	*Doronicum orientale*	6–10
Anacyclus depressus	7–10	*Astilbe japonica*	10–12	*Chasmanthium latifolium*	9–12	*Draba aizoon*	9–12
Andropogon scoparius	9–12	*Astilbe thunbergii*	10–12	*Cheiranthus cheiri*	6–9	*Erigeron speciosus*	6–10
Anemone multifida	11–15	*Astrantia major*	6–9	*Chrysanthemum coccineum*	7–10	*Eryngium alpinum*	7–10
Anemone vitifolia	11–15	*Aubrietia cultorum*	6–9	*Cimicifuga racemosa*	9–12	*Eryngium bourgatii*	7–10
Anemone x *hybrida*	11–15	*Aubrietia* x *hybrida**	5–8	*Coreopsis grandifolia**	9–12	*Eryngium planum*	7–10
Aquilegia alpina	6–9	*Aurinia saxatilis*	6–9	*Corydalis flexuosa*	5–8	*Euphorbia amygdaloides*	5–9
Aquilegia caerulea	6–9	*Bergenia cordiflolia*	5–8	*Corydalis lutea*	5–8	*Euphorbia epithymoides*	5–9
Aquilegia flabellata	6–9	*Brunnera macrophylla*	6–9	*Delosperma cooperi*	6–9	*Euphorbia myrsinites*	5–9
Aquilegia x *hybrida**	6–9	*Calamagrostis acutifolia*	9–12	*Delphinium elatum**	7–10	*Euphorbia polychroma*	5–9
Aquilegia vulgaris	6–9	*Calluna vulgaris*	6–10	*Delphinium zalil*	7–10	*Gaillardia grandiflora*	9–12
*Arabis albida**	4–6	*Campanula garganica*	10–15	*Dianthus barbatus**	6–10	*Galanthus nivalis*	8-10
Arabis caucasica	4–6	*Campanula glomerata*	10–15	*Dianthus caryophyllus*	6–9	*Galium odoratum*	12–15
Arabis sturii	4–6	*Campanula medium*	9–12	*Dianthus deltoides**	6–9	*Gentiana acaulis*	7–10

Gentiana septemfida	7–10	Inula orientalis	9–12	Pennisetum alopecuroides	10–12	Sedum floriferum	6–10
Geranium clarkei	6–9	Lathyrus latifolius	6–10	Penstemon barbatus	6–9	Sedum reflexum	6–10
Geranium dalmaticum	6–9	Laurentia axillaris	8–10	Penstemon campanulatus	6–9	Sedum spurium*	6–10
Geranium himalayense	6–9	Lavandula angustifolia*	10–15	Phlox divaricata	7–9	Sempervivum	7–10
Geranium x hybridum	6–9	Lavandula vera	10–15	Phlox subulata	7–9	Silene schafta	9–12
Geranium sanguineum	6–9	Leucanthemum superbum*	7–10	Physalis alkekengi	9–12	Stokesia laevis*	10–15
Geum chiloense	6–9	Lewisia cotyledon	6–10	Physostegia virginiana	10–12	Tanacetum coccineum	6–9
Geum coccineum*	6–9	Liatris spicata	9–12	Potentilla atrosanguinea	7–10	Tanacetum niveum	6–9
Geum hybridum	6–9	Lilium willmottiae	6–9	Potentilla megalantha	7–10	Teucrium chamaedrys	6–9
Gypsophila pacifica	10–12	Limonium caspium	6–9	Potentilla nepalensis	7–10	Thalictrum aquilegiifolium	6–9
Gypsophila paniculata	12–15	Limonium latifolium	6–9	Primula denticulata	6–9	Thalictrum kiusianum	6–9
Gypsophila repens	10–12	Limonium tataricum	6–9	Primula pubescens	6–9	Thymus serpyllum	7–10
Helianthemum nummularium	6–9	Linum flavum	7–10	Primula rosea	6–9	Trollius chinensis	6–10
Helichrysum thianshanicum	6–9	Linum perenne	7–10	Primula vialii	6–9	Trollius europaeus	6–10
Helictotrichon sempervirens	10–12	Lupinus x hybrida*	6–10	Primula vulgaris	6–9	Verbascum bombyciferum	6–9
Helleborus niger	4–8	Lupinus polyphyllus*	6–10	Primula polyanthus*	6–9	Veronica incana	10–12
Hemerocallis	9–12	Lychnis coronaria	6–9	Pulmonaria x hybrida	4–6	Veronica longifolia	10–12
Heuchera x hybrida	7–10	Lychnis flos-cuculi	6–9	Pulsatilla vulgaris	6–9	Veronica peduncularis	10–12
Heuchera sanguinea	7–10	Lychnis viscaria	6–9	Rodgersia aesculifolia	7–10	Veronica repens	6–9
Heucherella x hybrida	7–10	Miscanthus sinensis	10–12	Salvia officinalis*	6–9	Veronica spicata*	6–9
Hieracium villosum	9–12	Myosotis scorpioides	6–9	Santolina chamaecyparissus	9–12	Veronica teucrium	10–12
Hosta	4–8	Myosotis sylvatica*	6–9	Saponaria ocymoides	9–12	Viola odorata*	4–6
Hypericum calycinum	9–12	Nepeta faassenii*	6–9	Saxifraga arendsii	10–15	Viscaria alpina	9–12
Iberis sempervirens	8–10	Oenothera fruticosa	6–9	Saxifraga umbrosa	10–15		
Incarvillea delavayi	9–12	Papaver orientale	6–9	Sedum acre	6–10		

*Only select cultivars

them to receive a cold treatment. It should be noted that vernalization of non-cold-requiring perennials does not reduce the necessary forcing time after the cold treatment or provide any harmful side effects. There are a couple of exceptions, such as *Hibiscus moscheutos* 'Disco Belle Mixed', which may get injured at temperatures less than 41° F (5° C), and *Asclepias tuberosa,* which is prone to overwatering during cold treatments. It is recommended to start these varieties from seed the same year flowering is desired, keep them under long-day conditions, and avoid cold temperatures to prevent them from going dormant.

Cold required

Perennials with an obligate cold requirement will not flower unless they receive the necessary vernalization. Many perennials with an obligate cold requirement also have a juvenility requirement. These perennials must first reach a certain age or have produced the critical leaf number necessary for each variety before they are able to perceive the cold treatment. Plants that are still in the juvenile stage going into the cold period are not able to produce flowers after they have received a cold treatment. Some flowering may occur, but most likely it will be sporadic and not consistent enough for commercial production. It is recommended to verify, by counting the number of leaves, that the juvenility period is over before beginning the vernalization process. Take into consideration that not all cultivars within a species will have the same critical leaf number.

Cold beneficial

Plants that are classified in the cold beneficial category experience some benefits from receiving cold, although a cold treatment is not necessary for them to flower. Perennials found within the cold beneficial category usually require less time to reach flowering following vernalization as compared to plants that do not experience any cold. The production time for some perennials is reduced by one to three weeks following a cold treatment. Following exposure to cold, plants in this category often have more flowers per plant and flower more uniformly than plants that do not receive any vernalization.

Vernalization temperatures and methods

Up to now, I have made many references to cold, cold treatments, and vernalization, but what exactly are the requirements and how are they provided? The acceptable temperature range for most perennials is 28–45° F (-2–7° C). Temperatures below 28° F (-2° C) are detrimental to some perennials, particularly their root

Figure 10-4. ***Lychnis* 'Jenny' is a cold-beneficial plant and has limited flowering when cold is not provided, as seen in the image on the left. When fifteen weeks of cold are provided, it blooms very readily (right). The treatments to plants in both pictures are, left to right: short days, sixteen hours extended lighting with incandescent lighting, and sixteen hours extended lighting with high-pressure sodium lights.**

Michigan State University

systems. Vernalization temperatures above 45° F (7° C) may have some effect on flowering, but generally these temperatures will be less effective and not beneficial enough for commercial production. For growers who can control temperatures during the cold treatment, temperatures should be maintained at 38–44° F (3–7° C) with 41° F (5° C) being optimal.

Perennials should be well established and past the juvenility phase before the onset of vernalization. Perennials that are in plug trays should be near fully rooted, whereas larger size containers can be cooled once they are well established or have roots out to the sides and the bottom of the pot. Regardless of the container size, the more roots perennials have going into the cold period, the better.

Growers must note that some perennial varieties should not receive vernalization in the plug trays. Such perennials bloom rapidly when they are forced after the cold requirement has been satisfied. With quick bloom times, there is often an inadequate amount of time to transplant the plug and bulk it up prior to flowering. These perennials often bloom before there is enough foliage to fill out the pot and produce an attractive product. Some examples of perennials that should be bulked up in their final container before the cold period include *Ajuga reptans, Iberis sempervirens, Phlox subulata,* and *Pulmonaria* cultivars.

In the northern United States, commercial growers often vernalize perennials inside minimally heated greenhouses, unheated cold frames, or even outside. Growers using minimally heated structures usually keep soil temperatures at, or slightly above, freezing (32° F [0° C]) while maintaining air temperatures less than 40° F (4° C). Plug suppliers most often vernalize their plugs in minimally heated houses. Growers who vernalize perennials in cold frames or unheated structures may experience troubles at both ends of the temperature spectrum. Many perennial species can tolerate root zone temperatures much lower than 28° F (-2° C), but several perennials' root systems will become damaged at these temperatures. Additionally, maintaining air temperatures below 40° F (4° C) can be difficult on sunny days, as the ability to provide adequate ventilation may be limited.

Figure 10.5. **Many perennials are herbaceous, and the foliage completely dies back during the vernalization process. The original foliage of this *Echinacea purpurea* 'Magnus' has died back, and the new shoots are beginning to emerge during mid-spring.**

Another option for growers is to provide the required vernalization inside temperature-controlled coolers set at 38–41° F (3–5° C). This may be an efficient method for vernalizing plugs, but it may be too costly for vernalizing finished containers, as space and the capacity of the cooler may become a factor. Coolers are useful for growers who require vernalized plugs for late-season sales or summer plantings. For plants that do not go completely dormant while they are in the cooler, growers should provide nine hours of light (short days) with an intensity of 25–50 f.c. (269–538 lux). Growers really need to pay attention to the watering inside a cooler. The cooler acts as a dehumidifier, removing moisture from the air and growing medium. Be sure to monitor the moisture levels of the root zones regularly. Growers in southern states may not be able to properly vernalize their perennials unless they utilize coolers or purchase vernalized materials from other growers.

Table 10.4. Cold Beneficial Perennials

Cold beneficial plants often require less time to flower, produce more flowers per plant, and bloom more uniformly following exposure to cold, when the other requirements for flowering, such as the proper photoperiod, are provided. Many of these perennials can be forced successfully into bloom without providing a cold treatment. Generally, individual cultivars within the species listed below will have similar cold requirements. Cultivar-specific differences can and do exist. Use the information provided here only as a guideline. Providing more weeks of cold than the recommendations listed here is acceptable and may continue to increase the benefits with many perennial varieties. The cold duration is the recommended minimum amount of cold to provide for each species.

Variety	Weeks of Cold	Variety	Weeks of Cold
Achillea clypeolata	6–9	*Dianthus barbatus*	6–10
Achillea filipendulina	6–9	*Dianthus deltoides*	6–9
Achillea x *hybrida*	6–9	*Dianthus* x *allwoodii*	6–9
Achillea millefolium	6–9	*Dianthus gratianopolitanus*	6–9
Achillea ptarmica	6–9	*Digitalis grandiflora*	5-8
*Agastache foeniculum**	11–15	*Digitalis purpurea**	6–9
Agastache nepetoides	11–15	*Echinacea purpurea*	11-14
Anaphalis margaritacea	7–10	*Echinops ritro*	7–10
Anchusa capensis	6–10	*Erigeron glaucus*	6–9
Anemone hupehensis	11–15	*Erodium reichardii*	6–10
Anemone sylvestris	11–15	*Erysimum cheiri*	6–10
*Arabis albida**	4–6	*Gaillardia aristata*	9–12
Arabis blepharophylla	4–6	*Geranium* x *cantabrigiense*	6–9
Armeria x *hybrida*	10–15	*Geum coccineum**	6–9
Armeria latifolia	6–9	*Gypsophila paniculata*	12–15
Armeria maritima	6–9	*Gypsophila repens*	10–12
Aster alpinus	10–12	*Helenium autumnale*	6–9
Aster dumosus	10–12	*Heliopsis helianthoides*	6–10
Aster novae-angliae	10–12	*Inula ensifolia*	9–12
Aster novae-belgii	10–12	*Jasione laevis*	9–12
Aubrietia deltoidea	6–9	*Lavandula angustifolia*	10–15
Bellis perennis	6–9	*Leontopodium alpinum*	9–12
Calamintha nepetoides	6–10	*Leucanthemum vulgare*	7–10
*Campanula carpatica**	7–10	*Leucanthemum superbum**	7–10
Campanula cochleariifolia	10–15	*Linaria purpurea*	6–9
Campanula rotundifolia	10–15	*Lobelia siphilitica*	7–10
Catananche caerulea	6–10	*Lobelia* x *speciosa*	7–10
Centranthus ruber	6–9	*Lupinus* x *hybrida**	6–10
Clematis integrifolia	6–10	*Lychnis* x *haageana*	6–9
*Coreopsis auriculata**	6–9	*Myosotis sylvatica**	6–9
Coreopsis rosea	10–12	*Oenothera missouriensis*	6–9
Coreopsis verticillata	10–13	*Oenothera tetragona*	6–9
Cynoglossum amabile	11-14	*Oxalis crassipes*	7–10
Delphinium belladonna	5-8	*Papaver nudicaule*	6–9

Variety	Weeks of Cold	Variety	Weeks of Cold
Penstemon digitalis	6–9	*Salvia* x *superba**	6–9
Penstemon x *hybrida*	7–10	*Salvia* x *sylvestris**	6–9
Perovskia atriplicifolia	5-8	*Scabiosa caucasica*	10–15
*Platycodon grandiflorus**	7–10	*Scabiosa columbaria*	10–15
*Polemonium caeruleum**	6–9	*Sedum spectabile*	6–10
Polemonium reptans	6–9	*Sidalcea hybrida*	6–9
Primula x *polyantha**	6–9	*Silene orientalis*	10–12
Ranunculus repens	6–9	*Solidago rugosa*	7–10
Rudbeckia fulgida	10–12	*Stokesia laevis**	8--10
Rudbeckia triloba	10–12	*Tiarella* x *hybrida*	6–10
Sagina subulata	5-8	*Veronica spicata**	6–9
Salvia nemorosa	6–9		

*Only select cultivars

Many novice perennial growers often express unnecessary concern in regard to the appearance of their perennials after the vernalization period. Each variety of perennial behaves differently to the cold treatments. Some perennials are considered evergreen and have foliage that remains green throughout the year. Other perennials are semi-evergreen (semi-herbaceous) and can be described as having green new foliage at the base of the plant during the winter, but the majority of the previous year's growth dies back during the late fall. The majority of perennials are herbaceous, and their stems die back completely during the cold treatment. Regardless of how the top of the plant appears, it is important that a healthy root system is in place before, during, and after the vernalization process.

Duration of cold

The duration of cold necessary to vernalize perennials varies between three to sixteen weeks, with most commercially produced varieties requiring six to ten weeks. There are usually no detrimental effects from providing more than the minimum recommended amount of cold. In fact, many perennials exhibit a positive response, such as a shorter time to reach flowering, when a slightly longer cold treatment is given. It is perfectly acceptable, and even beneficial, to provide ten weeks of cold to perennials that have a six-week cold recommendation. For growers who do not have adequate facilities for providing differing durations of cold, providing a minimum of ten weeks of cold will satisfy the cold requirement for the majority of the commercially produced perennials.

For many growers, the space necessary to vernalize their product might be limited or unavailable due to other production needs. These growers should consider vernalizing perennials in plug flats to maximize the number of plants they can accommodate in their facilities. Vernalizing their own plugs will also help growers to ensure they have the right cultivars and are not subjected to the sometimes-limited availabilities of various plug suppliers. Commercial growers usually provide the cold treatment to 72-cell plug flats or larger plug sizes before transplanting them to larger sized containers. However, many perennial varieties that have a juvenility requirement or take a long time to fill out the container are transplanted into the final container and bulked up prior to receiving the cold treatment.

Another option would be for growers to purchase vernalized plugs, transplant them into the finished containers, and force them at the appropriate temperatures and photoperiods. Locating cold-treated plugs can be challenging at certain times of the year, as most growers only vernalize their materials during the winter months, when the temperatures are conducive to vernalization. Several growers vernalize plugs during the winter months but do not market or guarantee that they are vernalized. It is very beneficial to establish a relationship with a plug producer that has the ability to vernalize plugs any time of the year and can supply vernalized materials to fit your specific needs.

Growers must decide if they can provide the cold treatment to the finished containers or plug flats themselves or if they must purchase pre-cooled plugs from outside sources.

Light Requirements

The role of light in regard to flowering is getting more attention from researchers in recent years. Light may be the single most important factor that helps perennials overcome their juvenility. Elements of light growers need to consider include the duration, or photoperiod; the light intensity, or irradiance; the color, or light quality; and the total amount per day, or light integral.

For photosynthesis, growers are most interested in the combination of light intensity and the light integral, where light intensity is the instantaneous amount of light available and the light integral represents the total amount of light over a twenty-four-hour period. For flowering, researchers are finding that for many plants the light integral affects how soon some plants overcome juvenility and are capable of blooming. Unfortunately, not all plant species respond to the light integral or light intensities similarly.

Most plants fall into two categories, or response groups: irradiance indifferent and facultative (beneficial) irradiance. Plants within the irradiance indifferent response group essentially do not develop flowers quicker when extra lighting is provided. Conversely, plants within the facultative irradiance response group do develop flowers earlier. "Quicker" and "earlier" mentioned above should be interpreted as developmentally earlier. For example, for plants with a facultative irradiance response, the addition of light will not only cause flowers to form earlier on a time scale, but there will be fewer leaves below the flower as compared to plants grown at natural light intensities (flowers developmentally earlier as the juvenility phase is shortened). Conversely, plants within the irradiance indifferent group will not bloom developmentally early with additional lighting and will contain the same number of leaves below the flowers as compared with plants grown at natural light levels. When supplemental lighting is provided, growers may observe earlier development on a time scale due to the warmer temperatures created by the lights, but the juvenility phase has not been altered. Supplemental lighting is often used by growers to increase the quality characteristics and to reduce the production times of various crops.

Many perennials benefit from supplemental lighting during the forcing process. Certain perennials, such as *Asclepias, Artemisia, Aruncus, Coreopsis, Echinacea, Hibiscus, Rudbeckia,* and some *Sedum* cultivars, require additional lighting to allow them to grow and bloom properly. Other perennials, such as *Achillea, Campanula, Digitalis, Gaillardia, Lobelia, Lythrum,* and *Statice,* will benefit from supplemental lighting. Lighting will allow them to bloom slightly quicker than when grown at natural day lengths and light intensities.

It is recommended to use supplemental lighting on the perennial varieties listed above during times of the year when the natural day lengths are less than twelve hours. There are a number of sources of supplemental lights growers can use (see light sources, p. 319). A common method is to use one incandescent bulb for every 150–200 sq. ft. (13.9–18.6 m^2) of production space mounted 4–6 ft. (1.2–1.8 m) above the crop. The lights should be on for fifteen to eighteen hours per day from the time the crops are planted until the flower buds are visible.

It is easy for growers to justify using supplemental lighting for facultative-irradiance plants, where the duration of juvenility is decreased, the time to bloom is reduced, and the overall production time is decreased. Conversely, using lights to reduce crop production times of irradiance-indifferent plants is not justifiable and can be done more efficiently by adjusting the production temperatures. For perennial forcing, supplemental lighting generally increases the number of branches and flowers per plant, produces larger flowers, and often thickens plant stems; all of these benefits from lighting improve the overall quality characteristics of the crop.

Photoperiod

Most perennials have a certain day length, or photoperiod, that is necessary for them to flower. The photoperiod refers to the number of hours of light provided to, or perceived by, the plant each day. The photoperiod naturally perceived by plants is the duration between sunrise and sunset plus about thirty minutes to account for twilight. The natural photoperiod changes daily and varies depending on the geographic location. Perennials in nature use photoperiod as the environmental signal to trigger flowering, allowing them to flower when conditions are conducive for pollination and seed development, regardless of their geographic location or calendar date.

Researchers have discovered that for many perennials, a mere hour's variance in the photoperiod can mean the difference between flowering and non-flowering plants. How plants respond to the duration of light and darkness is commonly referred to as photoperiodism. By understanding the effect photoperiod has on flowering, growers can develop programs allowing them to bloom perennials nearly on demand.

Perennials have been split into three simple groups, defined by the photoperiod they need in order for them to produce flowers. Perennials within these categories are referred to as long-day plants, short-day plants, or day-neutral plants. The majority of commercially grown perennials are long-day plants requiring a minimum of thirteen hours of light per day in order to flower. Short-day plants contain a select group of perennials that flower when the day length is less than twelve hours per day. Day-neutral plants represent a wide variety of perennials that will flower either under long days or short days.

Not all of the perennials within each of these three categories will flower simultaneously for every photoperiod within each group. For example, long days (for most perennials) are considered to be any duration over thirteen hours of light; many perennials will bloom best at thirteen hours of day length while others will not begin to flower until the photoperiod has reached a minimum of fourteen hours. Many perennials will flower under a wide range of photoperiods, but will flower best, or faster, when they are exposed to their optimal photoperiod. It is important to consider that each plant species (and even cultivar) has its own unique optimal photoperiod.

The natural photoperiod growers experience varies dramatically throughout the year, depending on the location of their operation from the equator. The entire United States is located in the Northern Hemisphere, where the longest day of the year is on June 21 and the shortest day is December 21. From December 21 till June 21, the day length increases gradually, and following June 21 the day length decreases gradually. The magnitude of the day length fluctuation increases the farther away from the equator a greenhouse is located. For parts of the country located at or near 30° N latitude, the shortest day of the year (December 21) is slightly less than 11 hours and the longest day of the year (June 21) is over 14.5 hours. Moving north to a location at 45° N latitude, there is only 9.5 hours of day length on the shortest day and over 16 hours on the longest day.

Greenhouse growers are already familiar with producing crops that are photoperiodic. Chrysanthemums and poinsettias are two commercially produced crops that flower following the onset of short days. To more accurately schedule short-day crops, growers have also learned to keep them under long-day conditions by lighting, often referred to as mum lighting, until a certain number of weeks before the desired flowering date. To ensure the plants perceive short days during periods of naturally occurring long days, many growers pull black cloth over the production area or "black out" their plants for a minimum of thirteen hours to

Table 10.5. Long-day Obligate Perennials

Each perennial has a critical photoperiod necessary for flowering. Long-day plants typically require at least thirteen hours of light each day. Most perennials requiring long days will successfully flower when the photoperiod is greater than fourteen hours or if night interruption lighting is provided.

Achillea clypeolata	*Centranthus ruber*	*Gypsophila repens**	*Lupinus* x *hybrida*
*Achillea filipendulina**	*Ceratostigma plumbaginoides*	*Hakonechloa macra*	*Lupinus polyphyllus*
Achillea x *hybrida*	*Chasmanthium latifolium*	*Helenium autumnale*	*Lysimachia atropurpurea*
*Achillea millefolium**	*Chelone glabra*	*Helianthemum nummularium*	*Lysimachia clethroides*
*Achillea ptarmica**	*Cimicifuga racemosa*	*Helianthus angustifolius*	*Lysimachia punctata*
Agastache foeniculum	*Coreopsis grandifolia*	*Helichrysum thianshanicum*	*Malva alcea*
Agastache x *hybrida*	*Coreopsis rosea*	*Helictotrichon sempervirens*	*Melica transsilvanica*
*Agastache rugosa**	*Coreopsis verticillata*	*Heliopsis helianthoides*	*Miscanthus sinensis*
Anacyclus depressus	*Corydalis lutea*	*Heliopsis scabra*	*Monarda didyma*
Anaphalis margaritacea	*Cymbalaria muralis*	*Heuchera sanguinea*	*Nepeta* x *faassenii*
Andropogon scoparius	*Diascia barberae*	*Hibiscus x hybridus*	*Oenothera fruticosa**
Anemone hupehensis	*Digitalis purpurea*	*Hibiscus moscheutos*	*Oenothera missouriensis*
Anemone x *hybrida*	*Digitalis thapsi*	*Hieracium villosum*	*Oenothera speciosa*
Anemone vitifolia	*Echinacea purpurea*	*Hosta*	*Oxalis crassipes*
Arenaria montana	*Echinops ritro*	*Hypericum calycinum*	*Panicum virgatum*
Armeria pseudarmeria	*Erigeron aurantiacus*	*Hypericum polyphyllum*	*Papaver nudicaule**
Asclepias tuberosa	*Erigeron glaucus*	*Incarvillea delavayi*	*Papaver orientale*
*Aster alpinus**	*Erigeron karvinskianus*	*Inula ensifolia*	*Pennisetum alopecuroides*
Aster tongolensis	*Erigeron speciosus*	*Inula orientalis*	*Pennisetum setaceum*
Brunnera macrophylla	*Erinus alpinus*	*Jasione laevis*	*Penstemon barbatus**
Buddleia davidii	*Eryngium alpinum*	*Kniphofia uvaria*	*Penstemon campanulatus*
Calamagrostis acutifolia	*Eryngium bourgatii*	*Lavandula angustifolia**	*Phlox paniculata*
Calamintha nepetoides	*Eryngium planum*	*Lavandula stoechas**	*Physalis alkekengi*
Campanula carpatica	*Erysimum cheiri*	*Lavandula vera*	*Physostegia virginiana*
Campanula cochleariifolia	*Eupatorium rugosum*	*Leontopodium alpinum*	*Polemonium caeruleum*
Campanula garganica	*Euphorbia* x *hybrida*	*Leucanthemum* x *superbum**	*Polemonium reptans*
Campanula glomerata	*Gaillardia aristata*	*Leucanthemum vulgare*	*Polemonium viscosum*
Campanula longistyla	*Gaillardia* x *grandiflora*	*Liatris spicata*	*Potentilla megalantha*
Campanula medium	*Galium odoratum*	*Limonium bellidifolium*	*Potentilla nepalensis*
Campanula persicifolia	*Gaura lindheimeri*	*Limonium latifolium*	*Primula vulgaris*
Campanula punctata	*Gentiana acaulis*	*Limonium perezii*	*Primula japonica*
Campanula rotundifolia	*Gentiana septemfida*	*Limonium tataricum*	*Rodgersia aesculifolia*
Carlina acaulis	*Geranium* x *cantabrigiense*	*Linaria purpurea**	*Rudbeckia fulgida*
Catananche caerulea	*Geranium sanguineum**	*Linum flavum*	*Rudbeckia hirta*
Centaurea macrocephala	*Gypsophila pacifica*	*Lobelia siphilitica*	*Rudbeckia speciosa*
Centaurea montana	*Gypsophila paniculata*	*Lobelia* x *speciosa*	*Rudbeckia triloba*

Sagina subulata	*Sedum cauticolum*	*Sisyrinchium bellum*	*Verbascum bombyciferum*
Salvia greggii	*Sedum floriferum*	*Solidago canadensis*	*Verbascum chaixii*
Salvia hybrids	*Sedum maximum*	*Solidago sphacelata*	*Verbascum* x *hybrida*
Salvia nemorosa	*Sedum spectabile*	*Stokesia laevis**	*Verbascum phoeniceum*
Salvia officinalis	*Sedum telephium*	*Tanacetum coccineum*	*Verbena bonariensis*
Salvia x *superba*	*Sedum tetractinum*	*Tanacetum niveum*	*Verbena canadensis*
Salvia x *sylvestris*	*Sidalcea hybrida*	*Tanacetum parthenium*	*Verbena rigida*
Saponaria ocymoides	*Sidalcea malviflora*	*Teucrium chamaedrys*	*Veronica incana*
Scabiosa caucasica	*Silene maritima*	*Thalictrum aquilegiifolium*	*Veronica spicata**
Sedum aizoon	*Silene orientalis*	*Thymus serpyllum*	*Veronica subsessilis*
Sedum bithynicum	*Sisyrinchium angustifolium*	*Trollius chinensis*	*Veronica teucrium*

*Many long day perennials will flower under shorter photoperiods, but flowering is improved when long-day conditions are provided.

promote flower induction. Production of these crops is down to a science; growers can manipulate the photoperiod and temperatures to meet nearly any desired ship date.

Researchers have learned that just like the cold requirement, perennials have their own photoperiodic needs that are specific to each variety. Again, it was necessary to break these needs into different groups so that growers could manage their forcing programs more efficiently. These groups are: long day (LD) required, LD beneficial, short day (SD) required, and day neutral.

LD required

Perennials belonging to the LD-required group have an obligate requirement for long days and simply will not flower unless they are grown under long-day conditions. For perennial forcing, most long-day plants require a minimum of fourteen hours of light per day for flowering to occur, but each perennial has a critical photoperiod required for flowering. Plants will remain vegetative, and non-flowering, unless the photoperiod exceeds this critical duration.

As discussed above, many perennials have an obligate cold requirement and will not flower if they have not been exposed to a specific duration of cold. Interestingly, in some cases, satisfying the cold requirement also slightly reduces the critical photoperiod for some perennials. For example, *Rudbeckia*

Table 10.6. Long-day Beneficial Perennials
Many perennials will flower under various photoperiods but flower best when exposed to long days.

Achillea aegyptiaca	*Hemerocallis*
*Achillea filipendulina**	*Hosta**
*Achillea millefolium**	*Laurentia fluviatilis*
*Achillea ptarmica**	*Lavandula angustifolia*
Achillea tomentosa	*Leucanthemum superbum**
*Agastache rugosa**	*Lobelia speciosa**
Anchusa capensis	*Lychnis coronaria**
Anemone sylvestris	*Oenothera fruticosa*
Aquilegia x *hybrida*	*Oenothera tetragona*
Arabis caucasica	*Penstemon* x *hybrida*
Aster dumosus	*Penstemon mexicale*
Astilbe x *arendsii*	*Penstemon smallii*
Astilbe chinensis	*Phlox subulata*
Astilbe japonica	*Platycodon grandiflorus*
Astilbe thunbergii	*Pulmonaria* x *hybrida*
Campanula poscharskyana	*Salvia nemorosa**
Cerastium tomentosum	*Salvia* x *superba**
Delphinium grandiflorum	*Saxifraga arendsii*
*Gaillardia grandiflora**	*Saxifraga umbrosa*
Geranium dalmaticum	*Scabiosa columbaria*
*Gypsophila repens**	*Tiarella wherryi*

*Certain cultivars of these species express more benefits from long days than do others. Not all cultivars are considered to be long-day beneficial plants.

fulgida 'Goldsturm' is an obligate long-day, cold-beneficial plant. Without a cold treatment, the critical photoperiod is approximately fourteen hours. 'Goldsturm' plants that have been vernalized have a critical photoperiod of about thirteen hours. This example demonstrates that there is a critical photoperiod that has to be observed by the plant in order for it to bloom. It also points out that providing a cold treatment will reduce the critical photoperiod, which consequently reduces the time to flower by about three weeks.

LD beneficial

Many plants belong to the LD-beneficial group (also called LD facultative), which means that they exhibit some benefit from being produced under long-day conditions. Although LD-beneficial plants may flower under different photoperiods, producing them under LD conditions will usually increase the number of flowers they produce and/or decrease the time to flower.

In many instances, growers force LD-beneficial plants under long-day photoperiods to produce a more commercially appealing product. For example, *Lavandula angustifolia* cultivars may bloom under short days but will often have significantly fewer flowers per plant compared with plants grown under long-day photoperiods. Other perennials, such as *Salvia,* only exhibit a slight benefit to long-day conditions, but this response is often enough for growers to consider producing these plants under long-day conditions.

Table 10.7. Short-day Perennials

*Aster dumosus**
Aster novae-angliae
Aster novi-belgii
Chrysanthemum morifolium
*Eupatorium rugosum**
*Helianthus angustifolius**
Ranunculus repens
Sisyrinchium tinctorium
*Solidago canadensis**
*Solidago sphacelata**

*These perennials benefit from being produced under long day conditions prior to exposure to short days.

SD required

Many short-day perennials are naturally fall bloomers that remain vegetative under the long days of summer. Only a small group of commercially produced perennials fall into this category. These are plants that will flower only following the onset of short-day conditions, or day lengths of less than twelve hours. Like long-day plants, the actual day length necessary for flowering of short-day perennials varies with the plant species and cultivar. Flowering occurs once the plants are exposed to photoperiods shorter than some critical duration.

Similar to long-day plants, perennials requiring short days can be divided into two subgroups: obligate short-day plants, which must receive short days to flower, and facultative short-day plants, or short-day beneficial, which flower better under short-day conditions. Obligate short day plants, such as *Aster*, will flower only when the day length is less than some critical length. Facultative short-day perennials, such as *Eupatorium,* will flower eventually regardless of day length but will flower faster when the photoperiod is less than some optimal length.

Day neutral

Day-neutral perennials are those that will flower under either long or short days. Many of the perennials that flower naturally in the spring are day-neutral plants. In most cases, day-neutral perennials have an obligate cold requirement of five to fifteen weeks. As discussed previously, each perennial cultivar has its own optimal cold requirement. Regardless of the photoperiod, cold-requiring perennials will flower poorly, or not at all, unless the appropriate cold treatment has been provided. For cold-requiring, day-neutral plants, once the cold requirement has been satisfied, flower development is a function of temperature. Many perennials in this category

Table 10.8. Day Neutral Perennials

Perennials within this category will flower under any photoperiod.

Ajuga reptans	*Coreopsis auriculata*	*Geranium sanguineum*	*Phlox divaricata*
Alcea rosea	*Corydalis**	*Geum chiloense*	*Phlox subulata**
Alchemilla mollis	*Corydalis flexuosa*	*Geum coccineum*	*Physalis alkekengi*
Alyssum saxatilis	*Cynoglossum amabile*	*Gypsophila repens**	*Platycodon grandiflorus**
*Anchusa capensis**	*Delosperma cooperi*	*Helleborus niger*	*Potentilla atrosanguinea*
Anemone multifida	*Delphinium belladonna*	*Hemerocallis**	*Potentilla nepalensis*
*Anemone sylvestris**	*Delphinium* x *cultorum*	*Heuchera* x *hybrida*	*Primula denticulate*
Aquilegia alpina	*Delphinium elatum*	*Heuchera micrantha*	*Primula elatior*
Aquilegia caerulea	*Delphinium grandiflorum**	*Heuchera sanguinea*	*Primula* x *hybrida*
Aquilegia flabellata	*Delphinium nudicaule*	*Heucherella* x *hybrida*	*Primula polyanthus*
Aquilegia x *hybrida**	*Delphinium zalil*	*Iberis sempervirens*	*Primula pubescens*
Aquilegia vulgaris	*Dianthus* 'Allwoodii'	*Lamium maculatum*	*Primula rosea*
Arabis albida	*Dianthus barbatus*	*Lavandula angustifolia**	*Primula vialii*
Arabis blepharophylla	*Dianthus caryophyllus*	*Lavandula stoechas*	*Primula vulgaris*
*Arabis caucasica**	*Dianthus deltoides*	*Leucanthemum superbum**	*Pulmonaria* x *hybrida**
Arabis sturii	*Dianthus gratianopolitanus*	*Lewisia cotyledon*	*Pulsatilla vulgaris*
Armeria x *hybrida*	*Dianthus hybridus*	*Linaria purpurea*	*Rosa chinensis*
Armeria latifolia	*Dianthus plumarius*	*Linum perenne*	*Salvia nemorosa**
Armeria maritima	*Dianthus superbus*	*Lychnis* x *arkwrightii*	*Saxifraga arendsii**
Aster alpinus	*Dicentra eximia*	*Lychnis coronaria**	*Scabiosa caucasica*
Aster amellus	*Dicentra spectabilis*	*Lychnis* x *haageana*	*Scabiosa columbaria**
Aster dumosus	*Digitalis grandiflora*	*Lychnis viscaria*	*Silene schafta*
Astilbe x *arendsii**	*Digitalis obscura*	*Myosotis hybrida*	*Solidago rugosa*
*Astilbe chinensis**	*Digitalis purpurea*	*Myosotis scorpioides*	*Stokesia laevis*
*Astilbe japonica**	*Doronicum orientale*	*Myosotis sylvatica*	*Tanacetum coccineum*
*Astilbe thunbergii**	*Draba aizoon*	*Oenothera tetragona**	*Thalictrum kiusianum**
Aubrieta deltoidea	*Erodium reichardii*	*Papaver alpinum*	*Tiarella* x *hybrida**
Aubrieta x *hybrida*	*Erysimum suffruticosum*	*Papaver nudicaule*	*Tiarella wherryi*
Aubrieta cultorum	*Euphorbia amygdaloides*	*Penstemon barbatus*	*Trollius europaeus*
Aurinia saxatilis	*Euphorbia epithymoides*	*Penstemon campanulatus*	*Verbena canadensis*
Bellis perennis	*Euphorbia myrsinites*	*Penstemon digitalis*	*Veronica longifolia*
Bergenia cordiflolia	*Euphorbia polychroma**	*Penstemon hartwegii*	*Veronica peduncularis*
*Campanula portenschlagiana**	*Filipendula purpurea*	*Penstemon* x *hybrida**	*Veronica repens*
*Cerastium tomentosum**	*Geranium dalmaticum**	*Penstemon mexicale**	*Viola odorata*
Ceratostigma plumbaginoides	*Geranium himalayense*	*Penstemon smallii**	*Viola tricolor*
Cheiranthus cheiri	*Geranium* x *hybridum*	*Perovskia atriplicifolia*	*Viscaria alpina*
Chrysogonum virginianum			

*Long-day beneficial

have a juvenility requirement and/or should be bulked up prior to the cold period to ensure they are large enough for sales because in many cases, flowering occurs rapidly in the spring without much vegetative growth.

Manipulating photoperiod

Generally speaking, plants perceive light through small proteins called phytochromes found in their leaves. These phytochromes measure the duration of both the light and dark periods. When exposed to light, the phytochrome is converted into one form, and with dark exposure it converts into another. When this phytochrome conversion occurs for a certain number of days in a row, the flowering process is triggered within the plant. Each type of plant (LD or SD) has a critical photoperiod or a predetermined amount of phytochrome conversion that must be achieved in order for flowering to occur. Once this critical level is achieved, the meristem, or growing point, is initiated to flower, causing the plant to switch from a vegetative phase into a flowering or reproductive phase.

A more accurate way to describe how plants perceive the photoperiod is to explain that they perceive the dark period of the night rather than the light of the day. Short-day plants will only flower when exposed to a long dark period of a critical length, of at least twelve hours. Conversely, long-day plants will only flower when exposed to a short dark period of a critical length, which is less than twelve hours. To reduce confusion, I will continue this discussion referring to the day length as opposed to the dark period.

Growers often are concerned with keeping their plants vegetative, or non-flowering, in order to reach a particular marketing window or for stock plant production. To keep short-day plants from flowering, growers can interrupt the dark period, or the night, with night interruption lighting, which will be discussed shortly. In essence, this breaks up the night, causing the plant to perceive long-day conditions and remain vegetative. With long-day plants, creating short days using a blackout system is necessary to prevent these plants from flowering.

Creating short days

During naturally long days, it may be necessary for growers to create short-day photoperiods by blocking out all of the light. As mentioned previously, short-day conditions are created by pulling black cloth or black plastic over the production site, ensuring the plants only receive a maximum of twelve hours of light each day. The blackout area should be completely dark since some perennials can perceive even 1 f.c. (11 lux) of light, negating the full benefit of providing dark. Unwanted light reaching the production area is referred to as light pollution and often comes from nearby greenhouses or streetlights. To reduce potential light pollution, growers need to routinely check the blackout area to ensure there is no light entering the production facility. Growers can check visually or with light meters, targeting less than 0.5 f.c. (5 lux) of light.

Figure 10.6. **Growers often manufacture simple systems, such as these frames, to support the blackout material above the crop.**

Michigan State University

Besides light pollution, the next largest concern is heat buildup under the blackout system. To create short days, the blackout system is often closed before sunset. In doing so, heat from the sun is allowed to build up under the black cloth or black plastic. High temperatures during the dark period may cause an inhibition or delay of flowering, called heat delay. Poinsettia and mum growers are very familiar with heat delay and often experience it when the night temperatures are allowed to rise above 73° F (23° C) under a blackout system.

To help reduce the likelihood of heat buildup, there are a couple of steps growers should consider. Usually the outdoor temperatures are warmer in the evening than in the morning, and the sunlight is more intense as well. It may be beneficial to close the blackout at dusk or just after dark, once the sun's heating rays are gone, and then open it later in the morning, after sunrise. In areas where you do not have to worry about stray light sources, the plants can go into the night without being covered. Then you can apply the black cloth before any sign of morning light arrives (one hour before sunrise) and remove the black cloth later in the morning after a sufficient dark period (natural night, plus the blackout duration, should total at least twelve hours).

In addition to promoting flowering of short-day plants, creating short days is also a helpful tool some growers use to keep long-day plants from flowering early or to keep them in a vegetative state, such as for stock plant production.

Creating long days

Since the majority of perennials are classified as long-day plants, most growers will have to set up methods to create long days during periods where the day lengths are naturally short (autumn till spring). There are two methods commercial growers use to create long days. The first method is called day length extension, and the second is called night interruption lighting.

Day length extension

Growers accomplish day length extension by lighting their plants, extending the total number of hours each day the perennials perceive light, up to the desired day length. For example, if a grower is trying to achieve a fourteen-hour day length during a time of the year when the natural day length is only ten hours, the grower would provide lighting from about an hour before dusk up to the time the plant has received fourteen total hours of light. Day length extension is an effective method of delivering the desired photoperiod, but requires the lights to run slightly longer than other methods of providing long days (night interruption or cyclical lighting), which are discussed next.

The purpose of lighting is not to provide sufficient amounts of light useful for growth or for photosynthesis, but for the plants to only perceive enough light so that they are convinced it is still daytime. Researchers recommend growers supply only 10 f.c. (108 lux) of light in the darkest corner of the greenhouse. There are numerous sources of light that are effective for day length extension; they are discussed in the light sources section on page 319.

Night interruption

Night interruption, often referred to as night break lighting or mum lighting, is another method growers use to promote flowering of long-day plants. Contrary to what most growers believe, plants actually measure the length of the night as opposed to the length of the day. So interrupting their perception of night fools them into thinking they are experiencing long days.

The most common practice is to provide lighting continuously from 10:00 P.M. till 2:00 A.M. This breaks up the long continuous dark period into two short ones, or creates a long day as far as the plant is concerned. The intensity of the light over the production area should measure at least 10 f.c. (108 lux) at the furthest point from the light fixture. Every long-day plant I have studied has induced flowering when night interruption lighting is provided. Initially, growers used incandescent lighting for this procedure, but several additional light sources are being used today,

Figure 10.7. **Night interruption lighting can be used to promote flowering in outdoor production sites, such as with these metal halide lights. Growers cannot control temperatures outdoors, but they do observe that flowering occurs several weeks earlier when night interruption lighting is provided.**

including fluorescent, high-pressure sodium, and metal halide lamps.

Some perennials may respond appropriately to exposure to shorter periods of night interruption or to lower light intensities, but shorter night break durations have been shown to delay the flowering of many long-day plants. The duration of the night interruption lighting should always be four hours, unless specific plant recommendations state otherwise. For example, researchers have found that night interruption periods of less than four hours on *Echinacea purpurea* will not delay the time to flower and will significantly reduce the plant height. For *Echinacea purpurea* cultivars, it is recommended to provide between thirty minutes and four hours of night interruption lighting.

Growers have also found that using night break lighting when the day lengths are naturally long can reduce the time to produce flowering plants by as much as seven to fourteen days.

Night interruption lighting is used by some growers to keep plants from flowering. This method was developed to keep short-day plants from setting flowers, allowing them to bulk up properly during natural short days, or to allow growers to produce certain flowering crops year round. The night interruption breaks up the long night into two short ones, which prevents the plants from initiating flowers. Perennial growers utilize this method for both maintaining vegetative growth of short day plants, particularly for stock plant production, and for triggering flowering of long day varieties.

Cyclical lighting

Cyclical or intermittent lighting (also called flash lighting) is a slight variation of night interruption lighting. This method involves cycling on and off the lighting at certain intervals during the night, causing the perennials to perceive long days. Generally, the lights are turned on for six to ten minutes of every thirty-minute period for four to six hours during the night. This strategy is very effective at promoting flowering and saves electricity.

Growers using incandescent light sources can accomplish cyclical lighting by installing the lights in the same manner as if they were providing night interruption lighting, except that the fixtures should be on timers. The timers should be set for six minutes on, followed by twenty-four minutes off. If a grower runs intermittent lighting for four hours, the incandescent lamps would only be running a total of forty-eight minutes. It would reduce the total electricity needed to provide this lighting by over 75% as compared with running four hours of traditional night interruption lighting.

Growers utilizing high-pressure sodium lamps can still provide cyclical lighting, but because the ballasts take a few minutes to heat up and bring the lamps to their full intensity, intermittent lighting cannot be delivered in the same manner as done with incandescent lamps. Cycling on and off high-pressure sodium lamps will also dramatically reduce the life of the bulb. So growers and researchers have developed a couple of innovative methods to provide cyclical lighting to crops using high intensity discharge lamps.

The first method involves mounting high-pressure sodium lamps, or fluorescent lighting, onto a traveling irrigation boom. The watering boom is then run over the crop with the lights on from 10:00 P.M. till 2:00 A.M. (no water is applied). With lights mounted on booms, there are two important items to consider. The first is to ensure the delivery of at least 10 f.c. (108 lux) of light to the entire area the boom passes. And the second is to ensure the boom travels slow enough to deliver six minutes of light over each square foot during every thirty-minute time period. The boom can travel more than two times over the production space, but you must deliver at least six minutes of lighting to each square foot. The benefit of this type of lighting would be a reduced number of fixtures needed to apply light over the entire greenhouse space, as well as a reduction in the total amount of electricity needed. The main drawback results in extra wear and tear on the traveling irrigation boom, decreasing the longevity of some of the moving parts.

The second innovative method to deliver cyclical lighting over crops using high-pressure sodium lamps is to use a stationary bulb with a moving reflector that provides a moving light source within the greenhouse. Basically, the ballast is mounted in a fixed location separate from the lamp. The bulb is also mounted in a fixed location with a reflector that has an oscillating motor attached, causing the reflector to move around the bulb. As the reflector moves

Figure 10.8. **Beamflicker lights provide growers with a variation on cyclical lighting. The light contains a ballast, mounted on the top of the truss in this picture, a bulb and reflector (right), and an oscillating motor (left) mounted on the bottom of the truss.**

around the bulb, it causes the light to appear as if it were traveling back and forth through the greenhouse. These waves of light provide cyclical lighting and are very effective at promoting flowering of perennial crops. These fixtures are more expensive than the traditional high-pressure sodium lamps, but fewer fixtures are needed to cover the same production space. In a greenhouse in which I installed the oscillating lights I calculated that 60% less fixtures were needed over the traditional types, and the overall cost of these fixtures was slightly less than purchasing the appropriate number of stationary lamps. The main benefits would be that installation of these lights is slightly cheaper and the amount of electricity needed can be greatly reduced. The main drawback would have to be the potential maintenance of the oscillating motor compared with traditional lamps that do not have this feature. Oscillating high-pressure sodium light fixtures, named the Beamflicker, are available from PARsource Lighting Solutions (www.parsource.com).

Light sources

Historically, lighting has largely been used by greenhouse growers to supplement the natural light levels provided by the sun, particularly during the low-light winter months. The plants use this additional light to increase photosynthetic activity and increase their growth. Typically, the light supplied for this purpose is from high-intensity discharge sources such as high-pressure sodium or metal halide light fixtures. High-intensity discharge lighting provides the best quality light source available to growers, and it is relatively efficient for plants to convert the light energy into photosynthetic energy. The cost of the fixtures and the electricity to run them are often a limiting factor for many growers.

For photoperiodic lighting, researchers have shown that most light sources can be used to provide long-day conditions for perennials. Growers have successfully used incandescent, fluorescent, high-pressure sodium, and metal halide light fixtures to promote flowering of perennial crops. As with most everything, there are some distinct pros and cons to each type of lighting, and some light sources are better than others.

High-pressure sodium lights are the most commonly used form of high intensity discharge lighting (metal halide being the other). The most popular fixture utilizes 400W lamps, and depending on the configuration within the greenhouse and distance above the crop, can deliver up to 500 f.c. (5.4 klux) of light intensity. For photoperiodic lighting, growers typically supply 10–20 f.c. (108–215 lux) of light over the production area. High-pressure sodium lights deliver the best quality light of any of the commercially available supplemental light sources. For manipulating photoperiods, these lights provide a very reliable, high quality source of lighting.

Figure 10.9. **Incandescent lights are commonly used to provide photoperiodic lighting. Here, the grower has placed a shiny aluminum pie pan above the light to reflect the light back down onto the crop.**

Incandescent lamps have been used by growers to provide long-day conditions to short-day crops such as chrysanthemums and poinsettias to keep them from blooming. This type of lighting is most commonly referred to as mum lighting. Growers should follow the guidelines outlined in the night interruption section above. Incandescent lighting provides growers with an inexpensive and easy-to-install method to deliver long-day conditions to their crops. Generally, growers supply incandescent lighting using 60W bulbs spaced about 8 ft. (2.4 m) apart and 4 ft. (1.2 m) above the crop to deliver 10 f.c. (108 lux) of light over the crop.

The biggest disadvantage of using this source of lighting is the far-red light emitted, causing plant cells to elongate, which undesirably increases the plant's height. Plant stretch can be somewhat reduced by turning the lights off once the flower buds are present. To decrease plant stretch, growers should limit the use of incandescent lighting to only four hours. Most light sources cause some degree of stem elongation, but incandescent light causes at least 10% more stretch than other sources of lighting. The most increases in plant height are observed where incandescent lights are used to extend the day length rather than to interrupt the night. Regardless of the strategy, it may be necessary to provide additional height control whenever incandescent lighting is used.

Light meters

Growers using supplemental lighting of any source and for any reason should use a light meter to check the intensity and uniformity of the light being provided. Is the proper amount of light reaching all areas of the production site? For example, with night interruption lighting, growers are typically looking for 10 f.c. (108 lux) or more. But if they observe areas only receiving 3 f.c. (32 lux), the plants are not going to respond uniformly. Growers would need to modify the lighting by adjusting the placement and/or orientation of the fixtures or adding more lights to achieve the desired output over the entire production area. The light levels should always be measured at the furthest point away from the light source or any areas perceived to be the darkest. The intensity of light bulbs decreases over time, so their output should be checked at the start of every lighting season and randomly thereafter.

Temperature

After understanding the role of juvenility, vernalization, and photoperiod, growers should understand the role temperature plays with forcing perennials and the effects it has on crop quality and timing. Most growers are well aware that warmer temperatures will hasten plant development, but few growers take into consideration the role temperature plays on plant quality. Before discussing how temperature affects the quality attributes of perennials, let's quickly review the role temperature does play with crop development and finishing times.

The growth rate of perennials, and all plants, is essentially a function of temperature. Temperature affects the rate chemical reactions occur within plants and ultimately determines their rate of growth and development. The temperature must be above a certain level, or base temperature, or growth will not occur. For most plants the base temperature is slightly less than 50° F (10° C). As the temperature increases above the base temperature, the growth rate of the plant also increases until the optimal temperature is reached. The optimal temperature is the point where the growth rate is maximized and there is no additional increase in the growth rate with a subsequent increase in temperature. In fact, temperatures higher than the optimal temperature cause plant stress, and the growth rate actually begins to decrease. Each particular plant species and cultivar has its own base temperature and optimal temperature. In general, producing plants at warmer temperatures, near the optimal growth temperature, will reduce the crop time, causing them to bloom earlier, provided all of the other flowering requirements have been met. For many perennials the optimal temperature is around 80°F (27° C). If the optimal temperature is known for a particular plant species, growers can provide an environment where the growth rate can be maximized without causing undesirable plant stresses.

Every crop has a maximum growth temperature specific to it. However, to produce high-quality plants, it is often necessary to grow plants *below* temperatures that promote the fastest rate of growth. Therefore, the optimal growth temperature is not always the one growers should use for crop production.

One thing I've learned about temperature and plant development is that plant development is a function of average temperature. We need to consider this when determining our temperature settings. As long as the temperatures are maintained between the base temperature and the optimal temperature, plant growth will remain stable. However, when the day or night temperatures move outside of the normal temperature ranges (below the base or above the optimum), then the growth rate will slow down.

Keep in mind that when plants are behind schedule, it may make sense to raise the production temperatures to hasten crop development. When crops are behind schedule, it is best to maximize the growth rate by producing them at or near the optimal temperature constantly twenty-four hours per day. Growing plants at higher than optimal temperatures does not produce any better results and costs more money.

Remember, though, that temperature combined with light determines the developmental rate and affects the overall quality attributes of perennials. Plants grown under high light levels with low temperatures are generally sturdy but grow slowly, while plants produced with low light levels and high temperatures typically grow rapidly but are often thin or weak.

Determining growth rate

How can growers determine the growth rate? Perhaps the easiest method is to determine the leaf unfolding rate. Researchers and potted plant producers have developed leaf unfolding models for several commercially produced crops, including poinsettias and lilies. With certain crops, a particular number of leaves will be unfolded prior to flowering during the production period. The leaf unfolding rate is a function of temperature; leaves will unfold faster at warmer temperatures and slower at cooler temperatures. Counting leaves provides growers an indication whether crops are developing on schedule.

To determine the leaf unfolding rate, growers must first mark a recently matured leaf by using a marker, hanging a string over the petiole, or some other method where the leaf can be easily found and identified. After marking the leaf, growers wait seven days to let the plant grow and then return to the plant and determine how many leaves have developed during that time. If seven leaves have unfolded above the marked leaf, the grower can determine the current leaf unfolding rate to be one leaf per day (seven leaves divided by seven days equals one leaf per day).

To develop the leaf unfolding rate into something useful, growers should also calculate the average temperatures the plants were exposed to during the seven-day time frame. This would indicate the role temperature has on the rate of leaf unfolding. For the above example, assume the average temperature was 70° F (21° C). If the average temperature were reduced to 65° F (18° C), there would be a reduction in the number of leaves unfolded due to the lower temperatures. At 65° F (18° C), only five leaves may have unfolded, for a leaf unfolding rate of 0.7 leaves per day.

Combining these examples would show the role temperatures have with the leaf unfolding rates and ultimately the crop finishing times. Simply reducing the average temperatures by 5° F (3° C) reduced the leaf unfolding by nearly 29%, and the crop would need at least 29% more time to unfold all of the leaves before flowering could occur.

Temperature and crop timing

Growers must take into consideration the effect light and humidity levels have on the plant temperature, which affects the overall rate of growth and plant development. Plant temperatures are generally warmer than the air temperature during sunny conditions and cooler than the air temperatures during cloudy conditions. Growers can easily measure plant temperatures or leaf surface temperatures using inexpensive ($60–150) infrared temperature testers.

To maintain a consistent growth rate, or uniform leaf unfolding rates, growers will have to adjust the air temperatures based on the weather conditions. During sunny conditions, the leaf temperatures increase, causing the rate of development to increase (or decrease if the temperatures rise above the optimal temperatures); in most cases, growers will have to provide some form of cooling to maintain temperatures within the optimal range. Conversely, on cloudy days, the leaf temperatures may be below the air temperatures, decreasing the rate of development, resulting in the need for growers to increase the air temperatures to maintain leaf temperatures within the optimal range. Every day is different, requiring growers to constantly react to the conditions at hand when determining how to best maximize plant development and produce plants on schedule.

Growers should take into consideration the kind of growth they are looking for and determine the optimal temperatures necessary for that stage of growth. By providing optimal conditions for each stage of development, growers will improve the quality characteristics of perennials, producing plants with desirable stem and leaf mass, acceptable internode lengths, and earlier flowering.

The environment indigenous to the species being produced usually determines the optimal temperatures for a specific species. Temperate plants generally have lower optimal temperatures than ones from tropical origins. Generally, perennials are grouped into cool crops, requiring cooler temperatures (58–68° F [14–20° C]), and warm crops, requiring warmer temperatures (68–78° F [20–26° C]).

Low temperatures

Growing plants at temperatures below the optimal temperature range causes them to grow rather slowly, significantly delaying the time to reach flowering and, thus, the date they will be ready to be sold. Low temperatures can be used to slow the growth rate or to hold plants at a particular developmental state if they are coming too fast. Many growers place budded plants in coolers at 40° F (4° C) to hold them for short durations prior to shipping. Some tender perennials and tropical perennials will not tolerate exposure to temperatures less than 50° F (10° C) and may become seriously injured with exposure to these temperatures. Growers also observe an increased risk of root diseases and overwatering while plants are exposed to below-normal temperatures.

High temperatures

Growing plants at temperatures above the optimal range not only reduces the rate of plant development but may cause injury to plant tissues. Injury to the leaves often resembles leaf burn, or sun scald, that many plants get when produced under higher light levels. Plants grown near the top of the optimal temperature range usually have lower quality characteristics due to the "soft" growth; these plants often do not ship well nor do they have a satisfactory shelf life. Some other signs that plants have been grown at excessive temperatures include decreased flower number, aborted flowers, misshaped or smaller-than-normal flowers, and lighter or bleached coloration of the bloom. High temperatures during flower initiation can cause heat delay, most notably with short-day crops.

Temperature and plant quality

Flower size

Researchers have learned that production temperatures can have a dramatic effect on the size of the flowers of many perennials. As the production temperature increases, the flower size of many perennial crops decreases. However, the effect of temperature on flower size does not appear to take hold until after visible bud. Exceptions may occur when plants are produced above the optimal temperature before visible bud. In these cases, they often produce smaller flowers.

Coreopsis verticillata 'Moonbeam' provides a good example of how temperature affects the flower size. When 'Moonbeam' is grown at 63° F (17° C), the flower size is approximately 2 in. (5 cm) across. An increase in temperature up to 68° F (20° C) reduces the flower size by 20%, down to 1.6 in. (4 cm). Increasing the temperature from 63° F (17° C) to 73° F (23° C) results in a reduction of flower size by 30%, to 1.4 in. (3.5 cm). Really high production temperatures have a much more dramatic effect on the flower size. For example, temperatures of 84° F (29° C) will result in flowers measuring only 0.9 in. (2.3 cm), more than 50% smaller than the flowers on plants grown at 63°F (17° C).

Number of flowers

In addition to decreasing the flower size, warmer temperatures also tend to reduce the overall number of flower buds on some perennials. This reduction in flower buds is typically the result of reduced lateral branching that occurs at warmer temperatures. Some perennials have shown up to an 80% reduction in the number of flower buds when the temperatures are increased from 60° F (16° C) up to 80° F (27° C).

Temperature and stem elongation

Most growers are under the assumption that warmer production temperatures cause perennials to grow taller. Although the growth rate will increase until the optimal temperature is reached, an increase in the growth rate does not correlate to a corresponding increase in plant height. Actually, plants grown at warmer temperatures tend to be shorter when they are flowering than when produced under cooler temperature regimes. Temperatures can be manipulated to affect plant height; for more on limiting stem elongation with temperature, see chapter 8, Controlling Plant Height of Perennial Crops.

Temperature and flowering

As previously discussed, many perennials require a cold treatment (vernalization) or exposure to cool temperatures for certain durations before they will flower. For most perennials, successful vernalization occurs when exposed to temperatures of less than 44° F (7° C) for six to ten weeks. Each plant species responds differently to cold and has a specific optimal temperature and duration necessary for flowering.

Besides the often-necessary vernalization period, temperature affects flowering of perennials in differing ways. With many species, temperatures too warm or too cool will often reduce flowering or eliminate it altogether. Some perennials, such as *Fuchsia magellanica,* are day-temperature sensitive and the number of flowers is reduced as the day temperatures increase. Many perennials, such as garden mums, are night-temperature sensitive and often experience a delay of flowering (called heat delay) and/or a reduction of flowers from exposure to night temperatures greater than 73° F (23° C). Another temperature response occurs with plants that are day- and night-temperature sensitive, where the flowering is reduced, delayed, or eliminated when either the day or the night temperatures are above some critical level. There are other plants that are considered average daily temperature sensitive, meaning the flower number decreases as the average daily temperatures increase.

The rate of flower development, like plant development, is also a function of temperature up to some maximum temperature. Increasing the temperature above the optimal temperature for flower development will reduce the rate of flower development or delay flowering of many species.

Temperature summary

As the discussion above demonstrates, temperature is not only important for growers to deliver a perennial to market on schedule, but it also influences the final appearance and quality of the plant. Growers should consider the different roles temperatures have during crop development for each perennial species they are producing. Understanding the effects of temperature on such attributes as flower size, flower number, plant height, and plant quality can help growers to properly schedule and produce high-quality crops.

Stock Plant Management

Having an understanding about how to get perennials to bloom also gives growers the tools they need to maintain vegetative non-flowering plants. Many growers maintain stock and vegetatively propagate several of their perennial varieties. Unfortunately, most growers do not fully understand the mechanics behind producing an "ideal" perennial cutting or they do not have the ability to properly manage the stock plant environment to produce such cuttings. "Ideal" cuttings are vegetative propagules that are healthy, consist of uniform size and stem caliber, and have not initiated flower. These cuttings tend to root more readily during propagation and allow for better timing and uniformity of bloom for the end user.

Over the last decade, researchers have worked diligently to learn the mechanisms behind flowering perennials, allowing growers to produce perennials in bloom for nearly any foreseeable sale date. This same information can help growers to produce non-flowering stock plants, which can be used to produce high-quality vegetative propagation materials. Similar to forcing perennials, knowing the flowering response of each perennial to day length (photoperiod) is helpful for growers to properly provide and manage the photoperiods necessary to produce vegetative cuttings from perennial stock plants. Perennial stock plants should be maintained under photoperiods that will not induce flowering to ensure that they and the cuttings will remain vegetative.

Most commercially produced perennials fall into three flower inductive groups: cold required day neutral plants, cold required photoperiodic plants, and no cold required photoperiodic plants. Successful stock plant management entails understanding which flower inductive group a perennial belongs to and properly managing the environment to keep the stock in an actively growing vegetative state. For most perennials, the key is to maintain stock plants under non-inductive photoperiods to ensure they remain non-flowering.

Many of the day-neutral perennials, or perennials that will flower readily under any photoperiod, usually require exposure to a cold treatment to promote flowering. They tend to grow vegetatively until they are exposed to cool temperatures. For example,

Veronica longifolia 'Red Fox' is a perennial that once flowering has occurred can be kept vegetative all year, regardless of the photoperiod, provided it is not exposed to a cold treatment. Stock plants within this category should not be exposed to a cold treatment and can often be grown under natural day lengths to maintain the vegetative state.

Managing short-day stock plants with no cold can be done very similarly to how chrysanthemum growers keep their plants from flowering—by providing long days. When the natural day lengths are less than fourteen hours, growers can simply provide night break lighting to maintain vegetative stock plants of these short-day perennials. In most cases, growers supply night break lighting, achieving 10–20 f.c. (108–215 lux) between the hours of 10 P.M. and 2 A.M.

Most long-day perennials can be maintained at photoperiods just below the optimal photoperiod for flowering. For example, *Phlox paniculata* 'David' flowers when the photoperiods are at least thirteen hours; producing stock plants under twelve-hour day lengths will effectively keep 'David' in a vegetative state. This method will not work on all long-day perennials; *Gaura lindheimeri* 'Whirling Butterflies' is an exception. This cultivar is a facultative long-day plant, which flowers best when produced under long days. To maintain vegetative plants of 'Whirling Butterflies', growers have to reduce the day length dramatically, down to nine hours or less.

Some perennials become dormant when grown under short days. When producing stock plants, growers should control the photoperiod so that certain stock plant varieties will not go dormant from exposure to short photoperiods or flower prematurely when produced under long-day conditions.

Besides manipulating the photoperiod, growers might consider providing supplemental lighting to improve the quality of the cuttings produced and to increase the number of cuttings that can be harvested. Many growers successfully provide supplemental lighting during the winter months, when the light intensity is naturally low. Supplemental lighting is generally provided for eleven hours per day, with 350–575 f.c. (3.8–6.2 klux) being delivered to the crop. The challenge is to increase the light levels and the daily light integral without extending the day length; if the day length is extended, some perennial stock plants may form flowers.

Besides managing photoperiods and providing supplemental lighting, growers should also maintain ideal temperatures for stock plant and cutting production. For most perennials, stock plants should be grown at 65–72° F (18–22° C). Production of perennial stock at higher temperatures often leads to low-quality, leggy cuttings, and temperatures below this range adds significant time to produce each flush of cuttings. Growers should maximize the light quantity where applicable and consider providing supplemental lighting certain times of the year.

The stock plants should be maintained and cuttings should be harvested when they are ready. Frequent harvesting of the cuttings, every three to five weeks, keeps the stock plants from becoming overly grown, encourages lateral branching to produce more cuttings, and provides for uniform cutting production. Cuttings should be harvested whether they are needed or not; leaving them on the stock plants will result in lower cutting quality and poor propagation performance, and it will limit the cutting potential of the stock plants.

The goals of producing perennial stock plants should be to manipulate the photoperiod to prevent flowering, while providing all of the light possible to grow high-quality cuttings. Growers must weigh out the pros and cons of producing their own perennial stock plants. Many growers are not set up to produce their own stock plants. In many parts of the country, it is difficult to provide short days during the warm days of summer, while maintaining temperatures within the optimal ranges for plant development. Providing night-interruption lighting to manipulate the photoperiod is practical for many perennial producers, but it is often difficult for growers to justify the cost of supplemental lighting for perennial crops. Many vegetatively propagated perennials are not considered to be high-value crops, although many are, and unrooted cuttings can be purchased rather cheaply from outside sources.

Propagating for Forcing Programs

Just as it is important to manage the stock plant environment to ensure vegetative cutting production, it is important to also provide non-flower inductive conditions during propagation. Maintaining vegetative propagation materials allows growers the ability to accurately schedule perennials to flower uniformly.

In many cases, perennial cuttings are taken from mature stock plants that have already had flowering induced. This causes undesirable flowering during or slightly after propagation. It may be beneficial, if not necessary, to apply Florel to the cuttings to inhibit flowering or to pinch the cuttings and to move them to non-inductive conditions.

Perennial propagators should become familiar with the three flower inductive categories perennials fall into, as described above in the Stock Plant Management section, and provide the same environments (non-inductive conditions) to cuttings as recommended for stock plant production for each perennial being propagated. Most perennials with an obligate cold requirement for flowering can be successfully propagated without providing a particular photoperiod by omitting the cold treatment. For additional cultural recommendations for propagating cuttings, please refer to chapter 9.

Scheduling

The number one goal of scheduling should be to produce a crop on time, meaning that growers must focus on producing flowering perennials for specific dates. There are many challenging aspects to scheduling perennial crops, as perennials are slightly more difficult to schedule than most other commercially produced crops. Growers need a more complete understanding of crop biology and the mechanics of flowering than is needed for many bedding or potted crops. Each perennial will require its own schedule, as it has needs that may be unique to it.

Overview of scheduling factors

Following are some highlights of the critical factors relating to perennial growth and flowering, which have been discussed in detail throughout the book.

Light levels

Light levels vary with the geographic location and the season. In many instances, during the winter months, for example, the light levels are often too low, causing plants to grow slowly and at a reduced quality. Growers often use supplemental lighting to increase the growth rate and improve the quality of the crops. In most cases, growers must schedule extra time, often several weeks, for plant development to compensate for these low light levels. As spring approaches, the light levels increase and the day length gets longer, causing plants to grow more rapidly and plant quality to improve. Growers must compensate for the different seasons and build enough time into the schedules to allow for the worse possible conditions. Regardless, they need to constantly evaluate and adjust to the conditions at hand.

Photoperiod

For most perennials, the photoperiod plays a critical role with the flowering process. The photoperiod often regulates the type of growth a plant has, whether it remains vegetative or if it develops flowers. As discussed previously, perennials are typically divided into three photoperiod groups: short-day plants, day-neutral plants, and long-day plants. Each of these groups is subdivided into plants with obligate photoperiodic responses (the photoperiod is required) or facultative photoperiodic responses (the photoperiod is beneficial). It is necessary to understand that most perennials must be exposed to the appropriate photoperiod in order to properly schedule them to be in bloom for specific sales dates.

Temperature

Temperature plays many roles with successful perennial production, from ensuring a plant is properly vernalized, to controlling the rate of growth, to influencing the number and size of the flowers produced. Many perennials require a vernalization treatment to initiate flowering; without this cold treatment flowering will not occur or be greatly reduced with some species.

Temperature plays a critical role with the developmental rate or how quickly the crop will be salable. In regard to scheduling, temperature is everything—it has the most influence on determining whether a crop is early, on time, or late. In most cases, for every 5° F (3° C) increase in average temperature, there is approximately a one-week reduction in crop time. Growers should know what temperatures the crops are being grown at, and if possible, make any necessary adjustments to speed up or slow down crop development to ensure they are finished at the required time.

Growers should note that increasing production temperatures is not always in the best interest of the crop, as it may often reduce the quality or may delay crop development, causing growers to miss the intended sales window.

Fertilizer

Fertilizers can be used, to some extent, to influence the growth rate of many perennial varieties. Fertilizers high in phosphorous and ammonium nitrates generally promote more rapid plant growth, which can lower plant quality in some cases. Fertilizers low in phosphorus and high in nitrate nitrogen (low ammonium nitrogen) tend to promote stocky, more normal growth. Withholding nutrients is also an option, but growers should not rely on starvation to control plant growth, as this can cause nutritional deficiencies that could affect the quality attributes of the crop.

Watering

Watering is perhaps the most important aspect of growing perennials. The individuals handling the watering responsibilities on a daily basis have a great impact on the outcome of the crops. Watering is not only important to keep plants alive, but it can be a valuable tool to control certain quality characteristics, and it affects crop scheduling. Perennials kept moist tend to grow very rapidly. When perennials are slightly overwatered, they will often reach the intended ship date on schedule but taller than desired. Conversely, growing plants too dry often produces shorter-than-desired plants (often with stunted growth) and delays crop scheduling. Unless they are trying to turn a second crop, most growers prefer to grow on the slightly dry side, adding a small amount of time to the overall schedule.

Source material

The size of the starting material has a great effect on the overall timing of the finished crop. Starter materials for most growers are typically plugs or liners. Perennial growers also use bare-root plants, which are often quite large, mature, and offer quick finishing times. The maturity and stage of development of the starter materials varies greatly with the various sources. For example, small plugs such as 220s may be of transplantable size but may not be mature enough to produce flowers or to perceive the necessary treatments for flowering.

Large starting materials are generally more mature and can often finish the final container up to four weeks earlier than when smaller plug sizes are used. For example, growers using 72-cell plugs can finish a crop one to two weeks earlier than when using 128-cell plugs of the same variety. Many growers will use larger starter materials for the second or third turn of the production space, allowing them to grow and ship more plants before the spring season is finished. To finish a crop quicker or to produce the desired quality, growers often place multiple plugs per pot, especially when using smaller starter materials.

Multiple plantings

Many growers produce one variety to be shipped over an entire growing season. For example, a perennial grower might supply *Leucanthemum* 'Snow Cap' with each delivery over the entire shipping season, perhaps eight weeks or longer. For the best appearance and plant quality, it is not conducive to produce the entire crop during the same time frame. Growers often split the crop into multiple plantings with various start dates, allowing them to have differing ship dates, maintaining a steady supply of high-quality flowering plants. Depending on the crops being grown and the ship dates, it is quite common for growers to schedule plantings one to two weeks apart. Most operations do not provide a steady supply of equal volumes with each delivery throughout the shipping season, but rather provide more of a bell-shaped curve throughout the season, where volume usually starts small, increases to a peak, then reduces back to lesser amounts. Splitting a crop into multiple plantings allows growers to adjust the numbers of each planting to better match the volumes needed with each shipment while keeping the crop fresh and not overgrown.

Container size

The size of the container the crop is produced in influences the length of time it requires to finish the crop. As the container size increases, more time is needed to produce a "full" plant. Conversely, as the container size decreases the crop timing decreases, as the container fills out more quickly. To reduce crop times of larger containers, growers can also plant additional plugs into each pot. For each additional plug planted, the crop time is reduced by approximately one week, as it takes less time to produce a "full" plant.

How to schedule crops

Growers should have realistic expectations regarding their crop scheduling. Most growers receive cultural and scheduling advice from outside sources, taking this information as gospel. Use outside scheduling information only as a guideline, as someone else's recipe may not work in your production system. There are numerous cultivar differences; what works for one cultivar may not work for another similar cultivar. Growers often underestimate the power of the sales window or expect a schedule that worked for an April sales date will also work for a June sales window.

To successfully schedule perennial crops, growers need to start with the finish date and work backward. Without a clearly defined date of when the product needs to be finished, the crop cannot be properly scheduled. Crop schedules can be simple, only showing the number of weeks from transplanting to the time the crop is ready, or they can be very detailed, showing specific event dates such as when the crop needs to be pinched or when the photoperiodic lighting should begin. The more detailed the schedules are, the better the growers will be able to perform the necessary production activities when they need to be performed. If forcing perennials into bloom is the goal, the schedules become more complex, as various factors such as juvenility, vernalization, and photoperiod need to be considered and provided at the right times.

A detailed crop schedule for *Lavandula* 'Hidcote Blue', with a target flowering date of April 15, could read as follows:

- 8 weeks forcing at 68° F (20° C), begin on February 18
- 12 weeks vernalization at 41° F (5° C), begin on November 26
- 10 weeks bulking (transplant plug), begin on September 17
- 13 weeks plug, begin on June 18
- 6 weeks chill seed for germination, begin on May 7

From seed to finish, this crop can take nearly a year. If vernalization is omitted, it can be grown from seed in about thirty weeks. But flowering is inconsistent and delayed without cold treatment.

Crop schedules need to be adjusted to fit the season the plants are being grown. Producing a flowering perennial for an early April ready date will require a different schedule than producing the same plant for a June blooming date. There are seasonal variances, due largely to temperature, light levels, and photoperiod, that have a dramatic effect on crop timing. It takes longer to produce a plant for an April ship date than it does for a June ship date, so using the April schedule for both will result in the June plants being ready weeks before the intended sales window. Conversely, using the schedule to produce the June crop for an April ship date often does not allow a sufficient amount of time to grow the crop, resulting in product that is not ready by the intended sales date.

The type of weather conditions that are prevalent during the production season will have a great effect on how accurate the crop schedules will be. Most growers schedule crops based on average conditions; if the conditions are extreme one way or the other, the timing of the crop could be significantly different than intended. During cold, dark, wet springs, most crops will be significantly delayed due to reduced light levels and lower temperatures. Growers often adjust the production temperatures and consider supplemental lighting when prolonged cold periods are anticipated. The other extreme is when unusually warm and sunny periods occur, causing plants to develop much faster. During warm spells, growers often increase venting and use more growth regulators to curb plant growth. During either extreme, it is difficult to compensate properly and bring the growth rate back to where it should be. The best scenario is to anticipate that these conditions are going to occur and to make the necessary adjustments before it is too late. It is very difficult to make up lost time or to hold crops while maintaining quality when the schedule gets off course.

During the growing season, growers should record nearly everything and take notes regarding any factor that contributed to the outcome of the crop. This information should describe the actual events that occurred during production as opposed to the intended schedule. Cultural notes regarding the weather, temperatures, light levels, insect or disease problems, dates of activities such as pinching and spacing, heating and cooling set points, and any adjustments or changes that were made throughout the crop will all be helpful when formulating schedules for upcoming crops. Any information that is important and will likely be forgotten, if it is not recorded, should be documented. Notes regarding the flowering dates, when the product was ready to ship, and when the product actually shipped will all help to plan future crops.

Scheduling tables are found at the end of this chapter. Table 10.9 shows scheduling for 1 qt. (1 L) containers, and table 10.10 shows scheduling for 1 gal. (3.8 L) containers. When reading these, remember that they are guidelines and may need adjustment based on your location and your operation's specific practices.

Scheduling systems

There are numerous methods growers can use to compose crop schedules and production plans. Growers effectively use simple tables written on paper to very elaborate computer software. The type of scheduling method used depends on the complexity of the crops being grown and the size of the perennial production facility.

Scheduling by hand

Several growers are content to create simple tables on tablets of paper to meet the scheduling needs of their crops. These tables usually consist of one row for each cultivar and each container size, then a series of columns displaying several key headings including the sow date, transplant date, container size, photoperiod, temperature settings, and so on. It is also wise to leave a column for any pertinent comments regarding any necessary adjustments for the next season's schedule. Creating schedules in this manner often limits the amount of details that can be organized in advance as well as how much information can easily be collected for future reference.

Spreadsheet scheduling

Using spreadsheets to create crop schedules is a simple but effective method for growers to organize production plans and collect the information, which is useful for future scheduling modifications. Spreadsheets are assembled very similar to hand-made schedules, with the rows containing the cultivars and the columns containing the key headings. With computers, which most operations now have access to, growers can simply build a spreadsheet themselves to satisfy their scheduling needs. Spreadsheets allow growers to be organized, use the same information from year to year (without copying it by hand each year), sort the crop schedules any way they need to (chronological, alphabetical, by container size, etc.), and to conduct their mathematical calculations. Spreadsheets are a wonderful option for the small- to medium-size grower.

Software scheduling

There have been various types of scheduling software developed for growers, each designed for a particular function. For example, some software is developed to primarily help growers manage their inventory and to facilitate shipping activities, some only help growers with the production of the crops, and some combine these into production-inventory-shipping programs. There are several tools currently available, or being developed, to help growers plan greenhouse space utilization and/or coordinate production and sales activities.

Scheduling software can be easily used to manage diverse product lines (cultivars and container sizes), multiple starting dates, and multiple finishing times. Depending on the software, some of the activities growers can accurately schedule include when to sow seeds, stick cuttings, transplant plugs, pinch plants, space the crop, provide a cold treatment, adjust the photoperiod, and develop sales forecasts. Compared to spreadsheets, software can bring continuity and consistency to the process. Software also allows for forecasting of future needs and provides the necessary information for various departments (growers, production, shipping, and logistics) to more efficiently do their jobs or to know their expectations.

Growers are very interested in using software to help manage capacity and space utilization, as production space is the most limiting factor to the number of plants that can be produced. Many growers overestimate their capacity and produce more plants than is practical within the confines of their production space, often reducing the overall quality of the crop and resulting in crop losses. It is important for growers to turn their production space frequently and to keep it full as much as possible. Such scheduling is demanding and nearly impossible without a computer.

Here is a listing of some of the software companies designing production and scheduling software for commercial growers.

- ArcSoftware, www.arcgrowingsoftware.com
- Gart Plan (space planning), www.gartplan.dk/
- GroTime, www.grotime.com
- Plantware Inc., www.plantware.net
- Software Studio Ltd. (HortiPlan), www.smartschedules.com
- Starcom, www.starcomsoft.com

Growers should use the type of scheduling tools (spreadsheets or software) that works best for their operation. The size of the operation often determines the need to use software over spreadsheets, with larger businesses requiring more complex tools. However, small operations should not be deterred from using software, as software can be tailored to fit anyone's needs, and affordable software is on the market.

The grower's calendar

Many growers and floricultural businesses have adapted the practice of using a calendar based on weeks rather than specific days or dates. Numbering the weeks is used to schedule propagation, production, and shipping. It is easier to count the number of weeks on a calendar than it is days.

Many companies attempt to assign their own numbering system to the calendar year; however, this can lead to many businesses functioning under different week systems, which could create confusion and impact plant finish dates. For example, a grower ordered plugs to be delivered in week 7, but the plug supplier was using a different calendar—the customer's week 7 was the plug supplier's week 6. The customer received the plugs in week 8 (the supplier's week 7), one week later than anticipated, and the grower's schedule was now at least a week off (from the beginning of week 7 to the end of week 8 is actually a two-week period, or fourteen days), forcing them to make up for lost time and possibly missing their sales window.

The International Organization for Standardization (ISO) has established standard rules for numbering weeks (ISO 8601:1988) that was endorsed by Bedding Plants International (BPI) in 1993. This system ensures that all businesses are using the same numbering system and eliminates the need for businesses to develop their own calendars.

The following is a summary of the ISO week numbering system:

- All weeks are composed of seven days.
- The week starts on Monday (most North American calendars have Sunday as the first day of the week).
- Week number 1 of any year is the week that contains the first Thursday of the year. This is perhaps the most important rule, as it is used to determine which week is week 1.
- The first week of the year is the week that contains January 4.

To refer to a particular week, it is a common practice to use two numeric digits preceded by a W for week. For example, W01 is used for the first week, week 1, and W52 represents week 52. To refer to a specific week of a particular year, growers use a couple of methods. One example is to list the year first, followed by the week number; 2007W06 is week 6 of 2007. It is also commonly written as 07W06. Some growers omit the W and simply list the year and the week as 200706, where the first four numbers represent the year and the last two digits are for the week. The shorthand growers use varies from operation to operation; the important part is that every operation is using the same beginning point for week 1.

Source materials

Scheduling helps growers to better plan the production or purchase of the source materials needed to produce the final product. Early planning increases the likelihood growers can get the varieties they want when they want them, and it usually gets them the best pricing.

In most cases, growers should allow adequate time to acquire or produce the young starter plant materials. In addition to allowing adequate time to produce the finished product, it is necessary to allow enough time to produce the plugs or liners. If growers are obtaining starter materials from outside sources, they need to allow the supplier sufficient time to produce the grower's order. For example, many plug suppliers only grow what has been sold and won't have any "extra stock" on hand. Many suppliers will either maintain an availability of the items they have or will indicate the number of weeks before an item is needed for orders to be placed.

Orders for starter plants should be placed according to the following formula:

Order Date = Ship Date Final Container
– Final Container Weeks to Finish (WTF)
– Starter Material WTF

For example:

Ship Week	Final Container WTF	Starter Material WTF	Order Week
20	9	8	3

In order to produce a crop to be ready in week 20 that takes nine weeks from transplant to finish, the transplanting would have to occur during week 11. The starting materials, plugs for example, must be ready to be transplanted by week 11 and must be started earlier to ensure they are ready by then. In the above formula, the starter material requires eight weeks to finish; therefore, to be ready for week 11 it must be started in week 3. Week 3 would then be the order week for the starting materials the growers want to transplant in week 11 to finish a crop in week 20.

In some cases, when using seeds or unrooted cuttings, it is necessary to add even more time onto the above formula to obtain these items from seed distributors or cutting suppliers. This is the lead time necessary to obtain the starting materials. For seeds, it may only take two weeks on average from ordering until they are received. The lead time it takes to acquire the seeds needs to be subtracted from the order week. Using the above example, the seeds should be ordered in week 1 (week 3 – 2 weeks = week 1).

The new order date formula could be expressed as:

Order Date = Ship Date of Final Container – Final Container WTF – Starter Material WTF – Starter Material Lead Time

Ship Week	Final Container WTF	Starter Material WTF	Starter Material Lead Time	Order Week
20	9	8	2	1

When acquiring unrooted cuttings, it may also take an additional amount of lead time from the time the order is placed until the time the cuttings can be received. Always consider the location the starting materials are originating from and allow an adequate amount of lead time based on your supplier's recommendations or from your past experiences.

Scheduling conclusions

Scheduling is a very complex task that requires a combination of skills, abilities, and attention to details. Properly and predictably growing perennial crops in bloom requires more than simply planting a plug and growing it on. There are so many variables and requirements that must be met in order for flowering to even occur. It takes time to learn the needs of each crop and how

Figure 10.10. **Growers should know what the crop is supposed to look like at various production times. This is useful to gauge whether the crop is on schedule and if any adjustments need to be made in order to finish the crop on time. Shown here is the progression of a *Leucanthemum superbum* 'Snow Lady' crop: week 1 (top left), week 3 (top right), week 5 (bottom left), and week 9 (bottom right).**

to properly manage the environments to produce the plants for specific sales dates. The payoff is the ability to deliver the crop on time, satisfy your customers, and maximize the profitability, all while being in control of your crops.

Delivering More Bloom Naturally

Growers should have a good idea of what they can grow that will bloom naturally without a forcing program. Do not rely on catalog descriptions, which generalize spring or summer flowering. Catalogs will give good indications as to when plants will bloom in outdoor conditions, but not when flowering is likely to occur under greenhouse conditions. Due to warmer temperatures in greenhouses than outdoors, producing plants under various structures often moves the flowering season forward by a few weeks. As soon as flowering begins, growers need to consider shipping their perennials to ensure the best retail presence and shelf life.

Many perennial growers have several types of production facilities: cold frames, heated Quonsets, gutter-connect greenhouses, and even outdoor sites. Growers should know the flowering times from each of these production sites, as they will vary. I recommend that growers dedicate themselves to recording when each perennial will be flowering in order to better plan their production and shipping schedules. This flowering information can be collected while scouting for cultural problems, taking inventory, or preparing orders for shipping. The key is to have a strategy in place and to stick with it. It is best to collect information weekly from every location and for every container size. Growers should take notes as to when the product was of shippable size and when it came into bloom. I find it useful to download inventory by location into an Excel worksheet with the appropriate columns, print it out, and to use this as a scouting and flowering evaluation tool (see table 10.9). This information should be collected over at least three seasons to account for any unusual environmental conditions that may have occurred.

Growers should also have good records in regard to what the production temperatures were, the outdoor temperatures, light levels, the source of the starter materials, and when planting occurred. All of these factors, combined with the bloom times, will allow you to create production schedules that fit your environments and sales windows. Scheduling information from this book or other references only provides guidelines; they do not give good indications of how perennials will grow under your conditions and environments.

Over time, growers can develop a good mix of perennials for nearly any sales date by utilizing their knowledge and facilities. If there are not enough varieties flowering naturally to satisfy your customer base, then forcing a few perennials may be necessary. This information puts you in control of your production, your sales, and your profits.

References

Armitage, Allan M. "Spring into Forcing Perennials." *Greenhouse Grower.* July 1994.

Aylsworth, Jean D. "Selling Perennials in Flower." *Greenhouse Grower.* July 1995.

Barrett, Jim. "Temperature Effects on Plant Growth." *Greenhouse Product News.* September 2000.

Barrett, Jim, and Theo Blom. "The Grower's Calendar." *Greenhouse Product News.* March 1999.

Beytes, Chris. "What's 'Mum Lighting'?" *GrowerTalks.* September 2001.

Cameron, Art, Erik Runkle, et. al. "Herbaceous Perennials: Summary Tables." *Greenhouse Grower.* December 2003.

Enfield, Amy, Erik Runkle, et. al. "Herbaceous Perennials: Quick-cropping, Part I." *Greenhouse Grower.* January 2002.

Erwin, John, Neil Mattson, and Ryan Warner. "Fundamentals of Flowering, Part 1." *Greenhouse Grower.* July 2003.

————. "Fundamentals of Flowering, Part II." *Greenhouse Grower.* January 2004.

Fisher, Paul and Erik Runkle. *Lighting Up Profits: Understanding Greenhouse Lighting.* Willoughby, Ohio: Meister. 2004.

Greenhouse Grower Magazine and Michigan State University. *Firing Up Perennials: The 2000 Edition.* Willoughby, Ohio: Meister. 2000.

Healy, Will. "Know When to Schedule What Crops." *Greenhouse Product News*. May 2001.

Niu, Genhua, Erik Runkle, et. al. "Herbaceous Perennials: Light." *Greenhouse Grower*. January 2001.

Runkle, Erik. "Grower 101: Controlling Photoperiod." *Greenhouse Product News*. October 2002.

Simeonova, Neda. "Crop Scheduling: How Growers Operate with Software That Is and Is Not Currently Available." *Greenhouse Product News*. November 2003.

Figure 10.11. **Scouting and Flower Evaluation Tool**

Date: ____________________

Scout's Name: ____________________

Location	Variety	Size	Quantity	Date Shippable Size	Date in Flower	Insects Observed	Diseases Observed	PGRs Needed	Comments
Greenhouse 1	Dianthus deltoids ‘Arctic Fire’	1 Gal.	467						
Greenhouse 1	Dianthus gratianopolitanus ‘Firewitch’	1 Gal.	803						
Greenhouse 1	Dianthus gratianopolitanus ‘Star Pixie’	1 Gal.	843						
Greenhouse 1	Dianthus gratianopolitanus ‘Tiny Rubies’	1 Gal.	1,576						
Greenhouse 2	Aquilegia x hybrida ‘Songbird Cardinal’	1 Gal.	200						
Greenhouse 2	Doronicum orientale ‘Little Leo’	1 Gal.	66						
Greenhouse 2	Heucherella x hybrida ‘Viking Ship’	1 Gal.	26						
Greenhouse 2	Lamium maculatum ‘Orchid Frost’	1 Gal.	169						
Greenhouse 2	Lamium maculatum ‘Red Nancy’	1 Gal.	137						
Greenhouse 2	Tiarella x hybrida ‘Spanish Cross’	1 Gal.	270						
Greenhouse 3	Geranium pratense ‘Victor Reiter, Jr.’	1 Gal.	241						
Greenhouse 3	Salvia nemorosa ‘East Friesland’	1 Gal.	1,264						
Greenhouse 3	Salvia x sylvestris ‘Marcus’	1 Gal.	569						
Greenhouse 3	Salvia x sylvestris ‘May Night’	1 Gal.	560						
Greenhouse 4	Heuchera x hybrida ‘Amethyst Mist’	1 Gal.	156						
Greenhouse 4	Heucherella x hybrida ‘Chocolate Lace’	1 Gal.	172						
Greenhouse 4	Ranunculus ‘Bloomingdale Special Mix’	1 Gal.	250						
Greenhouse 4	Veronica spicata ‘Goodness Grows’	1 Gal.	201						
Greenhouse 4	Veronica spicata incana ‘Giles van Hees’	1 Gal.	100						

Table 10.9. Scheduling Flowering Perennials in 1 Qt. (1 L) Containers

The schedules shown below are developed to help growers estimate the approximate finishing times for producing various crops. Unless specified, these schedules are intended to produce flowering plants. Providing the proper treatments necessary for flowering is important to produce blooming plants using the information in this table. There are many variables that affect crop scheduling and performance including cultivar, type and size of the starting materials, production temperatures, photoperiod, vernalization, and production practices. Growers should use the information provided here as guidelines, not absolutes.

Variety^	Commonly Potted the Year Before Expected Sales[1]	Previous Season's Source Materials[2]	Spring Potted Weeks to Finish^^	Spring Source Materials[2]
Achillea filipendulina			5*	Vernalized plug
Achillea x *hybrida*			5*	Vernalized plug
Achillea x *hybrida 'Moonshine'*			8	Vernalized plug
Achillea millefolium			5*	Vernalized plug
Achillea tomentosa			7	Vernalized plug
Aegopodium podagraria	Late summer	Bare root	6*	Bare root
Ajuga pyramidalis	Early fall[1]	Plug	7	Bare root or vernalized large plug
Ajuga reptans	Early fall[1]	Plug	7	Bare root or vernalized large plug
Alcea rosea			5*	Vernalized plug
Alchemilla mollis			6*	Vernalized plug
Alyssum montanum	Early fall	Plug	6	Vernalized plug
Alyssum saxatile	Early fall	Plug	7	Vernalized plug
Anaphalis margaritacea			12	Vernalized plug
Anemone multifida	Late summer[1]	Plug		
Anemone sylvestris	Late summer[1]	Plug		
Anthemis tinctoria			5*	Vernalized plug
Aquilegia alpina	Early fall	Plug	7	Vernalized plug
Aquilegia caerulea	Early fall	Plug	7	Vernalized plug
Aquilegia flabellata	Early fall	Plug	5	Vernalized plug
Aquilegia x *hybrida*	Early fall	Plug	7	Vernalized plug
Aquilegia vulgaris	Early fall	Plug	7	Vernalized plug
Arabis blepharophylla			6	Vernalized plug
Arabis caucasica			4	Vernalized plug
Arenaria montana			5	Vernalized plug
Armeria x *hybrida*	Early fall[1]	Plug	9	Vernalized plug
Armeria latifolia			8	Vernalized plug
Armeria maritima	Early fall[1]	Plug	7	Vernalized plug
Armeria pseudarmeria			7	Vernalized plug
Artemisia x *hybrida*			5*	Plug
Artemisia ludoviciana			6*	Plug
Artemisia schmidtiana			5*	Bare root

(continued)

Table 10.9. Scheduling Flowering Perennials in 1 Qt. (1 L) Containers *(continued)*

Variety^	Commonly Potted the Year Before Expected Sales[1]	Previous Season's Source Materials[2]	Spring Potted Weeks to Finish^^	Spring Source Materials[2]
Aster dumosus			9	Plug
Aster novi-belgii/angliae			9	Plug
Aster tongolensis			7	Vernalized plug
Astilbe arendsii	Midsummer	Bare root	8	Bare root
Astilbe chinensis	Midsummer	Bare root	8	Bare root
Astilbe japonica	Midsummer	Bare root	7	Bare root
Astilbe x *rosea*	Midsummer	Bare root	7	Bare root
Astilbe simplicifolia	Midsummer	Bare root	10	Bare root
Astilbe thunbergii	Midsummer	Bare root	7	Bare root
Aubrieta cultorum			9	Vernalized plug
Aubrieta x *hybrida*			7	Vernalized plug
Aurinia saxatilis			5	Vernalized plug
Bellis perennis			6	Vernalized plug
Bergenia cordifolia	Midsummer	Plug		
Boltonia asteroides			7*	Vernalized plug
Briza media			10*	Plug
Buddleia davidii			5*	Plug
Buddleia x *weyeriana*			5*	Plug
Calluna vulgaris	Late summer*	Plug		
Campanula carpatica	Midsummer	Plug	8	Plug
Campanula cochleariifolia			8	Plug
Campanula garganica	Midsummer	Plug	7	Vernalized plug
Campanula poscharskyana			7	Vernalized plug
Carlina acaulis			14	Vernalized plug
Centaurea dealbata			6	Vernalized plug
Centaurea montana			5*	Vernalized plug
Cerastium tomentosum			5	Vernalized plug
Chasmanthium latifolium			7	Plug
Cheiranthus cheiri			12	Vernalized plug
Cheiranthus suffruticosum			14	Plug
Clematis x *hybrida*			13	Bare root or vernalized plug
Clematis viticella			12	Bare root or vernalized plug
Convallaria majalis			8	Bare root
Coreopsis auriculata			6	Vernalized plug
Coreopsis grandiflora			8	Vernalized plug
Coreopsis x *hybrida*			8	Vernalized plug
Coreopsis lanceolata			8	Vernalized plug

(continued)

Table 10.9. Scheduling Flowering Perennials in 1 Qt. (1 L) Containers *(continued)*

Variety^	Commonly Potted the Year Before Expected Sales[1]	Previous Season's Source Materials[2]	Spring Potted Weeks to Finish^^	Spring Source Materials[2]
Coreopsis rosea			8	Vernalized plug
Coreopsis verticillata			8	Bare root
Corydalis lutea			7	Plug
Delphinium belladonna			5*	Plug
Delphinium elatum			5*	Plug
Delphinium grandiflorum			9	Plug
Delphinium nudicaule			7	Plug
Dianthus allwoodii			8	Vernalized plug
Dianthus barbatus			7	Vernalized plug
Dianthus deltoides			7	Vernalized plug
Dianthus gratianopolitanus			7	Vernalized plug
Dianthus hybridus			7	Plug
Dicentra eximia			7	Bare root or vernalized large plug
Dicentra formosa			7	Bare root or vernalized large plug
Dicentra spectabilis			5	Bare root
Digitalis grandiflora			5*	Vernalized plug
Digitalis mertonensis			5*	Vernalized plug
Digitalis purpurea			5*	Vernalized plug
Doronicum orientale	Early fall[1]	Plug	7	Vernalized plug
Draba aizoon	Early summer	Plug		
Echinacea purpurea			7*	Vernalized plug
Echinops ritro			5*	Plug
Erianthus ravennae			7*	Plug
Erigeron glaucus			9	Vernalized plug
Erigeron karvinskianus			8	Plug
Erigeron speciosus			8	Vernalized plug
Erinus alpinus			20	Plug
Eupatorium maculatum			6*	Plug
Euphorbia polychroma			5*	Plug
Festuca amethystina			6*	Plug
Festuca glauca			6*	Bare root or plug
Festuca valesiaca			6*	Plug
Gaillardia aristata			7	Vernalized plug
Gentiana acaulis	Late summer	Plug		
Geranium sanguineum	Late summer	Plug	7	Vernalized plug
Geum chiloense			7*	Vernalized plug
Gypsophila repens			7	Vernalized plug

(continued)

Table 10.9. Scheduling Flowering Perennials in 1 Qt. (1 L) Containers *(continued)*

Variety^	Commonly Potted the Year Before Expected Sales[1]	Previous Season's Source Materials[2]	Spring Potted Weeks to Finish^^	Spring Source Materials[2]
Hedera helix	Midsummer	Plug	10	Plug
Helianthemum nummularium	Midsummer	Plug		
Helichrysum thianshanicum			10	Vernalized plug
Helleborus x *hybridus*	Late summer[1]	Plug		
Helleborus niger	Late summer[1]	Plug		
Helleborus orientalis	Late summer[1]	Plug		
Hemerocallis	Midsummer[1]	Bare root	8*	Bare root or large plug
Hemerocallis tetraploid	Midsummer[1]	Bare root	7*	Bare root or large plug
Heuchera x *hybrida*			7	Vernalized plug
Heuchera micrantha			6	Vernalized plug
Heuchera sanguinea			7	Vernalized plug
Hieracium villosum	Late summer	Plug	10	Vernalized plug
Hosta	Early summer[1]	Bare root	14*	Bare root
Hosta fortunei	Early summer[1]	Bare root	15*	Bare root
Hosta montana	Early summer[1]	Bare root	15*	Bare root
Hosta plantaginea	Early summer[1]	Bare root	13*	Bare root
Hosta sieboldiana	Early summer[1]	Bare root	15*	Bare root
Hosta tardiana	Early summer[1]	Bare root	14*	Bare root
Hosta tokudama	Early summer[1]	Bare root	15*	Bare root
Hosta undulata	Early summer[1]	Bare root	14*	Bare root
Hosta ventricosa	Early summer[1]	Bare root	13*	Bare root
Houttuynia cordata			6*	Plug
Iberis sempervirens			10	Vernalized plug
Imperata cylindrica	Late summer	Plug	7*	Large plug
Inula ensifolia			9	Vernalized plug
Iris, bearded standard dwarf	Late summer[1]	Bare root		
Iris x *siberica*			7*	Bare root
Kniphofia uvaria			9*	Vernalized plug
Koeleria glauca			6*	Plug
Lamiastrum galeobdolon			8	Vernalized plug
Lamium maculatum			6	Plug
Lathyrus latifolius			6*	Vernalized plug
Lavandula angustifolia	Midsummer	Plug	9	Vernalized plug
Lavandula x *intermedia*	Midsummer	Plug	11	Vernalized plug
Lavandula stoechas	Midsummer	Plug	10	Vernalized plug
Lavandula vera			8	Vernalized plug
Leontopodium alpinum	Late summer	Plug	10	Vernalized plug

(continued)

Table 10.9. Scheduling Flowering Perennials in 1 Qt. (1 L) Containers *(continued)*

Variety^	Commonly Potted the Year Before Expected Sales[1]	Previous Season's Source Materials[2]	Spring Potted Weeks to Finish^^	Spring Source Materials[2]
Leucanthemum superbum			8	Vernalized plug
Lewisia cotyledon	Late summer	Plug	8	Vernalized plug
Lilium, Asiatic			7–11	Bulb
Lilium, Oriental			9–13	Bulb
Linum flavum	Late summer	Plug	8	Vernalized plug
Linum perenne			7	Vernalized plug
Lobelia x *hybrida*			8*	Plug
Lobelia speciosa			7*	Plug
Lupinus x *hybrida*			6*	Plug
Luzula nivea			7*	Plug
Lysimachia nummularia			5*	Plug
Melica transsilvanica			7*	Plug
Monarda didyma			5*	Vernalized plug
Monarda x *hybrida*			5*	Vernalized plug
Myosotis hybrida			9	Plug
Myosotis sylvatica			6	Vernalized plug
Nassella tenuissima			7*	Plug
Nepeta faassenii			6	Vernalized plug
Oenothera missouriensis			8	Vernalized plug
Oenothera speciosa			5	Vernalized plug
Pachysandra terminalis	Midsummer	Plug	10*	Plug
Papaver alpinum			8	Vernalized plug
Papaver miyabeanum			8	Vernalized plug
Papaver nudicaule			8	Vernalized plug
Papaver orientale			5*	Vernalized plug
Penstemon barbatus			9	Vernalized plug
Penstemon digitalis			7	Vernalized plug
Penstemon x *hybrida*			7	Vernalized plug
Perovskia atriplicifolia			7*	Vernalized plug
Phlox subulata	Late summer[1]	Bare root or plug		
Platycodon grandiflorus			10	Plug
Polemonium boreale			6	Vernalized plug
Polemonium caeruleum			6	Vernalized plug
Primula denticulata	Late summer[1]	Plug		
Primula elatior			12	Plug
Primula pubescens	Late summer[1]	Plug		
Primula polyanthus			9	Vernalized plug

(continued)

Table 10.9. Scheduling Flowering Perennials in 1 Qt. (1 L) Containers *(continued)*

Variety^	Commonly Potted the Year Before Expected Sales[1]	Previous Season's Source Materials[2]	Spring Potted Weeks to Finish^^	Spring Source Materials[2]
Primula rosea	Early fall[1]	Plug	11	Vernalized plug
Primula vulgaris	Early fall[1]	Plug	10	Vernalized plug
Pulsatilla vulgaris			7	Vernalized plug
Ranunculus repens			8	Vernalized plug
Rosa chinensis			7	Plug
Rosa, miniature			5–7	Plug
Rudbeckia fulgida			9*	Vernalized plug
Rudbeckia hirta			7	Plug
Rumex sanguineus			6*	Plug
Sagina subulata			9	Plug
Salvia lyrata			9	Plug
Salvia nemorosa	Late summer	Plug	6	Bare root or vernalized plug
Salvia superba			8	Plug
Salvia x *sylvestris*	Late summer	Plug	6	Vernalized plug
Salvia verticillata			5	Vernalized plug
Santolina chamaecyparissus			9*	Plug
Saponaria ocymoides			8	Vernalized plug
Saxifraga arendsii	Late summer	Plug	7	Vernalized plug
Scabiosa caucasica			9	Vernalized plug
Scabiosa columbaria	Early fall	Plug	7	Vernalized plug
Sedum acre	Late summer	Plug		
Sedum cauticolum	Early fall	Plug	7*	Vernalized plug
Sedum x *hybrida*			8*	Bare root or vernalized plug
Sedum kamtschaticum			6	Vernalized plug
Sedum reflexum	Mid summer	Plug	10	Vernalized plug
Sedum sieboldii			6*	Vernalized plug
Sedum spectabile	Midsummer	Plug	6*	Bare root or vernalized plug
Sedum spurium			8	Bare root or vernalized plug
Sempervivum	Late summer[1]	Bare root	7*	Bare root or plug
Silene maritima			11	Plug
Silene schafta			14	Vernalized plug
Solidago canadensis			12	Plug
Stachys byzantina			7*	Plug
Stipa barbata			7*	Plug
Stipa capillata			7*	Plug
Stipa pennata			7*	Plug
Stokesia laevis			7	Vernalized plug

(continued)

Table 10.9. Scheduling Flowering Perennials in 1 Qt. (1 L) Containers *(continued)*

Variety^	Commonly Potted the Year Before Expected Sales[1]	Previous Season's Source Materials[2]	Spring Potted Weeks to Finish^^	Spring Source Materials[2]
Teucrium chamaedrys			9*	Plug
Thymus citriodorus			7*	Vernalized plug
Thymus serpyllum			9	Vernalized plug
Thymus vulgaris			6*	Vernalized plug
Tiarella wherryi			5	Vernalized plug
Tradescantia andersoniana			5	Bare root
Verbena rigida			10	Plug
Veronica x *hybrida*			10	Vernalized plug
Veronica incana			8	Vernalized plug
Veronica longifolia			8	Vernalized plug
Veronica prostrata			8	Vernalized plug
Veronica repens	Early fall	Plug	5	Vernalized plug
Veronica spicata			7	Vernalized plug
Veronica spicata incana	Midsummer[1]	Plug	8	Vernalized plug
Veronica subsessilis			9	Vernalized plug
Veronica teucrium	Late summer	Plug	8	Vernalized plug
Vinca minor	Midsummer[1]	Plug	10*	Plug
Viola cornuta			7	Plug
Viola labradorica			7	Vernalized plug
Viola odorata			7	Vernalized plug
Viola sororaria			7	Vernalized plug
Viola tricolor			6	Plug
Viscaria alpina			6	Vernalized plug

^There are often numerous cultivars within each variety listed. Cultural needs and scheduled times can and do vary from cultivar to cultivar. The information provided here is intended to provide growers with a general idea of how to begin making crop schedules. Growers will need to understand the specific needs of each cultivar to better schedule their crops.

[1]The best quality characteristics are achieved when these crops are planted the specified season prior to the spring sales window. In many cases, these crops can also be started during the spring of the same season of expected sales.

[2]The size of the starter material varies greatly from grower to grower and has a significant result on the number of weeks necessary to produce the crop. Planting multiple plugs in each container may also reduce crop timing and improve the quality characteristics.

^^Spring potted weeks to finish also represents the approximate number of weeks heat must be provided to overwintered crops before they reach flowering. The WTF are estimated based on producing each crop at its recommended temperature. Actual WTF will vary with actual production temperatures and growing conditions.

*Crops grown at these schedules will generally not produce flowers or are sold for the characteristics of the foliage.

Table 10.10. Scheduling Flowering Perennials in 1 Gal. (3.8 L) Containers

The schedules shown below are developed to help growers estimate the approximate finishing times for producing various crops. Unless specified, these schedules are intended to produce flowering plants. Providing the proper treatments necessary for flowering is important to produce blooming plants using the information in this table. There are many variables that affect crop scheduling and performance including cultivar, type and size of the starting materials, production temperatures, photoperiod, vernalization, and production practices. Growers should use the information provided here as guidelines, not absolutes.

Variety^	Commonly Potted the Year Before Expected Sales[1]	Previous Season's Source Materials[2]	Spring Potted Weeks to Finish^^	Spring Source Materials[2]
Achillea filipendulina			8	Vernalized plug
Achillea x *hybrida*			8	Vernalized plug
Achillea x *hybrida* 'Moonshine'	Midsummer	Plug	7	Vernalized plug
Achillea millefolium			9	Vernalized plug
Achillea ptarmica			7	Plug
Achillea tomentosa	Midsummer	Plug	7	Vernalized plug
Aconitum x *napellus*	Late summer	Plug	9	Bare root or vernalized plug
Aegopodium podagraria			8*	Bare root
Agastache cana			8	Plug
Agastache rugosa			9	Plug
Ajuga pyramidalis	Late summer[1]	Plug	9	Bare root or vernalized plug
Ajuga reptans	Late summer[1]	Plug	9	Bare root or vernalized plug
Alcea rosea			8*	Vernalized plug
Alchemilla mollis	Late summer	Plug	7	Vernalized plug
Ampelopsis brevipedunculata	Midsummer[1]	Plug	13	Plug
Anaphalis margaritacea			12	Vernalized plug
Andropogon gerardii	Midsummer[1]	Large plug	11*	Vernalized plug
Anemone x *hybrida*	Late summer	Bare root or plug	13	Bare root or vernalized large plug
Anemone multifida	Late summer	Bare root or plug	11	Bare root or vernalized large plug
Anemone sylvestris	Late summer	Plug		
Anemone tomentosa	Late summer	Bare root or plug	10	Bare root or vernalized large plug
Anthemis tinctoria			7	Vernalized plug
Aquilegia alpina	Late summer[1]	Plug	8	Vernalized plug
Aquilegia caerulea	Late summer[1]	Plug	8	Vernalized plug
Aquilegia flabellata	Late summer[1]	Plug	6	Vernalized plug
Aquilegia x *hybrida*	Late summer[1]	Plug	8	Vernalized plug
Aquilegia vulgaris	Late summer[1]	Plug	8	Vernalized plug
Arabis blepharophylla			8	Vernalized plug
Arabis caucasica			5	Vernalized plug

(continued)

Table 10.10. Scheduling Flowering Perennials in 1 Gal. (3.8 L) Containers *(continued)*

Variety^	Commonly Potted the Year Before Expected Sales[1]	Previous Season's Source Materials[2]	Spring Potted Weeks to Finish^^	Spring Source Materials[2]
Arenaria montana			6	Vernalized plug
Armeria x *hybrida*	Late summer[1]	Plug	10	Vernalized plug
Armeria latifolia	Early fall[1]	Plug	9	Plug
Armeria maritima	Late summer[1]	Plug	8	Vernalized plug
Armeria pseudarmeria	Early fall[1]	Plug	9	Plug
Artemisia ludoviciana			8*	Plug
Artemisia schmidtiana			7*	Bare root
Artemisia x *hybrida*			6*	Plug
Aruncus dioicus	Midsummer	Bare root or plug	9	Bare root or vernalized large plug
Asclepias curassavica			10	Plug
Asclepias tuberosa			9	Plug
Aster alpinus	Midsummer	Plug	12	Vernalized plug
Aster dumosus			12	Plug
Aster x *frikartii*			11	Vernalized plug
Aster novae-angliae	Midsummer	Plug	12	Vernalized plug
Aster novi-belgii			12	Plug
Aster tongolensis	Late summer[1]	Plug	8	Vernalized plug
Astilbe arendsii	Midsummer	Bare root	9	Bare root
Astilbe chinensis	Midsummer	Bare root	9	Bare root
Astilbe japonica	Midsummer	Bare root	8	Bare root
Astilbe x *rosea*	Midsummer	Bare root	8	Bare root
Astilbe simplicifolia	Midsummer	Bare root	11	Bare root
Astilbe thunbergii	Midsummer	Bare root	8	Bare root
Astrantia major	Early summer[1]	Plug		
Athyrium filix-femina	Midsummer[1]	Plug	12*	Plug
Athyrium nipponicum	Midsummer[1]	Plug	15*	Plug
Aubrieta x *hybrida*			10	Vernalized plug
Aurinia saxatilis			6	Vernalized plug
Baptisia australis			16	Vernalized plug
Bellis perennis			8	Vernalized plug
Bergenia cordifolia	Midsummer[1]	Plug		
Boltonia asteroides	Midsummer	Plug	9*	Vernalized plug
Brunnera macrophylla	Late summer[1]	Bare root or large plug	9	Bare root or vernalized large plug
Buddleia davidii	Midsummer	Plug	7	Plug
Buddleia x *weyeriana*	Midsummer	Plug	7	Plug

(continued)

Table 10.10. Scheduling Flowering Perennials in 1 Gal. (3.8 L) Containers *(continued)*

Variety^	Commonly Potted the Year Before Expected Sales[1]	Previous Season's Source Materials[2]	Spring Potted Weeks to Finish^^	Spring Source Materials[2]
Calamagrostis acutiflora	Midsummer[1]	Large plug	9*	Vernalized large plug
Campanula carpatica	Midsummer	Plug	9	Plug
Campanula cochleariifolia			9	Plug
Campanula garganica	Midsummer	Plug	8	Vernalized plug
Campanula glomerata	Midsummer	Plug	7	Vernalized plug
Campanula medium			7	Vernalized plug
Campanula persicifolia			9	Vernalized plug
Campanula poscharskyana	Midsummer	Plug	7	Vernalized plug
Campanula punctata	Midsummer	Plug	7	Vernalized plug
Campanula rotundifolia			7	Plug
Carex dolichostachya	Midsummer[1]	Large plug	9*	Large plug
Carex hachijoensis	Midsummer[1]	Large plug	10*	Large plug
Carex morrowii	Midsummer[1]	Large plug	10*	Large plug
Carex phyllocephala	Midsummer[1]	Large plug	10*	Large plug
Carex siderosticha	Midsummer[1]	Large plug	11*	Large plug
Carlina acaulis	Midsummer	Plug	14	Vernalized plug
Caryopteris x *clandonensis*	Midsummer	Plug	7	Vernalized plug
Centaurea dealbata			8	Vernalized plug
Centaurea macrocephala			13	Vernalized plug
Centaurea montana	Midsummer	Plug	7	Vernalized plug
Centranthus ruber			8	Plug
Cerastium tomentosum	Late summer	Plug	7	Vernalized plug
Ceratostigma plumbaginoides			10	Plug
Chasmanthium latifolium	Midsummer[1]	Large plug	9*	Large plug
Chelone lyonii			12	Plug
Cimicifuga racemosa	Midsummer	Bare root	11	Bare root
Clematis x *hybrida*			15	Bare root or vernalized plug
Clematis viticella			14	Bare root or vernalized plug
Coreopsis auriculata	Late summer	Plug	7	Vernalized plug
Coreopsis grandiflora	Late summer	Plug	9	Vernalized plug
Coreopsis x *hybrida*			8	Vernalized plug
Coreopsis lanceolata	Late summer	Plug	8	Vernalized plug
Coreopsis rosea	Midsummer	Plug	9	Vernalized plug
Coreopsis verticillata			10	Bare root
Cortaderia selloana	Midsummer[1]	Large plug	9*	Large plug

(continued)

Table 10.10. Scheduling Flowering Perennials in 1 Gal. (3.8 L) Containers *(continued)*

Variety^	Commonly Potted the Year Before Expected Sales[1]	Previous Season's Source Materials[2]	Spring Potted Weeks to Finish^^	Spring Source Materials[2]
Corydalis x *hybrida*	Early fall[1]	Plug	9	Vernalized plug
Corydalis lutea			8	Plug
Delosperma cooperi	Late summer	Plug	8	Vernalized plug
Delosperma floribunda	Late summer	Plug	8	Vernalized plug
Delphinium belladonna			13	Plug
Delphinium cultorum			12	Plug
Delphinium elatum			13	Plug
Delphinium grandiflorum			10	Plug
Delphinium nudicaule			9	Plug
Delphinium zalil	Late summer	Plug	11	Vernalized plug
Deschampsia caespitosa	Midsummer[1]	Large plug	11*	Vernalized plug
Dianthus 'Allwoodii'	Late summer[1]	Plug	9	Vernalized plug
Dianthus barbatus	Early fall[1]	Plug	8	Vernalized plug
Dianthus caryophyllus	Early fall[1]	Plug	9	Vernalized plug
Dianthus deltoides	Late summer	Plug	8	Vernalized plug
Dianthus gratianopolitanus	Late summer[1]	Plug	8	Vernalized plug
Dianthus hybridus			8	Plug
Dianthus plumarius			9	Plug
Dianthus superbus			11	Plug
Dicentra eximia	Early summer	Bare root or large plug	8	Bare root or vernalized large plug
Dicentra formosa	Early summer	Bare root or large plug	8	Bare root or vernalized large plug
Dicentra spectabilis	Early summer[1]	Bare root	6	Bare root
Digitalis grandiflora			11	Vernalized plug
Digitalis mertonensis			9	Vernalized plug
Digitalis purpurea			9	Vernalized plug
Doronicum orientale	Early fall[1]	Plug	7	Vernalized plug
Dryopteris affinis	Midsummer[1]	Plug	12*	Large plug
Dryopteris atrata	Midsummer[1]	Plug	12*	Large plug
Dryopteris erythrosora	Midsummer[1]	Plug	12*	Large plug
Dryopteris filix-mas	Midsummer[1]	Plug	12*	Large plug
Dryopteris spinulosa	Midsummer[1]	Plug	12*	Large plug
Echinacea paradoxa	Late summer[1]	Plug	12	Vernalized plug
Echinacea purpurea	Late summer[1]	Plug	13	Vernalized plug
Echinops ritro			11	Vernalized plug
Epimedium x *cantabrigiense*	Midsummer[1]	Bare root		

(continued)

Table 10.10. Scheduling Flowering Perennials in 1 Gal. (3.8 L) Containers *(continued)*

Variety^	Commonly Potted the Year Before Expected Sales[1]	Previous Season's Source Materials[2]	Spring Potted Weeks to Finish^^	Spring Source Materials[2]
Epimedium pinnatum	Midsummer[1]	Bare root		
Epimedium x *rubrum*	Midsummer[1]	Bare root		
Epimedium x *versicolor*	Midsummer[1]	Bare root		
Erianthus ravennae	Midsummer[1]	Large plug	9*	Bare root or large plug
Erigeron glaucus	Late summer	Plug	10	Vernalized plug
Erigeron karvinskianus			9	Plug
Erigeron speciosus	Late summer	Plug	9	Vernalized plug
Erinus alpinus			20	Plug
Eryngium alpinum			14	Vernalized plug
Eryngium bourgatii			14	Vernalized plug
Eryngium planum			14	Vernalized plug
Eupatorium maculatum	Midsummer	Plug	9*	Plug
Euphorbia myrsinites	Late summer	Plug	7	Vernalized plug
Euphorbia polychroma	Late summer[1]	Bare root or plug	6	Bare root or large plug
Fallopia japonica	Midsummer	Plug	12*	Plug
Festuca amethystina	Midsummer[1]	Large plug	8*	Large plug
Festuca glauca	Midsummer[1]	Large plug	8*	Bare root or large plug
Festuca valesiaca	Midsummer[1]	Large plug	8*	Large plug
Filipendula vulgaris	Midsummer[1]	Large plug	6*	Large plug
Fragaria x *frel*	Late summer	Plug	7	Plug
Fuchsia magellanica	Late summer	Plug	8	Plug
Gaillardia aristata			8	Vernalized plug
Gaillardia grandiflora			8	Vernalized plug
Galium odoratum	Midsummer	Plug	9	Vernalized plug
Gaura lindheimeri	Midsummer	Plug	9	Plug
Geranium x *cantabrigiense*	Late summer	Plug	9	Plug
Geranium cinereum	Late summer	Plug	7	Plug
Geranium x *hybridum*	Late summer	Plug	8	Plug
Geranium macrorrhizum	Late summer	Plug	6	Plug
Geranium x *oxonianum*	Late summer	Plug	8	Plug
Geranium phaeum	Late summer	Plug	7	Plug
Geranium pratense	Late summer	Plug	8	Plug
Geranium sanguineum	Late summer	Plug	7	Plug
Geum chiloense	Late summer	Plug	9	Vernalized plug
Geum coccineum			9	Vernalized plug

(continued)

Table 10.10. Scheduling Flowering Perennials in 1 Gal. (3.8 L) Containers *(continued)*

Variety^	Commonly Potted the Year Before Expected Sales[1]	Previous Season's Source Materials[2]	Spring Potted Weeks to Finish^^	Spring Source Materials[2]
Geum hybridum	Late summer	Plug	9	Vernalized plug
Gypsophila elegans	Late summer[1]	Plug	11	Vernalized plug
Gypsophila pacifica	Midsummer[1]	Plug		
Gypsophila paniculata	Late summer[1]	Plug	13	Vernalized plug
Gypsophila repens	Late summer[1]	Plug	8	Vernalized plug
Hakonechloa macra	Midsummer[1]	Large plug	11*	Vernalized plug
Helenium autumnale			12	Vernalized plug
Helichrysum thianshanicum	Late summer	Plug	10	Vernalized plug
Helictotrichon sempervirens	Midsummer[1]	Large plug	9*	Vernalized large plug
Heliopsis helianthoides	Midsummer	Plug	10	Vernalized plug
Heliopsis scabra			9	Plug
Helleborus x *hybridus*	Early summer[1]	Large plug		
Helleborus niger	Early summer[1]	Large plug		
Helleborus orientalis	Early summer[1]	Large plug		
Hemerocallis	Midsummer[1]	Bare root or large plug	10	Bare root or large plug
Hemerocallis, tetraploid	Midsummer[1]	Bare root or large plug	9	Bare root or large plug
Heuchera americana	Late summer	Plug	9	Vernalized large plug
Heuchera x *hybrida*	Late summer	Plug	9	Vernalized large plug
Heuchera micrantha	Late summer	Plug	8	Vernalized large plug
Heuchera sanguinea	Late summer	Plug	9	Vernalized large plug
Heucherella x *hybrida*	Late summer	Plug	7	Vernalized large plug
Hibiscus x *hybrida*			13	Large plug
Hibiscus moscheutos			13	Large plug
Hosta	Early summer[1]	Bare root or large plug	16*	Bare root or large plug
Hosta fortunei	Early summer[1]	Bare root or large plug	17*	Bare root or large plug
Hosta montana	Early summer[1]	Bare root or large plug	17*	Bare root or large plug
Hosta plantaginea	Early summer[1]	Bare root or large plug	15*	Bare root or large plug
Hosta sieboldiana	Early summer[1]	Bare root or large plug	17*	Bare root or large plug
Hosta tardiana	Early summer[1]	Bare root or large plug	16*	Bare root or large plug
Hosta tokudama	Early summer[1]	Bare root or large plug	17*	Bare root or large plug
Hosta undulata	Early summer[1]	Bare root or large plug	17*	Bare root or large plug
Hosta ventricosa	Early summer[1]	Bare root or large plug	15*	Bare root or large plug
Houttuynia cordata			7*	Plug
Hypericum calycinum	Midsummer	Plug	10	Vernalized plug
Hypericum x *moserianum*	Early summer[1]	Plug	10*	Large plug

(continued)

Table 10.10. Scheduling Flowering Perennials in 1 Gal. (3.8 L) Containers *(continued)*

Variety^	Commonly Potted the Year Before Expected Sales[1]	Previous Season's Source Materials[2]	Spring Potted Weeks to Finish^^	Spring Source Materials[2]
Hypericum polyphyllum			11	Plug
Iberis sempervirens	Late summer[1]	Plug		
Imperata cylindrica	Late summer[1]	Large plug	9*	Large plug
Incarvillea delavayi	Early summer	Plug		
Inula ensifolia	Midsummer	Plug	10	Vernalized plug
Inula orientalis	Midsummer	Plug	10	Vernalized plug
Iris, bearded intermediate	Late summer[1]	Bare root		
Iris, bearded tall	Late summer[1]	Bare root		
Iris ensata	Late summer[1]	Bare root	11	Bare root
Iris pallida	Late summer[1]	Bare root	8	Bare root
Iris pseudacorus	Late summer[1]	Bare root	8	Bare root
Iris x *siberica*	Late summer[1]	Bare root	8	Bare root
Jasione laevis	Late summer	Plug	9	Vernalized plug
Juncus effusus	Midsummer[1]	Large plug	11*	Large plug
Knautia macedonica	Midsummer	Plug	8	Vernalized plug
Kniphofia uvaria	Midsummer[1]	Plug	11	Vernalized plug
Koeleria glauca	Midsummer[1]	Large plug	8*	Large plug
Lamiastrum galeobdolon	Late summer[1]	Plug	9	Vernalized plug
Lamium maculatum	Late summer	Plug	7	Plug
Lathyrus latifolius	Midsummer[1]	Plug	9	Vernalized plug
Lavandula angustifolia	Midsummer[1]	Plug	9	Vernalized plug
Lavandula x *intermedia*	Midsummer[1]	Plug	11	Vernalized plug
Lavandula stoechas	Midsummer[1]	Plug	10	Vernalized plug
Lavandula vera	Midsummer	Plug	8	Vernalized plug
Leucanthemum superbum	Midsummer	Plug	10	Vernalized plug
Leucanthemum vulgare	Late summer	Plug	6	Vernalized plug
Liatris spicata	Midsummer[1]	Bare root or large plug	9	Bare root
Ligularia dentata	Midsummer[1]	Bare root	7*	Bare root
Lilium, Asiatic			7–11	Bulb
Lilium, Oriental			9–13	Bulb
Limonium latifolium	Late summer	Plug		
Limonium perezii	Late summer	Plug		
Limonium tataricum	Late summer	Plug		
Linum perenne	Midsummer[1]	Plug	7	Vernalized plug
Liriope spicata	Midsummer[1]	Bare root or large plug	10*	Bare root or large plug

(continued)

Table 10.10. Scheduling Flowering Perennials in 1 Gal. (3.8 L) Containers *(continued)*

Variety^	Commonly Potted the Year Before Expected Sales[1]	Previous Season's Source Materials[2]	Spring Potted Weeks to Finish^^	Spring Source Materials[2]
Lobelia x *hybrida*			9	Plug
Lobelia speciosa			8	Plug
Lupinus x *hybrida*	Midsummer	Plug	11	Vernalized plug
Lupinus polyphyllus	Midsummer	Plug	10	Vernalized plug
Luzula nivea	Midsummer[1]	Large plug	9*	Large plug
Lychnis chalcedonica			9	Vernalized plug
Lysimachia clethroides	Midsummer	Plug	10*	Plug
Lysimachia nummularia	Midsummer[1]	Plug	6*	Plug
Lysimachia punctata	Midsummer	Plug	8	Plug
Lythrum virgatum	Midsummer	Plug	9	Vernalized plug
Malva sylvestris	Midsummer	Plug	7	Vernalized plug
Matteuccia struthiopteris	Midsummer[1]	Plug	10*	Large plug
Melica transsilvanica	Midsummer[1]	Large plug	9*	Large plug
Miscanthus sinensis	Midsummer[1]	Large plug	9*	Large plug
Monarda didyma	Midsummer	Plug	10	Vernalized plug
Monarda x *hybrida*	Midsummer	Plug	11	Vernalized plug
Myosotis sylvatica	Late summer[1]	Plug	8	Vernalized plug
Nassella tenuissima	Midsummer[1]	Large plug	9*	Large plug
Nepeta faassenii	Late summer	Plug	6	Vernalized plug
Oenothera missouriensis			10	Plug
Oenothera speciosa			6	Vernalized plug
Ophiopogon planiscapus	Midsummer[1]	Large plug	11*	Large plug
Osmunda cinnamonea	Midsummer[1]	Plug	12*	Large plug
Osmunda regalis	Midsummer[1]	Plug	12*	Large plug
Panicum virgatum	Midsummer[1]	Large plug	8*	Large plug
Papaver alpinum	Late summer[1]	Plug	9	Vernalized plug
Papaver miyabeanum	Late summer[1]	Plug	9	Vernalized plug
Papaver nudicaule	Late summer[1]	Plug	9	Vernalized plug
Papaver orientale	Late summer[1]	Plug	9	Vernalized plug
Parthenocissus tricuspidata	Midsummer[1]	Large plug	16*	Vernalized plug
Pennisetum alopecuroides	Midsummer[1]	Large plug	8*	Large plug
Pennisetum orientale	Midsummer[1]	Large plug	8*	Large plug
Pennisetum setaceum			11*	Large plug
Penstemon barbatus	Midsummer	Plug	10	Vernalized plug
Penstemon digitalis	Midsummer	Plug	8	Vernalized plug

(continued)

Table 10.10. Scheduling Flowering Perennials in 1 Gal. (3.8 L) Containers *(continued)*

Variety^	Commonly Potted the Year Before Expected Sales[1]	Previous Season's Source Materials[2]	Spring Potted Weeks to Finish^^	Spring Source Materials[2]
Penstemon x *hybrida*	Midsummer	Plug	8	Vernalized plug
Perovskia atriplicifolia	Midsummer[1]	Plug	9*	Large plug
Persicaria affinis			7	Plug
Persicaria virginiana			6*	Plug
Phalaris arundinacea	Midsummer	Large plug	6*	Large plug
Phlox arendsii	Midsummer	Plug	11	Vernalized plug
Phlox carolina	Midsummer	Plug	11	Vernalized plug
Phlox divaricata	Midsummer[1]	Plug	10	Vernalized plug
Phlox maculata	Midsummer	Plug	11	Vernalized plug
Phlox paniculata	Midsummer	Plug	11	Vernalized plug
Phlox subulata	Late summer[1]	Bare root or large plug		
Physalis alkekengi	Late summer	Plug	8	Vernalized plug
Physostegia virginiana	Midsummer	Plug	9	Vernalized plug
Platycodon grandiflorus	Midsummer	Plug	10	Plug
Polemonium boreale	Late summer	Plug	8	Vernalized plug
Polemonium caeruleum	Late summer	Plug	6	Vernalized plug
Polemonium yezoense	Late summer	Plug	7	Vernalized plug
Polygonum aubertii	Midsummer[1]	Plug	6*	Plug
Polystichum acrostichoides	Midsummer[1]	Plug	12*	Plug
Polystichum polyblepharum	Midsummer[1]	Plug	12*	Plug
Primula denticulata	Late summer[1]	Plug		
Primula elatior			14	Plug
Primula polyanthus	Late summer[1]	Plug	11	Vernalized plug
Primula vialii	Late summer[1]	Plug		
Pulmonaria x *hybrida*	Late summer[1]	Plug	8	Vernalized plug
Pulmonaria saccharata	Late summer[1]	Plug	8	Vernalized plug
Pulsatilla vulgaris	Midsummer[1]	Plug	9	Vernalized plug
Ranunculus repens			10	Vernalized plug
Rodgersia aesculifolia	Midsummer[1]	Bare root	7*	Bare root
Rosa chinensis	Early fall	Plug	8	Plug
Rosa, miniature			9–11	Plug
Rudbeckia fulgida	Midsummer[1]	Plug	13	Vernalized plug
Rudbeckia hirta			8	Plug
Rudbeckia speciosa	Late summer	Plug	12	Vernalized plug
Rumex sanguineus			7*	Plug

(continued)

Table 10.10. Scheduling Flowering Perennials in 1 Gal. (3.8 L) Containers *(continued)*

Variety^	Commonly Potted the Year Before Expected Sales[1]	Previous Season's Source Materials[2]	Spring Potted Weeks to Finish^^	Spring Source Materials[2]
Salvia lyrata	Midsummer	Plug	7	Plug
Salvia nemorosa	Late summer	Plug	8	Bare root or vernalized plug
Salvia officinalis	Late summer	Plug	9	Vernalized plug
Salvia superba			10	Plug
Salvia x *sylvestris*	Late summer	Plug	7	Bare root or vernalized plug
Salvia verticillata	Late summer	Plug	7	Vernalized plug
Saponaria lempergii	Midsummer[1]	Plug	10	Plug
Saponaria ocymoides	Late summer	Plug	8	Vernalized plug
Scabiosa caucasica	Late summer[1]	Plug	10	Vernalized plug
Scabiosa columbaria	Early fall[1]	Plug	9	Vernalized plug
Sedum cauticolum	Late summer	Plug	9*	Vernalized plug
Sedum x *hybrida*	Late summer	Bare root or plug	10	Bare root or vernalized plug
Sedum kamtschaticum	Late summer	Plug	6	Vernalized plug
Sedum reflexum	Midsummer	Plug	11	Vernalized plug
Sedum sieboldii	Midsummer	Plug	7*	Vernalized plug
Sedum spectabile	Midsummer	Bare root or plug	12	Bare root
Sedum spurium	Late summer	Bare root or plug	10	Bare root or vernalized plug
Sempervivum	Late summer[1]	Bare root or plug	10*	Bare root or plug
Sidalcea hybrida			11	Plug
Silene maritima			12	Plug
Silene schafta	Late summer	Plug	15	Vernalized plug
Solidago canadensis			14	Plug
Solidago sphacelata			9	Vernalized plug
Stachys byzantina			8*	Plug
Stipa barbata	Midsummer[1]	Large plug	9*	Large plug
Stipa capillata	Midsummer[1]	Large plug	9*	Large plug
Stipa pennata	Midsummer[1]	Large plug	9*	Large plug
Stokesia laevis	Midsummer	Plug	9	Vernalized plug
Tanacetum coccineum (herb)			9	Vernalized plug
Tellima grandiflora			6	Vernalized plug
Thalictrum aquilegiifolium	Late summer[1]	Plug	10	Vernalized plug
Thymus citriodorus	Midsummer	Bare root or plug	12*	Bare root or vernalized plug
Thymus vulgaris	Midsummer	Bare root or plug	12*	Bare root or vernalized plug
Tiarella x *hybrida*	Midsummer[1]	Plug	7	Vernalized large plug
Tiarella wherryi	Midsummer[1]	Plug		

(continued)

Table 10.10. Scheduling Flowering Perennials in 1 Gal. (3.8 L) Containers *(continued)*

Variety^	Commonly Potted the Year Before Expected Sales[1]	Previous Season's Source Materials[2]	Spring Potted Weeks to Finish^^	Spring Source Materials[2]
Tradescantia andersoniana	Midsummer	Bare root	6	Bare root
Trollius chinensis	Late summer	Plug	9	Vernalized plug
Trollius x *cultorum*	Midsummer[1]	Plug	9	Vernalized plug
Verbascum bonariensis			9	Plug
Verbascum x *hybrida*			12	Plug
Verbascum phoeniceum			10	Plug
Verbena rigida			11	Plug
Veronica x *hybrida*	Midsummer	Plug	11	Vernalized plug
Veronica incana	Early fall	Plug	10	Vernalized plug
Veronica longifolia	Midsummer	Plug	9	Vernalized plug
Veronica prostrata	Midsummer	Plug	9	Vernalized plug
Veronica spicata	Midsummer	Plug	8	Vernalized plug
Veronica spicata incana	Midsummer[1]	Plug	8	Vernalized plug
Veronica subsessilis	Midsummer	Plug	9	Vernalized plug
Veronica teucrium	Late summer	Plug	10	Vernalized plug
Vinca minor	Midsummer	Plug	10*	Plug
Viola cornuta	Early fall	Plug	8	Plug
Viola labradorica	Midsummer[1]	Plug	9	Vernalized plug
Viola odorata	Midsummer[1]	Plug		
Viola sororaria	Midsummer[1]	Plug	9	Vernalized plug
Viola tricolor	Early fall	Plug	8	Plug
Viscaria alpina	Midsummer	Plug	6	Vernalized plug
Waldsteinia ternata	Midsummer[1]	Plug	14	Vernalized plug

^There are often numerous cultivars within each variety listed. Cultural needs and scheduled times can and do vary from cultivar to cultivar. The information provided here is intended to provide growers with a general idea of how to begin making crop schedules. Growers will need to understand the specific needs of each cultivar to better schedule their crops.

[1]The best quality characteristics are achieved when these crops are planted the specified season prior to the spring sales window. In many cases, these crops can also be started during the spring of the same season of expected sales.

[2]The size of the starter material varies greatly from grower to grower and has a significant result on the number of weeks necessary to produce the crop. Planting multiple plugs in each container may also reduce crop timing and improve the quality characteristics.

^^Spring potted weeks to finish also represents the approximate number of weeks heat must be provided to overwintered crops before they reach flowering. The WTF are estimated based on producing each crop at its recommended temperature. Actual WTF will vary with actual production temperatures and growing conditions.

*Crops grown at these schedules will generally not produce flowers or are sold for the characteristics of the foliage.

11

Overwintering Perennials

Many commercial perennial producers are faced with the challenge of carrying much of their containerized perennial inventory throughout the winter months, until growth resumes in the early spring. Container-grown perennials often require winter protection to survive low temperatures and temperature fluctuations typical of many regions of the United States and Canada.

There are many reasons growers consider and are willing to take the additional risk involved during the overwintering process. Due to long production times, many crops benefit from being produced the summer before they reach the market. Many growers overwinter perennials to satisfy the vernalization requirement necessary for many perennials to flower. Finally, many growers overwinter perennials for logistical reasons; in many cases, it is not very practical to plant all of the perennials required for spring sales during the late winter or early spring months.

Figure 11.1. **Growers overwintering perennials must provide adequate protection from cold, whether they are being overwintered in enclosed structures or at outdoor locations.**

The primary goal of the overwintering process is to protect the root system from cold temperatures and dramatic temperature fluctuations. The root system of a containerized plant does not receive the same buffering capacity of a plant that is growing in the ground. Without additional protection, the only insulation the root zone a containerized plant has is the thin wall of the plastic container it sits in. The temperature of the medium, when plants are grown aboveground, can be much colder than root zones underground. In fact, the soil temperature in containers is often comparable to the outside air temperature. For this reason, it is important to take measures to protect the root zones of containerized plants from cold temperatures.

Successful overwintering involves keeping the plants cold and alive but not actively growing. The amount of winter protection growers provide should be based on the root hardiness of each plant species, or the lowest temperature the roots will survive at. Root systems of many perennial varieties will become damaged at extended exposure to temperatures less than 20° F (-7° C). Short term exposure (a few days) to these or lower temperatures will not usually cause injury to the roots.

Root and crown hardiness for each perennial species is not clearly defined by researchers as of yet, which leaves many growers uncertain about the amount of protection to provide or which technique they should implement. Perennial producers should consider what the lowest expected minimum temperatures for their particular location are likely to be. As a guideline, growers use the USDA hardiness zones as an indication to the amount of cold a plant can tolerate. For example, a Zone 5 perennial can tolerate more cold than one from Zone 6, which implies that the Zone 6 plant would require more winter protection than the Zone 5 plant. Keep in mind that even though the USDA hardiness zones are derived from over forty years of climactic information, temperature extremes in any given year can range widely and may put your plants at risk.

Using the USDA hardiness zones does not guarantee that a particular plant will survive within the zone it is listed in. An example of this is a Zone 3 plant that might be overwintered successfully under heavy amounts of snowfall, but this plant may not successfully be overwintered if there is no significant snow cover or other means of protection. Regardless of a plant's anticipated hardiness, it is only capable of withstanding cold temperatures if is has been properly acclimated to those temperatures and provided adequate winter protection.

Occasionally growers experience unseasonably warm falls and early winters. These conditions do not allow a plant to become acclimated sufficiently enough to withstand the cold temperatures normal for the zone. When the temperatures drop dramatically, there is an increased possibility of injury occurring to the crop. If the temperatures fall gradually over time, then the likelihood of injury to the plant becomes less, as the plant is better acclimated to cold temperatures.

Winter hardiness is predetermined in the genetic make up of the plant and is influenced by several cultural factors, such as temperature, moisture, and light. Other factors such as plant stresses, diseases, and rodents can contribute to the success or failure of growers to overwinter their perennials. Plant stresses usually entail situations where plants are exposed to extreme conditions, such as drought, excessive moisture, or high nutrient levels, that affect the plant's ability to acclimate to colder weather conditions.

As we will discuss, there are a number of different methods commonly used to ensure plant survival over the winter months. Some experimentation may be necessary to determine which method works best for your specific production scenarios. The type of overwintering system used is first determined by the plant species and its ability to withstand cold temperatures, then by the type of production facilities available to the grower.

Overwintering Terminology

Before we explore the various types of overwintering systems, it is helpful to gain a better understanding of some of the terminology and physiological changes that occur during this process.

Hardiness

As mentioned earlier, cold hardiness refers to a particular perennial's ability to withstand cold temperatures. Hardiness is not constant and changes throughout the winter months as the plant becomes acclimated or de-acclimated to the current environmental conditions.

Acclimation

Acclimation, often called hardening off, is a process plants go through that allows them to tolerate cold temperatures. This process usually involves several weeks of exposure to gradually decreasing average daily temperatures. The temperatures during the acclimation process will determine a plant's specific ability to tolerate cold and may vary by location. For example, *Echinacea* acclimated in Minnesota will be able to withstand colder temperatures than will *Echinacea* produced and acclimated in Kentucky. Without the proper acclimation, plant injury and mortality will most likely occur.

Dormancy

Dormancy is a process plants undergo in response to an increasing length of darkness and exposure to colder temperatures. As a perennial becomes fully acclimated, it becomes dormant. The best method to acclimate plants toward reaching the dormant state is to subject them to short days followed by cool temperatures, then finally to temperatures just below freezing. After a short time, the cold temperatures are more important during the acclimation process than are short days.

Depending on average temperature and the point in time of the overwintering period, a plant can actually become de-acclimated, or more sensitive to cold temperatures and slip out of dormancy. As the temperatures increase during the late winter, plants begin to lose their ability to withstand cold temperatures. Even short periods—one or two days—of unusually warm temperatures can fool the plant into becoming de-acclimated. Therefore, to maintain perennials in the dormant state, it is important to maintain cold temperatures averaging less than 40° F (4° C). Once de-acclimation starts, plant injury or even death may occur following warm spells if sudden severely cold temperatures return. Most plant injury occurs either early or late in the overwintering cycle when plants are not fully acclimated to cold temperatures, or they have become de-acclimated during late winter warm spells.

Various plant parts become acclimated to different extents, allowing for the roots and stems of the plant to have differing amounts of tolerance to cold temperatures. In most cases, the stems of a plant can tolerate more cold than the root system can. Some plants may have hardy shoots for a given zone, but less hardy root systems that may not survive unless they are well protected. There are even differences between the hardiness of mature roots and young roots. For example, mature roots generally can tolerate temperatures 5–10° F (3–6° C) colder than young roots can. In most cases, the young roots become injured while the mature roots remain intact. Any root injury, to young or mature roots, affects the crop's development the following spring; plants often grow at a slower rate and are more susceptible to root rot pathogens.

Plants will gradually come out of dormancy as average temperatures increase and after their cold requirement has been met. Most perennials require a minimum of six weeks of cold to satisfy the vernalization requirement necessary for flowering. Fruit producers are accustomed to using a method called cold chilling unit accumulation to quantify how much cold their orchards have received. They essentially track the number of units of cold received during the winter months to establish when their orchards are likely to break dormancy and begin to grow. There has not yet been specific research conducted on a wide enough variety of perennial species for growers to effectively use this method.

The release from dormancy is a function of temperature, not photoperiod. Most perennials do not have any leaves present in the dormant form and are unable to perceive day length. It is even possible for plants to break dormancy when they have not received the proper amount of cold, such as after a very mild winter. When perennials do not receive an adequate cold period, they will not emerge or grow uniformly, will often have less blooms, and the flowering will be more sporadic.

Winter injury

There are various types of winter injury that commercial growers often observe. The most common types of injury include early frost injury, late frost injury, winter desiccation, and animal damage.

Early frost injury often occurs at the beginning of winter when perennials that have not been fully acclimated are exposed to temperatures below freezing. Late frost injury is likely to occur toward the end of the overwintering cycle when plants exposed to warm temperatures have begun to flush new growth and then freezing temperatures return.

Winter desiccation occurs when there is not enough water available for uptake. Desiccation can also occur as the result of exposure to both low temperatures and drying winds. Many evergreen perennials, such as *Heuchera,* still transpire or lose water through their leaves during the winter months. When the medium in the pot is frozen, the roots cannot take up water, causing plant cells to become desiccated, often resulting in plant injury.

Freeze injury occurs either by severe dehydration of plant cells or intracellular ice formation. In the previous example, transpiration is still going to occur with the *Heuchera*. However, instead of transpiring water, which normally is pulled up through the root system, the plant satisfies the evaporative demand by removing water from its plant cells. As water is drawn out of plant cells, the solute concentration (essentially sugars and salts) inside them increases, which reduces the cells' ability to withstand freezing temperatures. With intracellular ice formation, the free water located between the cells freezes as the temperatures drop below freezing. This formation of ice crystals causes further water to be pulled from within plant cells into the intracellular spaces, causing them to become dehydrated and more prone to freezing, as described above. The extent of freeze injury varies from small localized areas, such as growing points, to severe cases that may kill the entire plant.

Overwintering is one of those processes that we need to understand and watch carefully. There are a few things we can and should do to ensure plant survival, but in our efforts to ensure survival, it is quite easy to create the opposite effect- reducing cold hardiness and increasing the likelihood of plant injury. Producing perennials that are naturally hardy in your region and managing the environment around the root zone, specifically moisture, salts, and temperatures are critical to overwintering success.

Preparations for Winter

To successfully overwinter perennials, it is critical to have well-established plants before entering the cold winter months. Non-established plants will overwinter very poorly. Many growers start their perennials during the late summer or early fall for next spring's sales. On average, for most varieties, I would recommend allowing a minimum of six weeks before the first killing frost to allow adequate time for the root systems to become established. The better developed the root system, the greater the chance the plant will survive the winter.

To prevent the likelihood of harboring plant pathogens over the winter, which could become problematic during the next production cycle, it is highly beneficial to remove most of the existing foliage prior to the over-

Figure 11.2. **To prevent plant pests from overwintering, it is best to remove the foliage by trimming or mowing and removing the plant debris from the production site.**

wintering process. The most common practice is to remove any existing foliage of herbaceous and semi-evergreen varieties by cutting the plants back to 2–4 in. (5–10 cm) above to top of the container. Evergreen plants should only be trimmed slightly or not cut back at all. The perennials referred to as sub-shrubs are either not cut back at all or are cut back to 10–12 in. (25–30 cm) above the top of the container.

Foliage can be trimmed back using a variety of cutting tools, including hand shears, sickle bars, and mowing machines. After the plants are trimmed, the foliage that has been removed should be collected and moved away from the production site. This exercise should occur during the late fall, after autumn sales have subsided and before major cold and snow arrive.

The winter preparation involves cleaning the area of dirt and plant debris by sweeping or blowing it out of the production area and removing any existing weeds from the facilities. The production site should be cleaned and prepared before plants are placed in them. Quite often growers will cut plants from one location and move them to an adjacent location that has been prepared for the winter. This lateral move, from one site to the site right next to it, allows all of the overwintering areas to be properly prepared while keeping the cost of labor reasonable.

While preparing perennials for the winter months, it is a good time to move them, if necessary, to the locations where they are needed for spring forcing. For example, if a perennial is currently in an outdoor location and it needs to be placed in a heated location in order to produce it for an early spring ship date, now may be the most economical time to place the plant in the proper location. Moving plants before the plastic is placed on the structures will allow plants to be moved directly from their current location onto a wagon or other means of transportation. Otherwise moving them midway through the overwintering phase will require more labor, as protective covers may have to be temporarily removed.

Another consideration during winter preparation is to consolidate groups of the same varieties from multiple locations to one or two locations for the convenience of inventory management and efficiency during the shipping season. When shipping, it is very time consuming and costly to travel to multiple locations to find the same variety of perennial as opposed to a single location to fill the entire order. Fall is an excellent if not the best time to seriously look at the needs for next season, consolidate your inventory, or to move plants to the proper environments for spring forcing.

Many growers apply preemergent herbicides to the ground mats or growing surfaces to prevent the germination of weed seeds between late fall and the time of spring sales. If applying herbicides to areas that are going to be covered with plastic, do not use herbicides that may volatilize within the structure once the plastic is applied. Always refer to each herbicide's label to check for its potential to volatilize or cause injury to crops and for application guidelines. In addition to applying herbicides to the ground mats, several growers apply copper sulfate crystals to help control any algae that is present. After the herbicide and copper sulfate applications, I would recommend applying a small amount of irrigation, such as five minutes of overhead sprinklers, to transfer the chemicals into the production surface (ground, ground mat, or gravel). It is also important to wait at least ten days before the area where herbicides are applied is covered with plastic. This reduces the likelihood of volatilization and keeps the use of these products consistent with their labels, since many labels do not allow for applications within enclosed structures. Controlling the weeds on the production surfaces is a powerful method of controlling the weeds that will most likely end up on the crop.

After the plants are trimmed, most growers consolidate them to a pot tight configuration. By grouping them into blocks, they are providing some protection from freezing and thawing, preventing potential injury to the root systems. By configuring the pots in this manner, the amount of air movement between the pots is greatly reduced, which also reduces the amount of heat lost from within the root zone.

Figure 11.3. **To provide more protection from cold, it is beneficial to consolidate the containers into groups in a pot-tight configuration.**

If overwintering occurs within structures, the blocks of plants are usually placed toward the center of the house, leaving an area between the plants and the outside wall. The outside walls could be several degrees colder than the center of the house, and placing of the plants away from the outside walls provides additional protection and reduces potential plant losses. Cold-sensitive varieties should be placed in the center of the blocks and the cold-tolerant varieties around the edges. The size of the blocks within the structures is often determined by the size of the covering material, if any, used. Overwintering coverings and thermal blankets are discussed later in this chapter.

Growers should also consider placing the same number of plants within the structure as the structure will hold when the plants are spaced out in the spring. If a structure will hold five thousand pots when the plants are spaced, then five thousand pots should be set into the structure in the fall. An exception would be if spring planting into that facility is anticipated, then fewer pots overwintering inside the structure is acceptable. If too many containers are placed within a house, it may be necessary to move some out when spacing pots during the early spring. Overfilling the houses is only acceptable when the overwintering facilities are limited and the grower must utilize the facilities available to them; the costs associated with moving the excessive plant materials are then justified.

After the perennials are cut back and placed in their final destinations, it is recommended to apply a preventative fungicide drench. Soilborne diseases are often overlooked, and their presence in the fall will increase the likelihood of winter mortality. This application usually entails drenching fungicides that are effective on root rot diseases such as *Pythium*, *Phytophthora*, and *Rhizoctonia*. It may also be beneficial to include a fungicide that controls *Botrytis* or other fungal diseases that may attack the freshly cut stems of the perennials. This application is largely preventative in nature and is intended to help reduce the likelihood of root rot diseases during the overwintering cycle by reducing their presence going into the cold period and improving the overall health of the root system. A good combination drench that I have had successful results with is combining mefenoxam (Subdue Maxx) with a thiophanate-methyl product such as Cleary's 3336WP. The drench is usually applied within three days of mowing and well before the root zone becomes frozen.

Perennials should be fully acclimated before being moved into enclosed structures or having any protective covers applied. Any preemergent herbicides should be applied to the containers several weeks before placing them into enclosed structures or before covering with plastic.

Acclimation

Acclimating perennials to cold temperatures involves providing several external conditions simultaneously. The first consideration should be the plant's ability to become acclimated; not all perennials will acclimate successfully to below-freezing temperatures. Acclimation entails an interaction between day length (photoperiod) and the temperature; these combined with other cultural factors, such as managing fertility and irrigation, all affect how a plant is acclimated and its overall ability to withstand cold temperatures.

Light

In addition to the photoperiod, light, or more specifically, the light intensity, affects a plant's ability to become acclimated. Plants grown in shady conditions, such as under shade cloth, acclimate at a slower rate than those grown under higher light intensities. In warm climates where plants are intentionally grown under shade, it is important to leave the shade on during active growth while the light levels are high and remove the shading materials as the conditions become more favorable for acclimation.

Fertility

Fertilization is an often overlooked or misunderstood aspect of overwintering perennials. Generally speaking, it is important to have healthy, well-fertilized plants going into the winter. But what does well-fertilized mean? To different growers it could have various meanings. Basically, we want to have adequate nutrients within the plant going into the overwintering cycle to reduce the likelihood of any nutrient stress problems. The concept is that if they go into the winter healthy, they come out healthy. During the late summer and fall (late August to late September), growers should only apply enough nutrients to sustain normal growth and development. Heavily fertilized plants are often rapidly growing and may not acclimate to the cold temperatures as well as perennials that receive only nominal amounts of nutrients. As a result, they may not be able to withstand winter freezes and some injury could result.

Try to match the available nutrients to the current growth rate of the plant. As the fall progresses, cooler temperatures and shorter day lengths result in slower rates of plant development. It is a common practice to reduce the fertilizer rates by half the rates usually applied during the growing season. Many growers have adopted the practice of not applying fertilizer after September 15, which I do not recommend. Growers should not stop feeding plants altogether since perennials with nutrient deficiencies will be more prone to overwintering injury.

Many growers deliver nutrients to their perennial crops by incorporating or top-dressing controlled-release fertilizers (CRFs). CRFs can be a very effective means of delivering nutrients to perennial crops. The release of nutrients from all types of CRFs is a function of temperature. As the temperatures cool down in the fall, the rate of nutrients released slows down. Some of the CRF technologies will completely stop releasing at low temperatures (below 45° F [7° C]). Since the release of nutrients and the growth of plants are both functions of temperature, growers can feel comfortable that the release of nutrients from CRFs will mimic plant growth as the temperature increases or decreases.

Growers need to monitor their crops, using soil tests or pour-through testing to indicate the availability of nutrients at various points during the fall. If excessive nutrients are available, growers may need to consider leaching them from the potting substrates by applying a large volume of water to the medium. Conversely, if the nutrient levels are low, the perennials are still actively growing, and there are still a few weeks before they go dormant, growers need to consider fertilizing with a low rate, such as 50 ppm nitrate, of water-soluble fertilizer. I would recommend growers maintain EC levels at or below 0.7 during the overwintering process.

Irrigation

During acclimation, the amount of water perennials receive could affect their overall cold hardiness. Mismanagement of the irrigation practices can greatly reduce a crop's ability to withstand cold temperatures and affects the survival rate during the winter months.

When perennials are grown on the dry side slightly before and during the acclimation process, they can be induced into dormancy quicker than when an adequate water supply is available to the roots. Slight water stress prior to overwintering can increase the cold hardiness of many perennials. Slight water stress means that you should let them dry out slightly more than normal between irrigations. When irrigation is necessary, water them completely. The last watering of the season, just prior to covering, should be a good thorough irrigation to provide adequate moisture and humidity levels under the protective coverings, thus preventing freeze-drying of the soil.

Care should be taken to not provide drought-like conditions at this stage, as cold injury is likely to occur. Perennials produced under overly dry conditions do not become acclimated as well and do not reach the same level of cold hardiness as plants grown under adequate moisture regimes do. When grown too dry, many perennials do not store adequate amounts of carbohydrates in the root system, making them more susceptible to cold injury, and they often have poor shoot development in the following spring. Keep in mind that much of the freeze injury that occurs over the winter is some form of desiccation or drought injury (intercellular ice formation and cell dehydration).

Perennials that undergo complete dormancy (no visible vegetative parts) usually do not require irrigation until the pots thaw out in the spring (mid to late March), provided they are thoroughly watered before the medium becomes frozen (late November to early January). Evergreen perennials transpire, or lose water, throughout the winter months, but when the root zone is frozen they can become damaged due to their inability to acquire water. Therefore, it is important that they enter the overwintering period with adequate water availability and to irrigate, if necessary, when conditions allow for it during the winter months.

Many growers attribute nearly 50% of their plant losses to the medium being too wet and about 15% percent of their losses due to overly dry conditions. So in many cases, adequate cold protection is provided but the underlying moisture level plays a very significant role in overwintering successes or failures. Providing irrigation to frozen pots during the winter months will not harm the plants. Watering frozen pots will raise the temperature of the root zone slightly, and when ventilation is provided, a cooling affect can be observed.

Overwintering above freezing

The more the root zone is allowed to freeze and then thaw, the greater the risk of the roots to become injured, which could possibly lead to plant mortality. It is difficult to keep plants from thawing during climatic changes such as unseasonably warm weather patterns during the winter when temperatures often rise well above the freezing point. To reduce the risk of containers becoming frozen, then thawed, and frozen again, many growers have adopted the philosophy to keep plants thawed the entire duration of the dormancy period by providing "minimum heat" to that crop. The small amount of energy used results in less plant losses than when heat is not utilized.

The "minimum heat" concept entails providing minimal amounts of heat, maintaining soil temperatures just above freezing. Usually, soil temperatures of 35° F (2° C) are desirable. For vernalization (the necessary cold requirement for many perennials to flower) of most perennials, it is necessary to maintain temperatures less than 44° F (7° C) for the duration of the cold period. Besides providing minimum heat, growers using this method must also provide cooling to the plants during periods when temperatures rise above the desirable level (44° F [7° C]). By using vents, exhaust fans, or roll up sides, the temperature inside of these structures can be maintained at relatively the same temperatures as outside.

Perennials that are considered to be marginally hardy in your particular zone should probably be placed in houses with heaters to keep them from being exposed to extremely cold temperatures. Growers overwintering small container sizes, such as plug flats, should also consider providing minimum heat. The smaller the container size, the less hardy the root zone is and the more likely plants are to sustain winter injury. Most plug suppliers overwinter and vernalize their plugs by supplying minimum heat during this period. Many perennial producers choose to overwinter their perennials in minimally heated greenhouse structures.

When minimum heat is used, growers usually control the air temperature to keep the soil temperature above freezing. Most growers do not actually monitor the soil temperature except by visual observations. Within minimum heat structures, there should be at least a couple of temperature sensors placed in the medium to accurately show what the soil temperatures are. Depending on the size of the pot and its proximity to the heat source, the air temperature setting may need to be several degrees higher than the desired soil temperature. With 200 ft. (61 m) long Quonset polyhouses with a single heat source near one end, it may be necessary to have a heat setting of 40° F (4° C) to prevent the plug flats at the opposite end from reaching a soil temperature of less than 32° F (0° C).

Overwintering Structures

Quonsets

Quonset-type structures, often referred to as polyhouses, are typically 14–20 ft. (4.3–6.1 m) wide, 7–9 ft. (2.1–2.7 m) high, and range of 100–250 ft. (31–76 m) in length. They are simple to construct and relatively inexpensive. Throughout the years, many growers have used Quonset structures for overwintering purposes. The primary purpose of this type of structure for woody plant producers is to provide protection to their containerized trees and shrubs. For perennial producers, polyhouses are not just places to overwinter plants; they are valuable resources for them to grow plants for early spring sales.

Quonset structures are generally covered with either one or two layers of polyethylene (poly). If one layer is used, it is generally a white plastic. When two layers are applied, the top layer usually consists of white polyethylene and the bottom layer is a clear plastic. The white plastic is used to reflect solar radiation, reducing the amount of heat buildup within the structure. Structures covered with poly will offer plants good protection on cold days. During the winter months, poly-covered houses will maintain temperatures averaging 5–15° F (3–8° C) above the outdoor temperatures. When two layers of poly are used, it is beneficial to inflate air between the layers by installing and running small inflation fans. This air layer creates an insulation barrier between the outside environment and the inside of the structure.

Growers using two layers of clear poly film report several benefits over using a combination of white and clear plastics. Many peren-

Figure 11.4. **Growers commonly overwinter perennials in covered Quonset structures with heating and cooling capabilities. These structures also serve as areas to grow or force perennials for early spring sales.**

nials are semi-evergreen or evergreen and have some foliage during the winter months. Clear poly films allow the maximum amount of sunlight exposure to the crop during these dark months. The foliage stays drier, reducing the occurrence of foliar diseases, compared with houses covered with opaque coverings. Clear films allow for more heat to be absorbed by the floors and growing medium of the pots within the facility, which moderates the temperature drop in the evenings. Plants break dormancy earlier, allowing for early spring sales. Growers also report less stretch, or shorter plants, in houses covered with clear films compared with houses using white coverings.

During the early winter, when the outdoor temperatures are relatively low and it is often overcast outside, these structures provide adequate protection from the cold but do not heat up significantly during the day. On warm, sunny days, temperatures inside poly-covered houses can often be at least 10–15° F (6–8° C) warmer than the outside temperatures. As time moves more toward the spring season, the number of sunny days increases, the outside temperatures increase, and the day length increases, causing larger and larger increases with the air temperatures inside of the poly-houses as compared with outside. As the perennials inside the enclosed structures are exposed to warmer temperatures, they begin to de-acclimate and may begin to actively grow at some point. Increasing the opaqueness of the poly will reduce but not eliminate this temperature difference.

During these warm spells or sunny conditions, growers should provide ventilation to keep the crop as cool as possible before the plants become de-acclimated and are susceptible to cold and freezing conditions. Once temperatures have increased within these houses, growers should pay particular attention to the outdoor conditions and provide the appropriate protection over the crops, especially when the weather returns to freezing conditions.

Early in the spring or during a dark, cloudy spring, some plants may elongate, or stretch, due to the lower light levels and increased day temperatures. Normally, by the time they begin to grow, the light levels within the structure are suitable for plant development and do not cause significant increases in plant height. If height control is necessary, growers apply plant growth regulators to control stem elongation during this phase of production.

Within these structures, it is often necessary to provide some additional form of frost protection. At a minimum, all the plants within the structure should be placed pot tight to provide some insulation. Various types of coverings (described on pp. 366–368), particularly thermal blankets, are commonly applied over the crops.

Polyhuts

Many growers place plants in polyhuts, which resemble miniature polyhouses; they often measure 3–4 ft. (0.9–1.2 m) in height by 5–6 ft. (1.5–1.8 m) across. The plants are generally placed inside the structure at a pot-tight spacing, often times even stacked to several layers. This close configuration adds to the overall protection from cold and reduces the amount of space needed for growers to overwinter their perennials or groundcovers. The height of these structures does make it harder to access the materials on the inside and perform any necessary activities.

After the pots are placed in the polyhut, they need to be thoroughly watered, as irrigation under the plastic is not feasible during the winter months. It is also necessary to place rodent baits throughout the polyhut at this time to reduce the amount of damage caused by small critters throughout the winter. After the polyhuts are filled, rodent bait is applied, and the plants are watered in, the polyhuts are then covered with white polyethylene plastic. In early spring, if ventilation is necessary, growers often roll the plastic up one side of the polyhut or cut holes into the plastic. If the weather turns cold again, they roll back the plastic and seal it with Poly-Patch or duct tape. It is difficult, if not impossible, to monitor and address the plants' needs in small polyhut structures during the winter months. The opaque covers also make it difficult for growers to quickly observe if cultural problems exist beneath them.

These small structures trap temperatures better than the larger Quonset polyhouses mentioned above. The temperature within polyhuts increases rapidly in the late winter due to the decreased buffering capacity. There tends to be greater, often undesirable temperature fluctuations as compared with polyhouse structures. Plants within these structures may become de-acclimated and more susceptible to winter injury. Growers can somewhat reduce the dramatic increases in temperature during the late winter by constructing them with a north-south orientation. Polyhuts with north-south orientations are slightly cooler at midday than houses with an east-west orientation, which causes greater fluctuations of temperature throughout the day, which could lead to more cold injury.

Figure 11.5. **These dormant daylilies sit inside a greenhouse that has been modified with 4 x 4 in. (10 x 10 cm) posts to support the snow load on the structure.**

Other structures

There are numerous other structures or overwintering systems growers can use. Some of them are variations on the ones mentioned above. Gutter-connect greenhouses are often used to overwinter perennials, either providing minimum heat or no heat at all. Individual or gutter-connect structures with rolling sidewalls are also commonly used and allow growers to provide a great amount of cooling, often keeping temperatures equivalent to those outside. Some growers implement systems within structures that provide heat to the root zone in order to keep them at or near freezing, allowing the plants to remain dormant and keeping the root zone temperatures from falling to levels where injury is likely to occur. Retractable-roof greenhouses are an option more and more perennial growers are utilizing. These structures can provide the benefits of growing outside during the season with greater control of temperatures during the winter months.

Outdoor Sites

Many growers overwinter perennials in outdoor locations or in what are considered structureless systems. The pots are usually prepared in the same manner, as far as cutting them back and consolidating them into groups, except that they are placed in an open area. Protective thermal blankets are usually placed over the pots, and the edges are secured to the ground using ground mat staples or by placing heavy objects, such as cinder blocks, at 3 ft. (0.9 m) spacing around the perimeter of the covered site. Overwintering coverings and thermal blankets are discussed in more detail on pages 366–368. The covering traps residual heat from the soil surface, which provides substantial buffering from low temperatures and it holds moisture, often eliminating the need for supplemental irrigation during the overwintering period. In addition to covering, some growers surround the consolidated groups with bales of straw to provide an insulation barrier around the perimeter and to serve as a windbreak.

Before covering, it is important to ensure the crops have been watered thoroughly and any necessary rodent baits are broadcast. Once the perennials are covered, they most commonly will remain sealed for the duration of the winter. It is also difficult to provide adequate ventilation and cooling during the late winter. If weather conditions permit, the covers can be rolled to one side to provide the necessary ventilation and to apply water to the crops. However, growers should be prepared to reapply the covers if the climatic conditions turn cold.

When uncovering the blocks initially, uncover them leaving the windward side of the block still permanently attached to the ground. This would allow, if the need arises, a fairly quick means of reapplying the covers. The covers from leeward side should be pulled over to the windward side and secured to keep them from blowing in the wind. During the late winter, growers should watch the weather patterns throughout the day to determine if there is the need to cover or uncover their outdoor overwintering facilities. Once the perennials are actively growing, do not gamble on whether frost is likely to occur. If there is even a remote chance of frost, cover the plants at night. Unless the night temperatures are going to remain above 40° F (4° C), all structureless areas should be protected.

A method of overwintering often referred to as the coverless pot-to-pot method is often used in warmer parts of the country. This method entails many of the overwintering preparations described above, such as trimming the plants back and consolidating containers into pot-tight configurations but does not involve the use of coverings. Except for extremely hardy varieties, this coverless method is not recommended for northern growers. For those perennials that are root hardy enough to withstand the normal minimum lows for your area, this method would be acceptable. If you have any doubts or are unsure as to the root hardiness, do not gamble; use an overwintering method that involves some type of covering or placement within an enclosed structure.

Figure 11.6. **In warmer hardiness zones, containerized perennials can often be overwintered outside with no or minimal protection from the cold.**

Covering

There is no magical date that growers can write on their calendars indicating the most optimal time to cover perennials. The covering date will vary by geographic location and weather conditions each year. Depending on geographic location, the covers are most commonly applied from late November to early January. Outdoor overwintering locations should be covered before snowfall to ensure the coverings can be applied.

The root zone temperature should be allowed to drop to near freezing and anticipated to stay at or below these temperatures before the plants are covered. Perennials should be covered as late in the season as possible, allowing them enough time to be completely dormant and acclimated to cold temperatures. Covering plants too early can delay the acclimation, or hardening off, and may result in overheating and premature growth of new shoots.

Watering before the plants are covered is critical to ensure adequate protection of the root zone. Moist medium freezes slower than dry medium, and it releases heat, which helps to protect the roots. Dry medium does not offer this type of protection and often continues to desiccate the root zone, causing significant amounts of plant injury and/or death. Moisture in the root zone will also increase the relative humidity under the covers, which slows down transpiration and desiccation. It would be beneficial to check the moisture level of the medium during the winter and apply irrigation if necessary.

The majority of perennials are either cut back to the crown or go dormant, dying

Figure 11.7. Outdoor nurseries often protect plants within protective structures, such as polyhuts, or other customized covering systems.

Goris Passchier

completely back. Due to their dormant state, these types of plants generally do not experience any foliar diseases while under the covers. Evergreen perennials are more susceptible to foliar pathogens if they are placed under various insulating covers. If evergreen varieties are covered, monitor them frequently for diseases, paying particular attention toward the end of the overwintering cycle. At this point, it may be worthwhile to remove the covers from these types of perennials altogether. When covers are applied on top of evergreen varieties, many growers will provide some method of supporting the blankets above the top of the foliage, keeping the material from contacting or crushing the leaves, which could lead to disease problems.

Thermal blankets

Perhaps the most common technique growers use to maintain root zone temperatures is to cover their crops with thermal blankets. Generally, the containerized plants are placed in a pot-tight configuration and covered with a thin cloth material that has insulating qualities. Typically, the thermal blankets fully cover the block of pots and drape down to the ground. If the containers are in outdoor locations, it is necessary to secure the edges of the cloth by weighing down the edges using heavy objects such as cinder blocks or inserting ground staples through the cloth and into the ground, preventing the wind from blowing the cloth off of the containers. When thermal blankets are used inside covered structures, it is not necessary to secure them.

Thermal blankets moderate the temperatures of the root zone under the materials. A simple thermal blanket covering dormant plants can result in temperatures beneath the cloth 8–10° F (4–6° C) warmer than the air above the thermal blanket. The root zone temperature should be allowed to drop below freezing (25–30° F [-4–-1° C]) and expected to remain there before the plants are covered. During extended periods of temperatures above 40° F (4° C), the thermal blanket should be removed to avoid excessive moisture buildup under the cloth. When the temperatures decrease to below freezing again, cover the plants using the guidelines just mentioned.

There are several brands of thermal blankets with varying thickness available for growers. There are several trade names of thermal blankets available including Microfoam, Typar, and The Winter Blanket. These blankets are efficient, flexible, and lightweight. They are a relatively inexpensive and convenient method for growers to protect their crops during the winter months. Thermal blankets are sold by the weight per square yard they exert. Common trade weights used range from 0.5–6.50 oz. per square yard (14–184 g/0.84 m^2). These woven materials will also allow water to pass through them, which allows growers to provide irrigation if necessary without removing the covers. Thermal blankets can be reused from year to year and may last up to five years.

There are some disadvantages of using thermal blankets. Plant injury often occurs to evergreen varieties or perennials that have fragile stems. Plant size may restrict their use or reduce the benefit of using them. For example, tall plants such as ornamental grasses or small trees may have to be laid down sideways in order to be properly covered. Rodents love the habitat created beneath the covers. Air circulation is greatly reduced, which may lead to the occurrence of fungal pathogens. Temperatures may build up under the covers, which may cause the plants to de-acclimate. But the advantages of this type of protective system far outweigh these disadvantages.

Some growers apply more than one layer of thermal blankets over their crops. To provide a better insulation barrier, many growers will first apply the thermal blankets, then add a layer of white 4 mil plastic on top of the coverings. If this method is used in outdoor production areas, it is important to secure the plastic around the perimeter of the covered area to keep the wind from blowing away the covers. The combination of cloth and plastic helps to moderate temperatures and humidity levels while extending the longevity of the more expensive thermal blankets. It is recommended to use white covering materials to reflect light and to avoid excessive heat buildup.

Once the outdoor temperatures are consistently above freezing or when active growth resumes, the covers can be permanently removed for the season. The exception would be if a frost is expected, and then the covers should be replaced over the crop. When the covers are removed for the season, they can be stored in a dark, dry area.

Plastic/straw sandwich

One method growers use to protect perennials is to use a system often referred to as the plastic/straw sandwich, which utilizes both plastic and straw. After the pots are placed at a pot-tight configuration, an inexpensive wire or wood frame is built over the beds. An inexpensive sheet of clear plastic, such as 2 or 4 mil, is draped over the frame, completely covering the perennials below. Then, a 6–12 in. (15–30 cm) layer of straw is placed over the clear plastic. Finally, the straw is covered with a layer of white plastic that is also fastened to the frame.

This provides the best winter protection of any of the unheated thermal systems. It is generally used more in colder regions, such as in USDA Hardiness Zones 3 and 4. Growers who have had troubles with certain cold-sensitive varieties in the past should consider implementing this type of thermal protection at their operations. The downfall is that it involves a great deal of labor to set up and to take down (not to mention the mess), which makes it a more expensive option. An additional drawback to the plastic/straw sandwich is that the straw between the layers often becomes wet from condensation and may lose some of its insulation value or possibly cause injury to some of the varieties beginning to emerge below.

Other covering options

In many cases, growers will cover their consolidated plants with at least two different materials. A layer of white plastic is commonly placed on top of the plants, and then a second covering is applied. The material of the top layer can be poly blankets, straw, evergreen boughs, sawdust, or even another layer of plastic. This method tries to create a layer of dead air space between the plants and the outdoor environment. Using white plastic for the first layer covering helps to prevent heat from building up under the coverings.

Maintaining Dormancy

During any extended warm spells, or at the end of winter, in addition to removing the thermal blanket, some attention should be made in regard to temperature, ventilation, and disease management. During the warm spells, be sure to keep temperatures from building inside of the structures. High temperatures could cause plants to break dormancy prematurely, which may result in plant injury if cold weather returns. The humidity most likely will rise inside of the structure, potentially leading to unwanted plant pathogens. Providing some ventilation will help to keep diseases in check.

To keep plants from resuming growth too quickly, it is beneficial to provide some cooling. Perennials brought out of dormancy gradually will be under less stress and more tolerant of small temperature fluctuations. Any time temperatures rise above 40° F (4° C), some form of cooling should be provided. The doors should be opened if plants are being overwintered in traditional cold frames. Opening the doors during the day while temperatures are above 40° F (4° C) and closing them in the evening before temperatures fall is probably the best method of managing temperatures within the structures. If the night temperatures are expected to remain above 36° F (2° C), the doors can be left open. For houses with vents or exhaust fans, cooling should be provided when the criteria mentioned above apply.

Irrigation should be monitored and applied, as necessary, during any warm spells throughout the winter and early spring months. Allowing the plants to become overly dry may lead to cold injury and possibly plant loss. To avoid desiccation and other cold-related injuries, it is important to evaluate the need for irrigation routinely during the overwintering period and to apply water if necessary. Just because the perennials are not actively growing does not mean they can be allowed to dry out. The water serves as an insulation barrier to the roots, allowing some buffering between the medium and the air. Water also helps to reduce dramatic swings in air temperature, keeping the soil temperatures more constant and helping keep the plants dormant.

Controlling Plant Disease

Prevention is key to controlling plant diseases during the winter months. Prior to the winter, perennials should be as disease free as possible. Not only will disease-free plants be healthier, there will be less fungal spores around next spring. Even though the perennials are not actively growing, growers should routinely scout their overwintering facilities for symptoms and signs of pathogens. Besides looking for obvious symptoms, such as *Botrytis* spores or necrotic areas, pay attention to any particular odors, such as ammonium or methane gases, that may provide an indication that various fungi might be present.

As mentioned previously, avoid overwatering the containers. Excessive moisture may injure plant roots by reducing the oxygen levels available to them and creates a good environment for root rot pathogens. Many times, plant mortality occurs and growers associate the death to diseases such as *Pythium*, when in many cases the cause was due to cultural practices or inadequate protection during the cold winter months. In many instances, the roots initially become damaged from extremely cold temperatures, inadequate protection, or poor irrigation management, and then secondary pathogens (root rots) attack the already weakened roots.

Many growers successfully reduce the initial presence of root rot pathogens by the application of fungicide drenches prior to the overwintering period. Some growers have been successful using natural organisms from such products as RootShield (*Trichoderma harzianum*) and Companion (*Bacillus subtilis*) incorporated or drenched into the growing medium to suppress the development of root rot pathogens. These organisms should be applied during the growing season to allow them to become established prior to the winter months.

The evergreen perennials such as *Heuchera, Myosotis,* and *Phlox subulata* are the most prone to having foliar diseases during the overwintering period. These varieties should be checked regularly, especially during the warm spells, to determine if plant pathogens are present. A preventive fungicide spray application of a broad-spectrum fungicide such as Spectro 90 (chlorothalonil + thiophanate-methyl) may be beneficial during these warmer periods. Depending on plant hardiness, it may be best to not cover evergreen varieties to achieve maximum ventilation, less moisture on the leaf surfaces, and a reduced likelihood of foliar diseases.

Growers using mulches to insulate their crops should ensure that the mulch does not become compacted around the crown. This compaction inhibits air movement and increases the moisture level around the plant; both of these conditions create favorable environments for fungal pathogens.

Weed Control

Controlling various winter annual and perennial weeds may also be necessary prior to covering. There are only a few herbicides labeled for weed control on perennial crops within closed structures. Growers primarily use preemergent herbicides to control weed seeds. Many of these herbicides will cause some phytotoxicity to actively growing perennials; however, most of the perennials that exhibit herbicide injury do tend to grow out of it eventually. Many operations treat the overwintering area with either preemergent herbicides or a combination of a nonselective herbicide, such as glyphosate (Roundup Pro), mixed with a preemergent chemistry at least two weeks before covering. Treating the facility and the surrounding areas will both control the weeds and help to reduce the future weed contamination of the containers.

Controlling weeds within the individual containers of the crops may also be required. There are some preemergent chemicals on the market today that will do an excellent job at controlling weed seeds. A few even offer some postemergent control, killing small seedlings. These chemicals will provide excellent control provided they are safe to the crop they are being applied to. Some crops, such as *Phlox paniculata* and *Sedum spectibile* seem to be very sensitive to many of the herbicides commonly used. My recommendation would be to wait until the perennials are completely dormant and apply herbicides only to the varieties that completely die back. I am not comfortable putting these chemicals on semi-evergreen or evergreen varieties. Also, if you have never used herbicides at your operation before, do not apply them to all of your crops until you feel comfortable using them and can justify the risk. Be sure to trial a small amount of plants before trying any new herbicide.

Many growers will treat their pots with preemergent herbicides during the winter preparation. This application is often done at least two weeks before covering. I prefer to wait until all of the perennials are acclimated and dormant, often applying them in late December to mid January. Herbicides are commonly applied as broadcast applications of granules or as spray applications. Generally, better control of weeds can be achieved with spray applications, due to a more uniform application to the soil surface. Coverage is key to getting complete control.

Methods for controlling weeds in perennials are discussed more thoroughly in chapter 7.

Controlling Animal Pests

Another risk grower's face when overwintering perennials is the potential crop damage caused by small rodents, most often field mice and meadow voles. Operations that do not put control measures into place may observe nearly a third of all of their losses are due to small animal damage. These small critters, once inside the overwintering area, will eat many perennial varieties and often burrow into the pots if their presence goes undetected. They feed on bulbs, rhizomes, seeds, stems, and tubers, often causing significant crop damage and unsalable products.

The meadow vole's body measures 3–5 in. (8–13 cm) long, with the tail adding an additional 1.5–2.5 in. (4–6 cm). They are brown with small eyes and ears. Voles do not hibernate and are active both during the day and night. One sign that voles are present is the occurrence of the many tunnels and surface runways with numerous burrow entrances. Unfortunately, by the time these tunnels are observed, significant damage to the crop is already likely to have occurred.

The most obvious method of preventing rodent injury is through exclusion, preventing them from entering the production area. This can be accomplished by placing a fine mesh screen wire around the perimeter of the greenhouse or other structure. The screen should be buried at least 6 in. (15 cm) deep, and bent upward at a 90° angle, a couple of inches (5 cm) above the soil surface as a deflector. The presence of voles can also be reduced by mowing the grass around the structures as close to the ground as possible. Better yet, maintain a 4 ft. (1.2 m) grass-free zone around the overwintering site.

Most growers do not implement exclusionary measures, instead controlling rodents by using chemical repellents or baits. Repellents give off offensive odors or have repulsive flavors. They are most commonly applied as sprays directly to the perennials and the surrounding areas. Many growers use repellents containing capsaicin (hot pepper sauce) or thiram (Spotrete). Many of these products are not very residual and need to be reapplied with some frequency. Toxic baits include cracked corn or oats treated with zinc phosphide or anticoagulant baits, such as Ramik (diphacinone) and Ground Force (chlorophacinone). The zinc phosphide baits are single-dose toxicants and will kill rodents after consumption. The anticoagulants are slow acting, often taking five to fifteen days and require multiple feedings for them to be effective. Growers using baits have often observed the rodents develop what is called bait shyness, meaning they avoid the baits or only consume nonlethal doses.

Regardless of the type of bait being used, it is beneficial to begin baiting about a month before you expect to apply the covers. Great care must be used when using commercial baits, as they are toxic to all forms of animal life. Place all baits in containers rather than placing them in heaps, piles, or broadcasting them around the production site. Many growers use homemade containers, such as 1.5 in. (38 mm) diameter PVC pipe, inserting a pot of bait inside another pot, or utilizing commercial bait stations to place these lethal products around the nursery. Baits should be kept dry, as moisture reduces their effectiveness.

For plug production or specialty perennials with high crop values, I would recommend using a combination of zinc phosphide and anticoagulant baits. In these cropping scenarios, there is very little tolerance for crop damage and rodent control needs to be both aggressive and preventative. Larger containers, such as 1 gal. (3.8 L) perennial product lines, generally have a greater tolerance to rodent damage. Having said that, I would still always consider a preventative program, using anticoagulant baits. Wherever rodent baits are used, be sure to check the overwintering facilities to determine if additional baits are needed to supply sufficient control through the rest of the season.

Rabbits may cause damage to perennials being produced and overwintered in outdoor locations. Rabbits can be controlled through exclusion, such as wire mesh fencing around the plant materials. Repellents containing capsaicin, thiram, or denatonium saccharide (RoPel) are also effective at discouraging feeding, provided they are applied with some frequency. Habitat modification, live trapping, and hunting are also effective methods of controlling the presence of rabbits.

Outdoor overwintering sites may also get crop damage caused by the feeding of deer. Deer can be controlled using fence barriers, odor and taste repellents, dog restraint systems, scare devices, and can be hunted provided growers acquire the proper Department of Natural Resource permits. The best method to exclude deer from the production site is the use of deer barriers constructed of woven wire fences or walls that are at least 10 ft. (3 m) tall. There are numerous homemade and commercial odor and taste repellents available for growers to use to deter the presence of deer. Whichever repellents are used, they all need to be reapplied during the winter months since they last between three and ninety days.

Removing the Covers

As mentioned earlier, remove the covers when the temperatures remain above freezing for an extended amount of time. It is helpful to follow the weather predictions closely in the late winter to early spring so the decisions to remove or reapply the covers are timely enough that problems from diseases and cold injury can be prevented. Typically, this mid- to late-season removal is toward the end of the overwintering cycle, when some plants are beginning to break dormancy but may still require some protection if the cold weather returns. Removing the covers at this point will allow growers to monitor the irrigation needs of the crops and to provide the necessary ventilation to reduce moisture levels and disease incidence under the blanket. When low temperatures are expected to reach below 40° F (4° C), reapplying the covers over the crop is recommended. Depending on the weather patterns, it may be necessary to remove and recover the plants two or three times during this transitional period. Successfully managing the overwintering process requires some flexibility in the overwintering facilities, good instincts with predicting the weather patterns, and good timing.

In addition to managing the removal and application of the thermal blankets, growers should more routinely monitor the temperatures where the containers are located. Until spring arrives, a grower's overwintering goal is to keep the plants from experiencing large temperature fluctuations. High temperatures during the day can cause perennials to become de-acclimated and more sensitive to cold temperatures. Ventilation and cooling should be provided to reduce plant growth and to keep any actively growing plants acclimated to cool temperatures. When the temperatures are above 40° F (4° C), leave the polyhouse doors open or provide some form of ventilation throughout the day (and possibly even the night). If they are allowed to produce lush growth, they will be more sensitive to any abnormally cold temperatures that may occur.

At some point growers need to decide when they can remove the covers for the season. There is no set date or rule of thumb for the best time to remove the thermal blankets. And of course, this will vary due to different weather conditions each year. Due to the unpredictable nature of the weather and year-to-year variations, it is a good idea to keep the cloth within each structure, or next to the containers, until most of the probability of severe cold has passed. Toward the end of winter, the goal is to prevent early shoot growth while ensuring that cold temperatures do not cause injury to your perennials.

Frost Protection

Some growers use irrigation as a winter protection technique for outdoor production areas. This method is useful if the plants have shoot growth or tissues that are not fully hardened or acclimated to cold and the temperatures are expected to drop to or just below freezing. The irrigation must be applied before the air temperatures reach freezing (32° F [0° C]) and should be continued until the ice begins to melt. As strange as it may sound, the formation of ice generates heat. This heating process is called the heat of fusion. As the liquid water changes to a solid form, a small amount of heat is generated, which is enough to protect plant cells while the temperatures are near freezing. Applying irrigation to crops is not effective when temperatures fall dramatically past the freezing point or for long durations of time.

Overwintering Costs

There are several often-overlooked costs associated with overwintering perennials. Many operations do not consider the expenses incurred to build overwintering structures or the labor of trimming and moving plant materials. Other costs that should be considered are covering materials, fungicides, rodent control, heat sources, and energy costs. Potential plant loss is another cost of overwintering. Some plant loss is inevitable and extremely unpredictable, but the financial impact of plant mortality is real. Any costs incurred to overwinter perennials should really be looked at as if they were crop insurance. For example, let's assume that the normal mortality rate for a non-protected perennial is 20%. And let us also assume that by providing protection, the crop's survival rate is 95%, or a 5% loss. Consider when using a 1 gal. (3.8 L) container at a pot-to-pot configuration, the spacing is .292 sq. ft. (271 cm^2) per pot; therefore, 1 sq. ft. (929 cm^2) contains 3.42 pots. The cost of the covering material ($0.44/sq. ft. [929 cm^2]) and labor ($0.31/sq. ft. [929 cm^2]) amount to $0.75 per 1 sq. ft. (929 cm^2). The perennials could be sold for a wholesale price of $3.45 each. The value of the plants within 1 sq. ft. equates to $11.82 ($3.45 x 3.42 pots).

With a mortality rate of 20%, a grower could potentially lose $2.36 of sales ($11.82 x .20) per square foot, when no protection is provided. A grower's loss with a protection system is only 5%, resulting in a loss of $0.59 ($11.82 x .05) per square foot. This means the grower reduced the overall losses by 75%. To get a 75% reduction in plant mortality the grower had to invest $0.75 per square foot in materials and labor for crop insurance.

All the numbers start to look confusing, so what are the real costs and savings from providing adequate plant protection during the winter months. Using the example above, a system with no protection would result in a reduction of sales of $2.36 per square foot due to plant death over the winter. When a protection system is implemented, the loss is only $0.59 per square foot. At this point, we have created $1.77 per square foot of additional sales by reducing the number of plant deaths. A grower had to invest $0.75 per square foot (materials and labor) to acquire $1.77 more sales per square foot. The overall gain per square foot is $1.02.

Using the example above, for every acre of space dedicated to overwintering 1 gal. (3.8 L) perennials at pot-to-pot spacing, at least an additional $40,000 in sales could be generated where plant protection is provided.

References

Berghage, Robert. "How to Overwinter Perennials." *GMPro*. August 2003.

Berghage, Robert, and Jim Sellmer. "Over-wintering Herbaceous Perennials." *Penn State Ornamental Horticulture Monthly Newsletter*. Volume 2, No. 6. October 1999.

Fernandez, Tom. "Getting Through the Winter." *The Michigan Landscape*. September/October 2003.

———. "Keeping It Cool." *The Michigan Landscape*. January/February 2004.

Fisher, Paul. *Overwintering Perennials in Containers*. University of New Hampshire Cooperative Extension. January 2001. http://ceinfo.unh.edu/Agric/AGGHFL/Overwntr.pdf. Accessed January 31, 2006.

Overwintering Containerized Perennials. Amherst, Mass.: UMass Extension Greenhouse Management Factsheets. 2003.

Pealer, George. "Overwintering Container-Grown Perennials Using 'Minimum Heat' Polyhut Structures." *Perennial Plants*. Summer 2003.

Perry, Leonard P. *Herbaceous Perennials Production: A Guide From Propagation to Marketing*. NRAES-93. 1998.

Preventing Rodent Damage to Overwintering Perennials. Amherst, Mass.: UMass Extension Greenhouse Management Factsheets. 2003.

Schmidt, Steve R. "How to Overwinter Container Grasses." *NMPro*. February 2002.

Snyder, Jim. "Overwintering Container-Grown Perennials Using Retractable Roof Greenhouse Structures." *Perennial Plants*. Fall 2003.

Zinati, Gladis. "Weathering the Cold." *American Nurseryman*. April 1, 2004.

Table 11.1. Overwintering Costs and Savings

				Typical without Protection		Typical with Protection			Difference	
Square Feet per Gallon at Pot-tight Spacing	Gallons per Square Foot	Wholesale Price Each	Value per Square Foot	20% Loss	Value per Square Foot at 20% Loss	5% Loss	Value per Square Foot at 5% Loss	Protection vs. No Protection per Square Foot	Overwintering Expenses (Materials and Labor) per Square Foot**	Savings with Protection per Square Foot
0.292	3.42	$3.45	$11.82	$2.36	$9.45	$0.59	$11.22	$1.77	$0.75	$1.02
Square Feet per Gallon at Pot-tight Spacing	Gallons per Acre* (43,560 ft² per acre)	Wholesale price Each	Value per Acre	20% Loss	Value per Acre at 20% Loss	5% Loss	Value per Acre at 5% Loss	Protection vs. No Protection per Acre	Overwintering Expenses (Materials and Labor) per Acre	Savings with Protection per Acre
0.292	149,178.08	$3.45	$514,664.38	$102,932.88	$411,731.51	$25,733.22	$488,931.16	$77,199.66	$32,670.00	$44,529.66

*Assumes one acre of production space is covered at pot-tight spacing and does not imply that growers can finish the crop at this spacing.

**Overwintering expenses will vary for each operation and may be less expensive if the coverings can be used for several years.

12

Perennial Profiles

Ajuga

Bugleweed

The bugleweeds being grown by commercial growers primarily belong to the species *Ajuga reptans*, but also commonly include select cultivars of *A. genevensis*, *A. pyramidalis*, and hybrids of these or other species. *Ajuga* is a stoloniferous perennial commonly used as a groundcover, in mass plantings, perennial borders, patio pots, and as components of mixed containers.

Figure 12.1. ***Ajuga* 'Chocolate Chip' shipped in bloom makes an impressive color display.**

The characteristics of the foliage are attractive to gardeners and landscapers, providing season-long interest to those who utilize this perennial. Adding to its landscape value, bugleweed remains evergreen throughout the year in most climates. Depending on the cultivar, the leaves can be bronze, green, or variegated and can be smooth, wavy, or ruffled. The best leaf coloration is achieved when they are grown under full sun. Bugleweeds produced under shady conditions will appear to have green and bronze leaves. The clumps reach up to 6 in. (15 cm) in height when they are not flowering and spread to 24 in. (61 cm) across. In the landscape, they bloom in mid to late spring. The 8–10 in. (20–25 cm) flower spikes usually contain violet or blue flowers, but may also consist of pink, rose, or white colorations.

Ajuga cultivars tolerate a large range of growing conditions, from full sun to full shade throughout USDA Hardiness Zones 3–9 and AHS Heat Zones 9–1. There are hardiness and heat tolerance differences between cultivars. Many *Ajuga* are particularly sensitive to cold temperatures and may succumb to severe winter injury and even death when they are not adequately protected during the winter months.

Ajuga belongs to the Lamiaceae family, which includes many other popular perennials such as *Agastache, Calamintha, Lavandula, Monarda, Nepeta, Perovskia, Physostegia, Salvia, Stachys,* and *Thymus,* to name a few.

Propagation

Commercially, *Ajuga* cultivars are propagated by cuttings or division. As the plant grows, it forms stems called stolons, which grow horizontally at or just below the soil surface. Some species form underground rhizomes, which function very similarly to stolons. These rhizomes and stolons form small individual plantlets, which can be removed from the mother plant and used as propagules for propagation. These plantlets often have visible roots growing from them and can easily be rooted into a plug flat or larger container.

Commonly Grown *Ajuga* Cultivars

A. pyramidalis 'Metallica Crispa'	*A. reptans* 'Golden Glow'
A. pyramidalis 'Purple Crispa'	*A. reptans* 'Jungle Beauty'
A. reptans 'Black Scallop'	*A. reptans* 'Purple Brocade'
A. reptans 'Bronze Beauty'	*A. reptans* 'Silver Beauty'
A. reptans 'Burgundy Glow'	*A. reptans* 'Silver Queen'
A. reptans 'Catlin's Giant'	*A. reptans* 'Purple Torch'
A. reptans 'Gaiety'	*A.* x *tenorii* 'Chocolate Chip'

Stem cuttings with or without preformed roots are easiest to root in the midsummer after the flowering is finished but can be propagated successfully anytime of the year. *Ajuga* should be placed under a light misting regime for the first week of propagation, then mist is not necessary for the remainder of the rooting time. With their ease of rooting, it is not necessary to dip the base of the cuttings into a rooting hormone such as IBA (indole-butyric acid). Some growers have found it beneficial to apply a preventative fungicide drench, such as mefenoxam (Subdue Maxx) combined with fludioxonil (Medallion), at the time the cuttings are stuck. The rooting process usually takes three to four weeks to become fully rooted and ready for transplanting, provided the soil temperature is maintained at 69–73° F (21–23° C).

Division is another method growers use to propagate *Ajuga*. The crowns are commonly lifted from a field, or plants grown in containers are divided into several plants (including the root system), and transplanted into pots for finishing. For the quickest rooting, it is best to divide bugleweed in the spring or early fall.

Production

Ajuga are most often produced in 1 qt. (1 L) or 1 gal. (3.8 L) containers. The most common practice is to transplant rooted liners, such as 72-cell plugs into the final container. Bugleweed performs best when grown in a moist, well-drained growing mix. Most commercially available peat- or bark-based growing mixes work well, provided there is adequate drainage. After the plug has been transplanted, the growing medium of the pot should be even with the top of the plug. I recommend growers drench *Ajuga*, using a broad-spectrum fungicide, at the time of planting to reduce potential crop damage from plant pathogens such as *Pythium* or *Rhizoctonia*. For plant establishment, maintain average temperatures of at least 65° F (18° C). Once they are established, grow *Ajuga* at 55–70° F (13–21° C) during the day and 55–60° F (13–16° C) at night.

To produce a fuller appearing container and reduce the crop production time, plant more than one *Ajuga* plug in 6 in. (15 cm) or larger pots. For example, growers planting two plugs into 1 gal. (3.8 L) containers often produce *Ajuga* reaching the finished size three to four weeks earlier than when only one plug per pot is used. This allows growers to turn their production space quicker.

During production, the pH of the growing medium should be maintained at 6.0–6.5. *Ajuga* are light feeders. Fertility can be delivered using water-soluble or controlled-release fertilizers. Growers using water-soluble fertilizers either apply 100–125 ppm nitrogen as needed or feed with a constant liquid fertilization program using rates of 50 ppm nitrogen with every irrigation. Controlled-release fertilizers are frequently used to produce containerized bugleweed. Growers commonly incorporate time-release fertilizers into the growing medium prior to planting at a rate equivalent to 0.75–1 lb. nitrogen per cubic yard (340–454 g/0.76 m^3) of growing medium.

Ajuga require an average amount of irrigation, as they do not tolerate really wet or overly dry conditions. Growers should keep them moist but not consistently wet. Never let *Ajuga* dry out or significant crop losses will be likely. When irrigation is necessary, water them thoroughly, allowing the soil to dry slightly between waterings.

Insects

The occurrence of insects feeding on *Ajuga* is not uncommon but rarely becomes problematic. Whiteflies are the most troublesome insect pest of *Ajuga*, but occasional outbreaks of aphids, slugs, and snails may occur. Root knot nematodes, soft scales, and spider mites are also sometimes found. It is not necessary to implement preventative strategies to control most of these insect pests. Growers should have routine scouting programs to detect the presence of insects early and to determine if and when control strategies are necessary. With the likelihood of whiteflies being vectors, potentially moving viruses from plant to plant, I recommend growers set very low thresholds on this insect and consider implementing preventative control programs.

Diseases

Botrytis, Phytophthora, Pythium, Rhizoctonia, Sclerotinia, and *Sclerotium* are the most common diseases affecting *Ajuga*.

Growers should be aware of the frequent occurrence of crown rots before putting *Ajuga* into crop production. There are growers who seem plagued with the occurrence of these crown rot diseases. Conversely, there are others who seldom or even never experience them. Often times, growers can successfully produce *Ajuga* from spring to fall, only to experience difficulty with crown rots towards the end of the overwintering period. Crown rots become prevalent when the environmental conditions are conducive to their development. Overly moist or wet conditions are most often associated with the onset of these diseases. With *Ajuga*, overly dry conditions often cause the initial injury to the crown, which creates an entry point for these opportunistic diseases to begin their often-lethal infections.

To control *Ajuga* diseases, it is best to manage the environment by providing proper plant spacing and adequate air movement, controlling the humidity, and using sound irrigation practices. Growers should carefully watch the moisture levels during adverse times of the year and avoid over- and underwatering their plants. Routine scouting is useful and recommended to detect plant diseases early, allowing the appropriate control strategies to be implemented before significant crop injury or mortality occurs. Many growers find it beneficial to apply broad-spectrum fungicides as preventative sprenches or drenches to *Ajuga* at monthly intervals.

Growers must also be aware of potential viruses that may be observed when producing *Ajuga*. *Ajuga* are prone to several viruses, including cucumber mosaic virus (CMV) and tobacco mosaic virus (TMV). Virus symptoms often appear as dark or light green mosaic and mottling of the leaves, growth distortion and stunting, or as ring patterns or bumps on the plant foliage. Viruses are spread from plant to plant (vectored) by whiteflies. As mentioned above, controlling whiteflies can greatly reduce the occurrence of these viruses. There are no cures for plant viruses. Symptomatic plants should be discarded to prevent the spread of viruses to healthy, uninfected plants.

Plant Growth Regulators

Ajuga cultivars have naturally compact growth habits, often remaining as a rosette, and will usually not require height control. However, during the winter months, periods of low light levels, or when grown at high plant densities, excessive internode elongation might occur, requiring some type of height management strategy. Generally, the height of *Ajuga* can be controlled by providing adequate spacing between the plants and by withholding nutrients and irrigation. On occasion, some growers have found it beneficial to apply chemical plant growth regulators to manage stem elongation. Commercially, there has not been a lot of research conducted to determine which growth regulators control the height of bugleweed. They are extremely sensitive to uniconazole (Sumagic); the application of this product could lead to too much growth regulation and plant stunting. When using this growth regulator for toning and shaping purposes, growers should be cautious and only use low rates, such as 2.5 ppm (in a northern location).

Forcing

Ajuga are easy to produce as a foliage plant. When transplanting from 72-cell plugs, bugleweed grown at 65° F (18° C) typically take six to eight weeks to finish 1 gal. (3.8 L) pots in the summer or eleven weeks during the winter. Growers wishing to produce blooming plants need to understand and follow the guidelines discussed below.

Ajuga do not have a juvenility period and are often observed flowering while they are in the plug flat. Planting vernalized ajuga plugs in the late winter for spring sales may not be desirable, as they will most likely produce flowers before bulking up to fill out the pot. Bulking promotes lateral branching, fills out the container, and results in high-quality, attractive plants at the time of sale. For spring shipping of *Ajuga* cultivars in bloom, it is recommended to plant two 72-cell plugs into a 1 gal. (3.8 L) container or one plug per 1 qt. (1 L) pot during the late summer. This allows them to become established and fill out the pot prior to winter.

Bugleweed has an obligate cold requirement in order for them to flower. It is recommended to cool (vernalize) small containers of bulked-up *Ajuga* for a minimum of six weeks at 35–44° F (2–7° C). Since they remain evergreen throughout the cold treatment, provide 100–200 f.c. (1.1–2.2 klux) of light during this period. If they are overwintered in a greenhouse or cold frame, natural light levels should be sufficient. When coolers are used, supplemental lighting is best provided using cool-white fluorescent lamps. Growers may observe sporadic flowering throughout the year, even when no cold has been provided. This type of erratic blooming is not predictable and cannot be scheduled. To produce flowering plants predictably, cold is required.

After the cold requirement is achieved, *Ajuga* can be grown at any day length, as they are day-neutral plants. The length of the photoperiod does not have any effect on the time to flower or the number of blooms produced. The time to bloom after vernalization is a function of cultivar and temperature. Depending on the cultivars being produced, *Ajuga* will reach full bloom in approximately five to eight weeks. Producing them at cooler temperatures increases the time to flower but will improve the overall quality characteristics of the plant, such as the color intensity of the foliage and flowers. To achieve the highest plant quality, I recommend growers produce *Ajuga* using cool production temperatures, such as 60–65° F (16–18° C), and allow a couple additional weeks for crop production.

References

Nau, Jim. *Ball Perennial Manual: Propagation and Production*. Batavia, Ill.: Ball Publishing. 1996.

Pilon, Paul. Perennial Solutions: "*Ajuga* 'Black Scallop'." *Greenhouse Product News*. February 2004.

Aquilegia x *hybrida*

Columbine

Aquilegia, commonly known as columbine, is one of the best-known, old-fashioned perennials that still deliver great appeal in today's gardens. With breeding improvements, this genus keeps getting better. Genetic improvements have brought about a broader range of

Figure 12.2. ***Aquilegia* x *hybrida* 'Swan Rose and White'**

PanAmerican Seed

Commonly Grown *Aquilegia* x *hybrida* Cultivars

'Biedermeier Mixed'	'Origami' series
'Blue Shades'	'Red Hobbit'
'Crimson Star'	'Songbird' series
'McKana Giants Mixed'	'Spring Magic' series
'Music' series	'Swan' series
'Olympia' series	'Winky' series

flower colors and sizes, shorter plant habits, and reduced juvenility and vernalization periods, which all are conducive to the commercial perennial grower. There are many commercially produced series of *Aquilegia*, such as 'Origami', 'Songbird', 'Spring Magic', 'Swan', and 'Winkie', to name a few, that are great examples of how genetics have been used to create plants with superior garden performance, great flower power, and improved commercial production characteristics.

In the spring, most cultivars produce an abundance of large, spurred blooms above mounds of attractive medium green foliage. The flowers consist of various shades of blue, pink, purple, red, white, or yellow and may be bicolor or solid. The blooms may be used as cut flowers or to attract both butterflies and hummingbirds to the garden.

Columbines are clump-forming perennials often utilized in rock gardens, as border plants, or grown in containers. *Aquilegia* prefer full sun, although in the South they perform best when grown under partial shade. They are both heat and cold hardy in USDA Hardiness Zones 3–9 and AHS Heat Zones 8–1. There are hardiness and heat tolerance differences between cultivars. In general, *Aquilegia* are rather short lived in the garden, often only surviving a couple of years. It is not uncommon for seedlings to emerge from seeds that have fallen to the ground. Unfortunately, seedlings from hybrid *Aquilegia* are not true to type and will not resemble the parent.

The name *aquilegia* comes from the word *aquil,* meaning "like an eagle," and compares the spurlike flower petal to an eagle's beak. Interestingly, *Aquil* is also the name of a genus of eagles. *Aquilegia* belongs to the Ranunculaceae family. There are numerous perennials in the Ranunculaceae family being commercially produced. Some of them are *Aconitum, Anemone, Cimicifuga, Clematis, Delphinium, Helleborus, Pulsatilla, Ranunculus,* and *Trollius.*

Propagation

Aquilegia cultivars are commercially propagated by seed or division. Although using bare-root divisions provides for a quicker finishing time and removes the concerns of plant juvenility and proper vernalization, it often leads to crop variability and lower survival rates. Most commonly, growers propagate *Aquilegia* from seed.

To improve germination, many *Aquilegia* cultivars have a pre-chilling requirement. When propagating *Aquilegia* from seed, growers should determine if a stratification treatment is necessary by consulting with the seed supplier. Stratification is useful in overcoming seed dormancies and will increase the uniformity and rate of germination. Many of the hybrids available today do not require stratification to achieve good germination. If old seed is being sown, providing stratification is recommended.

The seeds should be sown in the intended plug flat and covered lightly with germination mix or medium-grade vermiculite. The covering helps to maintain a suitable environment around the seed during this phase. If stratification is necessary, the seed flats should be moistened and moved to an environment, such as a cooler, where the temperatures can be maintained at 40° F (4° C) for four weeks.

After the cold treatment, or if no cold is necessary, move the plug trays into a warm environment (70–75° F [21–24° C]) for germination. Many growers utilize germination chambers during this stage to provide uniform moisture levels and temperatures; this is optional, as *Aquilegia* will successfully germinate in the greenhouse. The seeds begin to sprout within a few days and, depending on the cultivar, may take up to two weeks for germination to be complete.

Following germination, they can be grown at 65–68° F (18–20° C) for eight to ten weeks until they are ready for transplanting. Fertilizers are usually applied once the true leaves are present, applying 100 ppm nitrogen every third irrigation or 50 ppm with every irrigation, using a balanced water-soluble source.

Production

For container production, *Aquilegia* hybrids are suitable for 1 qt. (1 L) to 1 gal. (3.8 L) containers. Most growers receive starting materials of *Aquilegia* as finished plug liners. Columbines perform best when they are grown in a moist, well-drained medium with good aeration and water-holding capacity. Most commercially available peat- or bark-based growing mixes work well, provided there is adequate drainage.

When planting, be careful not to plant the plugs too deeply, as this could lead to poor plant establishment, crop variability, crown rot, and losses. The top of the starter plug should be even with the soil line of the finished container. I would recommend applying a fungicide drench, using a broad-spectrum fungicide, such as etridiozole + thiophanate-methyl (Banrot) or the combination of mefenoxam (Subdue Maxx) and thiophanate-methyl (Cleary's 3336), after transplanting.

During production, the pH of the medium should be maintained at 5.8–6.4. *Aquilegia* are light to moderate feeders. Fertility can be delivered using water-soluble or controlled-release fertilizers. Growers using water-soluble fertilizers either apply 100–150 ppm nitrogen as needed or feed with a constant liquid fertilization program using rates of 50–75 ppm nitrogen with each irrigation. Growers commonly apply time-release fertilizers as a topdressing onto the medium's surface using the medium rate, or incorporated into the growing medium prior to planting at a rate equivalent to 0.75–1 lb. nitrogen per cubic yard (340–454 g/0.76 m^3) of growing medium.

Where water-soluble fertilizers are the primary source of nutrition, use high nitrate fertilizers. Ammonium-based fertilizers may promote excessive vegetative growth and internode elongation during cool, dark weather conditions.

Aquilegia require an average amount of irrigation. To avoid overly wet conditions that could lead to crown rot, keep plants that are not fully established slightly drier than fully rooted containers. When irrigation is necessary, water them thoroughly, then allow the soil to dry slightly between waterings.

Insects

Aphid	Grasshopper
Caterpillar	Leaf miner
Columbine borer	Slug
Columbine sawfly	Spider mite
Columbine skipper	Thrips
Foliar nematode	Whitefly

Aphids, leaf miners, two-spotted spider mites, and whiteflies are the most common insects that may be observed feeding on *Aquilegia* cultivars. The presence of leaf miners is often hard to detect until the larvae begin their feeding between the leaf surfaces, leaving behind white trails often referred to as mines. Adult leaf miners can be caught on yellow sticky cards. They can be identified as small 2 mm bright yellow and black flies, resembling small bees. Recognition of the adults and treating the hot spots are keys to reducing excessive crop injury. Applying sprays of abamectin (Avid) or spinosad (Conserve) when the adults are present is effective at reducing the number of larvae mines per plant.

Spider mites are usually found feeding on the undersides of the leaves. Damaged leaves are stippled with small yellowish to silvery-gray speckles. Controlling spider mites is not easy because it is difficult to deliver the chemicals to the lower leaf surfaces, where they are feeding. There are several life stages present at any time, and they build resistance quickly to pesticides. Spider mite control is more successful when growers rotate chemical classes at every application, tank mix ovicides such as hexythiazox (Hexygon) and clofentezine (Ovation) with adulticides such as abamectin (Avid) and bifenazate (Floramite), and by ensuring good coverage of the sprays to the crop.

Diseases

Alternaria blight	*Rhizoctonia*
Ascochyta (leaf spot)	Rust
Botrytis	*Sclerotinia*
Powdery mildew	*Thielaviopsis*

Under the right circumstances, certain diseases may become prevalent when growing *Aquilegia* cultivars. Powdery mildew is the disease most problematic to growers. At first, powdery mildew appears as small, white, talcum-like colonies on the upper leaf surfaces, but under the right conditions they may engulf the plant with a powdery appearance. To control this disease, it is best to manage the environment by providing the proper plant spacing and adequate air movement and by controlling the humidity. Or, if desired, follow a preventative spray program using the appropriate chemicals. *Aquilegia* hybrid cultivars vary in plant height, from 14–24 in. (36–61 cm), when they are in bloom. Controlling the plant height may be required while producing certain cultivars in containers. Before using chemicals to reduce plant height, it is usually beneficial to provide adequate space between each plant, which will reduce the competition between plants for light and prevent the plants from growing taller. Several of the commercially available PGRs are effective at controlling plant height when they are applied using the appropriate rates, frequency, and timing. Depending on your geographic location, apply foliar applications of PGRs beginning with the rates listed in the table below. Apply growth regulators as the flower stalks are just beginning to elongate above the foliage. It will usually require two to three applications at seven-day intervals to provide adequate height control.

Forcing

To improve marketability, *Aquilegia* cultivars can be forced to bloom throughout the year. Growers wishing to produce blooming plants need to understand and follow the guidelines discussed below.

All *Aquilegia* have a juvenility phase and are not capable of flowering unless they are mature enough to perceive the other factors that promote flowering. Immature plants will not flower even after being exposed to the proper conditions necessary for flower formation. Researchers have used the number of leaves unfolded as a measurement of plant maturity. For example, the 'Swan' series is not capable of flowering until twelve to fifteen leaves are present. See the adjoining table *Aquilegia* Juvenility to determine the number of leaves necessary for many of today's popular *Aquilegia* cultivars to become mature enough to flower. For cultivars not listed, consult with the seed supplier for the juvenility requirement, or, if the duration of juvenility is unknown, plan to bulk plants up until at least twelve leaves are present.

Aquilegia need to be bulked or grown for a period of time until they have reached a

Figure 12.3. **Spring-planted vernalized *Aquilegia* plugs can be forced to bloom in five to eight weeks.**

PGRs for *Aquilegia* x *hybrida* Cultivars

Trade Name	Active Ingredient	Northern Rate
A-Rest	Ancymidol	25 ppm
B-Nine or Dazide	Daminozide	2,500 ppm
Bonzi, Piccolo, or Paczol	Paclobutrazol	30 ppm
Cycocel	Chlormequat chloride	1,250 ppm
Sumagic	Uniconazole-P	5 ppm
A-Rest + B-Nine or Dazide	Ancymidol + daminozide	10 ppm + 2,000 ppm
B-Nine or Dazide + Cycocel	Daminozide + chlormequat chloride	2,500 ppm + 1,000 ppm
B-Nine or Dazide + Bonzi, Piccolo, or Paczol	Daminozide + paclobutrazol	2,500 ppm +15 ppm
B-Nine or Dazide + Sumagic	Daminozide + uniconazole-P	2,000 ppm + 3 ppm

These are northern rates and need to be adjusted to fit the location where the crop is being produced.
These rates are for spray applications and geared to growers using multiple applications.

mature stage and are capable of perceiving the additional factors necessary for flowering. Columbines reach maturity better when grown in containers rather than in plug flats. In many instances, *Aquilegia* plugs that have been bulked have a lower flowering percentage than plants that have been bulked in larger container sizes.

Most growers have the best success when planting large plug liners into the final container during the late summer or early fall of the year prior to the desired spring sales and overwintering them as a potted plant. It is most beneficial to vernalize plants in the final container when the pot size is greater than 5 in. (13 cm). Spring planting of vernalized plugs in large container sizes does not allow the plants to produce a "full" pot prior to them blooming. However, large vernalized plugs can be useful for planting in small container sizes, such as 1 qt. (1 L) pots, during the spring.

Aquilegia generally require cold in order for them to bloom. Through rigorous breeding efforts, the vernalization requirement is slowly being bred out of some of the newer cultivars. Currently, many of today's cultivars will flower without a cold treatment, but cold is beneficial and increases the flowering percentages, hastens the time to bloom, and/or increases the number of flowers produced. Without a cold treatment, many cultivars will flower sporadically the first year from seed.

In most cases, the vernalization requirement has been reduced over the needs of the older cultivars. The cold duration varies widely with the cultivars being produced and may be as little as three weeks or up to twelve weeks. The cold treatment should consist of temperatures at 35–45° F (2–7° C) for the recommended duration for each cultivar being produced. Please refer to the table *Aquilegia* Cold Requirement for the cold recommendations of many *Aquilegia* cultivars. For cultivars not listed, consult with the seed supplier for the vernalization requirement or, if the duration of the cold treatment is unknown, plan to provide a minimum of six to nine weeks of cold.

Columbines are day-neutral plants and will flower under either short or long days. Plants forced under long days tend to grow taller than those grown under short days and may require additional applications of plant growth regula-

Aquilegia Juvenility

Cultivar or Series*	Leaf # Necessary for Maturity
'Crimson Star'	≥12
'McKana Giants Mixed'	≥12
'Music' series	≥14
'Origami' series	7–9
'Songbird' series	12–16
'Swan' series	12–15
'Winkie' series	9–12

*Individual cultivars within a series often vary slightly.

tors. It is recommended to force *Aquilegia* under natural photoperiods. For cultivars that will flower without a cold treatment, it is important—after they have become mature—to grow them under long days and at temperatures below 60° F (16° C) to initiate flower buds.

The time to bloom varies widely with cultivar and production temperatures. As mentioned previously, *Aquilegia* will flower quite rapidly following vernalization. Some cultivars flower within three to four weeks after the appropriate temperatures are provided. The time to bloom generally decreases as temperatures increase. High temperatures (above 78° F [26° C]) are detrimental to flowering, causing delayed flowering and a reduction in flower number and size. Most cultivars are forced into bloom at temperatures of 57–68° F (14–20° C). I would recommend growers use production temperatures of 60–65° F (16–18° C) to force *Aquilegia* cultivars; this produces plants with the largest flower size and best overall plant appearance. At these temperatures, most cultivars will be blooming in approximately five to eight weeks.

***Aquilegia* Cold Requirement**

Cultivar or Series*	Weeks of Cold for Vernalization
'Crimson Star'	6
'McKana Giants Mixed'	≥12
'Music' series	12
'Origami' series	5
'Songbird' series	6
'Swan' series	3
'Winkie' series	10

*Individual cultivars within a series often vary slightly.

References

Pilon, Paul. Perennial Solutions: "*Aquilegia* x *hybrida* 'Swan'." *Greenhouse Product News.* March 2004.

Whitman, C., A. Cameron, E. Runkle, and R. Heins. "Herbaceous Perennials: *Aquilegia* (Columbine)." *Greenhouse Grower.* October 2003.

Aster

Hardy Aster

The increased popularity of perennials has created season-long demand for flowering perennial crops. It is not uncommon for perennial growers to offer perennials from frost to frost. As the season progresses, it becomes increasing difficult to provide flowering perennials into the marketplace because the number of flowering perennial varieties decreases as the days become shorter. Historically, garden mums and fall pansies have dominated autumn plant sales.

Figure 12.4. ***Aster dumosus* 'Woods Purple'**

With their ease of production and natural bloom times, *Aster* crops are well suited for late-season sales. Depending on the cultivar, naturally blooming *Aster* can be sold from late August through October. *Aster* can be easily forced into bloom out of season, creating sales opportunities from late spring through early autumn. *Aster* plants are available in a variety of growth habits and colors including shades of blue, lavender, pink, purple, raspberry, and white. They are very floriferous, forming mounds of daisy-like blooms. The flowers consist of rows of thin, petal-like ray flowers surrounding a yellow disk consisting of many tightly packed, petalless flowers. The name *Aster* is derived from the Greek word referring to the starlike shape of the flower.

Aster species perform well in sunny locations throughout USDA Hardiness Zones 4–9 and AHS Heat Zones 8–1. In southern locations, it

Commonly Grown *Aster* Varieties

A. dumosus 'Sapphire'	*A. novi-belgii* 'Jenny'
A. dumosus 'Wood's Pink'	*A. novi-belgii* 'Magic'
A. dumosus 'Wood's Purple'	*A. novi-belgii* 'Newton Pink'
A. novae-angliae 'Alma Potschke'	*A. novi-belgii* 'Pink Bouquet'
A. novae-angliae 'Purple Dome'	*A. novi-belgii* 'Professor Kippenburg'
A. novae-angliae 'Red Star'	*A. novi-belgii* 'Snow Cushion'
A. novi-belgii 'Alert'	*A. novi-belgii* 'White Swan'
A. novi-belgii 'Dragon'	*A. novi-belgii* 'Winston Churchill'

is recommended to grow them under partial shade. They are versatile perennials that can be used in a number of ways, including container, border, and mass plantings. Depending on the variety, hardy *Aster* reach 2–8 ft. (0.6–2.4 m) in height; most of the varieties produced as flowering containers reach 2–4 ft. (0.6–1.2 m) in the landscape. Growers produce *Aster* in a wide range of container sizes, from 4 in. (10 cm) up to 1 gal. (3.8 L) containers, and utilize them in combination pots or color bowls. With these characteristics, *Aster* can be marketed alongside of fall bedding plants, extending the growing season for many operations.

Hardy asters belong to the Asteraceae family (formerly Compositae). There are numerous perennials in the Asteraceae family being commercially produced. Some of them are *Achillea, Bellis, Centaurea, Coreopsis, Doronicum, Echinacea, Echinops, Erigeron, Gaillardia, Helenium, Heliopsis, Leucanthemum, Ligularia, Rudbeckia, Stokesia,* and *Veronica.*

Propagation

Aster plants are vegetatively propagated from tip cuttings or bare-root divisions. Many cultivars are patented, and unlicensed propagation is prohibited. To avoid illegal activities and the consequences, always check for plant patents before attempting to propagate *Aster.*

Propagation by vegetative tip cuttings usually occurs in the late spring and early summer. Tip cuttings should be taken while they are vegetative; cuttings that are already in bloom will take longer to root and have a lower survival rate than purely vegetative starting materials. During the late winter or early spring, growers should provide night interruption lighting (10:00 P.M. till 2:00 A.M.) to keep the stock plants vegetative for ideal cutting production and quality.

Tip cuttings should measure approximately 2 in. (5 cm) and contain several nodes. They can be stuck into plug liners or directly into the final container for rooting. Regardless, the rooting medium should be moistened prior to sticking. The cuttings can be dipped into a rooting hormone, such as a solution of indolebutyric acid (IBA) at rates between 500 and 1,000 ppm prior to sticking. *Aster* can successfully root without rooting compounds but tend to root slightly faster when these treatments are provided.

Cuttings should be placed under low misting regimes for the first two weeks of propagation. The misting can gradually be reduced as the cutting forms calluses and root primordia. The cuttings are usually rooted in less than three weeks with soil temperatures at 68–74° F (20–23° C). For best results, the air temperature during rooting should be maintained above 60° F (16° C) and below 80° F (27° C). Long days should be provided throughout the rooting process. It is beneficial to begin constant liquid feeding with 200 ppm nitrogen at each irrigation beginning ten days from sticking.

Production

Growers producing 6 in. (15 cm) or smaller containers usually plant one liner or direct stick two or three unrooted cuttings per pot. When larger sizes are grown, planting multiple liners or sticking additional cuttings per pot is recommended. *Aster* prefers to be grown in a moist, well-drained medium with a slightly acidic pH, 6.0–6.5. It is beneficial to use growing mixes containing at least 20–35% peat moss to help retain moisture, particularly once the plants reach a finished size and tend to use more water. *Aster* require frequent irrigation, particularly as they become more developed. When irrigation is necessary, I recommend watering thoroughly, allowing the soil to dry slightly between waterings.

Aster are heavy feeders and perform best when a constant liquid fertilization program is used, feeding at rates of 150–200 ppm nitrogen. Use complete N-P-K fertilizers with the majority of the nitrogen being in the nitrate form. Fertilizing with micronutrients is also recommended. Controlled-release fertilizers can be incorporated into the growing mixes at medium to high rates; applying 1.25 to 1.50 lb. nitrogen per cubic yard (567–680 g /0.76 m^3) of growing medium is recommended. It may be necessary to leach with clear water or to reduce the fertility rate on occasion if the soluble salts accumulate. Once the flowers begin to open, growers often switch to applying only clear water.

Insects

Aphid	Slug
Caterpillar	Spider mite
Chrysanthemum lace bug	Spittle bug
Japanese beetle	Tarnished plant bug
Leafhopper	Thrips
Root knot nematode	Whitefly

Aphids, spider mites, thrips, and whiteflies occasionally will become problematic. Of these insect pests, aphids and whiteflies are the most prevalent. Growers should have scouting programs in place to detect these pests early before the populations build up to excessive levels. Some growers implement preventative spray programs using systemic chemicals such as pymetrozine (Endeavor), thiamethoxam (Flagship), imidacloprid (Marathon II), or acetamiprid (TriStar) to keep *Aster* free of aphids and whiteflies for up to one month following application.

Diseases

Anthracnose
Aster yellows
Botrytis blight
Cercospora leaf spot
Coleosporium (rust)
Crown gall
Downy mildew
Entyloma (white smut)
Fusarium wilt (crown and root rot)
Phomopsis
Phytophthora
Powdery mildew
Pseudomonas leaf spot
Pythium
Puccinia (rust)
Rhizoctonia
Sclerotinia
Septoria leaf spot
Verticillium wilt
Xanthomonas

Botrytis, powdery mildew, *Rhizoctonia* foliage blight, and rust are common diseases of *Aster* crops. All of these diseases can be reduced by using adequate crop spacing, avoiding overhead irrigation late in the day, providing adequate air circulation, and controlling the humidity. Routine scouting is recommended to detect these problems early. Growers should also consider implementing preventative spray programs using the appropriate chemicals to control these pathogens.

Botrytis often causes the lower leaves to become yellow and brown. It is most prevalent when plants are grown at high densities and with poor air circulation. In most instances, only the lower leaves are affected. Growers can control this pathogen with foliar applications of iprodione (Chipco 26019), thiophanate-methyl (Cleary's 3336), trifloxystrobin (Compass), chlorothalonil (Daconil), azoxystrobin (Heritage), or chlorothalonil + thiophanate-methyl (Spectro 90). Simply providing wide crop spacing can greatly reduce the occurrence of *Botrytis*.

Powdery mildew appears as small, white, talcum-like colonies, but under the right conditions it may engulf the plant with a powdery appearance. To control this disease, it is best to manage the environment by providing the proper plant spacing and adequate air movement and by controlling the humidity. Many growers follow a preventative spray program using the appropriate chemicals. I have observed good results when rotating trifloxystrobin (Compass), azoxystrobin (Heritage), piperalin (Pipron), myclobutanil (Systhane), triadimefon (Strike), and triflumizole (Terraguard) with my preventative powdery mildew programs.

Rhizoctonia foliage blights form brown-black necrotic spots on leaf surfaces and stems, with the leaves dying back from the tips in advanced stages, often leading to plant death. It is imperative to monitor the crops routinely for this pathogen and begin control measures immediately upon detection. Chemical controls include iprodione (Chipco 26019), thiophanate-methyl (Cleary's 3336), flutolanil (Contrast), azoxystrobin (Heritage), and fludioxonil (Medallion).

Rust causes yellow spots to appear on the upper leaf surfaces and bright orange to brown pustules to form on the undersides. Once rust occurs, it spreads quickly through air movement and splashing water. Certain chemicals are effective at controlling rust either after damaged foliage is removed or on a preventative basis. Some of the suggested chemicals for controlling rust include triadimefon (Bayleton), pymetrozine (Banner Maxx), flutolanil (Contrast), azoxystrobin (Heritage), triadimefon (Strike), and myclobutanil (Systhane). Rotate chemical classes to reduce the development of fungicide resistance.

Plant Growth Regulators

Controlling the plant height may be necessary when producing *Aster*. In many cases, the height of the plant can be manipulated using proper crop scheduling and pinching practices. Providing adequate spacing between the plants will reduce plant stretch caused by competition. Under certain growing conditions or high plant densities, it may be necessary to use chemical plant growth regulators. In the northern parts of the country, I would recommend applying daminozide at 2,500 ppm. Paclobutrazol sprays at 30 ppm have also been effective at controlling plant height. Applying one to two PGR applications seven to ten days apart should provide adequate height control. The first application should occur after pinching, once the new shoots are about 1 in. (2.5 cm) in length.

Forcing

Forcing *Aster* into bloom out of season is relatively easy. They can be programmed into bloom throughout the spring and summer months. In most instances, growers schedule their *Aster* crops, paying close attention to propagation and pinching dates. The combination of these helps growers to dial in their crop scheduling in a predictable or, more important, repeatable manner.

Aster grown in controlled environments or forced to bloom out of season, as discussed below, often require pinching to promote more

PRGs for *Aster* Cultivars

Trade Name	Active Ingredient	Northern Rate
B-Nine or Dazide	Daminozide	2,500 ppm
Bonzi, Piccolo, or Paczol	Paclobutrazol	30 ppm
B-Nine or Dazide + Sumagic	Daminozide + uniconazole-P	2,000 ppm + 3 ppm
Bonzi, Piccolo, or Paczol drench	Paclobutrazol	6 ppm

These are northern rates and need to be adjusted to fit the location where the crop is being produced. Unless noted otherwise, these rates are for spray applications and geared to growers using multiple applications. Drench applications are usually applied once per crop.

Sample *Aster* Schedules*

	Quart	1 gal.
Flowering/ship date	July 18	July 18
Start of short days (6 weeks before ship)	June 6	June 6
Second pinch (21 days from first pinch)	----	May 23
First pinch (21 days from stick/transplant)	May 30	May 2
Night interruption lighting	May 9	April 11
Direct stick or transplant	May 9	April 11

*The scheduling of *Aster* will vary by cultivar; use these schedules only as guidelines. Check with your plant supplier for cultivar-specific recommendations.

branching and create a fuller pot. Pinching usually occurs within two weeks of planting a rooted cutting into the final container or three weeks from sticking a cutting, after the roots reach the sides and the bottom of the pot. Typically, growers pinch off the tip or growing point of the plant (soft pinch), leaving four to six leaves. Small container sizes, such as 4 in. (10 cm) pots are usually pinched one time. Larger container sizes may require two pinches to create fuller pots and more bloom. If a second pinch is needed, it should occur three to four weeks following the first, leaving three to four leaves above the initial pinch. When producing *Aster* for natural-season blooming, the final pinching should occur between July 25 and August 10. Later pinches delay flowering and may result in inadequate plant size at flowering.

Timely spacing of the crop is important to help control plant pathogens, reduce plant stretch caused by competition, and improve overall crop quality. Most growers produce *Aster* at pot-tight spacing until they receive the final pinch, then plants are placed at wider spacing. Adequate and on-time spacing is important for producing a quality plant. I recommend producing *Aster* at the following minimum spacing: 4 in. (10 cm) pots at 6 x 6 in. (15 x 15 cm) centers, and 1 gal. *Aster* should be spaced at 12 x 12 in. (30 x 30 cm) centers.

Aster requires short days for flower bud initiation and development. Production under long days causes them to remain vegetative. To keep plants vegetative and to avoid early bloom, growers often provide night interruption lighting, even during the summer months when the photoperiods are naturally long, until short days are desirable.

During naturally long days, it may be necessary for growers to create short-day photoperiods by blocking out all of the light. Short-day conditions are created by pulling black cloth or black plastic over the production site, ensuring the plants only receive twelve hours of light each day. The blackout area should be completely dark, as even 1 f.c. (11 lux) of light can be perceived by *Aster*, negating the full benefit of the conditions. Short days typically should begin one week after the final pinch for 4 in. (10 cm) containers or two weeks from the final pinch for 6 in. (15 cm) or larger pots. *Aster* respond quickly to short days and can be flowering within six weeks from the onset of the short photoperiods. *Aster* should be grown with night temperatures of 60–65° F (16–18° C) and day temperatures of 65–75° F (18–24° C). During the last two weeks of production, as the crop nears open bloom, lowering the night temperature to 55–60° F (13–16° C) will help intensify the color of the bloom.

References

Jeagar, Jim. "How We Grow Asters." *GrowerTalks.* November 1999.

Pilon, Paul. Perennial Solutions: "*Aster dumosus* 'Sapphire'." *Greenhouse Product News.* July 2005.

Yoder Brothers, Inc. http://www.yoder.com/inside_frameset.asp?body=%2Fgrower%2Fasters%2Ftips%2Fcontent%2Easp. Accessed January 5, 2006.

Astilbe

False Goat's Beard, False Spirea

False spireas are very popular shade and woodland garden perennials. Any landscape with shady, moist conditions would welcome the presence of *Astilbe* to brighten up the often-colorless scenery. *Astilbe* form beautiful mounds of fernlike foliage bearing tiny flowers on erect to arching plumelike flower panicles that rise above the foliage on slender upright stems. Through breeding and plant selection, today's *Astilbe* cultivars consist of a wide range of flower colors, including carmine, coral, lavender, pink, rose, salmon, scarlet, and white, plus numerous shades in between.

Figure 12.5. ***Astilbe chinensis* 'Visions' flower.**

The cultivars commercially grown today were crossbred within each species or between species and are classified into five main groups: Arendsii, Chinensis, Japonica, Simplicifolia, and Thunbergii. These groups help to more accurately identify the specific traits of each cultivar.

Cultivars within the Arendsii group make up over 95% of all the *Astilbe* sold in the United States. The group is named after the German nurseryman George Arends (1862–1952) who bred seventy-four new *Astilbe* cultivars in fifty years. Almost every hybrid *Astilbe* can be traced directly or indirectly back to Arends. Numerous cultivars were bred from various *Astilbe* species, including *A. astilboides, A. chinensis* var. *davidii, A. japonica,* and *A. thunbergii.* Arendsii hybrids reach various heights, depending on the cultivar, ranging from 18–36 in. (46–91 cm) tall in bloom and spreading 15–24 in. (38–61) wide.

The Chinensis group contains hybrids within the *A. chinensis* species. Cultivars are known to extend the blooming season into the summer months. The foliage is deeply incised, coarsely textured, and clear to bronze-green in color. The flowers are very showy and borne on narrow-branched panicles. The cultivars within this group are moderately drought tolerant but still perform best in moist soils. Chinensis cultivars reach 12–24 in. (31–61 cm) in height when they are blooming and have a 15–18 in. (38–46 cm) spread.

The Japonica group produces flowers in early summer on dense, pyramidal clusters. The foliage is glossy green, narrow, toothed, and is often tinged with red. They are basically short cultivars that have been crossbred.

The Simplicifolia group contains compact, slow-growing cultivars. The glossy fernlike foliage has leaves that are lobed, not divided. The starlike flowers are borne on open, airy panicles, providing a less formal appearance than cultivars from other *Astilbe* groups. When blooming, Simplicifolia cultivars tend to reach 12–18 in. (31–46 cm) tall with a 15–20 in. (38–48 cm) spread.

Cultivars within the Thunbergii group are late-blooming *Astilbe*. The foliage is glaucus and loosely branched. They have distinctive open flower clusters cascading downward from tall stems. These hybrids are crosses of an *A. thunbergii* species originating in Japan and Korea.

Astilbe can be easily produced in average, medium-wet, well-drained soils across USDA Hardiness Zones 4–9 and AHS Heat Zones 8–2. Most *Astilbe* species require partial to heavy shade. However, cultivars of *Astilbe*

Commonly Grown *Astilbe* Cultivars

Cultivar	Group	Flower Color
'Amethyst'	Arendsii	Bright lavender
'August Light'	Arendsii	Scarlet-red
'Bella'*	Arendsii	Red to rose
'Bressingham Beauty'	Arendsii	Salmon-rose
'Bridal Veil'	Arendsii	Lacy white
'Cattleya'	Arendsii	Lilac to pink
'Erica'	Arendsii	Light heather pink
'Etna'	Arendsii	Dark red
'Fanal'	Arendsii	Bright red
'Flamingo'	Arendsii	Pink
'Glow'	Arendsii	Dark red
'Grande'*	Arendsii	Mixed
'Grete Pungel'	Arendsii	Light pink to white
'Nikky'	Arendsii	Light and dark pink
'Pink Pearl'	Arendsii	Light pink
'Rhythm and Blues'	Arendsii	Raspberry-pink
'Rock and Roll'	Arendsii	White
'Rotlicht'	Arendsii	Deep red
'Showstar'*	Arendsii	Mixed
'Sister Theresa'	Arendsii	Salmon pink
'Snowdrift'	Arendsii	White
'Spinell'	Arendsii	Red
'White Gloria'	Arendsii	Snow white
'Maggie Daley'	Chinensis	Lavender-purple
'Pumila'*	Chinensis	Lilac-pink
'Purple Candles'	Chinensis	Violet-red
'Taquetii' ('Superba')*	Chinensis	Lilac
'Veronica Klose'	Chinensis	Dark purplish rose
'Vision'	Chinensis	Raspberry red
'Vision in Pink'	Chinensis	Pale pink
'Vision in Red'	Chinensis	Purplish red
'Deutschland'	Japonica	Pure white
'Elisabeth van Veen'	Japonica	Raspberry lilac
'Ellie'	Japonica	Pure white
'Europa'	Japonica	Pink
'Montgomery'	Japonica	Dark red
'Peach Blossom'	Japonica	Pink
'Red Sentinel'	Japonica	Scarlet red
'Rheinland'	Japonica	Rose pink
'Vesuvius'	Japonica	Reddish pink
'Washington'	Japonica	White
'Aphrodite'	Simplicifolia	Rosy red
'Hennie Graafland'	Simplicifolia	Light pink
'Jacueline'	Simplicifolia	Pink
'Pink Lightning'	Simplicifolia	Rose-purple to pink
'Sprite'	Simplicifolia	Shell pink
'Moerheims Glory'	Thunbergii	Light pink
'Ostrich Plume'	Thunbergii	Coral rose
'Straussenfeder'	Thunbergii	Salmon pink

*Propagated from seed

chinensis prefer partial shade and can tolerate more sun and drier conditions. All *Astilbe* cultivars are sensitive to moisture stress and should not be allowed to dry out, as leaf scorch is likely to occur.

In Greek, the word *astilbe* translates to "without brilliance," probably coined from the appearance of a single flower on the panicle. However, as a group of flowers on plumes, the cultivars currently being grown produce a brilliant and spectacular display in shady landscapes. These slow-spreading, rhizomatous plants belong to the family Saxifragaceae, which contains several commonly grown perennials, including *Bergenia, Heuchera, Heucherella, Rodgersia, Saxifrage,* and *Tiarella*. *Astilbe* are often used for cut flowers, in containers, mass plantings, or small groups, and as border plants or groundcovers in shade gardens.

Propagation

Most cultivars are propagated vegetatively by division. Divisions of *Astilbe* are best done during the spring or fall. Divisions are made by cutting the thick, woody crowns into portions containing several eyes (shoots) and roots using a sharp, sterilized knife. The size of the division usually depends on the size of the container in which it is being potted. For 1 qt. (1 L) production, a crown consisting of one or two eyes is commonly used. For larger containers, such as a 1 gal. (3.8 L), divisions with at least two or three eyes are typically used.

Some *Astilbe* cultivars are propagated by seed. They can be difficult to germinate and often have a low germination rate, 70% or less. The seeds are small and difficult to handle. The best results occur when growers sow fresh seed rather than seed that has been in storage. Sow them on the surface of the growing medium and do not cover with vermiculite or plug mix. If the seeds are covered, germination will likely be reduced, as light is necessary for germination to occur. Temperatures should be kept at 62–67° F (17–19° C) for germination. It is important to keep the medium very moist during this stage, or the germination rates are likely to decrease. Seedlings will emerge over a period of time ranging from fourteen to twenty-one days after sowing. Once germination has occurred, the temperatures can be reduced to 55–60° F (13–16° C) until they are ready for transplanting. It takes fourteen to sixteen weeks for them to finish in a 128-cell plug tray. If old seed is used, it is beneficial to stratify them prior to germination by providing a moist-cold period for at least four weeks prior to placing them at temperatures suitable for germination. Since *Astilbe* require a cold period for flowering, seeds should be sown in early summer the year prior to the sale date for a finished flowering product.

In most cases, container growers do not propagate *Astilbe* cultivars; rather, they purchase bare-root divisions or large plug liners from growers who specialize in *Astilbe* propagation.

Production

Astilbe are available to the marketplace as bare-root plants or large plug liners. They are commonly planted in 1 qt. (1 L) or 1 gal. (3.8 L) containers immediately after they have been received. Growers commonly plant them in bark- and peat-based growing mixes with good drainage and a fair amount of water-holding ability. When planting bare-root divisions, the woody crown should be placed in the center of the pot with the crown slightly covered (no more than 0.5 in. [13 mm]) or just exposed after they are thoroughly watered in. Planting the crown too deep will lead to crop variability and losses. After planting, I recommend drenching with a broad-spectrum fungicide such as etridiozole + thiophanate-methyl (Banrot) or the combination of mefenoxam (Subdue Maxx) and thiophanate-methyl (Cleary's 3336).

The main cultural problem of *Astilbe* is improper water management. They should be kept evenly moist throughout production. Water stress causes the leaf margins to turn brown and become crispy, often rendering the plant unsalable. Moisture stress is likely to occur at two points during crop production. The first being a few weeks after potting; the foliage is flushed, but there are only a few roots (or no roots) to support the transpiration rate of the leaves. Under certain circumstances the root system simply cannot keep up with the water demands of the top growth, and the leaves become scorched. In this situation, many growers observe that the growing mix had an adequate moisture level, but with an insufficient root system, the plant is not able to take up water fast enough. The second point at which growers frequently observe moisture stress is when the plants are fully flushed and too much time has lapsed between irrigations, causing the plant to become scorched.

I recommend watering thoroughly when irrigation is required, allowing the substrate to dry slightly between waterings. Overlooking the water status and needs of an *Astilbe* crop just one time can seriously impact the quality and salability of the entire crop.

Growers should also take steps to reduce water stress, particularly when caused by high light levels. During the late spring and summer months, most growers produce *Astilbe* in a shade facility such as a retractable roof structure or under a 30% shade cloth.

During production, the pH of the medium should be maintained at 5.8–6.2. Feeding is generally not necessary during the first few weeks of production. When actively growing, *Astilbe* are moderate feeders, requiring a controlled-release fertilizer incorporated at a rate equivalent to 1 lb. nitrogen per cubic yard (454 g/0.76 m^3) of growing medium or 50–100 ppm nitrate delivered under a constant liquid fertilizer program. They are sensitive to high salts and may experience root injury and become scorched if the soluble salt levels are allowed to build up. *Astilbe* are particularly sensitive to high salt levels before they become established—the combination of water stress and high salts at this stage can wreck havoc on the crop. I recommend growers monitor the soluble salt levels routinely and leach them out with clear water if the EC rises above 2.0 using a 2:1 extraction method.

Insects

Aphid	Leafhopper
Black vine weevil	Root-knot nematode
Caterpillar	Soft scale
Foliar nematode	Spider mite
Grasshopper	Whitefly

Astilbe are not susceptible to many insect pests. Aphids, spider mites, and whiteflies are the insects that are the most prevalent.

Diseases

Botrytis	*Fusarium wilt*
Erwinia	Powdery mildew
Discosia (leaf spot)	*Pythium*

Astilbe are not susceptible to many diseases. Crown rot caused by *Fusarium* or *Pythium* is likely to occur when crops are overwatered or if the growing medium has insufficient drainage. None of these diseases or pests requires preventative control strategies. Growers should have routine scouting programs to detect their presence early and to determine if and when control strategies are necessary.

Many *Astilbe* cultivars do not require height control strategies if they are grown with adequate spacing between the plants. When grown under high plant densities, plant height can be controlled by applying daminozide at 2,500 ppm or uniconazole-P at 5 ppm to the foliage as the plant canopy begins to enclose or once the inflorescences begin to elongate. Two applications seven days apart will provide an adequate reduction of the flowering stalk without altering the overall appearance of the plant. These are considered northern rates and should be adjusted accordingly.

Overwintering *Astilbe* is relatively simple. In the late fall, trim the plants back to 2 in. (5 cm) above the top of the container. Once plants are trimmed, group the pots together inside a cold frame, greenhouse, or outdoor

PGRs for *Astilbe* Cultivars*

Trade Name	Active Ingredient	Northern Rate
A-Rest	Ancymidol	25–50 ppm
B-Nine or Dazide	Daminozide	2,500 ppm
Bonzi, Piccolo, or Paczol	Paclobutrazol	30 ppm
Cycocel	Chlormequat chloride	1,250 ppm
Sumagic	Uniconazole-P	5 ppm
B-Nine or Dazide + Cycocel	Daminozide + chlormequat chloride	2,500 ppm + 1,250 ppm
B-Nine or Dazide + Sumagic	Daminozide + uniconazole-P	2,000 ppm + 3 ppm

*Most of the PGR research has been conducted on *Astilbe arendsii* hybrids. Although all cultivars of the various *Astilbe* groups have not been tested, similar results have been observed using the PGRs in this table. These are northern rates and need to be adjusted to fit the location where the crop is being produced. These rates are for spray applications and geared to growers using multiple applications.

production bed. To improve the health of the root zone going into the winter months, I find it beneficial to drench the plants after trimming with a broad-spectrum fungicide. In colder zones, it is also beneficial to cover them with a protective frost fabric during the winter months. During this season, if there are periods where the temperatures remain above freezing, open the cold frame doors or provide ventilation during the day to keep the temperatures inside as cool as possible. Provide adequate ventilation any time the outdoor temperatures are above 40° F (4° C). It is important to spread mouse baits, such as diphacinone (Ramik), throughout the production area, as mice are likely to feed on the crowns during the overwintering period.

Forcing

Astilbe can be planted and successfully forced into bloom either by planting them in the late summer and overwintering the established plants in finished containers or by planting overwintered bare-root divisions or plugs in the early spring. Using established plant materials will ensure the plants are not juvenile and they can perceive the other factors that promotes flowering. *Astilbe* require cold in order for them to produce bloom. Providing ten to twelve weeks of cold at 35–45° F (2–7° C) should be sufficient. The time to bloom varies widely with cultivar and production temperatures.

The photoperiod has very little effect on flowering. Some cultivars will reach flowering approximately one week earlier when long days are provided. They can be successfully forced into bloom using natural photoperiods or by providing night interruption lighting. *Astilbe* grown under long days tend to be slightly taller than when natural photoperiods are used.

Generally, early to mid-season cultivars will flower using the following temperature recommendations. Late-season cultivars usually require two to three additional weeks of forcing time. *Astilbe* grown at average temperatures of 65° F (18° C) will reach a finished size in six to eight weeks and will be flowering in twelve to fourteen weeks. Forcing at warmer average temperatures (68–72° F [20–22° C]) will only slightly decrease the time to flower (by approximately one week). Temperatures above 72° F (22° C) are likely to lead to heat and water stress, possibly causing injury to the leaves. Cool

Figure 12.6. ***Astilbe chinensis* 'Vision' prior to shipping. False spireas are often shipped just before the plumelike flowers open.**

production temperatures (60–65° F [16–18° C]) increase the number of flowers produced and intensify their color.

References

Freitas, Stina. "Propagating Astilbes." *GrowerTalks.* July 2002.

Greenhouse Grower Magazine and Michigan State University. *Firing Up Perennials: The 2000 Edition.* Willoughby, Ohio: Meister Publishing. 2000.

Nau, Jim. *Ball Perennial Manual: Propagation and Production.* Batavia, Ill.: Ball Publishing. 1996.

Pilon, Paul. Perennial Solutions: "*Astilbe chinensis* 'Visions'." *Greenhouse Product News.* August 2005.

Campanula carpatica

Carpathian Bellflower

Bellflowers have been used in gardens for over five hundred years. With over three hundred species in the genus, a number of forms and textures are available for landscapers to incorporate into today's landscape. *Campanula carpatica* is one of the popular perennial species being grown across the country. The common names, Carpathian bellflower and Carpathian harebell, are derived from their native habitat, the Carpathian mountains of Eastern Europe. In Latin, the name *Campanula* means "little bell" and refers to the bell shape of most *Campanula* flowers.

Figure 12.7. *Campanula carpatica* 'Pearl Deep Blue'

Commonly Grown *Campanula carpatica* Cultivars

'Alba'	'Pearl Deep Blue'
'Bliss Blue'	'Pearl White'
'Bliss White'	'Uniform Blue'
'Blue Clips'	'Uniform White'
'Dark Blue Clips'	'White Clips'
'Light Blue Clips'	

With a compact habit forming small floriferous mounds reaching 9–12 in. (23–30 cm) in height, gardeners and landscapers commonly use *Campanula* cultivars as an attractive and showy perennial in today's gardens. Cultivars of *Campanula carpatica* are commonly used as accent plants, border plants, in rock gardens, container plantings, and as potted house plants. *Campanula* flowers are primarily blue or white, cup-shaped, face upwards, and are held above the foliage. Carpathian bellflowers naturally bloom continuously from June through August. Many cultivars are suitable for production in USDA Hardiness Zones 3–9 and AHS Heat Zones 9–1; there may be slight hardiness and heat tolerance differences between cultivars. Bellflowers prefer to be grown in full sun, although locations receiving partial sun are often acceptable.

Campanula belongs to the Campanulaceae family. There are several perennials in the *Campanulaceae* family being commercially produced, including *Adenophora, Codonopsis, Jasione, Lobelia,* and *Platycodon.*

Propagation

Campanula carpatica cultivars are propagated from seed or by cuttings. Due to economics and ease of production, they are almost always propagated by seed. Sow *Campanula* seeds in the intended plug flat and do not cover the seed with germination mix or vermiculite. Light is required for germination; covering reduces the amount of light reaching the seed and may adversely affect the germination process. The seed flats should be moistened and moved to a warm environment, where the temperatures can be maintained at 65–70° F (18–24° C) for germination. Many growers utilize germination chambers during this stage to provide uniform moisture levels and temperatures. It is important to keep the medium very moist but not saturated during this stage or the germination rates are likely to decrease.

Seedlings will emerge over a period of time ranging from ten to eighteen days after sowing. Following germination, reduce the moisture levels somewhat, allowing the growing medium to dry out slightly before watering to help promote rooting. Some growers have found it beneficial to lightly cover the emerging seedlings with medium-grade vermiculite following germination to help maintain moisture levels. Fertilizers are usually applied once the true leaves are present, applying 100 ppm nitrogen every third irrigation or 50 ppm with every irrigation, using a balanced water-soluble source. Maintain photoperiods of less than thirteen hours, throughout all plug stages to keep plants in the vegetative state. When plugs are grown at 65° F (18° C), they are usually ready for transplanting in nine to eleven weeks.

Growers often have difficulty germinating bellflowers during the summer months. I recommend to avoid starting them during this time of year, unless cool conditions and good moisture management can be provided until the first true leaves are present.

Production

For container production, *Campanula carpatica* is suitable for 1 qt. (1 L) to 1 gal. (3.8 L) containers. Most growers receive starting materials as finished plug liners. When planting large containers, such as 1 gal. (3.8 L) or larger pots, it is best to plant at least two plug cells per container to properly fill out the pot. Bellflowers perform best when they are grown in a moist, well-drained medium with a slightly acidic pH, 5.8–6.2. Most commercially available peat- or bark-based growing mixes work well, provided there is adequate drainage.

When planting, the plugs should be planted so the original soil line of the plug is even with the surface of the growing medium of the new container. Planting the crown too deep will lead to crop variability and losses. After planting, I recommend drenching with a broad-spectrum fungicide such as etridiozole + thiophanate-methyl (Banrot) or the combination of mefenoxam (Subdue Maxx) and thiophanate-methyl (Cleary's 3336). The best quality is achieved when plants are grown in full sun or in greenhouses with high light intensities.

Campanula can be grown using light to moderate fertility levels. Fertility can be delivered using water-soluble or controlled-release fertilizers. Growers using water-soluble fertilizers either apply 100–150 ppm nitrogen as needed or feed with a constant liquid fertilization program using rates of 50–75 ppm nitrogen with every irrigation. Grower's commonly apply time-release fertilizers as a topdressing onto the medium's surface using the medium labeled rate or incorporate them into the growing medium prior to planting at a rate equivalent to 1 lb. nitrogen per cubic yard (454 g/0.76 m^3) of growing medium.

Campanula require an average amount of irrigation, as they do not tolerate very wet or overly dry conditions. Root zones that remain waterlogged tend to get root rot pathogens and can quickly lead to crop losses. Overly dry growing conditions greatly reduce crop quality and delay flowering. When irrigation is necessary, water them thoroughly, then allow the soil to dry slightly between waterings.

Insects

Aphid	Slug
Caterpillar	Snail
Foliar nematode	Spider mite
Leafhopper	Thrips
Root knot nematode	Whitefly

Bellflowers are relatively free of serious problems associated with insects. Occasionally, aphids, caterpillars, leafhoppers, spider mites, thrips, and whiteflies may appear, causing only a minimal amount of crop injury. None of these insect pests require preventative control strategies. Growers should have routine scouting programs to detect the presence of insects early and to determine if and when control strategies are necessary.

Diseases

Aster yellows	*Pythium*
Botrytis	*Rhizoctonia*
Botrytis blight	Rust
Phyllosticta leaf spot	*Sclerotium*
Powdery mildew	*Verticillium* wilt

Campanula carpatica can generally be grown free of plant pathogens. The primary diseases growers should watch for is *Pythium* and *Rhizoctonia* crown and root rots. Plants are most susceptible to these diseases when they are grown under cool and wet conditions, such as going into or coming out of winter dormancy. *Botrytis* is another disease that could become problematic. *Botrytis*, like *Pythium* and *Rhizoctonia*, often occurs around the overwintering process but is also likely to occur under dense plant canopies.

To control these diseases, it is best to manage the environment by providing proper plant spacing and adequate air movement and controlling the humidity. Growers should carefully watch the moisture levels during adverse times of the year and avoid overwatering their plants. Many growers apply a broad-spectrum fungicide drench before and just after the overwintering period to reduce the population of these pathogens.

Campanula carpatica cultivars have naturally compact growth habits and will usually not require PGRs to control plant height. However, during the winter months, during periods of low light levels, or when grown at high plant densities, excessive plant stretching may occur, requiring some type of height management strategy. Before applying these chemicals, the height of bellflowers can often be effectively controlled by providing adequate spacing between the plants and by withholding water and nutrients. Most commercially available PGRs are effective at controlling plant height of bellflowers. In fact, *Campanula* cultivars are sensitive to several of them, paclobutrazol and uniconazole-P in particular. The application of these products could lead to too much growth regulation and plant stunting. For toning and shaping purposes, one application is often adequate. When additional height control is necessary, make a second application seven to ten days following the first.

Forcing

For container production, *Campanula carpatica* cultivars are well suited for forcing into bloom throughout the year. When producing blooming plants is the goal, a few requirements should be met in order to produce uniform, consistent, high-quality flowering plants.

Bellflowers do not have a cold requirement that must be met prior to forcing. They will easily bloom from plants started by seed during the first growing season without a cold treatment.

PGRs for *Campanula carpatica* Cultivars

Trade Name	Active Ingredient	Northern Rate
A-Rest	Ancymidol	25 ppm
B-Nine or Dazide	Daminozide	2,500 ppm
Bonzi, Piccolo, or Paczol	Paclobutrazol	15 ppm
Cycocel	Chlormequat chloride	750 ppm
Sumagic	Uniconazole-P	2.5 ppm

These are northern rates and need to be adjusted to fit the location where the *Campanula* crop is being produced. These rates are for spray applications and geared to growers using multiple applications.

Figure 12.8. **To expand sales opportunities, *Campanula* can be marketed and sold as houseplants that can be planted into the landscape. Shown: *Campanula carpatica* 'Pearl Deep Blue' and 'Pearl White'.**
Benary

Bellflowers do not have a juvenility period and have been observed flowering with as few as ten leaves present. In some cases, flowering occurs before the containers are bulked up, forming small unsalable blooming plants. Bulking promotes lateral branching, fills out the pot, and results in high-quality, attractive plants with numerous flowers at the time of sale.

The primary factor for flowering of *Campanula carpatica* crops is the day length. Flower induction and formation begins when day lengths are over fourteen hours. Growers should allow a period during production where the days are naturally short and the plants will remain vegetative, allowing them to bulk up prior to exposing them to conditions that promote flowering. During the times of the year where the day lengths are naturally long, growers should consider providing short days by blocking out all of the light. Short-day conditions are created by pulling black cloth or black plastic over the production site, ensuring the plants only receive twelve hours of light each day.

They are obligate long-day plants and will not flower under short days. As long as the photoperiod is less than twelve hours, *Campanula* will remain as compact, non-flowering rosettes. Growers should provide a minimum of fourteen-hour photoperiods or four-hour night interruptions until the flower buds are visible. Growers aim to deliver a minimum light intensity of 10 f.c. (108 lux) using cool-white fluorescent, high-pressure sodium, or metal halide lamps to provide photoperiodic lighting. To improve the quality and appearance of the crop, long days should not be provided until the plants are bulked up and have at least fifteen leaves.

Once flower buds are visible, growers can remove the lighting and produce them under natural day lengths to finish the remainder of the forcing. After visible bud, production under naturally short days decreases the overall height of the plant by reducing plant stretch. Incandescent light sources will also cause undesirable internode elongation, increasing the plants' overall height and appearance.

The time to bloom after the proper photoperiod is provided is a function of temperature. *Campanula carpatica* cultivars grown at 68° F (20° C) will take eight to nine weeks to reach flowering, while plants grown at 60° F (16° C) will flower in approximately twelve weeks. The size of the flowers is larger when they are forced at cooler temperatures. To obtain the best plant quality, I recommend producing bellflowers at 65–68° F (18–20° C) where possible.

References

Greenhouse Grower Magazine and Michigan State University. *Firing Up Perennials: The 2000 Edition.* Willoughby, Ohio: Meister Publishing. 2000.

Nau, Jim. *Ball Perennial Manual: Propagation and Production.* Batavia, Ill.: Ball Publishing. 1996.

Coreopsis grandiflora

Tickseed

Coreopsis, also referred to as tickseed, is a very popular perennial. The name *Coreopsis* means bug-like and comes from the Greek words *koris* (bug) and *opis* (like). The seed of *Coreopsis* resembles a tick, which obviously led to its common name.

Figure 12.9. ***Coreopsis grandiflora* 'Early Sunrise'**

Coreopsis belongs to the Asteraceae family (formerly Compositae). There are numerous perennials in the Asteraceae family being commercially produced. Some of them include *Achillea, Aster, Bellis, Centaurea, Doronicum, Echinops, Echinacea, Erigeron, Gaillardia, Helenium, Heliopsis, Leucanthemum, Ligularia, Rudbeckia,* and *Stokesia.*

Coreopsis grandiflora cultivars perform well across a wide portion of the United States, throughout USDA Hardiness Zones 3–9 and AHS Heat Zones 9–1. They prefer full sun, although in the South they perform best when some partial shade is provided. In the landscape, tickseed reaches 12–24 in. (30–61 cm) in height. These American natives are used as accent plants, border plants, in mass plantings, patio containers, and as cut flowers.

Commonly Grown *Coreopsis grandiflora* Cultivars

'Baby Sun'	'Rising Sun'
'Cutting Gold'	'Sunburst'
'Domino'	'Sunfire'
'Early Sunrise'	'Sunray'
'Flying Saucers'	x 'Tequila Sunrise'

Propagation

Depending on the cultivar, *Coreopsis* are started from seed or vegetatively propagated by tip cuttings or division. The method of propagation is often dependent on cultivar, availability, and cost. Although using bare-root divisions provides for a quicker finishing time, it often leads to crop variability and lower survival rates. Propagation from seed is the most common method used by commercial growers, since it is less expensive than bare-root divisions or tip cuttings.

Sow *Coreopsis* seeds in the intended plug flat and cover them lightly with germination mix or medium-grade vermiculite. The covering is optional but helps to maintain a suitable environment around the seed during this phase. The seed flats should be moistened and moved to a warm environment, where the temperatures can be maintained at 65–75° F (18–24° C) for germination. Many growers utilize germination chambers during this stage to provide uniform moisture levels and temperatures. Using germination chambers is optional, though, as *Coreopsis* cultivars will successfully germinate in the greenhouse.

The seeds should be germinated in seven to ten days. Following germination, reduce the moisture levels somewhat, allowing the growing medium to dry out slightly before watering; this will help promote rooting. The light levels can gradually be increased as the plug develops, from 500 f.c. (5.4 klux) following germination to 2,500 f.c. (27 klux) at the final stage. Fertilizers are usually applied once the true leaves are present, applying 100–150 ppm nitrogen every third irrigation or 75 ppm with every irrigation, using a balanced water-soluble source. When plugs are grown at 65° F (18° C), they are usually ready for transplanting in five to seven weeks.

Tip cuttings should be taken while they are vegetative, or non-flowering. The cuttings should measure approximately 2–2.5 in. (5–6 cm) and contain several nodes. The well-drained rooting medium should be moistened prior to sticking. The base of the cuttings can be dipped in a rooting hormone, such as a solution of indolebutyric acid (IBA) at rates of 750–1,000 ppm prior to sticking. *Coreopsis* can successfully root without rooting compounds but tend to root slightly faster and more uniformly when these treatments are provided.

Cuttings should be placed under low misting regimes for the first two weeks of propagation. Prolonged exposure to mist may cause the leaves to rot during propagation. The misting can gradually be reduced as the cuttings form calluses and root primordia. It is beneficial to begin constant liquid feeding with 200 ppm nitrogen at each irrigation beginning ten days from sticking. The cuttings are usually rooted in less than three weeks with soil temperatures of 70–75° F (21–24° C). For best results, the air temperature during rooting should be maintained above 60° F (16° C) and below 80° F (27° C).

Production

Coreopsis cultivars are commonly produced in 1 qt. (1 L) to 2 gal. (7.6 L) containers. They perform best when grown in a moist, well-drained medium with the pH maintained at 5.8–6.4. Many commercially available peat- or bark-based growing mixes work well, provided there is good water-holding ability and adequate drainage.

Tickseed prefers to be grown in a moist but not wet growing medium. Water as needed when the plants are young and becoming established. Once the plants are large, they will require more frequent irrigations, as they will dry out rather quickly. Under stressful growing conditions, such as warm temperatures and high light levels, *Coreopsis* wilt very easily. Generally, if they are watered within a reasonable amount of time after they have begun to wilt, tickseed will recover quickly. When irrigation is needed, water them thoroughly, ensuring the entire growing medium is wet or nearly saturated. It is best to only allow the growing medium to dry slightly between irrigations.

Coreopsis cultivars are moderate feeders. Fertility can be delivered using water-soluble or controlled-release fertilizers. Growers using water-soluble fertilizers either apply 200 ppm nitrogen as needed or feed with a constant liquid fertilization program using rates of 75–100 ppm nitrogen with every irrigation. Controlled-release fertilizers are commonly applied as a topdressing on the medium's surface using the medium recommended rate on the fertilizer label or incorporated into the growing medium prior to planting at a rate equivalent to 1 lb. nitrogen per cubic yard (454 g/0.76 m^3) of growing medium. Plants grown under high fertility regimes generally become very lush and may take longer for them to flower.

Insects

Aphid	Slug
Beetle	Soft scale
Cucumber beetle	Spider mite
Leafhopper	Spittlebug
Leaf miner	Thrips
Plant bug	Whitefly
Root knot nematode	

Generally, *Coreopsis* can be produced relatively insect free. Aphids and whiteflies occasionally will become problematic. These pests can be controlled after they are detected or prevented using a proactive strategy. A preventative application of imidacloprid (Marathon 60WP) as a drench will generally ensure aphid-free plants from spring planting until the plants are shipped. Other preventative strategies include monthly spray applications of systemic chemicals such as pymetrozine (Endeavor), thiamethoxam (Flagship), imidacloprid (Marathon II), dinotefuran (Safari), or acetamiprid (TriStar). These strategies will also prevent the occurrence of whiteflies. Other insect pests listed in the table above are often observed feeding on tickseed but rarely become problematic. Most insects can be detected with routine crop monitoring. Control strategies may not be necessary unless the scouting activities indicate actions should be taken.

Diseases

Alternaria	Powdery mildew
Aster yellows	*Pseudomonas*
Botrytis	*Rhizoctonia*
Botrytis blight	Rust
Cercospora leaf spot	*Sclerotium*
Downy mildew	*Streptomyces* (scab)
Erwinia	*Verticillium*

Plant diseases may be observed when environmental conditions are favorable for their development. The most common diseases observed attacking *Coreopsis* crops are downy mildew and powdery mildew. As with many perennials, the occurrence of plant diseases can be negated or greatly reduced when the proper cultural practices are followed. To control foliar diseases, it is best to manage the environment by providing proper plant spacing and adequate air movement, controlling the humidity and, if desired, following a preventative spray program using the appropriate chemicals.

Downy mildew is often mistaken for powdery mildew and is more difficult to control. Downy mildew usually appears first on the undersides of the leaves as a mass of white or gray spores, and often the upper leaf surface (directly above where the spores are observed) will appear mottled, discolored, or blistered. Powdery mildew, on the other hand, appears as small white, talcumlike colonies on the upper leaf surfaces, but under the right conditions may engulf the plant with a powdery appearance. To control downy mildew, it is best to manage the environment by providing proper plant spacing and adequate air movement, controlling the humidity, and watering early in the day, which allows the foliage to be dry before night. Many chemicals work well on a preventative basis, but none will clean up an established downy mildew infestation. Preventative fungicide applications act as a barrier, not allowing the disease to infect the plant. The best strategy is to rotate products such as fosetyl-aluminum (Aliette), trifloxystrobin (Compass), azoxystrobin (Heritage), copper hydroxide + mancozeb (Junction), mancozeb (Protect T/O), dimethomorph (Stature DM), and mefenoxam (Subdue), making applications every seven to ten days beginning at the onset of favorable conditions for this disease.

Plant Growth Regulators

Coreopsis may require height control when grown under greenhouse and nursery conditions. Providing adequate spacing between the plants will reduce plant stretch caused by competition. When produced under low light

PGRs for *Coreopsis grandiflora*

Trade Name	Active Ingredient	Northern Rate
A-Rest	Ancymidol	25–50 ppm
B-Nine or Dazide or Dazide	Daminozide	2,500 ppm
Bonzi, Piccolo, or Paczol	Paclobutrazol	45 ppm
Cycocel	Chlormequat chloride	1,250 ppm
Sumagic	Uniconazole-P	5 ppm
B-Nine or Dazide + Bonzi, Piccolo, or Paczol	Daminozide + paclobutrazol	2,000 ppm + 15 ppm
B-Nine or Dazide + Cycocel	Daminozide + chlormequat chloride	2,500 ppm + 1,000 ppm
B-Nine or Dazide + Sumagic	Daminozide + uniconazole-P	2,000 ppm + 3 ppm

These are northern rates and need to be adjusted to fit the location where the crop is being produced. These rates are for spray applications and geared to growers using multiple applications.

levels or at high plant densities, it may be necessary to use chemical plant growth regulators. Several of the commercially available PGRs are effective at controlling plant height when they are applied using the appropriate rates, frequency, and timing. A chart containing the effective height control products and northern rate recommendations is provided on page 399. For foliar applications, it usually requires two or three applications at seven-day intervals to provide adequate height control. Begin the PGR applications when the leaves from adjacent plants are just beginning to touch one another.

Forcing

Producing flowering *Coreopsis grandiflora* cultivars out of season is relatively easy, provided a few guidelines are followed. There are differences regarding the specific flower requirements of certain cultivars, namely 'Early Sunrise', which are discussed below.

Research has shown that *Coreopsis* cultivars must reach a particular size, or maturity, before they are capable of flowering. Juvenile plants are incapable of flowering and will not flower in response to vernalization or day length. It is recommended to begin the forcing process using mature plant materials, consisting of at least sixteen leaves. Therefore, it is recommended to grow *Coreopsis* at photoperiods of thirteen hours or less until at least sixteen leaves have formed. Temperatures of 70–75° F (21–24° C) will promote rapid development during this growth phase.

Most *Coreopsis* cultivars have an obligate cold requirement in order for them to flower. It is recommended to cool (vernalize) plugs or small containers of tickseed for a minimum of ten weeks at 35–44° F (2–7° C). Regardless of the container size, growers must ensure that plants are fully rooted and past the juvenile stage prior to exposing them to cold temperatures. The cultivar 'Early Sunrise' is the exception, as it does not have an obligate cold requirement and will readily flower without a cold treatment.

All cultivars, including 'Early Sunrise', are considered to be obligate long-day plants, absolutely requiring long days for them to flower. With photoperiods of less than fourteen hours, plants will not flower. Following vernalization, it is recommended to provide at least fourteen-hour photoperiods or night interruption lighting when the natural photoperiod is less than fourteen hours.

During naturally short days, use cool-white fluorescent, high-pressure sodium, incandescent, or metal halide lights to extend the day length or to deliver four-hour night interruption; lights should deliver 10 f.c. (108 lux) of light to promote flowering. Using incandescent light sources will promote flowering but will cause the internodes to stretch excessively, reducing the overall quality attributes of the plant.

Once flower buds are visible, growers can remove the lighting and produce them under natural day lengths to finish the remainder of the production schedule. After visible bud, production under naturally short days decreases the overall height of the plant by reducing plant stretch. The drawbacks of removing plants from long photoperiods are a reduction in the number of flower buds produced and it will take a few extra days for them to flower. If the photoperiodic lighting is removed, it is best to wait until at least seven to ten days past visible bud to return the plants to natural day lengths. This will promote the development of additional flowers buds and improve the overall appearance of the plant while it is blooming.

The time to bloom after vernalization and the proper photoperiod is provided is a function of temperature. The following temperature recommendations fit most varieties, but each cultivar will have a slightly different finishing time at each of the temperatures illustrated. *Coreopsis grandiflora* grown at 68° F (20° C) will take eight to nine weeks to reach flowering, while plants grown at 60° F (16° C) will flower in eleven to twelve weeks. Temperatures above 70° F (21° C) will hasten plant and flower development but also reduce the number of flowers produced. To obtain the best plant quality, I recommend producing *Coreopsis* at 65–68° F (18–20° C).

To obtain the highest quality finished plants, particularly for large container sizes, growers should consider planting plugs in the desired pot during the late summer of the year prior to the intended date of sale. Established containers should be overwintered in a protected area prior to spring forcing. Planting liners of *Coreopsis* in this manner will allow them to bulk up, produce more flowers per plant, and bloom earlier than when they are planted and grown only in the spring. When small container sizes, such as 1qt. (1 L) pots, are desired, it is usually best to transplant vernalized plugs into the final container and immediately provide the proper photoperiod and temperatures.

References

Greenhouse Grower Magazine and Michigan State University. *Firing Up Perennials: The 2000 Edition.* Willoughby, Ohio: Meister Publishing. 2000.

Delphinium grandiflorum

Larkspur

Delphinium grandiflorum cultivars are early blooming, well-branched varieties that have ideal growth habits for today's commercial grower, retailers, and consumers. With an abundance of blue, lavender, or white flowers, combined with their compact habit, reaching only 8–18 in. (20–46 cm) in height, *Delphinium grandiflorum* cultivars are excellent choices for containers, mixed containers, and perennial borders.

Figure 12.10. *Delphinium grandifolium* 'Butterfly Blue'

Delphinium grandiflorum prefer full sun and grow well in USDA Hardiness Zones 3–7. They are classified by American Horticultural Society (AHS) as Heat Zone 6–1 plants, meaning they do not tolerate extreme summer temperatures. In southern portions of the country (Hardiness Zones 8 and higher), they are often grown as annuals due to the warm climate and should be produced under partial shade.

The name *Delphinium* in Greek translates to "dolphin flower," describing the shape of the spur on the upper part of the flower. *Delphinium* belongs to the Ranunculaceae family. There are numerous commercially produced perennials in this family, including *Aconitum, Anemone, Aquilegia, Cimicifuga, Clematis, Helleborus, Pulsatilla, Ranunculus,* and *Trollius.*

Commonly Grown *Coreopsis grandiflora* Cultivars

'Blue Butterfly'	'Summer Blues'
'Blue Mirror'	'Summer Nights'
'Blue Pygmy'	'Summer Stars'

Delphinium, also commonly referred to as larkspur, is rather short lived in the garden, often only surviving a couple of years. Regardless of where they are grown or how many years they live, gardeners and landscapers embrace the incomparable color and value *Delphinium* bring to the landscape.

Propagation

Delphinium grandiflorum cultivars can easily be started from seed. Growers commonly sow the seed into small cell sizes, such as 288-cell or 200-cell plug trays, filled with a germination medium that provides both good aeration and water-holding capacity. The germination percentages of some cultivars are slightly lower than many perennials, often ranging between 60–80%; these lower numbers can be overcome to a certain extent by multiple sowing two or three seeds per cell. At sowing, cover the seed lightly with germination mix or fine-grade vermiculite to help preserve moisture around the seed.

The seed flats should be moistened and moved to a warm environment, where the temperatures can be maintained at 68–72° F (20–22° C) for germination. Many growers utilize germination chambers during this stage to provide uniform moisture levels and temperatures. Starting larkspur inside of a germination chamber will increase both the germination rate and percent germination, while decreasing the time necessary for all of the seeds to sprout. Using germination chambers is optional, as they will successfully germinate in the greenhouse. It is important to keep the medium moderately moist (not saturated) during this stage, or the germination rates are likely to decrease. Providing high humidity levels during the germination process is more advantageous than maintaining a particular moisture level in the medium.

At these temperatures, germination will occur in fourteen to eighteen days. Following germination, reduce the moisture levels somewhat, allowing the growing medium to dry out slightly before watering to help promote rooting. Once germinated, the temperatures can be reduced to 60–68° F (16–20° C). Fertilizers are usually applied once the true leaves are present, applying 100 ppm nitrogen every third irrigation or 50 ppm with every irrigation, using a balanced water-soluble source. During the plug stage, it is important to keep the medium uniformly moist but not wet. *Delphinium grandiflorum* cultivars may vary slightly but are usually ready for transplanting in six to eight weeks.

Production

For container production, cultivars are well suited for 1 qt. (1 L) to 1 gal. (3.8 L) containers. Most growers receive starting materials as finished plug liners. When planting large containers, such as 1 gal. or larger, growers should plant at least two plug cells per container to properly fill out the pot. Larkspurs perform best when they are grown in a moist, well-drained medium with a slightly basic pH, 6.4–6.8. Most commercially available peat- or bark-based growing mixes work well, provided there is adequate drainage. The best quality is achieved when plants are grown in full sun or in greenhouses with high light intensities.

When planting, the plugs should be planted so the original soil line of the plug is even with the surface of the growing medium of the new container. Planting the crown too deep will lead to crop variability and losses from crown rot. After planting, I recommend drenching with a broad-spectrum fungicide such as etridiozole + thiophanate-methyl (Banrot) or the combination of mefenoxam (Subdue Maxx) and thiophanate-methyl (Cleary's 3336).

Larkspurs can be grown using moderate fertility levels. Fertility can be delivered using water-soluble or controlled-release fertilizers. Growers using water-soluble fertilizers either apply 100–150 ppm nitrogen as needed or feed with a constant liquid fertilization program using rates of 50–75 ppm nitrogen with every irrigation. Controlled-release fertilizers are commonly used to produce containerized larkspurs. Growers commonly apply time-release fertilizers as a topdressing onto the medium's surface using the medium rate or incorporated into the growing medium prior to planting at a rate equivalent to 1 lb. nitrogen per cubic yard (454 g/0.76 m^3) of growing medium. When they are underfed, they will take on an overall chlorotic appearance, and the lower leaves will turn yellow as nitrogen is translocated up the plant. When detected early, nitrogen deficiencies can easily be remedied with one or more drench applications of a water-soluble fertilizer, such as 20-20-20, using high rates (300 ppm nitrogen).

Although they are moderate feeders, they are sensitive to high salts and may experience root injury and root rots if the soluble salt levels are allowed to build up. *Delphinium grandiflorum* is particularly sensitive to high salt levels before it becomes established. Growers should monitor the soluble salt levels routinely and leach them out with clear water if the EC rises above 2.0 using a 2:1 extraction method.

Delphinium require an average amount of irrigation, as they do not tolerate very wet or overly dry conditions. They are susceptible to root rots when over-irrigated and should be grown on the dry side. When irrigation is necessary, I recommend watering them thoroughly, allowing the soil to dry slightly between waterings.

Insects

Aphid	Root knot nematode
Beetle	Slug
Borer	Snail
Cutworm	Sow bug
Cyclamen mite	Spider mite
Leafhopper	Thrips
Leaf miner	

Generally, *Delphinium* can be commercially produced in containers relatively free of insects. Aphids, thrips, and whiteflies may occasionally be observed feeding, but they rarely become problematic. None of these insect pests require preventative control strategies. Growers should have routine scouting programs to detect the presence of insects early and to determine if and when control strategies are necessary. Several of the insect pests listed in the table above are more likely to be apparent when larkspurs are in the landscape. However, in most cases, the damage is minimal and control strategies are seldom required.

Diseases

Aster yellows	*Pseudomonas*
Botrytis	*Pythium*
Botrytis blight	Rust
Crown gall	*Sclerotinia*
Erwinia	*Sclerotium*
Fusarium	Smut
Phytophthora	*Verticillium*
Powdery mildew	

As mentioned above, when the root zone remains wet for extended periods of time or is exposed to high salt levels, root rots, namely *Pythium* and *Phytophthora*, are likely to occur. *Botrytis* is occasionally a problem on the lower foliage, where air movement is limited and the foliage often stays wet after irrigation for extended periods of time. To control this disease, it is best to manage the environment by controlling the humidity and providing proper plant spacing and adequate air movement. Watering early in the day will allow the foliage to dry before night, which will further decrease the likelihood of *Botrytis* infestations.

Plant Growth Regulators

Since *Delphinium grandiflorum* cultivars are fairly compact, it is usually not necessary to control the plant height. Under certain growing conditions or under high plant densities, it may be necessary, although not common, to use chemical PGRs. In the northern parts of the country, I would recommend applying uniconazole-P at 5 ppm. Apply growth regulators as the flower stalks are just beginning to elongate above the foliage. Two applications applied seven days apart will provide an adequate reduction of the flowering stalk without altering the overall appearance of the plant.

Forcing

Delphinium grandiflorum cultivars are easy to force into bloom any time of the year. Although there is not a juvenility phase, it is best to transplant plugs that have at least seven leaves in the final container.

They will easily bloom from plants started by seed during the first growing season without providing a cold treatment. Larkspur does not have a cold requirement, nor is cold beneficial to flowering. In fact, larkspurs exposed to a cold treatment are often somewhat delayed and exhibit sporadic flowering. Prolonged exposure to cold temperatures often results in plant mortality. Whenever possible, avoid putting them through a cold period and use fresh plug materials. If cold is provided, shorter dura-

PGRs for *Delphinium grandiflorum* Cultivars

Trade Name	Active Ingredient	Northern Rate
A-Rest	Ancymidol	25–50 ppm
B-Nine or Dazide	Daminozide	2,500 ppm
Bonzi, Piccolo, or Paczol	Paclobutrazol	30 ppm
Sumagic	Uniconazole-P	5 ppm
B-Nine or Dazide + Bonzi, Piccolo or Paczol	Daminozide +paclobutrazol	2,000 ppm + 15 ppm
B-Nine or Dazide + Cycocel	Daminozide + chlormequat chloride	2,500 ppm + 1,000 ppm
B-Nine or Dazide + Sumagic	Daminozide + uniconazole-P	2,000 ppm + 3 ppm

These are northern rates and need to be adjusted to fit the location where the crop is being produced. These rates are for spray applications and geared to growers using multiple applications.

tions of time, such as six weeks, are better than providing longer periods.

The cultivar 'Blue Mirror' is the exception; it is considered to be cold beneficial. This cultivar will flower without a cold treatment but flowers up to two weeks faster when a five-week cold treatment has been provided. There may be additional cultivars that exhibit advantages or disadvantages to providing a cold treatment. Unless you have seen research specifying the benefits of cold on current or future cultivars, err on the side of caution and omit vernalization from your *Delphinium grandiflorum* forcing programs.

Larkspurs will flower under any photoperiod and can be forced into bloom under natural day lengths. If the photoperiods during forcing are less than twelve hours, it will take longer for them to bloom, plant height is dramatically reduced, and most cultivars produce fewer flowers. In many cases, production under short days results in poor plant quality and unsalable product. As the day length increases, growers can expect the plant height and flower number to increase. For the best plant quality, I recommend to produce them under photoperiods of twelve hours or longer. Plant quality can also be improved during short days or periods where the light levels are naturally low by providing 400–500 f.c. (4.3–5.4 klux) of supplemental lighting using high-pressure sodium lamps.

Depending on the cultivars being produced and production temperatures, *Delphinium grandiflorum* will reach full bloom in approximately six to ten weeks. Most cultivars will take six weeks to reach flowering when grown at temperatures averaging 70° F (21° C), while plants grown at 60° F (16° C) will flower in ten weeks. The size of the individual flowers will be larger and more intensely colored when they are grown at cooler temperatures (60–65° F [16–18° C]). Conversely, forcing with warm production temperatures (76° F [24° C] or warmer) adversely affects the quality attributes of the crop by dramatically reducing the number of flower buds formed and decreasing plant height. To obtain the best plant quality, produce larkspurs at 63–68° F (17–20° C).

When the days are naturally short and the production temperatures are less than 55° F (13° C), larkspurs tend to remain as rosettes and have delayed flowering. To avoid rosette formation with cool production temperatures, it is beneficial to provide long days by extending the day length or using night interruption lighting.

References

Greenhouse Grower Magazine and Michigan State University. *Firing Up Perennials: The 2000 Edition.* Willoughby, Ohio: Meister Publishing. 2000.

Pilon, Paul. 2004. Perennial Solutions: "*Delphinium grandiflorum* 'Summer Nights'." *Greenhouse Product News.* April 2004.

Whitman, C., A. Cameron, E. Runkle, and R. Heins. "Herbaceous Perennials: Delphinium." *Greenhouse Grower.* July 2003.

Dianthus gratianopolitanus

Cheddar Pinks

Dianthus gratianopolitanus cultivars bear numerous solid or bicolor flowers of differing hues of pink, red, or white that are held above glaucous blue-green foliage. The flowers of most cultivars measure 1 in. (2.5cm) across and have a velvety appearance, highlighted by subtle to jagged margins. They form small compact mounds measuring about 7 in. (18 cm) tall by 12 in. (30 cm) wide and remain evergreen throughout the year.

Figure 12.11. ***Dianthus gratianopolitanus* 'Shooting Star'**

Plant Haven

Dianthus was named by the famous botanist Theophrastus, from the Greek words *dios,* meaning "divine," and *anthos,* which translates into "flower." The common name most often used for this genus is pinks, referring to the ragged edges of the flowers, which most often appear to have been cut by pinking shears. *Dianthus gratianopolitanus,* which is commonly called cheddar pinks, is one of the most widely grown and marketed species and is well suited for landscapes across much of the country. The popularity of cheddar pinks is likely to continue based on its garden performance and desirable characteristics. Based on these attributes, the Perennial Plant Association named *Dianthus gratianopolitanus* 'Firewitch' as the 2006 Plant of the Year.

Commonly Grown *Dianthus gratianopolitanus* Cultivars

'Bath's Pink'	'Mountain Mist'
'Double Spotty'	'Spotty'
'Firewitch'	'Star' series
'Grandiflorus'*	'Tiny Rubies'

*Propagated from seed

The genus *Dianthus* contains over three hundred species, many of which are very suitable for commercial production as bedding plants, biennials, and perennials. *Dianthus* belongs to the Caryophyllaceae family. There are numerous commercially produced perennials in the Caryophyllaceae family, including *Arenaria, Cerastium, Gypsophila, Lychnis, Minuartia, Moehringia, Sagina, Saponaria, Silene,* and *Stellaria.*

Cheddar pinks are widely produced and used in landscapes throughout USDA Hardiness Zones 3–10 and AHS Heat Zones 8–1. They prefer to be grown in full sun, although locations receiving partial sun are often acceptable. Cheddar pinks bloom prolifically in early summer and will continue to bloom sporadically throughout the season. Deadheading, or removing spent blooms, is highly recommended to promote a reflush of flowers.

Propagation

Depending on the cultivar, *Dianthus gratianopolitanus* cultivars are started from seed or vegetatively propagated by tip cuttings or bare-root divisions. They are primarily vegetatively propagated by cuttings. There are a few cultivars commercially grown from seed, 'Grandiflorus' being the most common. Many cultivars are patented, and unlicensed propagation is prohibited. To avoid illegal activities and the consequences, always check for plant patents before attempting to propagate *Dianthus.*

Tip cuttings should be taken while they are vegetative. Cuttings that are already in bloom will take longer to root and have a lower survival rate than purely vegetative starting materials. Propagators have the most success when propagating during the early spring or late summer, when the plants are not flowering and are growing vegetatively. Tip cuttings should measure approximately 2 in. (5 cm) and contain several nodes. They are generally stuck into plug liners for rooting. The well-drained rooting medium should be moistened prior to sticking. The base of the cuttings can be dipped into a rooting hormone, such as a solution of indolebutyric acid (IBA) at rates of 750–1,000 ppm prior to sticking. *Dianthus* can successfully root without rooting compounds but tend to root slightly faster and more uniformly when these treatments are provided.

Cuttings should be placed under low misting regimes for about the first two weeks of propagation. When possible, it is best to propagate under high humidity levels (90% relative humidity) with minimal misting. Prolonged exposure to mist may cause the leaves to rot during propagation. The misting can gradually be reduced as the cuttings form calluses and root primordia. They are usually rooted in approximately three to four weeks with soil temperatures of 68–73° F (20–23° C). For best results, the air temperature during rooting should be maintained above 60° F (16° C) and below 80° F (27° C). It is beneficial to begin constant liquid feeding with 200 ppm nitrogen at each irrigation beginning ten days from sticking.

Production

For container production, *Dianthus gratianopolitanus* cultivars are suitable for 1 qt. (1 L) to 1 gal. (3.8 L) sized containers. Most growers receive starting materials of cheddar pinks as finished plug liners. They can be started using bare-root materials; however, the quality of the finished product is often greatly reduced. When bare-root materials are used, it is not uncommon for the center of the plant to fall open and appear brown and empty, greatly reducing the aesthetic appearance and marketability of the plant. Growers will achieve superior performance and quality attributes when plug liners are used for the starting materials.

When planting large containers, such as 1 gal. (3.8 L) pots, growers should plant two plug cells per container to properly fill out the pot and reduce the production time. Spring-planted cheddar pinks often remain small and produce bloom before the plants have reached a marketable size. For spring sales, it is highly beneficial to plant them during the late summer of the previous year, allowing them to bulk up prior to overwintering. When planted in this manner, only one plug per pot is necessary in most instances. For small containers, such as 1 qt. (1 L) pots, growers can successfully finish them with only one plug liner planted into each container, regardless of when they are planted.

Dianthus perform best when they are grown in a well-drained, porous growing medium with a slightly acidic pH, 6.0–6.5. Most commercially available peat- or bark-based growing mixes work well, provided there is adequate drainage. When planting, the plugs should be planted so the original soil line of the plug is even with the surface of the growing medium of the new container. Planting the crown too deeply will lead to crop variability and losses. After planting, particularly if bare-root starting materials are used, I recommend drenching with a broad-spectrum fungicide such as etridiozole + thiophanate-methyl (Banrot) or the combination of mefenoxam (Subdue Maxx) and fludioxonil (Medallion).

Dianthus gratianopolitanus require an average amount of irrigation, as they do not tolerate very wet or overly dry conditions. When moist or wet conditions occur, they are very susceptible to root rots. When irrigation is necessary, water them thoroughly, and then allow the soil to dry slightly between waterings.

Cheddar pinks are moderate feeders; fertility can be delivered using water-soluble or controlled-release fertilizers. Growers using water-soluble fertilizers either apply 100–150 ppm nitrogen as needed or feed with a constant liquid fertilization program using rates of 50–75 ppm nitrogen with every irrigation. Growers commonly apply time-release fertilizers as a topdressing onto the medium's surface using the medium rate, or incorporated into the growing medium prior to planting at a rate equivalent to 1 lb. nitrogen per cubic yard (454 g/0.76 m^3) of growing medium.

The best quality is achieved when plants are grown in full sun or in greenhouses with high light intensities, 3,000–5,000 f.c. (32–54 klux) is sufficient. When produced under lower light levels, the stems will become leggy and overall plant quality will be reduced. During the winter months, crop quality can be greatly improved when 400–500 f.c. (4.3–5.4 klux) of supplemental lighting is provided.

Insects

Aphid	Root knot nematode
Cabbage looper	Slug
Caterpillar	Spider mite
Cutworm	Spittlebug
Cyst nematode	Thrips
Fungus gnat larva	

Aphids, caterpillars, and thrips occasionally will become problematic. The primary insect observed feeding on *Dianthus gratianopolitanus* is aphids. Although *Dianthus* is an acceptable food source for aphids, it does not seem to be their favorite. With aphids and other insect pests only occurring occasionally, growers do not have to implement preventative programs for them, but can detect their presence through weekly scouting activities to determine if control measures are necessary.

Diseases

Alternaria	*Pellicularia* (stem canker)
Botrytis	*Phyllosticta* (leaf spot)
Botrytis blight	*Phytophthora*
Cladosporium (leaf spot)	*Pythium*
Colletotrichum (anthracnose)	*Rhizoctonia*
Crown gall	Rust
Erwinia	*Sclerotinia*
Fusarium	

Dianthus can generally be grown free of plant pathogens. The primary diseases growers should watch for is *Pythium* and *Rhizoctonia* crown and root rots. They are most susceptible to these diseases when they are grown under cool and wet conditions, such as going into or coming out of winter dormancy. Leaf spots caused by the fungal pathogens *Alternaria*, *Cladosporium*, and *Phyllosticta* are other diseases that could become problematic. Similar to crown and root rots, leaf spots often appear while overwintering and are also likely to occur under dense plant canopies.

As with many perennials, the occurrence of plant diseases can be negated or greatly reduced when the proper cultural practices are followed. To control these diseases, it is best to manage the environment by providing proper plant spacing and adequate air movement, and controlling the humidity. Growers should carefully watch the moisture levels during adverse times of the year and avoid overwatering their plants. Many growers apply a broad-spectrum fungicide drench before and just after the overwintering period to reduce the population of these pathogens. Routine scouting is useful and recommended to detect plant diseases early, allowing the appropriate control strategies to be implemented before significant crop injury or mortality occurs.

Plant Growth Regulators

Due to its naturally small growth habit, controlling plant height with chemical growth regulators should not be necessary. However, during the winter months, during periods of low light levels, or when grown at high plant densities, excessive plant stretching might occur, especially as they begin to develop flowers, requiring some type of height management strategy. The height of *Dianthus* can often be effectively controlled by providing adequate spacing between the plants. Cheddar pinks can be grown at high plant densities throughout most of the production cycle. As the flower buds develop and the stems begin to elongate, typically the last few weeks of production, it is beneficial to space the pots slightly further apart.

If excessive plant stretch is occurring, the height can be controlled using the combination of daminozide at 2,000 ppm plus uniconazole-P at 3 ppm. These are northern rates and need to be adjusted to fit the location where the crop is being produced. For toning and shaping purposes, one application is often adequate. When additional height control is necessary, make a second application seven to ten days following the first.

Forcing

When producing blooming plants is the goal, a few requirements should be met in order to achieve, consistent, high-quality flowering plants. Growers should begin the forcing process using plant materials that have been bulked up to nearly fill out the container they are being produced in. For plant establishment, it is best to provide natural photoperiods and maintain temperatures of at least 65° F (18° C).

To produce full plants with numerous blooms, most cultivars of *D. gratianopolitanus* have an obligate cold requirement for flowering. Many cultivars will flower sporadically and have low bud counts when vernalization is omitted. Growers should cool large containers of cheddar pinks that have been bulked up for six to nine weeks at 35–44° F (2–7° C). Plugs for spring transplanting into small containers can successfully be vernalized prior to planting.

Since they remain evergreen throughout the cold treatment, growers should provide 100–200 f.c. (1.1–2.2 klux) of light during this period. If plants are overwintered in a greenhouse or cold frame structure, the natural light levels should be sufficient. When coolers are used, supplemental lighting is best provided using cool-white fluorescent lamps.

After the cold requirement is achieved, cheddar pinks can be grown at any day length, as they are day-neutral plants. The length of the photoperiod does not have any effect on the number of blooms produced. *Dianthus gratianopolitanus* flowers faster when grown under higher light intensities and/or long days. Providing supplemental lighting is not necessary but will reduce the time to flower, especially when grown under low light levels. Therefore, cheddar pinks are sometimes considered to be long-day beneficial plants.

The time to bloom after vernalization is a function of cultivar, light intensity, photoperiod, and temperature. Depending on the cultivars being produced, cheddar pinks will reach full bloom in approximately seven to eight weeks when they have been grown at temperatures averaging 65° F (18° C). Producing them at cooler temperatures increases the time to flower but will improve the overall quality characteristics of the plant, such as the color intensity of the foliage and flowers. To achieve the highest plant quality, I recommend growers produce *Dianthus gratianopolitanus* using cool temperatures, such as 60 to 65° F (16–18° C). When they are grown under naturally short day lengths and/or low light levels, growers should expect an additional two to three weeks of production time for them to reach flowering.

References

Pilon, Paul. Perennial Solutions: "*Dianthus* 'Eastern Star'." *Greenhouse Product News*. June 2004.

Digitalis purpurea

Foxglove

Digitalis, also referred to as foxglove, belongs to the Scrophulariaceae family, which contains numerous commercially produced perennials, including *Antirrhinum, Chelone, Linaria, Penstemon, Verbascum,* and *Veronica.* The word *digitalis* means "a finger" in Latin and refers to the numerous fingerlike, nodding flowers. Foxglove is native to Britain and has become a very popular perennial in the American landscape.

Figure 12.12. ***Digitalis purpurea* 'Apricot'**
Jelitto Perennial Seeds

Foxglove forms large 2–3 ft. (61–91 cm) wide clumps with dark green foliage, and produces 2–6 ft. (0.6–1.8 m) tall vertical flower spikes in the early summer. When blooming, they create bold and colorful displays with their huge lavender, pink, purple, rose, yellow, or white tubular florets. Most cultivars have the characteristic maroon spotting and speckling on the lower lip of the corollas. In addition to the main flower stalk, several varieties form secondary flower spikes, extending the bloom time to over four weeks in many instances. Many cultivars will flower reliably the first year from seed and will bloom heavily the second year, while others will only flower following a cold treatment.

Foxgloves grow best in partial shade but can tolerate full sun provided they are kept moist and not allowed to dry out. They are hardy in USDA Hardiness Zones 4–9 and AHS Heat Zones 9–1. There are slight heat and cold tolerance differences between cultivars. They are commonly used as accent plants and in background or mass plantings. Additionally, gardeners use foxgloves as cut flowers and to attract butterflies and hummingbirds into the garden.

Foxgloves are old-fashioned perennials, grown less today than they were in the past, but they seem to be on the comeback trail and gaining popularity once again. *Digitalis purpurea* cultivars are biennials, not true perennials, and are usually treated as annuals or biennials in the garden. Clumps of *Digitalis* seem to persist for years due to the ability to self-seed. 'Foxy' has been grown and marketed as an annual in recent years since it was an All-America Selections winner and it flowers the first year from seed. New F_1 hybrids, such as the 'Camelot' series, are being developed to achieve reliable first-year flowering, extended bloom times, and uniform growth habits. With its bold and impressive flower displays and improved plant and flowering characteristics, *Digitalis purpurea* will be embraced by perennial enthusiasts for years to come.

Propagation

Digitalis purpurea cultivars are propagated from seed. The seed is small and difficult to handle, though some of the new cultivars, such as 'Camelot', are pelleted, making it easier to handle and more suitable for automated seeders. Sow foxglove seeds in the intended plug flat and do not cover the seed with germination mix or vermiculite. Light is required for germination; covering reduces the amount of light reaching the seed and may adversely affect the germination process. The seed flats should be moistened and moved to a warm environment, where the temperatures can be maintained at 65–70° F (18–21° C).

Many growers utilize germination chambers during this stage to provide uniform humidity levels and temperatures. Starting foxglove inside of a germination chamber will increase both the germination rate and percent germination while decreasing the time necessary for all of the seeds to sprout. It is important to keep the medium very moist (not saturated) during this stage or the germination rates are likely to decrease. Providing high humidity levels during the germination process is more advantageous than maintaining a particular moisture level within the medium. Using germination chambers is highly beneficial but optional, as foxglove can be successfully germinated in the greenhouse when the proper humidity and moisture levels can be maintained.

Seedlings will emerge within four to nine days after sowing. Following germination, reduce the moisture levels somewhat. Allowing the growing medium to dry out slightly before watering will help promote rooting. Once germinated, the temperatures can be reduced to 60–65° F (16–18° C). At these temperatures, it takes six to eight weeks from sowing for 128-cell plugs to reach transplantable size.

During the plug stage, it is important to keep the medium uniformly moist but not wet. Do not fertilize the seed flats prior to germination, as *Digitalis purpurea* cultivars are sensitive to high salt levels during this stage. Fertilizers are usually applied once the true leaves are present, applying 100 ppm nitrogen every third irrigation or 50 ppm with every irrigation, using a balanced water-soluble source.

Production

Digitalis purpurea performs best when grown in a moist, well-drained medium with a pH of 5.8–6.2. It performs well in many commercially available peat-based growing mixes, and a number of growers have success using peat- and bark-based growing mixes. The most important principal of the growing mix is that it provides the appropriate physical properties. *Digitalis* performs best when an adequate amount of moisture is present at all times. Therefore, it is important to provide a growing medium that has an above average water-holding ability and still provides sufficient drainage, allowing some air to be in the root zone at all times. Due to their large size when flowering, it is best to produce *Digitalis* in 1 gal. (3.8 L) or larger containers.

Generally, growers plant one plug per container. When planting, the plug should be planted so the original soil line of the plug is even with the surface of the growing medium of the new container. Planting them too deep will lead to crop variability and losses.

Foxgloves are moderate to heavy feeders and perform well when constant liquid fertilization programs are used, feeding at rates of 75–100 ppm (ppm) nitrogen with every irrigation, or using 150–200 ppm nitrogen as needed. Controlled-release fertilizers are commonly incorporated into the growing medium prior to planting at a rate equivalent to 1.0–1.25 lb. nitrogen per cubic yard (454–567 g/0.76 m^3) of growing medium, or top-dressed onto the medium's surface using the medium recommended rate on the fertilizer label.

Digitalis purpurea prefers to be grown in a moist growing medium. Plants should be kept moist, but not saturated, the entire time they are being grown. Proper irrigation management is even more important when they are grown under high light conditions. When irrigation is necessary, water them thoroughly, ensuring that the entire growing medium is wet or nearly saturated. It is best to only allow the growing medium to dry slightly between irrigations.

Insects

Aphid	Mealybug
Beetle	Spider mite
Bulb nematode	Stem nematode
Caterpillar	Thrips
Japanese beetle	Whitefly

Generally, *Digitalis purpurea* is relatively insect free. Aphids, whiteflies, and thrips occasionally will become problematic. Of these insect pests, aphids are the most prevalent. A preventative application of imidacloprid (Marathon 60WP) as a drench will generally ensure aphid-free plants from spring planting until the plants are shipped. Other preventative strategies include monthly spray applications of systemic chemicals such as pymetrozine (Endeavor), thiamethoxam (Flagship), dinotefuran (Safari), or acetamiprid (TriStar). These strategies will also prevent the occurrence of whiteflies.

Diseases

Acremonium wilt	*Pythium*
Botrytis	*Rhizoctonia*
Downy mildew	*Sphaceloma* (anthracnose)
Fusarium	*Verticillium* wilt
Leaf spot	
Powdery mildew	

Under the right circumstances, certain diseases may become prevalent when growing *Digitalis purpurea*. *Botrytis* is likely to occur late in the crop cycle once the canopy closes in and plants begin to bloom. Spent flowers generally fall down on the foliage below, where *Botrytis* infections are most likely to arise. In most cases, *Botrytis* can be prevented or reduced by providing adequate spacing and good air circulation at all times, maintaining a relative humidity below 70%, selling plants when the lower flower buds begin to open, and, if necessary, implementing a preventative fungicide spray program using products such as fenhexamid (Decree) and chlorothalonil (Daconil). Root rots from the fungal pathogens *Pythium* and *Rhizoctonia* are also likely to occur, especially when the root zone remains saturated for extended periods. Practicing sound irrigation practices, ensuring moist but not overly wet conditions, is the best preventative strategy against root rot pathogens.

Plant Growth Regulators

Foxgloves are naturally tall, reaching up to 6 ft. (1.8 m) in height when they are blooming. Controlling the plant height may be necessary when producing them in containers in the greenhouse. Before using PGRs, it is beneficial to provide adequate space between each plant, which will reduce the competition between plants for light and prevent them from growing taller. Several of the commercially available PGRs are effective when they are applied using the appropriate rates, frequency, and timing. Depending on your geographic location, apply foliar applications of PGRs beginning with the rates listed in the table on page 412. Apply PGRs as the flower stalks are just beginning to elongate above the foliage. It will usually require two to three applications at seven-day intervals to provide adequate height control.

Forcing

Digitalis purpurea cultivars can be successfully forced into bloom throughout the year. There may be differences between cultivars in regard to their forcing requirements, and plant performance may vary slightly from the recommendations described below.

Although they seem to reach a certain age or size before flowering, they are not considered to have a juvenility requirement. When the proper conditions for flowering are provided, plants tend to flower after twelve to fifteen leaves have formed.

Figure 12.13. ***Digitalis* cultivars grow quickly and do not have a juvenility requirement. This is *D. purpurea* 'Camelot' three weeks after potting.**

PGRs for *Digitalis purpurea* Cultivars

Trade Name	Active Ingredient	Northern Rate
A-Rest	Ancymidol	25 ppm
B-Nine or Dazide	Daminozide	2,500 ppm
Bonzi, Piccolo, or Paczol	Paclobutrazol	30 ppm
Sumagic	Uniconazole-P	5 ppm

These are northern rates and need to be adjusted to fit the location where the crop is being produced. These rates are for spray applications and geared to growers using multiple applications.

Digitalis purpurea is considered to be cold beneficial, meaning the flowering process benefits from exposure to a cold treatment. In many cases, it may flower without a cold treatment, but flowering is often hastened and is usually more uniform following vernalization. It is recommended to cool (vernalize) plugs or small containers of foxgloves for five to ten weeks at 35–44° F (2–7° C). Some cultivars, such as 'Foxy', will flower completely without a cold treatment, while other varieties may flower more sporadically and unpredictably.

Foxgloves are considered to be quantitative long-day plants. This means they will flower faster under long photoperiods compared to short day lengths. With this quicker flowering response under long photoperiods, they are also referred to as long-day beneficial plants. During naturally short days, I recommend providing *Digitalis purpurea* with sixteen-hour photoperiods by extending the day if necessary, or use a four-hour night interruption during the middle of the night, using supplemental lighting to provide a minimum of 10 f.c. (108 lux) at plant level.

Without a cold treatment, most foxglove cultivars will only flower when they are grown under long photoperiods. In most cases, long durations of cold can be provided as a substitute for long days. When plants are going to be produced under short photoperiods, the longer the cold exposure prior to forcing, the more flowering is hastened. For example, when a crop is being produced into bloom for April 1 sales and the day lengths are naturally short, the longer the cold treatment is prior to forcing, the quicker they will be flowering.

Interestingly, in addition to the day length, the light intensity non-cooled crops receive affects the flowering response of foxglove. At low light levels, flowering is often incomplete and delayed. To improve the flowering percentage, decrease the time to flower, and increase the quality characteristics of the crop, supply 300–500 f.c. (3.2–5.4 klux) of supplemental lighting when the light levels are naturally low.

As described above, there are numerous variables affecting the amount of time necessary to produce a flowering crop of *Digitalis*. The time to bloom after providing a cold treatment and/or long photoperiods is primarily a function of temperature. Foxgloves are considered to be cool-season crops and perform best when the production temperatures remain below 72° F (22° C). Production at warmer temperatures may result in deformed flowers or flower bud abortion.

There may be slight differences between cultivars, but from the onset of long day photoperiods, *Digitalis purpurea* cultivars grown at 68°F (20° C) will take nine to ten weeks to reach flowering, while plants grown at 60°F (16° C) will flower in eleven to thirteen weeks. Plants grown under naturally short day lengths take approximately two weeks longer than the above schedules unless supplemental lighting is provided. Providing a cold treatment can reduce the time to flower by two to four weeks when plants are produced under short days, but it has less of an effect on crop timing when plants are grown under long days.

References

Fausey, B., A. Cameron, R. Heins, and E. Runkle. "Herbaceous Perennials: Digitalis (Foxglove)." *Greenhouse Grower.* September 2003.

Pilon, Paul. Perennial Solutions: "*Digitalis purpurea* 'Camelot'." *Greenhouse Product News.* December 2004.

Echinacea purpurea
Purple Coneflower

Purple coneflower is one of the most recognizable perennials used in the landscape today. The long-lasting display of intensely colored ray florets surrounding the raised bronze-colored central cone is always a showstopper. In the past, various shades of red-purple petals were most commonly observed. Today, breeding breakthroughs within the *E. purpurea* species and between other species has allowed for a wide range of flower colorations. Some of the popular colors being produced today include various shades of orange, pink, rose, yellow, and white. 'Magnus' is the most widely grown cultivar of *Echinacea* and was selected by the Perennial Plant Association as the Plant of the Year in 1998 for its landscape performance and desirable characteristics.

Figure 12.14. *Echinacea purpurea* 'Double Decker'

Jelitto Perennial Seeds

Echinacea is native to the prairies and dry plains of the central United States. The name *Echinacea* is derived from the word *echinos,* which means "hedgehog," referring to the bristly cone in the center of the flower. It belongs to the Asteraceae family (formerly Compositae) and is closely related to the genus *Rudbeckia,* which consists of yellow and orange coneflowers. There are numerous perennials in the Asteraceae family being commercially produced, including *Achillea, Aster, Bellis, Centaurea, Coreopsis, Doronicum, Echinops, Erigeron, Gaillardia, Helenium, Heliopsis, Leucanthemum, Ligularia, Rudbeckia,* and *Stokesia.*

This American native is used as an aromatic border plant to attract hummingbirds and butterflies into the gardens. Coneflowers are also used as accent plants and for use as cut flowers. *Echinacea* is widely used as an herbal medicine to stimulate the immune system and fight off various viral and bacterial infections.

Most *Echinacea purpurea* cultivars reach 24–36 in. (61–91 cm) in height. Many cultivars, such as 'Dwarf Star', 'Kim's Knee High', and 'Little Giant', have been selected with shorter growth habits. These varieties often reach 12–18 in. (30–46 cm) while they are in bloom and are all propagated vegetatively. Purple coneflower varieties prefer full sun, although in the South they perform best when some partial shade is provided.

Commonly Grown *Echinacea purpurea* Cultivars

***Echinacea purpurea* cultivars commonly propagated by seed**	
'Alba'	'Primadonna Deep Rose'
'Bravado'	'Ruby Star'
'Bright Star'	'White Swan'
'Magnus'	
***Echinacea purpurea* cultivars commonly propagated vegetatively or tissue culture**	
'Double Decker'	'Prairie Frost'
'Fragrant Angel'	'Ruby Giant'
'Kim's Knee High'	'Sparkler'
'Little Giant'	
***Echinacea* hybrid cultivars**	
'Big Sky Sunrise'	'Orange Meadowbrite'
'Big Sky Sunset'	'Razzmatazz'
'Mango Meadowbrite'	

Echinacea purpurea performs well across a wide portion of the United States, throughout USDA Hardiness Zones 3–10 and AHS Heat Zones 9–1. There are hardiness and heat tolerance differences between cultivars. Many of the new cultivars may not have the appropriate hardiness trials conducted prior to their release, which may lead growers to initially produce them in locations not suitable for their survival.

Propagation

Growers use seeds, basal cuttings, root cuttings, or divisions to propagate *Echinacea purpurea.* The method of propagation is often dependent on cultivar, availability, and cost. Although using bare-root divisions provides for a quicker finishing time, it often leads to crop variability and lower survival rates.

The seeds should be sown in the intended plug flat and covered lightly with germination mix or medium-grade vermiculite. Covering is optional, as seed can be exposed to light during germination but the covering helps to maintain a suitable environment around the seed during this phase. The seed flats should be moistened and moved to a warm environment, where the temperatures can be maintained at 70–80° F (21–27° C) for germination. Many growers utilize germination chambers during this stage to provide uniform moisture levels and temperatures. The seeds should germinate in three to seven days. Following germination, they can be grown at 70° F (21° C) for six to eight weeks, until they are ready for transplanting. Most commercially grown *Echinacea purpurea* is propagated from seed.

With recent advances in plant breeding and selection, it is becoming more common to propagate purple coneflower vegetatively using tissue culture or by cuttings. Vegetative propagation improves crop uniformity, as every plant of a given cultivar is genetically identical. Seed propagation has been reliable but offers some genetic variability and slight differences in each plant's appearance and performance.

Basal cuttings are usually taken in the spring, when the shoots are 4–6 in. (10–15 cm) in length before the plants have shifted to a reproductive mode. The cuttings are usually dipped in 1,000 ppm IBA (indolebutyric acid) to promote root development, increase the rooting percentage, provide uniformity, and decrease the rooting time. Place them under a light misting regime. For ideal rooting, maintain soil temperatures of 70–75° F (21–24° C), relative humidity of 85–90%, and photoperiods of less than fourteen hours. When optimum conditions are provided, they root in approximately three weeks.

Tissue culture has allowed many *Echinacea purpurea* cultivars and many new perennial varieties to reach the market in a relatively short period of time. Many cultivars available from tissue culture are patented; unlicensed propagation of these cultivars is prohibited. Most growers will purchase cultivars from tissue culture as 72-cell or larger sized plugs from licensed propagators.

Production

Echinacea purpurea performs best when grown in a moist, well-drained medium with a slightly acidic pH, 5.5–6.5. They perform well in many commercially available peat- or bark-based growing mixes, provided there is adequate drainage. Due to their large size when flowering, it is best to produce coneflowers in 1 gal. (3.8 L) or larger containers.

Coneflowers do not tolerate excessive amounts of water; when moist or wet conditions occur they are very susceptible to root rots. Unless the growing mix is excessively porous or holds onto too much moisture, normal irrigation practices should suffice. When irrigation is needed, water thoroughly and allow the medium to dry between waterings.

They are moderate feeders; fertility can be delivered using water-soluble or controlled-release fertilizers. Growers using water-soluble fertilizers either apply high rates (200–300 ppm) of nitrogen as needed or feed with a constant liquid fertilization program using rates of 75–150 ppm nitrogen with every irrigation. Using water-soluble fertilizers containing high amounts of ammonium nitrate may promote excessive leaf elongation and weak stems. Controlled-release fertilizers are commonly applied as a topdressing onto the medium's surface using the medium rate or incorporated into the growing medium prior to planting at a rate equivalent to 1.0–1.25 lb. nitrogen per cubic yard (454–567 g/0.76 m^3) of growing medium.

The best quality coneflowers are produced with night temperatures of 50–60° F (10–16° C) and day temperatures of 65–75° F (18–24° C). Temperatures above 76° F (24° C) near the time of bloom decrease flower quality and the uniformity of bloom. Conversely, cooler temperatures at this stage improve flower quality and promote blooms to take on a deeper coloration. If sales of non-flowering *Echinacea purpurea* are desired, they can reach a shippable size in six to eight weeks following planting when they are grown at 60–65° F (16–18° C). Production temperatures will be discussed under Forcing.

High light levels (3,000–5,000 f.c. [32–54 klux]) are desirable for green-leaved cultivars. However, variegated cultivars such as 'Prairie Frost' should be grown at lower light intensities (2,500–3,500 f.c.[27–38 klux]) to reduce to the likelihood of leaf scorch.

Insects

Aphid	Mealybug
Caterpillar	Mite
Four-lined plant bug	Slug
Fungus gnat	Spittlebug
Grasshopper	Shore fly
Japanese beetle	Thrips
Leafhopper	Vine weevil
Leaf miner	Whitefly

Generally, coneflower can be produced relatively insect free. Aphids and whiteflies occasionally will become problematic. These pests can be controlled after they are detected or can be prevented using a proactive strategy. A preventative application of systemic products, such as thiamethoxam (Flagship 25WG), imidacloprid (Marathon II), or acetamiprid (TriStar 70 WSP), will generally provide pest-free plants for up to one month following the application. Other insect pests listed in the table above are often observed feeding on coneflower but rarely become problematic. Most insects can be detected with routine crop monitoring. Control strategies may not be necessary unless the scouting activities indicate they are necessary.

Diseases

Alternaria leaf spot	Powdery mildew
Aster yellows	*Pseudomonas*
Botrytis	*Pythium*
Botrytis blight	*Rhizopus* (blight)
Cercospora rudbeckii leaf spot	*Sclerotinia*
Downy mildew	*Sclerotium*
Fusarium (wilt) crown and root rot	*Septoria lepachydis* leaf spot
Myrothecium leaf spot	*Verticillium*
Phytophthora	*Xanthomonas*

Plant diseases may be observed when environmental conditions are favorable for their development. The most common diseases observed attacking *Echinacea purpurea* crops are listed in the table above. As with many perennials, the occurrence of plant diseases can be negated or greatly reduced when the proper cultural practices are followed. To control the foliar diseases, it is best to manage the environment by providing proper plant spacing and adequate air movement, controlling the humidity, and, if desired, following a preventative spray program using the appropriate chemicals. The onset of root rot diseases can often be prevented by avoiding high salts and overly moist or wet conditions. When using bare-root starter materials, it is recommended to drench the containers after planting with a broad-spectrum fungicide, such as etridiozole + thiophanate-methyl (Banrot) or thiophanate-methyl (Cleary's 3336), to reduce pressure from pathogens and to hasten plant establishment.

PGRs for *Echinacea purpurea* Cultivars

Trade Name	Active Ingredient	Northern Rate
A-Rest	Ancymidol	25 ppm
B-Nine or Dazide	Daminozide	2,500 ppm
Bonzi, Piccolo, or Paczol	Paclobutrazol	30 ppm
Cycocel	Chlormequat chloride	1,250 ppm
Florel	Ethephon	500 ppm
Sumagic	Uniconazole-P	5 ppm
Topflor	Flurprimidol	22 ppm
B-Nine or Dazide + Cycocel	Daminozide + chlormequat chloride	2,500 ppm + 1,250 ppm
B-Nine or Dazide + Sumagic	Daminozide + uniconazole-P	2,500 ppm + 5 ppm
Bonzi, Piccolo, or Paczol Drench	Paclobutrazol	6 ppm
Sumagic Drench	Uniconazole-P	1 ppm

These are northern rates and need to be adjusted to fit the location where the crop is being produced. Unless noted otherwise, these rates are for spray applications and geared to growers using multiple applications. Drench applications are usually applied once per crop.

Plant Growth Regulators

Except for the dwarf varieties, many cultivars require height control when they are grown under greenhouse and nursery conditions. Providing adequate spacing between the plants will reduce plant stretch caused by competition. Withholding water and nutrients can also reduce the height to some extent. Under certain growing conditions or at high plant densities, it may be necessary to use chemical plant growth regulators. Most of the commercially available PGRs are effective at controlling plant height when they are applied using the appropriate rates, frequency, and timing. A table containing the effective height control products and northern rate recommendations is provided above. For each of these products applied as foliar sprays, it usually requires two or three applications at seven-day intervals to provide adequate height control. Begin the applications when the flower stalks are near the leaf canopy and are beginning to elongate or bolt.

Forcing

To improve marketability, *Echinacea purpurea* can be forced to bloom throughout the year. Forcing coneflowers to flower out of season involves following a few key guidelines. Although I have not forced or seen research on every *Echinacea purpurea* cultivar, I feel it is safe to make a few assumptions based on my experiences and research conducted by Michigan State University on other cultivars of *Echinacea purpurea.*

Echinacea purpurea cultivars do not require a cold treatment for flowering. However, they are considered to be cold-beneficial plants, as flowering will occur two to three weeks earlier following a cold period. It is recommended to cool (vernalize) small containers of coneflower for a minimum of ten weeks at 41° F (5° C). *Echinacea* can be vernalized as a plug or in the final container. Regardless of the container size, it is recommended to ensure they are fully rooted prior to exposing them to cold temperatures. Juvenility is not an issue, as they do not need to reach a particular size prior to exposing them to cold or other flower-inductive treatments. Overwatering during the cold treatment could result in root rots and possibly plant losses.

Echinacea purpurea cultivars are considered to be intermediate day plants, requiring between twelve to sixteen hours of light for the best flowering. Light durations of less than twelve hours or greater than sixteen hours will cause them to flower poorly or not at all. It is recommended to provide fourteen-hour photoperiods or night interruption lighting when the natural photoperiod is less than fourteen hours.

During naturally short days, using high-pressure sodium lights to deliver a four-hour night interruption between 10 P.M. and 2 A.M. is effective in promoting flowering. Using incandescent light sources will promote flowering but will cause the internodes to stretch excessively, reducing the overall quality attributes of the plant. Research at MSU has shown that reducing the duration of the night interruption down to thirty to sixty minutes effectively promotes bloom and reduces plant height significantly.

The time to bloom after vernalization and the proper photoperiod is provided is a function of temperature. *Echinacea purpurea* grown at 68° F (20° C) will take twelve to fourteen weeks to reach flowering, while plants grown at 60° F (16° C) will flower in approximately sixteen weeks. The flower quality and uniformity of flowering is greatly reduced at high temperatures (greater than 76° F [24° C]). To obtain the best plant quality, I recommend producing *Echinacea purpurea* cultivars at 65–68° F (18–20° C). Coneflowers not receiving a cold treatment typically take two to four weeks longer (depending on temperature) than the durations listed above.

To obtain the highest quality finished plants, growers should consider planting rooted liners in the desired pot during the late summer of the year prior to the intended date of sale. Established containers should be overwintered in a protected area prior to spring forcing. Planting liners of coneflower in this manner will allow them to bulk up, produce more flowers per plant, and bloom earlier than when they are planted and grown only in the spring.

References

Greenhouse Grower Magazine and Michigan State University. *Firing Up Perennials: The 2000 Edition.* Willoughby, Ohio: Meister Publishing. 2000.

Nau, Jim. *Ball Perennial Manual: Propagation and Production.* Batavia, Ill.: Ball Publishing. 1996.

Pilon, Paul. Perennial Solutions: "Echinacea 'Little Giant'." *Greenhouse Product News.* June 2005.

Schoellhorn, R., and A. Richardson. *Warm Climate Production Guidelines for Echinacea.* University of Florida- IFAS Commercial Floriculture Update Bulletin ENHFL04-008. 2004.

Ferns

Ferns are an increasingly popular group of plants that do not receive enough recognition for their contribution in today's landscapes. Even though they are widely grown, ferns are still underutilized by today's gardeners and landscapers. The ornamental contributions ferns offer to the landscape include a great array of colors, sizes, shapes, and textures. Ferns also contribute significantly to floral arrangements as backing materials, and they are often used in dried arrangements.

Ferns belong to a major division of the plant kingdom called Pteridophyta. There are over 12,000 species of pteridophytes found throughout the world, only a handful of these have made their way to commercial production. Ferns do not produce flowers and are grouped with other plants with similar reproductive characteristics; these plants are commonly called cryptogams and include algae, mosses, and liverworts. One of the distinctive differences between ferns and these other "simpler" plants is that they have internal vascular structures associated with the movement of water and nutrients within the plant.

Characteristics of Ferns

Ferns take on many types of growth habits. The ferns commercially produced and marketed as perennials all have a distinctive set of fronds (leaves) that radiate to form a crown. The fronds vary in architecture, color, size, and venation. Young fronds are characteristically coiled or hooked before they unfurl (they are called fiddleheads at this stage) and expand into what resembles a compound leaf. These expanded fronds have various types of divisions or lobes (pinnae), giving them their fernlike appearance, which is useful for proper identification.

Unlike most perennials, ferns do not flower. They reproduce by tiny spores carried in sporangia, which are usually clustered in a group known as a sorus (sori is the plural). Sori are almost always located on the underside of the fronds. Ferns go through a process known as an alternation of generations, which entails passing from the sporophyte (plant) stage to the gametophyte stage in their development. A very general description of this process goes as follows:

The spore contains half the number of chromosomes as the sporophyte. It germinates into a structure called a prothallus, which contains the same number of chromosomes as the spore. As the prothallus develops, it produces male and female sex organs on its surface. The female organs are known as archegonia and contain the female eggs. The male sex organs are called antheridia and produce spermatozoids (sperm). The sperm swim by means of flagella to the archegonia in a film of water present on the surface of the prothallus and fertilize the egg. Two cells unite to form a zygote with a complete set of chromosomes. The zygote develops and grows into the sporophyte, the fern plant we recognize and commercially produce.

Ferns are mostly indigenous to moist, wet climates. Most of the species in commercial production today are native to marshes, woodlands, and open forests and perform best under shady conditions.

As alluded to previously, ferns offer a great deal of aesthetic appeal and are commonly used in public and private landscapes throughout the world. There are a number of easy to grow and reliable ferns available to perennial producers. Some of the newest hybrids and cultivars offer new colorations for growers to consider. Instead of the various shades of green provided by the old genetics, today's varieties are becoming more vibrant, offering various shades of pink, red, or purple in addition to copper tones and metallic hues. Some fern varieties have fronds that change coloration with maturity.

Growers should note that there are hardy ferns and tropical ferns. Hardy ferns are commonly grown as perennials and used for landscape purposes. Tropical or tender ferns are commonly grown in greenhouses and used for household decoration or for cut flower displays. Care should be taken when producing ferns for the perennial trade to ensure they are indeed hardy varieties.

Eight Popular Commercially Produced Ferns

The five genera discussed here (*Athyrium, Dryopteris, Matteuccia, Osmunda,* and *Polystichum*) represent some of the most commonly produced ferns grown by commercial growers, but they only represent a portion of the fern species in production today. Although the cultural requirements are often similar, there are some variations that if not provided, could lead to unsatisfactory crop performance or losses. These differences will be discussed below where appropriate.

Athyrium filix-femina (lady fern)

Athyrium filix-femina, commonly called lady fern, is a popular species used in gardens and landscapes in North America, Europe, and Asia. When reproduced naturally from spores, there tends to be a great deal of variation across the population. After a couple years of observation, careful selection for desirable characteristics in the young sporlings can lead to first-rate varieties that can be reproduced by means of tissue culture to ensure the stability of these traits.

Figure 12.15. *Athyrium filix-femina,* commonly known as lady fern

Lady ferns are considered a decorative species and are valued for their delicate and finely divided fronds. There are numerous cultivars of lady ferns that have been selected with various characteristics, including dwarf forms and differing frond characteristics. With plant selection, there are numerous plant heights available today, ranging from 12–36 in. (30–91 cm) tall. The frond shape can be highly variable, depending on the cultivar, with various pinnae characteristics. All lady ferns have solid green frond colorations. They perform well throughout USDA Hardiness Zones 3–9 and prefer locations with partial to full shade and soils with sufficient moisture levels. The fronds become tattered if the plants dry out or are exposed to too much direct sunlight. Gardeners commonly use lady ferns in shady borders and rock gardens.

Athyrium nipponicum (Japanese painted fern)

The Japanese painted fern is one of the showiest, most colorful ferns used in today's landscapes. The delicate tricolor metallic fronds with maroon midribs offer great landscape appeal. There are numerous frond colorations, ranging from almost pure silvery gray to deep metallic red. Growers commonly use this fern as a garden accent, in group plantings, or in combination pots with other shade perennials.

Despite the fragile appearance of the Japanese painted fern, it is actually tough and easy to grow in the landscape or in containers. It survives and performs well across much of the country, thriving in shady locations throughout USDA Hardiness Zones 3–9. *Athyrium nipponicum* has a great landscape habit, with its clumplike, noninvasive form and fronds that reach 12–24 in. (30–61 cm) in height.

With its desirable characteristics, ease of production, and wide availability, *Athyrium nipponicum* 'Pictum' was selected by the Perennial Plant Association as the 2004 Perennial Plant of the Year. This recognition has not only helped to increase the popularity of the Japanese painted fern, but has sparked awareness and interest in all types of ferns used in today's landscapes.

Figure 12.16. ***Athyrium nipponicum* 'Pictum'**

Dryopteris affinis (scaly male fern)

Dryopteris affinis thrives in cool, moist, and shady climates throughout USDA Hardiness Zones 4–8. They are considered to be semi-evergreen, as they generally go dormant during the winter months in cold climates but may remain evergreen in warm areas. The new fronds that flush in the spring are very decorative and are densely covered with coppery scales. Scaly male ferns are easy to grow and are utilized in a number of landscape scenarios, including mixed borders and woodland schemes.

Clumps of scaly male ferns are considered to be medium sized, but certain cultivars could reach 4 ft. (1.2 m) tall and wide over time. There are a number of cultivars and variations of this species being grown today. The overall frond shape is described as being ovate-lanceolate with the pinnules evenly oblong, and it usually has flat, spreading crests that taper to a point. A distinctive characteristic they all share is a conspicuous sooty patch at the junction of the pinna midrib and rachis (stem).

Dryopteris erythrosora (autumn fern)

The autumn fern is one of the most colorful ferns in the genus *Dryopteris*. The most distinguishing characteristic of this evergreen fern is the distinctive coppery-red coloration of the new fronds, which contrasts with the older green fronds. As autumn approaches, the fronds turn to a russet coloration, providing additional seasonal interest. This is one of the few ferns that provide both color and texture into the shady landscape throughout the growing season.

It performs well in partial to full shade across USDA Hardiness Zones 5–9. This is a very adaptable and easy-to-grow fern that performs exceptionally well in the shade garden. A single plant reaches 18 in. (46 cm) tall and approximately 24 in. (61 cm) across.

Figure 12.17. **Autumn fern foliage**

Autumn ferns are truly no maintenance and easy to grow. They can tolerate a couple hours of direct sunlight daily but perform best in shady conditions. *Dryopteris erythrosora* also can be grown in dry shade and once established are somewhat drought tolerant.

Dryopteris filix-mas (male fern)

The male fern is a tough, reliable, clump-forming fern found in woodlands and gardens throughout USDA Hardiness Zones 3–9. The species and common names refer to the fern's vigorous nature. They have narrow, fringed, light green, deciduous fronds and grow 2–4 ft. (0.6–1.2 m) tall.

Dryopteris filix-mas cultivars are easy to cultivate and collectively are some of the best performing ferns gardeners and landscapers can grow. They tolerate sun better than most fern species and can be tolerant of relatively dry growing conditions. Male ferns do perform best when these adverse conditions are avoided but will hold their own when unfavorable circumstances are present. However they are used, as specimen plants or mass plantings, they always make a grand statement in the landscape.

Matteuccia struthiopteris (ostrich fern)

Ostrich ferns are large, colony-forming ferns that are commonly used for naturalizing, foliage backdrops, and as foundation plantings. They reach 2 ft. (61 cm) wide and 3–5 ft. (0.9–1.5 m) tall at maturity and are grown throughout USDA Hardiness Zones 3–8. These elegant, robust ferns are considered to be easy to grow and require little maintenance. Since they spread by underground rhizomes, ostrich ferns could become invasive under the right conditions.

Ostrich ferns perform best in moist, shady sites, but locations with partial shade are acceptable. The more direct sunlight they are exposed to, the moister the site will need to be. There is a tendency for the fronds to yellow and scorch while exposed to hot, direct sunlight, or overly dry conditions. They should be grown with consistent moisture levels and not be allowed to dry out. *Matteuccia struthiopteris* prefers cool climates and is generally intolerant of the hot and humid summers of the Deep South.

The fronds are featherlike and somewhat resemble the plumes of an ostrich. The fronds of *Matteuccia struthiopteris* differ from most

Figure 12.18. Ostrich fern

ferns as they produce both vegetative (sterile) and fertile fronds. The large, green, sterile fronds are oblanceolate (widest in the middle and inversely tapering at each end), grow up to 5 ft. (1.5 m) in length, and form symmetrical vaselike clusters. The dense and rigid fertile fronds arise from the center of the clump in mid- to late-summer, grow about 2 ft. tall (61 cm), and turn brown with the many spore cases. The sterile fronds wither with the first frost, leaving the brown fertile fronds that last throughout the winter when the spores are released.

Osmunda cinnamonea (cinnamon fern)

Cinnamon ferns are popular, easy-to-grow ferns with showy fertile fronds containing cinnamon-colored spores. The fertile fronds stand above the vaselike cluster of sterile fronds surrounding them. *Osmunda cinnamonea* are often seen growing naturally in moist habitats. Gardeners commonly use these ferns in moist borders or as garden accents to add a lush, tropical look to the landscape. Cinnamon ferns spread by rhizomes and are typically used to naturalize landscapes.

In the early spring, the fertile fronds are the first to appear, arising bright green and then turning cinnamon-brown as the spores mature. The spore-bearing fertile fronds are modified fronds that grow separately from the sterile or vegetative fronds. These modified fronds are stiff and erect and consist of numerous pairs of spore cases, which look like stunted leaflets. After the fertile fronds are developed, the plant forms the familiar green arching vegetative fronds. The base of each frond contains tufts of cinnamon-colored fibers useful for proper identification of plants without fertile fronds.

Osmunda cinnamonea are large ferns and grow 3 ft. (0.9 m) wide and 3–5 ft. (0.9–1.5 m) tall. They are deciduous, with the fronds turning golden yellow and then bronze before succumbing to frost in the autumn. They are relatively heat and sun tolerant when adequate moisture levels are provided. Even with their heat tolerance, cinnamon ferns still do not perform well in the Deep South and are best produced in cooler portions of the country. They are widely grown throughout USDA Hardiness Zones 3–10.

Polystichum acrostichoides (Christmas fern)

Polystichum acrostichoides, commonly called Christmas fern, is one of the most commonly grown ferns. They are found in woodlands and landscapes throughout USDA Hardiness Zones 3–9. Christmas ferns have dark green, shiny, evergreen leaves. The common name refers the evergreen leaves that are often used for Christmas decoration.

They typically grow in fountain-like clumps reaching 18–24 in. (46–61 cm) wide and tall and feature leathery, lance-shaped,

Figure 12.19. *Polystichum acrostichoides,* Christmas fern

evergreen fronds. Spores are produced on the undersides of the pinna on the upper third of the fertile fronds. Any spores located on the lower two-thirds of the fertile fronds are sterile. Although they form rhizomes, this fern will not spread or naturalize; however, the clumps will gradually increase in size over time.

Christmas ferns are easy to grow and are considered to be tough and undemanding. They can tolerate more sun than many other ferns if they are provided with adequate soil moisture. However, the best growth occurs in moist shady locations. *Polystichum acrostichoides* are often utilized in woodland gardens, shade gardens or shady borders, and for wild or native plant gardens. Growers value this fern year-round, as the evergreen fronds provide good winter interest in the landscape.

Propagation

Propagation of ferns is commonly done by tissue culture, division, or by planting spores. To maintain the desirable characteristics of the original fern, only vegetative forms of propagation (tissue culture or division) should be used. Spore propagation is a form of sexual propagation and does not ensure that the desirable characteristics, such as the growth habit and uniformity of color, will carry over into the new offspring.

Tissue culture

Commercially, tissue culture is the most common method of producing fern species and cultivars. Tissue culture allows growers to produce plants with the identical and desirable characteristics of the mother plant. Since every plant propagated is genetically identical, it allows growers to produce fern crops with great uniformity. Most growers will purchase ferns propagated from tissue culture as finished plug liners from growers specializing in this type of propagation.

Division

Ferns grow from rhizomes that are usually just beneath the surface of the soil. The rhizome is the foundation from which the roots and fronds originate. Division basically entails cutting the underground stem (the rhizome) into sections. Since roots grow continuously along the rhizome, these sections do not require any specific propagation methods or procedures to promote rooting when this procedure is practiced. For a division that results in a nice propagule, it is important to ensure that each piece of rhizome contains at least two or three growing points (fiddleheads).

Commercially, division is not performed too frequently due to the efficiency and cost effectiveness of purchasing young plant materials propagated via tissue culture. The advantage of dividing ferns is that growers can get larger ferns in a shorter period of time. The disadvantage is they are limited by the number of new ferns that can be propagated at one time.

It is best to divide most ferns in the early spring, just as they are beginning to actively grow. The methods of division vary slightly by species but can be generalized as follows. Gently move the roots and fiddleheads (small unrolling fronds) out of the way to reveal the rhizome. Count two or three fiddleheads in each hand and make the division between them by cutting or gently prying apart the rhizome. Both sections can be replanted into containers or back into the landscape. Plant the new clump of ferns with the rhizome just below the surface of the soil.

There are numerous types of rhizomes ferns produce. The method of division is dependent on the type of rhizome and growth habit of each particular fern. *Athyrium* and *Dryopteris* produce multiple crowns and can be divided very similarly to herbaceous perennial plants. After lifting the mother plant, cut the rhizome into sections, using a sharp knife or spade. These sections should each contain two or three young fronds and a corresponding amount of roots. *Polystichum acrostichoides* form a long thin rhizome with a loose network of roots, which is easily pulled apart or separated. *Osmunda cinnamomea* develops a rhizome that resembles a small bulblike structure the size of a tangerine. This type of rhizome can also be cut or divided to produce more plants. Cinnamon ferns tend to take a long time to recover following division. Creeping ferns, such as *Matteuccia struthiopteris,* which spread by underground rhizomes, can be divided using edge division or cutting the rhizome where the roots and fronds are adjacent to one another.

Spores

For commercial production, propagating ferns from spores is not very practical. This method is very rewarding for the fern enthusiast or hobbyist, but it simply does not work well for large-scale producers. The advantage of propagating ferns in this manner is that a large number of new ferns can be obtained; the disadvantage is that it will take at least two years to have these ferns grow to a marketable size. Spore propagation also does not result in true-to-type offspring; they are not genetically identical to the mother plants. If the traits of the mother plant are desired, the fern must be propagated vegetatively by division or by means of tissue culture.

Production

Ferns are available to growers primarily as rooted plug liners started via tissue culture. Although some small operations may start fern crops using bare-root divisions, it is relatively uncommon for most growers to start them in this manner. They are most commonly planted in 1 qt. (1 L) or 1 gal. (3.8 L) containers immediately after they have been received.

Ferns perform best when grown in a moist, well-drained medium with a pH of 5.5–6.5. They perform well in many commercially available peat-based growing mixes. A number of growers have success growing ferns using peat- and bark-based growing mixes. The most important principal of the growing mix is that it provides the appropriate physical properties. Most ferns perform best when an adequate amount of moisture is present at all times. They do not like to remain waterlogged; therefore, it is important to provide a growing medium that has an above average water-holding ability and still provides sufficient drainage, allowing some air to be in the root zone at all times.

Generally, growers plant one plug per container, unless they wish to finish large containers slightly faster. When planting, the plugs should be planted so the original soil line of the plug is even with the surface of the growing medium of the new container. When planting bare-root divisions, the top of the rhizome should be placed in the center of the pot, with the crown slightly covered (no more than 0.5 in [13 mm]) or just exposed after it is thoroughly watered in. Planting the crown too deep will lead to crop variability and losses. It is very important to not let the ferns dry out at this time, or serious crop losses could result. I find it beneficial to apply a preventative fungicide drench of thiophanate-methyl (Cleary's 3336, OHP 6672, FungoFlo) after planting.

Good irrigation management is important throughout crop production, from the time they are planted until the day they are shipped. Ferns should be kept moist the entire time they are being grown. Water stress in the beginning of the crop, before they are established, could be detrimental to ferns. Inadequate moisture levels on established crops could cause the margins of the fronds to turn brown and become crispy. I recommend watering thoroughly, when irrigation is required, and allow the substrate to dry slightly between waterings. To avoid potential foliar problems associated with sun scorch or *Botrytis,* it is best to apply water early in the day to allow the fronds plenty of time to dry before night. Overlooking the water status and needs of a fern crop just one time can seriously impact the quality attributes and salability of the entire crop.

Most ferns perform best when they are grown under low light levels. Most growers find it beneficial to use a 50% shade cloth over the production area to reduce exposure to direct sunlight. Optimum light levels for fern production are from 1,000–2,500 f.c. (11–27 klux). Production of ferns in containers provides more stressful conditions, such as higher root zone temperatures and more dramatic swings in the moisture level surrounding the roots than when plants are grown in the ground. These stresses make ferns even more sensitive to exposure to high light levels and more prone to scorched leaves. The coloration of the fronds will be darker at lower light intensities and appear light green to yellow when they are grown at high light levels. Keeping the light levels below 2,500 f.c. (27 klux) will maintain the quality attributes of the fronds and keep most fern varieties from becoming overly stressed.

Ferns require relatively small amounts of fertilizer compared with other perennials. In fact, they are sensitive to high salt levels and may become scorched or experience root injury, which could lead to root rot pathogens such as *Pythium*. Fertility can be delivered using low rates of water-soluble or controlled-release fertilizers. Growers using water-soluble fertilizers normally use constant liquid feed programs at low concentrations, such as 50–75 ppm nitrogen with every irrigation. The rate of time-release fertilizer to incorporate should be about half of that used for most perennials or between 0.5 and 0.75 lb. of elemental nitrogen per cubic yard (227–340 g/0.76 m^3) of growing medium. Growers can also top-dress controlled-release fertilizers on the medium's surface using the low rates recommended on the product's label.

In the past, fern growers were wary of using time-release fertilizers to provide nutrients to their crops. Historically, the main concern using these types of fertilizers on ferns was the fertilizers' tendency to continue releasing nutrients during the winter months, causing salts to build up and leading to plant injury and even loss. Today, there are still some suppliers whose formulations may behave in this manner, but there are many formulations and technologies currently available that do completely shut

Insects of Ferns

Athyrium	*Dryopteris*	*Matteuccia*	*Osmunda*	*Polystichum*
Aphid	Aphid	Aphid	Aphid	Aphid
Army worm	Black vine weevil	Army worm	Armyworm	Armyworm
Black vine weevil	Japanese beetle	Black vine weevil	Black vine weevil	Black vine weevil
Caterpillar	Caterpillar	Caterpillar	Caterpillar	Caterpillar
Cricket	Cricket	Cricket	Cricket	Cricket
Foliar nematode	Foliar nematode	Foliar nematode	Foliar nematode	Foliar nematode
Grasshopper	Grasshopper	Grasshopper	Grasshopper	Grasshopper
Grub	Passion vine hopper	Japanese beetle	Japanese beetle	Japanese beetle
Japanese beetle	Scale	Mealybug	Mealybug	Mealybug
Mealybug	Snail	Passion vine hopper	Scale	Passion vine hopper
Slug	Thrips	Scale	Slug	Scale
Snail	Whitefly	Snail	Snail	Snail
Whitefly		Thrips	Thrips	Thrips
			Whitefly	Whitefly

Diseases of Ferns

Athyrium	*Dryopteris*	*Matteuccia*	*Osmunda*	*Polystichum*
Botrytis	*Botrytis*	*Botrytis*	*Botrytis*	*Botrytis*
Damping off	Damping off	Damping off	Damping off	*Cylindrocladium* leaf blotch
Foliar *Rhizoctonia*	Foliar *Rhizoctonia*	Foliar *Rhizoctonia*	Foliar *Rhizoctonia*	Damping off
Leaf spots	Leaf spots	Leaf spots	Leaf spots	Foliar *Rhizoctonia*
Pythium	*Pythium*	*Pythium*	*Pythium*	Leaf spots
Rust	Sooty mold	Rust	Rust	*Pythium*
Sooty mold	Rust	Sooty mold	Sooty mold	Rust
	Taphrina leaf blister		*Taphrina* leaf Blister	Sooty mold
				Taphrina leaf blister

down during the winter months and do not release nutrients. Check with your fertilizer distributor about their formulations, technologies, and winter-release patterns.

Ferns are not susceptible to many plant pathogens or insect pests. The table above does list numerous insects or diseases for each type of fern, but under normal circumstances none of these become problematic or devastating to fern crops. Growers should have routine scouting programs to detect the presence of insects or diseases early and to determine if and when control strategies are necessary.

Of the insects, aphids, Japanese beetles, slugs, and snails are observed feeding most frequently. Fern producers most often contend with controlling slugs and snails. Growers may experience *Botrytis*, *Pythium*, and foliar *Rhizoctonia*. *Pythium* can often be prevented by properly managing the fertility levels and ensuring the crops do not become too stressed from overly dry or excessively wet conditions. These stresses lead to root injury and provide entry points for root rot diseases. *Botrytis* and *Rhizoctonia* often occur when the foliage remains wet for long durations, especially during cool, cloudy weather. The presence of both of these pathogens can be reduced by irrigating early in the day (allowing adequate time for the fronds to dry), providing adequate spacing between each plant, and ensuring there is good air movement in the production facility.

Figure 12.20. **Dormant *Dryopteris affinis* prior to forcing in the spring.**

Overwintering

Overwintering perennial ferns is easy, provided a few simple steps are followed. Most growers overwinter containerized ferns within enclosed structures such as cold frames or greenhouses. In the fall, let the plants go completely dormant. Deciduous ferns, such as *Athyrium*, will completely die back, much like herbaceous perennials do. Evergreen varieties, such as Christmas fern, will have foliage that will remain green or may turn yellow over the winter months. Evergreen ferns are not necessarily evergreen in all climates. In many cases, northern growers observe that evergreen varieties may be semi-evergreen or even behave like deciduous ferns, depending on the type of climate they are in. Regardless, most growers leave the fronds alone until late winter. Group containers in a pot-to-pot configuration, and cover them with a thermal blanket to help provide additional insulation. It is important to not overwater or have saturated root zones during the winter months, or there could be a greater incidence of root rots and possibly crop losses.

During the late winter but before the new fronds begin emerging, the dead foliage can be easily removed from the top of the containers of deciduous ferns and the outermost fronds can be trimmed off from evergreen varieties, if necessary. If the new fronds (fiddleheads) have begun to actively grow, use extreme caution when removing the old foliage, or omit this step altogether, reducing any potential damage to the new growth.

Forcing

Ferns produced in large containers, such as 1 gal. (3.8 L) pots, for spring sales (between April and June) are best when planted during the previous summer to allow adequate time for bulking. Growers should also plant all of their ferns before September 1 (earlier in cold climates); ferns need an adequate amount of time to become established prior to overwintering. The optimal temperatures for rooting are 65–70° F (18–21° C). These temperatures are difficult to achieve in outdoor environments or structures where most ferns are placed during the fall. Planting in the fall will not allow proper time for plant establishment and could result in poor overwintering and plant losses.

In the spring, it takes eight to ten weeks to force, or flush, ferns growing at 60–70° F (16–21° C). Initially, I like to flush ferns using warmer temperatures, 65–70° F (18–21° C), for several weeks until the fronds are expanded and near the edge of the pot. Then I lower the production temperatures down to 58–63° F (14–17° C) to finish the crop.

Smaller container sizes, such as 4 in. (10 cm) pots, can be grown in a similar manner to the large containers, starting in the summer and overwintering them, or they can be planted using fresh plug materials during the same season they are to be shipped. To flush small container sizes, it requires six to ten weeks with an average temperature of 68° F (20° C) to obtain ferns that are shippable. The amount of forcing time depends largely on the variety of fern being produced and, more important, on the size of the starting materials. Large 72-cell plugs will finish small containers considerably faster than will small 128-cell plug sizes.

References

Armitage, Allan M. *Herbaceous Perennial Plants: A Treatise on Their Identification, Culture, and Garden Attributes.* Champaign, Ill.: Stipes. 1997.

Jones, D. L. 1987. *Encyclopedia of Ferns.* Portland, Ore.: Timber Press. 1987.

Pilon, Paul. Perennial Solutions: *"Athyrium nipponicum* 'Pictum'." *Greenhouse Product News.* September 2003.

Gaillardia x *grandiflora*, *G. aristata*

Blanket Flower

Gaillardia x *grandiflora* and *G. aristata* are often used interchangeably, although they do not represent the same group of plants. *G. aristata* is one of the parents of the *G.* x *grandiflora* hybrids and is not itself widely grown

Figure 12.21. *Gaillardia* x *grandiflora* 'Fanfare'

Walter's Gardens

(contrary to catalog listings, which often list many of the *G.* x *grandiflora* cultivars as belonging to *G. aristata*). *G. aristata* is native to the high plains of North America and is believed to give *G.* x *grandiflora* cultivars their cold hardiness characteristics. The other parent, *G. pulchella,* is native to the southern United States and Mexico and is said to provide *G.* x *grandiflora* with its ability to withstand heat and humidity.

Once established, the native American blanket flower becomes heat and drought tolerant. It thrives in full sun and can be produced throughout the country in USDA Hardiness Zones 3–10 and AHS Heat Zones 12–1, though there are hardiness and heat tolerance differences between cultivars. *Gaillardia* cultivars begin flowering in midsummer and continue through early fall. The daisylike blooms have deep yellow to rich burgundy petals surrounding the buttonlike centers. When growers cut back the plants during the growing season, or deadhead them, I have seen *Gaillardia* flower into October. With the differing height characteristics of the various cultivars, ranging from 8–30 in. (20–76 cm), blanket flowers can be used for numerous landscape applications. Gardeners and landscapers commonly use them as accent

Commonly Grown
***Gaillardia* x *grandiflora* Cultivars**

'Arizona Sun'	'Indian Yellow'
'Baby Cole'	'Monarch Mix'
'Bijou'	'Red Goblin'
'Bremen'	'Single Mix'
'Burgundy'	'Summer's Kiss'
'Dazzler'	'The Sun'
'Dwarf Goblin'	'Tokajer'
'Fanfare'	'Torchlight'
'Goblin' ('Kobold')	'Yellow Queen'
'Golden Goblin'	

plants, border plants, cut flowers, and in mass and container plantings.

The common name, blanket flower, came about since the striking three-colored flowers of *Gaillardia* resemble the colors commonly found in Indian blankets. Blanket flowers are one of the most popular perennials found in today's gardens. Their popularity is based on their ease of production, weather tolerance, and long blooming season. Recently, new introductions 'Fanfare' and 'Arizona Sun' have brought a new attention and awareness to an already well-known group of perennials.

Gaillardia belongs to the Asteraceae family (formerly Compositae). There are numerous perennials in the Asteraceae family being commercially produced, including *Achillea, Aster, Bellis, Centaurea, Coreopsis, Doronicum, Echinops, Echinacea, Erigeron, Helenium, Heliopsis, Leucanthemum, Ligularia, Rudbeckia,* and *Stokesia.*

Propagation

Depending on the cultivar, *Gaillardia* is started from seed or vegetatively propagated by tip cuttings. Some cultivars, such as 'Goblin', can be propagated either vegetatively or by seed. To maintain the desirable characteristics of a superior form, it is necessary to propagate it vegetatively, as the exact characteristics will not be carried over when propagating by seed. Seed propagation results in some crop variability but is still the most economical way to grow these cultivars. The method of propagation is often dependent on cultivar, availability, and cost.

Blanket flowers can be successfully started in various plug trays sizes from 288-cell up to 72-cell plug flats. Sow seeds in the intended plug flat and do not cover the seed with germination mix or vermiculite. Light is required for germination, and covering reduces the amount of light reaching the seed and may adversely affect the germination process. The seed flats should be moistened and moved to a warm environment, where the temperatures can be maintained at 70–75° F (21–24° C) for germination. During the germination process, it is important to keep the growing medium uniformly moist but not wet, or the germination rates are likely to decrease.

Many growers utilize germination chambers during this stage to provide uniform moisture levels and temperatures. Using germination chambers is optional, as *Gaillardia* will successfully germinate in the greenhouse.

At these temperatures, the seeds should be germinated in four to eight days. Following germination, reduce the moisture levels somewhat, allowing the growing medium to dry out slightly before watering to help promote rooting. A light covering of medium-grade vermiculite or germination mix can be applied to the emerged seedlings to help maintain moisture levels around the tender young plants. The temperatures can be reduced down to 68–72° F (20–22° C) for the remainder of the plug stages. At these temperatures, most cultivars will finish the plug stage in approximately five to seven weeks.

The light levels can gradually be increased as the plug develops, from 500 f.c. (5.4 klux) following germination to 2,500 f.c. (27 klux) at the final stage. Fertilizers are usually applied once the true leaves are present, applying 100–150 ppm nitrogen every third irrigation or 75 ppm with every irrigation, using a balanced water-soluble source.

Tip cuttings should be taken while blanket flowers are vegetative, or non-flowering. Cuttings that are already in bloom will take longer to root and have a lower survival rate than purely vegetative starting materials. When taking cuttings from cultivars that are also commonly propagated from seed, it is best to only obtain cuttings from plants that have the most desirable characteristics.

Tip cuttings should measure approximately 2–3 in. (5–8 cm) and contain several nodes. The well-drained rooting medium should be moistened prior to sticking. The base of the cuttings can be dipped into a rooting hormone, such as a solution of indolebutyric acid (IBA) at rates of 1,000–1,500 ppm prior to sticking. *Gaillardia* can successfully root without rooting compounds but tend to root slightly faster and more uniformly when these treatments are provided. They are generally stuck into plug liners for rooting, although they can be directly stuck in the final containers.

Cuttings should be placed under low misting regimes for about the first two weeks of propagation. When possible, it is usually best to propagate under high humidity levels (90% relative humidity), with minimal misting. Prolonged exposure to mist may cause the leaves to rot during propagation. The misting can gradually be reduced as the cuttings form calluses and root primordia.

The cuttings are usually rooted in less than three weeks with soil temperatures of 68–75° F (20–24° C). For best results, the air temperature during rooting should be maintained above 60° F (16° C) and below 80° F (27° C). It is beneficial to begin constant liquid feeding with 200 ppm nitrogen at each irrigation beginning ten days from sticking.

Production

Gaillardia x *grandiflora* cultivars are well suited for commercial container growers and are commonly produced in 1 qt. (1 L) to 1 gal. (3.8 L) containers. They perform best when they are grown in a moist, well-drained medium with a slightly acidic to neutral pH, 5.8–6.4. Many commercially available peat- or bark-based growing mixes work well, provided there is good water-holding ability and adequate drainage.

Gaillardia cultivars are moderate feeders. Growers using water-soluble fertilizers either apply 150 ppm nitrogen as needed or feed with a constant liquid fertilization program using rates of 75–100 ppm nitrogen with every irrigation. Controlled-release fertilizers are commonly applied as a topdressing onto the medium's surface using the medium recommended rate on the fertilizer label or incorporated into the growing medium prior to planting at a rate equivalent to 1.0–1.25 lb. nitrogen per cubic yard (454–567 g/0.76 m^3) of growing medium.

Blanket flowers prefer to be grown in moist growing medium. Water as needed when the plants are young and becoming established. Once they are large, they will require more frequent irrigations, as they will dry out rather quickly. Due to their large leaf surface area, it is not uncommon for some cultivars to wilt before the growing medium dries out. When irrigation is needed, water them thoroughly, ensuring the entire growing medium is wet or nearly saturated. During container production, they are susceptible to dry conditions; once they are established in the landscape, they are considered to be drought tolerant.

Gaillardia prefers to be grown at high light levels. Where possible, growers should place them in an environment with no less than 4,000 f.c. (43 klux) of light. Plants grown at low light levels tend to be of lower quality, as they become elongated and do not flower as profusely as those produced at ambient levels. During the winter months, when the light levels are naturally low, providing 400 f.c. (4.3 klux) of supplemental lighting helps to improve crop quality.

Insects

Aphid	Root knot nematode
Beetle	Slug
Caterpillar	Spider mite
Leafhopper	Spittlebug
Leaf miner	Thrips

Aphids, leaf miners, whiteflies, and thrips will occasionally become problematic. Of these insect pests, aphids are the most prevalent. All of these pests, with the exception of leaf miners, can be controlled after they are detected or prevented using proactive strategies. The presence of leaf miners is often hard to detect until the larvae begin feeding between the leaf surfaces, leaving behind white trails often referred to as mines. Adult leaf miners can be caught on yellow sticky cards. They can be identified as small (2 mm) bright yellow and black flies, resembling the appearance of small bees. Recognizing the adults and treating the hot spots are keys to reducing excessive crop injury. Applying sprays of abamectin (Avid) or spinosad (Conserve) when the adults are present are effective at reducing the number of larvae mines found per plant. Most insects can be detected with routine crop monitoring. Control strategies may not be necessary unless the scouting activities indicate actions should be taken.

Diseases

Aster yellows	*Pythium*
Botrytis	Rust
Phytophthora	*Septoria*
Powdery mildew	Smut
Pseudomonas	

Plant diseases may be observed when environmental conditions are favorable for their development. Under various circumstances, *Gaillardia* may be attacked by various plant pathogens: aster yellows, bacterial leaf spots and wilts, powdery mildew, rust, and smut. Usually these diseases are not detrimental. As with many perennials, the occurrence of plant pathogens can be negated or greatly reduced when the proper cultural practices are followed. To control foliar diseases, it is best to manage the environment by providing proper plant spacing and adequate air movement, controlling the humidity, and, if desired, following a preventative spray program using the appropriate chemicals. The onset of root rot diseases, such as *Pythium* or *Phytophthora,* can often be prevented by avoiding high salts and overly wet conditions.

Plant Growth Regulators

Blanket flowers may require height control when they are grown under greenhouse and nursery conditions. Providing adequate spacing between the plants will reduce plant stretch caused by competition. Under certain growing conditions or at high plant densities, it may be necessary to use chemical PGRs. Several of the commercially available PGRs are effective at controlling plant height when they are applied using the appropriate rates, frequency, and timing. Compared to many perennials, the PGR rates needed to achieve sufficient control are relatively high. Individual cultivars respond differently to PGR applications. For example, 'Burgundy' has a moderate response to applications of uniconazole-P while 'Goblin' shows no response. For foliar applications, it usually requires two or three applications at seven-day intervals to provide adequate height control. To achieve the greatest results, these applications should begin as the stems are rapidly elongating and before the flower buds appear.

Forcing

To improve marketability, *Gaillardia* x *grandiflora* cultivars can be forced into bloom throughout the year. Growers wishing to produce blooming plants need to understand and follow the guidelines discussed below.

All blanket flowers propagated by seed have a juvenility phase and are not capable of flowering unless they are mature enough to perceive the other factors that promote flowering. Immature plants will not flower even after being exposed to the proper conditions necessary for flower formation. The primary method used to establish a plant's maturity is to count the number of leaves unfolded. The exact number of leaves needed to determine plant maturity and its ability to flower varies among cultivars. Most seed-propagated cultivars are considered mature after they have unfolded twelve to sixteen leaves. Vegetatively propagated cultivars often do not have a juvenility phase and may be flowering in the plug flat they were propagated in.

PGRs for *Gaillardia* x *grandiflora* Cultivars

Trade Name	Active Ingredient	Northern Rate
A-Rest	Ancymidol	50 ppm
B-Nine or Dazide	Daminozide	3,750 ppm
Bonzi, Piccolo, or Paczol	Paclobutrazol	45 ppm
Sumagic	Uniconazole-P	10 ppm
B-Nine or Dazide + Bonzi	Daminozide + paclobutrazol	2,500 ppm + 30 ppm
B-Nine or Dazide + Cycocel	Daminozide + chlormequat chloride	3,000 ppm + 1,250 ppm
B-Nine or Dazide + Sumagic	Daminozide + uniconazole-P	2,500 ppm + 5 ppm

These are northern rates and need to be adjusted to fit the location where the crop is being produced. These rates are for spray applications and geared to growers using multiple applications.

Most cultivars are considered to be cold-beneficial plants. Cold beneficial plants will flower faster, more uniformly, and produce more flowers per plant when cold treatments are provided. Some cultivars, such as 'Arizona Sun', will flower reliably without a cold treatment, while others, such as 'Goblin' ('Kobold'), have an obligate cold requirement for consistent flowering. Several cultivars may flower without receiving a cold treatment, but the entire population may not flower or the flowering is sporadic and unpredictable. Unless a specific cultivar is known to flower reliably without providing a cold treatment, it is recommended to vernalize plugs or small containers of blanket flowers for a minimum of ten weeks at 35–44° F (2–7° C). Longer durations of cold will increase the percentage of flowering plants and the number of flower buds when the population is marginally mature.

Most cultivars are considered to be long-day beneficial plants, meaning they flower best under long days. Following the cold treatment, most blanket flowers will flower under any photoperiod but will flower faster and produce more blooms when grown under long-day conditions. Providing more than ten weeks of cold is recommended when long days cannot be provided, but increasing the cold duration has no effect on flowering times when forcing with long-day photoperiods. A few cultivars, including 'Fanfare', are very floriferous and will flower under any photoperiod. For most cultivars, flowering can occur under short photoperiods but there are fewer flowers per plant, the percentage of blooming plants is reduced, the flowering is less uniform, the time to flower is delayed, and the blooming is more sporadic. I recommend providing photoperiods of sixteen hours, extending the day if necessary, or using a four-hour night interruption during the middle of the night, providing a minimum of 10 f.c. (108 lux) of light at plant level.

After the cold treatment and long days are provided, the time to flower depends on the temperature the plants are grown at. The following temperature recommendations fit most varieties, but each cultivar will have a slightly different finishing time at each of the temperatures illustrated. Blanket flowers grown at 60° F (16° C) will flower in about ten weeks, while plants grown at 70° F (21° C) will flower in as little as six weeks. Forcing blanket flowers at warmer temperatures (above 75° F [24° C]) will reduce plant quality and decrease lateral branching. To obtain the best plant quality, I recommend producing *Gaillardia* cultivars at 65–68° F (18–20° C).

References

Greenhouse Grower Magazine and Michigan State University. *Firing Up Perennials: The 2000 Edition.* Willoughby, Ohio: Meister Publishing. 2000.

Pilon, Paul. Perennial Solutions: *"Gaillardia compaositae* 'Fanfare'." *Greenhouse Product News.* January 2004.

———. Perennial Solutions: *"Gaillardia aristata* 'Arizona Sun'." *Greenhouse Product News.* April 2005.

Schoellhorn, Rick. "Vegetative Matters: Gaillardia." *Greenhouse Product News.* February 2004.

Gaura lindheimeri

Whirling Butterflies, Wand Flower, Bee Blossom

Gaura lindheimeri has been known to be a plant that offers growers and gardeners great versatility and resiliency throughout a wide range of the country. Due to the popularity of perennials and increased interest in *Gaura,* breeders across the world have been working diligently to improve certain characteristics of this plant species.

Figure 12.22. *Gaura lindheimeri* 'Ballerina Blush'

The flowers resemble small charming pink or white butterflies whirling around on slender stalks. Unlike some of the earlier cultivars, which reached 3–4 ft. (0.9–1.2 m) tall, many of today's cultivars grow to a manageable 12–18 in. (30–46 cm) when blooming. They bloom consistently throughout the summer, especially when the spent flower spikes are removed. These characteristics make whirling butterflies excellent for perennial forcing programs, container production, mass plantings, and as accent plants.

Gaura lindheimeri is native to North America, originating in Texas and Mexico. It prefers full sun and can tolerate a great deal of heat and humidity. Whirling butterflies perform well across a wide portion of the United States, throughout USDA Hardiness Zones 5–10 and AHS Heat Zones 9–5. With their resiliency and tolerance of heat and humidity, *Gaura* are well suited for production in the South, where conditions are not always suitable for many perennial species.

Gaura belongs to the Onagraceae family. There are not many perennials in this family being commercially produced. *Gaura, Fuchsia,* and *Oenothera* are the most popular perennials being grown from this family, while others include *Calylophus, Chamerion, Circaea, Epilobium, Ludwigia,* and *Zauschneria.*

Propagation

Gaura lindheimeri cultivars are started from seed or vegetatively propagated by tip cuttings or bare-root divisions. Commercially, they are primarily propagated by cuttings. *Gaura* 'The Bride' is the most common cultivar propagated from seed. Growers often observe crop variability, and plants that do not appear to be true to type when attempting to propagate named cultivars by seed. Many cultivars are patented, and unlicensed propagation is prohibited. To avoid illegal activities and the consequences, always check for plant patents before attempting to propagate.

Tip cuttings can be taken while they are in either the vegetative or reproductive stages. Cuttings already in bloom will root successfully, but they will take longer to root and will have a lower survival rate than purely vegetative starting materials. Whenever possible, provide short days (nine-hour photoperiods) to both stock plants and during propagation to maintain vegetative growth. Throughout most of the United States, this would require providing short days by applying a blackout period for most of the year. Providing short days

Commonly Grown *Gaillardia* x *grandiflora* Cultivars

'Ballerina Blush'	'Pink Cloud'
'Ballerina Rose'	'Pink Fountain'
'Blushing Butterflies'	'Siskiyou Pink'
'Corrie's Gold'	'Sunny Butterflies'
'Crimson Butterflies'	'The Bride'*
'Passionate Pink'	'Whirling Butterflies'

*Propagated from seed

for this duration is not practical and almost never practiced commercially. Fortunately, *Gaura* cultivars will root even if they are flowering. Growers producing stock plants under long photoperiods take cuttings every four to five weeks to encourage continued branching and production of young cuttings.

Tip cuttings should measure approximately 2 in. (5 cm) in length and contain several nodes. They are generally stuck into plug liners for rooting. The well-drained rooting medium should be moistened prior to sticking. The base of the cuttings can be dipped into a rooting hormone, such as a solution of indolebutyric acid (IBA) at rates between 500–1,000 ppm. *Gaura* cultivars can successfully root without rooting compounds but tend to root slightly faster and more uniformly when these treatments are provided.

Cuttings should be placed under low misting regimes for the first ten days of propagation. It is usually best to propagate under high humidity levels (90% relative humidity) with minimal misting. Prolonged exposure to mist may cause the leaves to rot during propagation. Most losses growers experience during propagation occur because the cuttings were misted too frequently or for too many days after the cuttings were stuck. Misting can gradually be reduced as the cuttings form calluses and root primordia. They are usually rooted in approximately three to five weeks with soil temperatures of 68–75° F (20–24° C). For best results, the air temperature during rooting should be maintained above 65° F (18° C) and below 80° F (27° C). It is beneficial to begin constant liquid feeding with 150 ppm nitrogen at each irrigation beginning ten days from sticking.

Production

For container production, *Gaura lindheimeri* is suitable for 1 qt. (1 L) to 1 gal. (3.8 L) containers. Most growers receive starting materials of whirling butterflies as finished plug liners. When planting large containers, such as 1 gal. (3.8 L) pots, I recommend planting at least two plug cells per container to properly fill out the pot. For the best performance, plant in a well-drained medium, preferably a nursery-type mix (bark based) rather than traditional greenhouse (peat-vermiculite) medium. Not only is the porosity (drainage) of the medium beneficial during crop production, it is essential to successfully overwinter *Gaura,* as it does not tolerate wet soils for long durations.

When planting, the plugs should be planted so that the original soil line of the plug is even with or just below the surface of the growing medium of the new container. Crops are susceptible to root rots, particularly after transplanting or when overly wet conditions occur.

After planting, it is recommended to apply a preventative fungicide drench using a broad-spectrum fungicide such as etridiozole + thiophanate-methyl (Banrot) or the combination of mefenoxam (Subdue Maxx) and thiophanate-methyl (Cleary's 3336).

It is beneficial to pinch most *Gaura* cultivars prior to or shortly after they are planted in the final container. Pinching increases lateral branching and the total number of flowers produced on each plant. Flowering will be delayed approximately two to three weeks when they are pinched; therefore, growers should modify their crop schedules or ready dates accordingly.

The best quality is achieved when plants are grown in full sun or in greenhouses with high light intensities. Where possible, growers should place them in an environment with no less than 4,000 f.c. (43 klux) of light. Plants grown at low light levels tend to be of lower quality, as they become elongated and do not flower as profusely as those produced at ambient levels. Plant quality can also be improved during short days or periods where the light levels are naturally low by providing 400–500 f.c. (4.3–5.4 klux) of supplemental lighting using high-pressure sodium lamps.

Whirling butterflies are moderate feeders and perform best when the pH is maintained at slightly acidic levels, 5.8–6.5. Fertility can be delivered using water-soluble or controlled-release fertilizers. Growers using water-soluble fertilizers either apply 200 ppm nitrogen as needed or feed with a constant liquid fertilization program using rates of 75–100 ppm nitrogen with every irrigation. Growers commonly apply time-release fertilizers as a topdressing on the medium's surface using the medium rate or incorporate them in the growing medium prior to planting at a rate equivalent to 1 lb. nitrogen per cubic yard (454 g/0.76 m^3) of growing medium.

Whirling butterflies generally prefer to be grown under slightly dry irrigation regimes but will also perform well under average watering scenarios. They do not tolerate overly wet conditions. Root zones that remain waterlogged tend to get root rot pathogens, which can quickly lead to crop losses. They are tolerant of drought and heat stress, but perform best when watered regularly. When irrigation is necessary, water them thoroughly, allowing the soil to dry moderately between waterings.

Insects

Generally, *Gaura lindheimeri* cultivars can be produced with relatively few insect problems. Aphids, leaf miners, and whiteflies may occasionally be observed but rarely cause significant injury to the crop. Of these insect pests, aphids are the most prevalent and often can be found feeding on the growing tips of the newest shoots. None of these insect pests require preventative control strategies. Growers should have routine scouting programs to detect the presence of insects early and to determine if and when control strategies are necessary.

Diseases

Botrytis	*Pythium*
Downy mildew	Rust
Phytophthora	*Synchytrium* (leaf gall)
Powdery mildew	

Botrytis is occasionally a problem on the lower foliage, where air movement is limited and the foliage often stays wet after irrigation for extended periods of time. To control *Botrytis*, it is best to manage the environment by providing proper plant spacing and adequate air movement and watering early in the day to limit the length of time the foliage remains wet.

Root rots from such pathogens as *Pythium* and *Phytophthora* are often observed when improper cultural conditions are present. For example, improper drainage of the growing medium and overwatering are the most common conditions that promote these pathogens. In many cases, these conditions occur just prior to or while overwintering plants. Injury to the root system from high salt levels or other stresses is also conducive to these diseases. Providing and managing the proper environment around the root system are the most important methods growers can use to prevent root rot diseases. There are several fungicides effective at controlling root rots, if chemical controls are necessary.

Plant Growth Regulators

Gaura cultivars have various growth habits. Cultivars such as 'Ballerina Rose' have naturally compact growth habits and will usually not require PGRs to control plant height. Other varieties, such as 'Siskiyou Pink', have taller growth habits and may need height control while they are being commercially produced. During the winter months, periods of low light levels, or when grown at high plant densities, excessive plant stretching might occur, requiring some type of height management strategy.

Most commercially available growth regulators are effective at controlling plant height of whirling butterflies. Before applying these chemicals, the height of *Gaura* can often be effectively controlled by providing adequate spacing between the plants and by withholding water and nutrients. Each cultivar responds differently to the various growth regulators. For example, uniconazole-P provides sufficient control of plant height with most *Gaura lindheimeri* cultivars but provides no reduction in plant height when it is applied to 'Siskiyou Pink'. Some cultivars may only require one application for toning and shaping purposes, while others will require multiple applications to provide acceptable control.

PGRs for *Gaura lindheimeri* Cultivars

Trade Name	Active Ingredient	Northern Rate
B-Nine or Dazide	Daminozide	2,500 ppm
Bonzi, Piccolo, or Paczol	Paclobutrazol	30 ppm
Cycocel	Chlormequat chloride	1,250 ppm
Sumagic	Uniconazole-P	5 ppm
B-Nine or Dazide + Cycocel	Daminozide + chlormequat chloride	2,000 ppm + 1,000 ppm
B-Nine or Dazide + Bonzi, Piccolo, or Paczol	Daminozide + paclobutrazol	2,000 ppm + 30 ppm
B-Nine or Dazide + Sumagic	Daminozide + uniconazole-P	2,000 ppm + 5 ppm

These are northern rates and need to be adjusted to fit the location where the crop is being produced. These rates are for spray applications and geared to growers using multiple applications.

Forcing

For container production, *Gaura* cultivars are well suited for forcing into bloom throughout the year. When producing blooming plants is the goal, a few requirements should be met in order to produce uniform, consistent, high-quality flowering plants.

Gaura lindheimeri does not have a juvenility period and has been observed flowering with as few as six leaves present. Flowering is often observed while they are in the plug flat and may bloom before the containers are completely filled out or bulked up. Bulking promotes lateral branching, fills out the pot, and results in high-quality, attractive plants with a lot of flowers at the time of sale.

Gaura lindheimeri does not require cold for flowering, but is considered to be cold-beneficial. They will flower without vernalization, but the overall flower number and quality attributes will be reduced. Providing a cold treatment for a period of six weeks at temperatures of 35–44° F (2–7° C) will decrease the time to flower, increase the flower number, and greatly improve overall plant quality and appearance. I recommend bulking cultivars for four to six weeks prior to providing a cold treatment. Many growers trim or pinch them back at the beginning of the bulking period to promote lateral branching, which provides fuller plants with more flowers. Other growers have been successful soft pinching the plants approximately two weeks after the start of forcing. Methods for maintaining vegetative growth were discussed previously in the propagation section on page 431.

Gaura is a facultative long-day plant, which means it will flower under any photoperiod but will flower faster under long days. When forcing during periods of the year where the natural day length is less than thirteen hours, provide long-day conditions using a four-hour night interruption or extending the day length to fourteen hours using supplemental lighting. Growers should provide long days using high-pressure sodium or metal halide lights, as incandescent light sources will cause the internodes to stretch excessively, reducing the overall quality attributes of the plant.

Following the recommended cold treatment and long-day photoperiods, the time it takes whirling butterflies to bloom is a function of temperature. Plants grown at 65°F (18° C) will take approximately seven weeks to reach flowering, while plants grown at 75°F (24° C) will flower in five weeks. Temperatures above 75° F (24° C) will delay flowering and produce plants with smaller flowers and reduced vigor compared with those grown under cooler temperature regimes. The best flower size is achieved by growing at temperatures averaging 64–68° F (18–20° C).

References

Enfield, A., E. Runkle, R. Heins, A. Cameron, and W. Carlson. "Herbaceous Perennials: Quick-cropping Part I." *Greenhouse Grower.* March 2002.

Greenhouse Grower Magazine and Michigan State University. *Firing Up Perennials: The 2000 Edition.* Willoughby, Ohio: Meister Publishing. 2000.

Pilon, Paul. Perennial Solutions: "Gaura 'Ballerina Series'." *Greenhouse Product News.* October 2004.

Grasses, Ornamental

Ornamental grasses represent members of the grass family, Gramineae; the rush family, Juncaceae; and the sedge family, Cyperaceae. They are becoming increasingly popular as garden and landscape plants. There are numerous types of grasses; many are suitable to a broad range of growing conditions and landscape applications, while others require more specific environments and have limited use by gardeners, growers, and landscapers.

There is a place in almost every landscape design and situation for ornamental grasses. They are easy to grow and can be utilized in a number of ways. They are often used alone, in mass plantings, as small groupings in perennial borders, as groundcovers, or in rock gardens. Many grasses provide structure and draw interest to the landscape during the winter months, which often appears barren and desolate otherwise.

Characteristics of Ornamental Grasses

Most ornamental grasses are true perennials, growing year after year; some are extremely long lived, surviving for decades. Other ornamental grasses are perennials in some environments but must be handled as annuals in others. In most parts of the country, *Pennisetum setaceum* 'Rubrum', with USDA Hardiness in Zones 9–11, is perhaps the most widely grown ornamental grass growers and gardeners must treat as an annual. Each type and cultivar of grass has its own cold and heat tolerances and will react differently from one environment to the next.

Grasses are often classified into two groups, warm season or cool season, based on when they are actively growing. The seasonal growth groupings are largely influenced by a plant's origin or where it is native. Warm-season grasses grow best at warm temperatures (75–90° F [24–32° C]). Growth begins in mid-spring as the temperatures begin to rise, flowering occurs in the summer or fall, and they become dormant prior to the onset of winter. Cool-season grasses grow best at cool temperatures (60–75° F [16–24° C]). Growth begins during the late winter or early spring, and they flower from spring to early summer. Growth slows during the heat of summer, often becoming dormant. Cold-season grasses often resume growth with cool fall temperatures and may continue growing throughout the winter months. Seasonal classifications may vary slightly between species, and some species may have characteristics of both classifications. For example, many cool-season grasses remain evergreen in the South but go completely dormant when produced in northern environments.

Grasses generally take on one of two types of growth habits: clumping or running. Clumping grasses, also referred to as bunch grasses, grow in tufts, or clumps, which gradually increase in width each year. The majority or ornamental grasses grown are clumping grasses and are considered highly ornamental. Running grasses, also referred to as creeping or spreading grasses, spread by means of underground stems (rhizomes) or aboveground stems (stolons). Many running grasses are considered invasive, as they rapidly form dense mats, often covering large areas.

There are many forms, or shapes and sizes, characteristic to ornamental grass foliage. The foliage is often classified into one of the following categories. Arching foliage arches up and out in fairly equal proportion. Mounded foliage forms mounds of somewhat weeping foliage where the top or new growth covers the lower leaves. Tufted foliage appears spiky with fine textured upright leaves arising from the basal clump. Upright foliage appears erect, growing vertically in a uniform manner. Upright arching foliage ascends vertically, then the top half of the foliage appears to become arching. Upright divergent foliage grows up and out in an erect or stiffly ascending manner.

The leaves of ornamental grasses take on many forms, appearing broad, curly, fine, flat, folded, wiry, or anything in between.

Depending on the genus and cultivar, grass foliage comes in numerous colorations, including nearly every shade of green and various colors of blue, brown, purple, red, and yellow. Grass leaves can be solid or contain variegation in the form of strips or spots. The leaves contain various textures, such as fuzzy and soft, smooth or pleated, dull or shiny, or possess razorlike edges. The foliage creates seasonal interest as it turns various shades of orange, purple, red, and yellow during the fall months and provides accents to the winter landscape.

The flower structures of ornamental grasses are useful for proper identification. What is commonly referred to as the flower of grasses is actually an inflorescence, or a group of flowers. Each individual grass flower is held in a structure called a spikelet, which remains hidden until flowering. The inflorescences of grasses take on one of three forms: spike, raceme, or panicle. A spike is the simplest type of inflorescence, as the individual flowers are attached directly to the main stem with no branches. Racemes contain spikelets held on branches directly connected to the main stem. The panicle is the most complex form of inflorescence, bearing spikelets on branches or stalks off of a main stem. Panicles may also contain flowers arranged in racemes on its branches. Regardless of the type of inflorescence, grass flowers may be arranged in dense or loose formations, providing for a wide range of ornamental appearances.

The spikelets typically appear as one color, usually various shades of bronze, green, pink, red, or silver, and then with maturity they change colors to brown, gold, gray, or tan. Many spikelets contain needlelike awns that appear bristly or hairy, while others are hairless or do not have pronounced awns.

The differing sizes and shapes of grasses, combined with the differing leaf colorations, flowering structures, and seasonal characteristics have all contributed to the popularity of ornamental grasses today. Depending on how they are utilized, grasses offer gardeners and landscapers a sense of structure or texture, softness or boldness, and grace that few plants can offer.

Six Popular Ornamental Grasses

The six genera discussed here (*Calamagrostis, Carex, Festuca, Miscanthus, Panicum,* and *Pennisetum*) represent some of the most commonly produced grasses grown by commercial growers, but they are just a small portion of the grass species in production today. These grasses are versatile and fit nearly every desired landscape application. Although the cultural requirements are often similar, there are some variations that if not acknowledged and provided for could lead to unnecessary crop performance or losses. These differences will be discussed below where appropriate.

Calamagrostis x *acutiflora* (feather reed grass)

Calamagrostis, commonly called feather reed grass, is a popular ornamental grass utilized by gardeners and landscapers across the country. The popularity of this grass skyrocketed following the award of Perennial Plant of the Year 2001 to *Calamagrostis* x *acutiflora* 'Karl Foerster' from the Perennial Plant Association.

Calamagrostis cultivars are considered to be attractive, carefree, and versatile. The foliage is upright, reaching 2–3 ft. (61–91 cm) tall, forming narrow clumps that reach 18 in. (46 cm) wide in the landscape. They are considered

Figure 12.23. ***Calamagrostis* x *acutiflora* 'Karl Foerster'**

cool-season grasses, as they flush best and flower under cool spring temperatures. The loose, feathery flower inflorescences appear in late spring (May to June). The inflorescences are light pink in color initially, turning to golden brown as the seed heads mature. The flower stems last throughout the growing season and into the winter months. With its vertical growth habit and long-lasting flower stalks, feather reed grass provides ornamental value throughout the year.

C. x *acutiflora* cultivars are long-lived perennial grasses commonly used as specimen plants, cut flowers, vertical accents to the landscape, or in patio pot containers. They perform well throughout USDA Hardiness Zones 4–9 and prefer locations with partial to full sun and soils with sufficient moisture levels. Feather reed grass is somewhat drought tolerant but tends to cease growth and go dormant during the summer months when temperatures are hot and moisture levels have decreased. The seeds are sterile, and the plants remain in clumps, keeping them easy to maintain and non-invasive.

Commonly Grown
***Calamagrostis* x *acutiflora* Cultivars**

'Avalanche'
'Karl Foerster'
'Overdam'

Carex (sedge)

Carex belongs to the sedge family, Cyperaceae, and it is not a true grass. The main differences between sedges and grasses are often difficult to distinguish, but with close observation, characteristics of the flowers and foliage can lead to a proper identification. The flowering stems of sedges are triangular and without nodes, while the flowering stems of grasses are most often round, hollow, and with nodes. Sedges have separate male and female flowers on the same plant, whereas grasses have perfect flowers with both male and female flower parts on the same flower. Most grasses have conspicuous ligules, or appendages, at the junction of the sheath and blade; ligules are absent on sedges. Sedges are most often evergreen, while most ornamental grasses go through a complete dormancy.

Figure 12.24. *Carex hachijoensis* 'Evergold'

There are over two thousand species of sedges found throughout the world. Only a handful of these are being commercially produced today, but more are being added annually. *Carex* cultivars are relatively easy to grow and have a variety of leaf shapes, sizes, and colorations. The color of the leaves can be brown, copper, orange, red, silver, or numerous shades of green. The foliage is often variegated and evergreen, adding to the ornamental value. The flowers are often inconspicuous, and most cultivars are sold for their foliage.

Commonly Grown
***Carex* Species and Cultivars**

C. buchananii 'Red Rooster'
C. caryophyllea 'Beatlemania'
C. comans 'Bronze Curls'
C. comans 'Frosted Curls'
C. comans 'Island Brocade'
C. comans 'Milk Chocolate'
C. dolichostachya 'Gold Fountains'
C. elata 'Bowles Golden'
C. hachijoensis 'Evergold'
C. morrowii 'Ice Dance'
C. morrowii 'Silver Sceptre'

The plant height of *Carex* cultivars can be as little as 6 in. (15 cm) or up to 3 ft. (91 cm) tall. They generally perform best in moist areas under light shade, but some cultivars perform well under drier conditions. Although most cultivars are hardy in USDA Hardiness Zones 5–9, cultivar-specific hardiness differences do occur. Growers should check the winter hardiness prior to starting a crop. The cultivars of most species are clump forming. Some cultivars of *C. flacca*, *C. glauca*, *C. muskingumensis*, *C. nigra*, *C. pansa*, and *C. siderosticha* are creeping varieties that form non-invasive colonies.

Festuca (fescue)

Festuca, commonly known as fescue, contains several species being commercially produced. There is great confusion regarding the parentage, nomenclature, and synonyms of *Festuca* species. Most cultivars in the trade today are classified as the species *F. glauca.* Synonyms of *F. glauca* include *F. arvensis*, *F. cinerea*, and *F. ovina*.

Figure 12.25. *Festuca glauca* 'Select'
Jelitto Perennial Seeds

Commonly Grown *Festuca glauca* Cultivars

'Boulder Blue'	'Sea Urchin'
'Elijah Blue'	'Select'
'Golden Toupee'	

Most *Festuca* cultivars in commercial production are clump-forming, cool-season grasses with blue-gray evergreen foliage. The size, color, and texture of the foliage varies not only by cultivar but with the environment where it is produced. Blue to purplish flowers emerge from May to June, drying to a golden coloration as they mature, and then taking on a bleached grayish color at maturity.

Cultivars range from 6–16 in. (15–41 cm) in height and have fine to medium textured foliage. They are grown in containers, combination pots, rock gardens, borders, mass plantings, and as groundcovers throughout USDA Hardiness Zones 4–9. *Festuca* is considered a short-lived grass and may need to be divided every two to three years to keep them looking good.

Festuca is generally an easy-to-grow plant but is somewhat particular about temperature and moisture. They do not perform well during hot muggy summers, extremely dry and hot conditions, periods of excessive rainfall, or in heavy soils. They perform best when grown in full sun, and, once established, they show some level of drought tolerance. With their ease of production and desirable characteristics, growers often use fescues to provide color, form, and texture to the landscape.

Miscanthus sinensis (maiden grass)

The most commonly grown ornamental grass is *Miscanthus.* The species and cultivars within this genus provide gardeners and landscapers with a variety of blooming characteristics, growth habits, and hardiness ranges. The most notable feature of *Miscanthus* is the beautiful display of flowers. The spectacular inflorescences emerge as silky bronze, pink, silver, tan, or multicolored tassels; the flowers mature into puffy plumes that persist throughout the winter.

Figure 12.26. *Miscanthus sinensis* 'Morning Light'

Commonly Grown *Miscanthus sinensis* Cultivars

'Adagio'	'Kirk Alexander'
'Andante'	'Little Nicky'
'Arabesque'	'Little Zebra'
'Autumn Light'	'Morning Light'
'Blue Wonder'	'Nippon'
var. *condensatus* 'Cabaret'	'Puenktchen'
var. *condensatus* 'Cosmopolitan'	var. *purpurescens*
'Dixieland'	'Silberfeder'
'Ferner Osten'	'Strictus'
'Gracillimus'	'Variegatus'
'Graziella'	'Yaku Jima'
'Huron Sunrise'	'Zebrinus'

Miscanthus sinensis cultivars are the most widely produced of the various *Miscanthus* species, and are some of the most desirable ornamental grasses grown today. They are clump-forming, warm-season grasses commonly grown throughout USDA Hardiness Zones 4–9, and there are slight variances of plant hardiness between cultivars. Being warm-season grasses, they do not begin to flush in the spring until the temperatures are conducive for plant growth, usually in late March or April.

The height of the clump varies with the cultivar and ranges from 3–8 ft. (0.9–2.4 m) tall. The foliage comes with many forms and colorations. During the growing season, the foliage varies with cultivar and may consist of various hues of green or different variegations, with white, yellow, or silver stripping or vertical banding. The showy whisk-like plumes begin to appear in late summer and through the fall. The foliage provides astounding color displays during autumn, with brilliant shades of red-orange and yellow to neutral colorations of tan and brown.

These grasses prefer moist growing conditions and thrive near water gardens or ponds. They are often used as accent plants or in mass plantings along highways or on golf courses. The foliage and flowers of most cultivars hold up rather well during the winter months, providing additional architectural interest during this time. *Miscanthus* cultivars are easy to grow and always deliver an impressive display, whether they are in bloom or not.

Panicum virgatum (switch grass)

Panicum virgatum, or switch grass, is a very popular native American ornamental grass. Switch grass is a versatile, clump-forming, warm-season grass that tolerates a wide range of

Figure 12.27. *Panicum virgatum* in the landscape

growing conditions. *Panicum* cultivars are widely grown and perform well across USDA Hardiness Zones 4–9. They are commonly used as accent plants, in background and mass plantings, or planted near water gardens and ponds.

Panicum virgatum forms large upright clumps reaching 4–7 ft. (1.2–2.1 m) tall. Depending on the cultivar, the deep green to gray-green flat bladed upright leaves are 0.5–0.75 in. (13–19 mm) wide and 18–24 in. (46–61 cm) long. Most switch grass cultivars put on an impressive display of fall color, turning various shades of yellow, but some turn red or orange. As winter approaches, they fade to a beige coloration and remain upright throughout the winter. They flower in midsummer. The inflorescence, which rises 1–2 ft. (30–61 cm) above the foliage, is a tightly branched panicle at first and becomes more open and airy as it spreads open. The flowers are pink, red, or silver and become brown or grayish white at maturity.

Commonly Grown *Panicum virgatum* Cultivars

'Cloud Nine'	'Prairie Sky'
'Dallas Blues'	'Shenandoah'
'Heavy Metal'	'Trailblazer'
'Northwind'	

Switch grass is a versatile ornamental grass valued for its erect upright form, showy inflorescences, and the year-round contributions it provides to the landscape, including brilliant fall colors and the winter effects it provides while dormant.

Pennisetum alopecuroides (fountain grass)

Fountain grasses are some of the most dependable, showy, and versatile ornamental grasses of all the grasses in production. *Pennisetum alopecuroides* is used synonymously with the species *P. japonicum*. There are dwarf cultivars, such as 'Little Bunny', that only reach 6–12 in. (15–30 cm) in height, or tall varieties, such as 'Foxtrot', that reach 4–5 ft. (1.2–1.5 m) tall. They are clump-forming, warm-season grasses

Figure 12.28. ***Pennisetum alopecuroides* in the landscape**

that deliver handsome foliage and showy flowers to the landscape.

The glossy, bright green leaves are 0.25–0.5 in. (6–13 mm) wide and, depending on the cultivar, from 6–30 in. (15–76 cm) in length. The dwarf cultivar 'Little Honey' has variegated foliage with a white stripe down the midrib. The clumps form dense, upright mounds and often grow as wide as the foliage gets in height. During the fall, the leaves become streaked with yellow and brown, becoming completely almond color in late fall and fading to a straw color as winter approaches.

The flowers of *Pennisetum* are usually described as feathery, bottlebrush-like foxtails. The inflorescences are 1–3 in. (3–8 cm) wide and 4–10 in. (10–25 cm) long. The creamy white to light pink, coppery bronze to deep purple, or tan inflorescences emerge on arching stems during midsummer, creating a cascade of flowers. The flower coloration changes generally to a reddish brown as they mature.

Most cultivars thrive in USDA Hardiness Zones 5–9. They are extremely adaptable, thriving in a wide range of soil conditions. Fountain grass performs best when grown in sunny locations; some partial shade is acceptable. With a range of plant forms, shapes, and

Commonly Grown *Pennisetum alopecuroides* Cultivars

'Cassian'	'Little Bunny'
'Foxtrot'	'Little Hunny'
'Hameln'	'Moudry'

heights, *Pennisetum* cultivars can be utilized in a variety of landscape applications. Gardeners and landscapers commonly use fountain grass as container plants, specimen or border plants, and in mass and background plantings.

Growers should avoid seed-propagated species, such as the straight species *Pennisetum alopecuroides,* as they produce a lot of viable seeds and can become a nuisance in the landscape.

Propagation

Ornamental grasses are propagated by division or seed. Propagation is primarily done by division, as it is the only form of propagation that reliably perpetuates a plant's unique characteristics. In many instances division is the only method to propagate grasses, as many cultivars produce sterile seed or seed that is not viable and will not germinate.

Many cultivars do not grow true to type when propagated from seed, meaning the seedlings are not genetically identical to the parent plant and slight variations may be observed from plant to plant. Only a handful of grasses are commercially propagated by seed.

Division

Most grasses in commercial production are propagated by division. Division entails dividing or splitting the crown into smaller sections containing at least one stem, also commonly referred to as a culm or tiller, and several adjoining roots. Division of warm-season grasses (*Miscanthus, Panicum, Pennisetum*) is best when done in the late winter or early spring, while the plants are still in a dormant state. Cool-season grasses (*Calamagrostis, Festuca*) are best to propagate in the fall, winter, or early spring. Sedges (*Carex*) are best when propagated by division during the spring. A general rule of thumb regarding the proper propagation timing is to perform the division before or just as new growth is beginning. Most grasses that have been actively growing for some time will not propagate as successfully, or will result in a higher percentage of plant losses, as compared with those propagated at the right stage of development. Several grasses will tolerate division at other times of the year, but will root and survive better when it occurs at the optimum time for that particular grass.

When making divisions, the best results are achieved when the foliage is cut back one fourth or one third to reduce the loss of moisture through transpiration. The divisions should be made immediately after the crown of the mother plant is harvested, and they should be transplanted quickly into the desired container. This must be done immediately so the divisions do not dry out. The size of the container ranges widely from a 21-cell sized plug up to a 2 gal. (7.6 L) container.

Care should be taken when planting the division into the finished container. When planting, pay particular attention to ensure the potting substrate comes into good contact with the propagule. If the proper conditions exist, ornamental grasses will root easily after the division.

After they are planted, it is important to keep the growing medium moist, not saturated, to prevent further desiccation of the plant materials. It is beneficial initially to keep them protected from high light and windy conditions. Once they are beginning to root into the new growing medium, they can be supplied with more ambient growing conditions.

The size of the division will vary, depending on the size of the desired finished container. For example, the starting material for a 21-cell plug will vary dramatically from the material needed for a 1 gal. (3.8 L) container. For 1 qt. (1 L) production, a crown consisting of two or three stems is commonly used. For larger containers, such as 1 gal. (3.8 L), divisions of four to six tillers are commonly used. This varies greatly by variety. For example, *Festuca* will require many more stems per division to properly fill out the final container than many other grass varieties. Use past experience and the needs of the customer to determine what the proper division size needs to be. Failure to use large enough starting materials will result in longer production times and decreased quality, as the final product may appear thin and not proportionate to the size of the container.

The length of time necessary for propagation varies widely with the ornamental grass variety being propagated, the size of the propagation materials, the stage of development of the stock plant, and the propagation temperatures. It typically takes a couple of weeks for the divisions to begin rooting, and the length of time to become fully rooted often depends on the size of the container they are in. Since larger divisions are typically used for larger containers, the length of time needed is often offset and may not be terribly significant. A rough estimate of the propagation time necessary for most grasses is between eight to twelve weeks.

Seed

Plants started from seed will take much longer to finish compared with grasses propagated by division. They will also show great variability of plant color, habit, and vigor, as the seedlings are generally not true to type. Depending of the variety, starting ornamental grasses from seed can be very easy to very challenging. Some grasses emerge within a few days, some a few weeks, and yet others may take several months to germinate. In some cases, growers have found various seed treatments such as presoaking in warm water or stratification to be effective at increasing the germination success.

For several grasses, such as numerous *Carex* species, it is beneficial to stratify the seed prior to germination. Stratification is useful to overcome seed dormancies and will increase the uniformity and rate of germination. The seeds should be sown in the intended plug flat and covered lightly with germination mix or medium-grade vermiculite. The covering helps to maintain a suitable environment around the seed during this phase. The seed flats should be moistened and moved to an environment, such as a cooler, where the temperatures can be maintained at 40° F (4° C) for four weeks. After the cold treatment, move the plug trays into a warm environment (75–85° [24–29° C]) for germination. Many growers utilize germination chambers during this stage to provide uniform moisture levels and temperatures. Germination begins within a few days, but it often takes a couple of weeks for all of the seeds to germinate. Following germination, they can be grown at 70° F (21° C) until they are ready for transplanting.

The seeds of ornamental grasses are often small and difficult to handle. Sow them on the surface of the growing medium. The seeds can be covered lightly with germination mix or vermiculite, or they can remain uncovered, provided sufficient moisture levels can be maintained. After sowing, moisten the seed flats and move them to a warm environment for germination. Most grasses, including *Festuca, Miscanthus,* and *Pennisetum,* can be successfully germinated at 67–72° F (19–22° C) temperatures. Some grasses, such as *Panicum,* prefer to be germinated at warmer temperatures (75–80° F [24–27° C]). Many growers utilize germination chambers during this stage to provide uniform moisture levels and temperatures, though this is optional. It is important to keep the medium very moist but not saturated during this stage or the germination rates are likely to decrease.

Seedlings will emerge over a period of fourteen to twenty-one days after sowing. If rapid germination does not occur with *Carex, Miscanthus,* or *Panicum* species, growers should consider placing them through a stratification process such as the method described above. Following germination, reduce the moisture levels somewhat, allowing the growing medium to dry out slightly before watering to help promote rooting. Once germination has occurred, the temperatures can be reduced to 60–70° F (16–21° C) for most species until they are ready for transplanting.

Following germination, fertilizers are usually applied once two true leaves are present. Most plug growers apply 100 ppm nitrogen every third irrigation or 50 ppm with every irrigation, using a balanced water-soluble source. The light levels can gradually be increased as the plug develops, from 500 f.c. (5.4 klux) following germination to 2,500 f.c. (27 klux) at the final stage. When 128-cell plugs are grown at the above temperatures, they are usually ready for transplanting in twelve to fourteen weeks. Grass seedlings should not be transplanted until they have formed several sets of leaves and are fully rooted.

Production

The type of starting materials utilized by growers often depends on personal preference, availability, and the time of the year the planting occurs. Divisions are most commonly used during early spring, while plugs are used for transplanting throughout the production season.

Ornamental grasses will perform well in a wide range of potting mediums. It is recommended to use a medium with adequate drainage and water-holding capacity. Grasses generally prefer to be kept on the moist side. However, saturated conditions or extended periods of being overly wet will lead to root rots and possibly result in plant loss.

Growers can manipulate the growing mix to match the season the plants are being produced. For example, when producing grasses for summer sales it is often advantageous to use a growing medium with good water-holding capacity, such as a peat-based mix, allowing the mix to hold onto the water better between irrigations. Conversely, production of grasses in the fall may warrant a mix with less water-holding ability, reducing the likelihood the roots will be exposed to wet conditions for extended periods, which reduces the potential for root rot pathogens to attack the crop.

Growers transplant grass divisions or rooted plug liners into the desired container. Some grasses, such as *Carex* and *Festuca,* are suitable for production in small container sizes (1 qt. [1 L]), but most of the ornamental grasses discussed here are best when produced in larger containers (1 gal. [3.8 L] or larger).

When planting, it is important for the roots to be in good contact with the potting medium to prevent unnecessary drying out of the root system. This is especially important when bare-root divisions are used. Try to avoid air pockets between the division and the growing medium. While transplanting plugs, try to avoid planting them too high or too low; always plant to match the original soil line of the plug with the growing mix of the final container. Bare-root divisions should be planted with the crown being even with the soil line of the new container. If the crowns of ornamental grasses are planted below the soil surface, crown rots are likely to occur.

In many cases, bare-root divisions are only available for a limited time during the spring. Fortunately, there are several perennial producers offering large plugs (21-cell or larger) of various ornamental grasses to the industry. Growers often find that using plugs allows them ease of transplanting and more flexibility as to when the crop must be planted as compared with using only bare-root starting materials.

For successful establishment and plant growth, growers should properly manage the moisture levels of the growing mix for the first couple of weeks and avoid conditions that lead to excessive water stress. It is important to keep the root zone of newly potted grasses moist but not wet until they become established. Once they are fully rooted, grasses can be allowed to dry out more fully between waterings.

No ornamental grasses like to be grown overly dry. Many of them can tolerate short periods of dry conditions, but perform best when an adequate amount of irrigation is provided. *Calamagrostis, Miscanthus, Panicum,* and *Pennisetum* require average amounts of irrigation and often need to be watered daily when they are actively growing. *Carex* can be grown using average to above average watering regimes since they naturally grow in moist environments. *Festuca* prefer to be grown with average to below average irrigation practices; they do not tolerate moisture extremes and should not be grown at excessively moist or overly dry conditions for any length of time.

Most ornamental grasses (*Calamagrostis, Miscanthus, Panicum,* and *Pennisetum*) are moderate to heavy feeders. Some grasses (*Carex* and *Festuca*) are considered light to moderate feeders and require less fertilizer to grow the crop. The pH of the growing medium should be maintained within the 6.0–6.5 range. Growers typically deliver nutrition to grasses using liquid fertilizer programs or time-release fertilizers.

When constant liquid feed programs using water-soluble fertilizers are used, grasses that are considered to be moderate to heavy feeders are fed using rates of 100–200 ppm nitrogen, and light to moderate feeders require rates of 75–125 ppm nitrogen with each irrigation. When intermittent fertilizer applications are applied as needed, or on a weekly basis, growers often use higher rates of water-soluble fertilizers, such as 300–400 ppm nitrogen as needed.

Controlled-release fertilizers are commonly top-dressed onto the medium's surface using the medium or high rate recommended on the fertilizer label. More commonly, time-release fertilizers are incorporated into the growing mix prior to planting at a rate equivalent of 1.25–1.5 lb. nitrogen per cubic yard (567–680 g/0.76 m^3) of growing medium for moderate to heavy feeders or 1 lb. nitrogen incorporated per cubic yard (454 g/0.76 m^3) for light to moderate feeders.

Some of the most successful grass producers have found it beneficial to combine water-soluble and time-release fertilizer strategies as an effective nutrient management program for ornamental grasses. In many cases, these growers intentionally provide most of the fertilizer requirement using controlled-release fertilizers and supplement with liquid fertilizers as needed. This allows them to grow a healthy crop without providing too many nutrients up front and gives them the ability to tone the crop prior to sales or shipping.

Growers overwintering grasses have also found it beneficial to top-dress them while they are dormant to supplement any nutrients already released by the time-release fertilizers during the previous growing season. If growers anticipate needing additional fertilizer to properly grow the crop during the upcoming production season, this is the best time to apply it. Once the crop has flushed, it is difficult to broadcast fertilizers over the top of the crop and achieve uniform application.

Growers should also avoid placing the entire amount of topdressing material into the crown of the plant. Even though these are controlled-release fertilizers, there will be too much fertilizer and salts released into the crown, causing significant injury to the plant, possibly leading to mortality. When top-dressing pot-to-pot, it is best to place the fertilizer away from the crown, preferably in at least two spots on opposite sides of the medium's surface. Time-release fertilizers broadcasted over the top of a crop do not seem to cause any injury to the crown since only a small amount of fertilizer actually comes into contact with this portion of the plant.

For plant establishment of warm-season grasses, I recommend maintaining average temperatures of 65–75° F (18–24° C). When these temperatures are maintained throughout crop production, 1 gal. (3.8 L) pots can be produced from divisions or 21-cell plugs in seven to nine weeks. Cool-season grasses should be produced at slightly lower average temperatures; 55–65° F (13–18° C) is adequate for most of these grasses. Grasses that require cooler production temperatures generally call for slightly longer production times; many of these grasses can reach finished size in nine to twelve weeks. Keep in mind that most producers of ornamental grasses do not market them as flowering plants. In fact, unless they are produced and sold on site, it is often very expensive to ship a tall flowering grass to retail sites. The amount of time required to grow the crop also depends greatly on the size of the starter materials used, as larger divisions or plugs usually finish quicker than when smaller sizes are used.

In terms of forcing ornamental grasses into bloom, nearly all of them have an obligate cold requirement for flowering. There are a few exceptions, such as some *Festuca* varieties started from seed are capable of flowering the first year. Even these varieties will flower better once cold has been provided. There has been relatively little research conducted to determine the exact requirements for flowering of a broad range of ornamental grass species. The cold requirements are usually long. I recommend growers wishing to produce flowering grasses provide a minimum of twelve weeks of temperatures less than 40° F (4° C). Spring-planted bare-root divisions usually have already received adequate amounts of cold while in the dormant stage in the field prior to harvesting and will normally flower after the other conditions necessary for flowering have been met.

Most grasses are long-day plants and will not flower while they are grown under short days. In fact, several of them, including *Miscanthus, Panicum,* and *Pennisetum,* will not grow or go dormant under short days. Most growers do not market flowering grasses. To produce non-flowering grasses that are fully flushed, use the schedules mentioned above. If flowering grasses are desired, produce them at 68–72° F (20–22° C) and use the following schedules as guidelines.

Schedule for Flowering Ornamental Grasses

Variety	Cold Requirement	Photoperiod	Weeks to Flower
Calamagrostis	Obligate	Obligate LD	4–7
Miscanthus	Obligate	Obligate LD	16+
Panicum	Beneficial	Obligate LD	8–10
Pennisetum	Beneficial/ obligate	Obligate LD	10–14

Ornamental grasses should be produced under high light levels, with a minimum of 5,000 f.c. (54 klux). Lower light levels or areas of partial shade are still adequate for plant growth and development but may cause the foliage to take on a lighter green coloration. Plants grown under shade with consistently low light levels tend to become floppy and have lower quality characteristics.

Ornamental grasses are easy to overwinter when provided minimum amounts of protection. In the late fall, after they have gone dormant, trim the plants back to 2–3 in. (5–8 cm) above the top of the container. Many of the short evergreen cultivars, such as those of *Festuca* and *Carex,* may not require trimming prior to overwintering, depending on crop age and size. It is often beneficial to not trim these varieties prior to winter and cut them back as needed at the end of the overwintering period just before they begin to flush in early spring. Once they are trimmed, group the pots together inside a cold frame, greenhouse, or outdoor production bed. In many parts of the country, grouping them together is the only protection necessary, especially if they are located in covered structures. In colder zones, I recommend covering them with a protective frost blanket during the winter months. It is very important to not let the root zones dry out during the winter months, as overly dry conditions during this time will usually result in crop losses.

Insects

Calamagrostis	*Carex*	*Festuca*
Aphid Soft scale	Aphid Mealybug Slug	Aphid
Miscanthus	***Panicum***	***Pennisetum***
Aphid Japanese beetle Mealybug Slug Spider mite	Japanese beetle Spider mite Thrips	Caterpillar Japanese beetle Spider mite Spittlebug Whitefly

Insect problems are generally a rare occurrence with commercial grass production. Aphids, Japanese beetles, mealybugs, slugs, snails, spider mites, spittlebugs and whiteflies may occasionally be observed feeding on ornamental grasses. Most insects can be detected with routine crop monitoring. Control strategies may not be necessary unless the scouting activities indicate actions should be taken. These pests seldom cause significant injury to the crop and can be controlled relatively easily using the appropriate contact insecticides.

Diseases

Calamagrostis	*Carex*	*Festuca*
Crown and root rots Rust	Crown and root rots Rust	Crown and root rots *Sclerotium*
Miscanthus	***Panicum***	***Pennisetum***
Crown and root rots *Miscanthus* blight Rust	Crown and root rots	Crown and root rots *Helminthosporium* leaf spot

Plant pathogens are also not very common when producing ornamental grasses. The most common disease, crown rot, often occurs within a couple of weeks of transplanting. Improper planting practices or poor irrigation management are often responsible for the onset of root or crown rot pathogens. Other potential causes of root rots of ornamental grasses include high salt levels in the growing medium, poor physical properties of the medium (namely too much water-holding capacity and decreased aeration), or the crop has been grown in the same container and growing mix for too long. Any of these conditions could lead to plant stress and the onset of root rot pathogens. Try to pick a medium that has good water-holding and drainage characteristics and will not deteriorate or settle over time. Monitor the irrigation practices and the fertility levels on a regular basis, making adjustments accordingly. When possible, do not hold grasses in the same container for an extended period of time (twelve months or more). When these measures are taken, most crown and root rots can be prevented.

The occurrence of rust diseases may be observed under certain circumstances. Rust is a fungal disease caused by the pathogen *Puccinia*. It appears as orange spots on the stems and leaf blades of ornamental grasses. If it goes undetected, rust can rapidly cover the entire plant, quickly rendering the plant unsalable. Culturally, the presence of rust can be reduced by allowing plenty of space between the plants, allowing for good air circulation, and applying irrigation directly to the growing medium (not the foliage) using drip or subirrigation systems. If overhead irrigation must be applied, irrigate early in the day so the water droplets on the leaves can dry before the end of the day. Localized areas of rust can be removed from the production area by discarding infected leaves or plants. Some growers have found it beneficial to apply fungicides, such as triadimefon (Bayleton, Strike), or myclobutanil (Eagle, Systhane), on a preventative basis to reduce the occurrence of this disease.

As discussed in chapter 5, plant pathogens are most commonly introduced by cultural problems and in many instances can be prevented. One common problem that occurs is observed within two weeks of planting. It is essential for the division to be in good contact with the potting medium when it is planted and for the moisture levels to be properly monitored. Many times after planting, growers will observe the plants are not actively growing and actually appear to be turning brown. Many of the clumps continue to deteriorate and ultimately die. Using fungicide drenches to slow down or stop the deterioration of the clumps will most often only have a limited effect. The extra time spent properly planting and maintaining new crops of grasses is well worth the effort.

Another common observation is very similar to what often happens at planting but occurs after the plants are well established. The clumps within the pot often become lopsided, dying out in spots, or dying out completely. This problem is often caused by one or more of the following factors: high salt levels in the medium, too much irrigation, poor physical properties of the medium, or the crop has been grown in the same pot/medium for too long. Any of these conditions could lead to plant stress and the onset of root rot pathogens.

Plant Growth Regulators

Several growers have expressed a need to reduce the plant height of ornamental grasses when they are grown in containers. Growers producing plants for the landscape industry are rarely concerned with reducing plant height, but those growers producing grasses for box stores or shipping them long distances are very interested in reducing plant height, as they can fit fewer tall plants than small plants per load. Of course, it is undesirable to turn a 3 ft. (91 cm) tall grass into a 6 in. (15 cm) specimen. A realistic reduction of plant height by 25–35% is often desirable and still allows an ornamental grass to be an ornamental grass instead of a clump of turf.

The height of grasses can be maintained by routinely trimming them back, which both reduces the height of the plant and promotes more shoots to develop from the crown of the plant. When trimming grasses, it is important to remove no more than one-third of the existing

PGRs for *Calamagrostis, Miscanthus, Panicum,* and *Pennisetum* Cultivars

Trade Name	Active Ingredient	Northern Rate
Bonzi, Piccolo, or Paczol drench	Paclobutrazol	10 ppm
Sumagic drench	Uniconazole-P	2 ppm
Topflor drench	Flurprimidol	10 ppm

These are northern rates and need to be adjusted to fit the location where the crop is being produced. These rates are for drench applications and are usually applied once per crop. Use these rates as beginning rates and adjust them accordingly to provide the desired level of height reduction.

foliage or the crop may become stunted or reflush unsatisfactorily. Trimming can effectively be used to maintain plant height of tall-growing grasses such as *Calamagrostis, Miscanthus, Panicum,* and certain *Pennisetum* cultivars throughout the growing season. It is not usually necessary to maintain plant height of short grasses such as *Festuca* and most *Carex* cultivars.

PGRs can also be used to reduce the height of ornamental grasses. Foliar applications are rather ineffective, as the chemical has difficulty getting good contact with the stems, which are most often covered by the leaf sheath. When these products are applied as one-time drenches, growers commonly will observe a reduction in plant height. The most effective PGRs are those taken up by the root system, namely flurprimidol, paclobutrazol, and uniconazole-P. The rates necessary to achieve sufficient levels of control are higher than drench rates used for most perennials. Since there has only been limited research using PGRs on grasses, it is best to conduct small trials before making large-scale applications to the entire crop. Drenches should be applied by the time the plants are 6–12 in. (15–30 cm) tall. Later applications seem to be less effective and will not provide the desired results.

References

Armitage, Allan M. 1997. *Herbaceous Perennial Plants: A Treatise on Their Identification, Culture, and Garden Attributes.* Second edition. Champaign, Ill.: Stipes Publishing. 1997.

Greenlee, John. *The Encyclopedia of Ornamental Grasses.* Emmaus, Penn.: Rodale Press. 1992.

Pilon, Paul. Perennial Solutions: *"Pennisetum orientale* 'Tall Tails'." *Greenhouse Product News.* May 2004.

———. Perennial Solutions: *"Festuca glauca* 'Boulder Blue'." *Greenhouse Product News.* November 2003.

Still, Steven. "*Calamagrostis* x *acutiflora* 'Karl Foerster'." *Perennial Plants.* Autumn 2000.

Hemerocallis

Daylily

Daylilies are one of the most widely grown perennials in the United States. In fact, *Hemerocallis* is the second most popular perennial grown today, second only to *Hosta.* A great deal of breeding and selection work has led to the new varieties being produced today. Breeding improvements include better growth habits, foliage and flower textures, flower color and bloom periods, and sun and heat tolerance. There are currently over 50,000 named cultivars of *Hemerocallis* being produced.

Figure 12.29. *Hemerocallis* 'Daring Deception'

Eric Olson

Like the common name daylily implies, the individual flowers usually only last for one day. Even the scientific name *Hemerocallis* translates into "beauty for a day" from the Greek words *hemera* ("day") and *kallos* ("beauty"). Most established daylilies produce many buds, forming clumps that bloom continuously for thirty to forty days. Daylilies perform well throughout USDA Hardiness Zones 3–10, where they are commonly used as groundcovers, border plantings, mass plantings, and for naturalizing.

In the landscape, daylilies prefer full sun to partial shade and are very tolerant to a wide variety of growing conditions. They can withstand heat and drought better than most commonly grown perennials. Although daylilies do not prefer water logged soils, they have an excellent ability to survive floods. These attributes give *Hemerocallis* the reputation for being one of the toughest, most adaptable perennials used in the landscape.

Daylily clumps have smooth, grasslike foliage. The leaves are long, narrow, heavily ribbed, and two ranked, forming regular fans. The lily-like flowers come in clusters atop slender flower stalks, called scapes. Originally, daylilies were limited to only orange, red, and yellow blooms. Today, through extensive breeding, *Hemerocallis* varieties consist of nearly every color, including different hues of carmine, gold, lavender, maroon, orange, pink, red, rose, scarlet, and yellow.

Daylilies classified as rebloomers are often referred to as repeat bloomers. Daylilies in this category produce an initial flush of flowers in late spring to early summer and will have one or two repeated blooming cycles later that growing season. Any additional flushes of blooms produce fewer flowers than the original flush. Daylilies are also categorized as extended bloomers, meaning each flower will last for sixteen hours or more per day.

Foliage Classification and Landscape Performance

Even though we often consider all daylilies as equals, growers should know that certain daylilies perform differently in various climates. Daylilies can be classified into one of three categories based on foliage type (dormants, evergreens, and semi-evergreens) to help determine hardiness and plant performance throughout various parts of the country. There are no hard and fast rules to foliage classification and landscape performance. These classifications were more accurate in the past, before several generations of breeding have developed hardy evergreen and non-hardy dormant cultivars. Understanding these classifications and individual cultural characteristics can lead to better production practices as well as improved customer satisfaction.

Dormant

Daylilies within this category are deciduous, completely losing their foliage during the fall and winter months. Some daylilies in this category gradually die back as a response to photoperiod, while cold temperatures kill others. Next year's new shoots, or fans, develop from the crown and remain slightly below the soil line until conditions are conducive for growth.

Generally, dormant varieties, such as the very popular 'Stella de Oro', are totally hardy in the North, requiring a sustained period of cold to enter and remain dormant. In mild climates, such cold periods do not occur, causing the daylily to perform unsatisfactorily. In frost-free areas, dormant varieties struggle, often only blooming for a season or two before they eventually die. Several dormant varieties do perform very well in southern climates. Southern growers should research dormant varieties and their performance before placing them into production, though growers in the Deep South should avoid *Hemerocallis* varieties that fall within the dormant category.

Evergreen

Daylilies in this category have the foliage present year round. They continue to grow in mild climates, grow slightly in transitional climates, and often experience leaf injury in cold climates. The amount of injury progresses as temperatures become lower. In many cases, only the foliage tips express injury symptoms. Cold injury can cause dieback down into the clump. Cold injury often causes the foliage of evergreen varieties to become mushy and is slow to dry, appearing as greenish-brown mush. In most cases, the injury occurs to the upper portions of the leaves, leaving the bottom couple of inches of live, green tissues.

New growth in the spring resumes from the center of the foliage clump, sloughing away any damaged leaves over time. In most cases, the foliage of evergreen varieties is slow to green up in the spring, often appearing yellowish-green until it has returned to full active growth.

Evergreen varieties are often difficult to successfully overwinter in northern climates. To survive the cold temperatures, it is often necessary to provide excessive winter protection, such as placing them in overwintering structures and using insulating materials, such as mulches or thermal blankets. There are some exceptions; several evergreen varieties do perform very well in northern climates. Northern growers should research evergreen varieties and their performance before placing them into production.

Semi-evergreen

Roughly speaking, semi-evergreen daylilies behave like dormant varieties in the North but like evergreen varieties in southern frost-free climates. Semi-evergreen foliage dies back slower than dormant varieties, and during mild winters they often remain evergreen. Unlike evergreen varieties, the main body of old foliage dies completely, and the new growth resumes from the center of each fan. They flush quicker than dormant types in the spring. Semi-evergreen daylilies usually perform equally well in northern and southern climates.

Diploids versus tetraploids

Most plants and perennials are diploid; they have two complete sets of chromosomes in every cell. Polyploids are plants that have more than two sets of chromosomes in every cell. For example, triploids have three sets, tetraploids have four sets, and so on. There are more diploid daylilies than tetraploid. Diploids provide growers with more ever-blooming varieties, more pink-blooming cultivars, more spider- and double-flowering varieties, and are easier for breeders to work with.

Tetraploid daylilies often have several desirable characteristics over diploids. Some of these include: large flower sizes, more intense color of the blooms, stronger scapes, higher plant vigor, and more chromosomes mean more breeding opportunities for future improvements. Some tetraploids exhibit thick, corm-like stems, while others have brittle stems that can break in high winds. Some tetraploids exhibit premature bud drop. With a few exceptions, tetraploids are considered to be improved cultivars over the diploid types.

Propagation

Daylilies are most commonly propagated by division. Division entails dividing, or splitting, the crown into smaller sections containing at least one fan, or shoot, and several adjoining roots. Division of daylilies is best done during the spring or fall, but they may be propagated any time of the year. When making divisions, the best results are achieved when the foliage is cut back to 4–6 in. (10–15 cm) prior to lifting and dividing.

Production

Daylilies consisting of three to five fans are mostly used in 1 gal. (3.8 L) or larger pots, although for some markets they can be produced successfully in smaller containers, such as 1 qt. (1 L) or 5 in. (13 cm) pots, using bare-root divisions with one to three fans. Bare-root divisions or large plugs should be planted in a well-drained potting medium. Many growers use mixes containing various components, including bark, peat moss, coir, sand, or perlite. When planting, spread the roots over a central cone of root medium in the center of the pot. Fill the pot, covering the crown with about 1 in. (25 mm) of potting medium. Water them in thoroughly after potting and place them in the production area.

During production, the pH or acidity of the medium should be maintained at 6.2–6.7. Feeding is generally not necessary during the first few weeks of production. When actively growing, daylilies are moderate feeders, requiring a controlled-release fertilizer incorporated at a rate equivalent to 1 lb. nitrogen per cubic yard (454 g/0.76 m^3) of growing medium or 50–100 ppm nitrogen delivered under a constant liquid fertilizer, program. Avoid over-fertilization, as it may lead to soft, spindly growth. Until daylilies are established, keep the rooting medium evenly moist. Once established, I recommend watering thoroughly, when it is required, allowing the substrate to dry slightly between waterings. Routine watering during bud development will improve the quality of the flowers.

Figure 12.30. **Perhaps the most widely grown and recognizable daylily, 'Stella de Oro' is a reliable performer for commercial growers and home gardeners.**

Daylilies planted in the early spring are usually grown at an average temperature of 60° F (16° C). At this temperature, most daylilies will reach a finished size in six to eight weeks and will be flowering in about ten weeks. Optimal growing temperatures for established (overwintered) plants are slightly higher, with a day temperature of 75° F (24° C) and 65° F (18° C) at night. These warmer temperatures are similar to conditions naturally occurring outdoors in late spring and will decrease the time needed to reach flowering. Many operations plant their daylilies in the late summer or early fall for next year's spring sales. These summer/fall plantings produce better plants with more flowers than daylilies that are potted the same year they are to be sold. Be sure to allow six to eight weeks for rooting before the temperatures remain below freezing. I would also recommend withholding nutrients in the early fall to allow the plants to "shut down" as they prepare for their winter dormancy.

Overwintering daylilies is relatively simple. In the late fall, trim the plants back to 2 in. (5 cm) above the top of the container. Trim as late in the fall as reasonably possible for your operation, as daylilies will produce a flush of new growth until several hard frosts occur. Once they are trimmed, group the pots together inside a cold frame, greenhouse, or outdoor production bed. In colder zones, I would also recommend covering them with a protective thermal blanket during the winter months. During this season, if there are periods where the temperatures remain above freezing, open the cold frame doors or provide ventilation during the day to keep the temperatures inside as cool as possible. Provide adequate ventilation any time the outdoor temperatures are above 40° F (4° C).

Insects

Aphid	Mealybug
Caterpillar	Slug
Cutworm	Snail
Gall midge	Southern root nematode
Grasshopper	Spider mite
Japanese beetle	Spittlebug
Leafhopper	Thrips

Daylilies are not susceptible to many insect pests. Aphids, spider mites, and thrips are the insects that are the most prevalent. Aphids can be detected by the presence of cast skins left behind from molting. Aphid injury causes damage to the foliage and distorts the flower buds. Growers usually control aphids using various contact or systemic insecticides labeled for aphid control. Spider mite injury usually goes undetected unless the damage is severe and the mite population is high. Mite injury results in russeting or speckling of the foliage, reduced plant vigor, and causes the leaves to turn tan and even die under severe mite pressure. Usually a preventative program for spider mites that utilizes ovicides such as hexythiazox (Hexygon) or clofentezine (Ovation), and adulticides such as abamectin (Avid) or bifenazate (Floramite) can nearly eliminate their presence. Thrips will most likely be observed at or near flowering, causing misshapen and discolored blooms. Use chemicals such as spinosad (Conserve) or methiocarb (Mesurol) when the presence of thrips becomes apparent. Be sure to read the labels of each of these products for their recommended uses and rates.

Figure 12.31. **Daylily rust, *Puccinia hemerocallidis,* has become problematic for many growers.**

Diseases

Alternaria	*Fusarium*
Armillaria (root rot)	*Kabatiella* (leaf blight)
Aureobasidium (leaf streak)	*Puccinia hemerocallidis* (rust)
Botrytis blight	*Rhizoctonia*
Collecephalus (leaf streak)	*Sclerotium*
Colletotrichum (leaf spot)	
Erwinia	

There are only a couple of diseases associated with the production of daylilies. Crown rot sometimes occurs shortly after overwintering; usually this tends to be observed on evergreen cultivars. Although seldom a concern, inadequate aeration of the growing medium during cool wet weather may lead to soilborne fungal infections from *Fusarium* or *Rhizoctonia.*

Leaf streak is a fungal disease caused by the pathogen *Aureobasidium microstictum* affecting the leaves of daylilies. Injury to the plant from climatic factors such as frost damage or pest injury usually precedes a leaf streak infection. Leaf streak is described as yellow streaks along the central leaf vein, followed by browning and reddish spots that are often mistaken for rust. Damaged areas in the leaf often join, causing the leaf to appear streaked. Secondary pathogens such as *Alternaria alternata* and *Fusarium* may also attack infected leaf tissues.

The last couple years have brought a lot of concern and attention to a new pathogen, daylily rust, caused by the fungal pathogen *Puccinia hemerocallidis,* which is spreading throughout the country. Growers are at a greater risk of observing rust when purchasing plant material from outside of the United States or when daylilies are shipped out of the southern states. The rust does not appear to overwinter in the northern U.S., where temperatures during the winter are below 50° F (10° C), and growers carrying over daylilies in these regions may be less susceptible to this fungal disease. Daylily rust causes lesions on the leaves that kill the surrounding foliage. The most distinguishing characteristic is the raised yellowish orange powdery pustules on the infected leaf surfaces. The pustules easily rub off onto your fingers if the leaf surface is rubbed.

Rust is best controlled on a preventative basis by making weekly spray applications. Chlorothalonil (Daconil Ultrex), azoxystrobin (Heritage), and mancozeb (Protect T/O) have all provided effective control of rust when they are applied prior to inoculation. Some of the suggested chemicals for controlling this rust after infections are detected include propiconazole (Banner Maxx), triadimefon (Bayleton), flutolanil (Contrast), triadimefon (Strike), and myclobutanil (Systhane). Rotate between chemical classes to reduce the development of fungi-

PGRs for *Hemerocallis* Cultivars

Trade Name	Active Ingredient	Northern Rate
A-Rest	Ancymidol	50 ppm
B-Nine or Dazide	Daminozide	3,750 ppm
Bonzi, Piccolo, or Paczol	Paclobutrazol	45 ppm
Sumagic	Uniconazole-P	10 ppm
B-Nine or Dazide + Cycocel	Daminozide + chlormequat chloride	3,750 ppm + 1,250 ppm
B-Nine or Dazide + Sumagic	Daminozide + uniconazole-P	2,500 ppm + 5 ppm
A-Rest Drench	Ancymidol	5 ppm
Bonzi, Piccolo, or Paczol Drench	Paclobutrazol	6 ppm
Sumagic drench	Uniconazole-P	1 ppm

These are northern rates and need to be adjusted to fit the location where the crop is being produced. Unless noted otherwise, these rates are for spray applications and geared to growers using multiple applications. Drench applications are usually applied once per crop.

cide resistance. Although daylily rust has not been observed on all varieties, growers should still be cautious and implement preventative programs where applicable. Further information regarding daylily rust can be obtained at: www.aphis.usda.gov/npb/daylily.html.

Plant Growth Regulators

Many *Hemerocallis* cultivars require height control when they are grown under greenhouse and nursery conditions. Providing adequate spacing between the plants will reduce plant stretch caused by competition. Under certain growing conditions, such as low light levels or production at high plant densities, it may be necessary to use chemical plant growth regulators. Compared to many perennials, the PGR rates needed to achieve sufficient control are relatively high. For each PGR applied as a foliar spray, it usually requires two or three applications at seven-day intervals to provide adequate height control. Begin making foliar applications when the plants reach 3 in. (8 cm) tall when producing small containers and 6 in. (15 cm) for larger pots. One-time drench applications of ancymidol, paclobutrazol, or uniconazole-P are highly effective at reducing plant height. Drench applications should be applied at approximately the same time foliar applications are typically made.

References

Kroll, Arthur M. "Are You Growing the Right Daylilies?" *NMPro.* July 2002.

Pilon, Paul. Perennial Solutions: *"Hemerocallis* 'Little Missy'." *Greenhouse Product News.* April 2003.

Viette, André. "From Start to Finish: Daylily." *Greenhouse Grower.* July 1996.

Heuchera

Coral Bells

The native American *Heuchera* has become one of the most popular perennial plants in today's gardens. In the past, there were only a handful of cultivars being commercially produced. The Perennial Plant Association named *Heuchera* 'Palace Purple' as the Plant of the Year in 1991

Figure 12.32. *Heuchera sanguinea* 'Chocolate Ruffle'

Commonly Grown *Heuchera* Cultivars

'Amber Waves'	'Gypsy Dancer'	'Regina'
americana 'Dales Strain'	'Harmonic Convergence'	'Rose Majesty'
americana 'Green Spice'	'Jade Gloss'	*sanguinea* 'Bressingham Hybrids'
'Amethyst Myst'	'Lime Ricky'	*sanguinea* 'Chatterbox'
'Black Beauty'	'Mardi Gras'	*sanguinea* 'Firefly'
'Can Can'	'Marmalade'	*sanguinea* 'Ruby Bells'
'Cappuccino'	*micrantha* 'Palace Purple Select'	'Silver Lode'
'Cathedral Windows'	'Midnight Burgundy'	'Silver Scrolls'
'Champagne Bubbles'	'Monet'	'Starry Night'
'Cherries Jubilee'	'Obsidian'	'Stormy Seas'
'Chocolate Ruffles'	'Petite Pearl Fairy'	'Strawberry Candy'
'Chocolate Veil'	'Pewter Moon'	'Velvet Night'
'Crimson Curls'	'Pewter Veil'	'Venus'
'Ebony & Ivory'	'Plum Pudding'	'Vesuvius'
'Fireworks'	'Purple Mountain Majesty'	*villosa* 'Caramel'
'Frosted Violet'	'Purple Petticoats'	
'Geisha's Fan'	'Raspberry Ice'	

for its landscape performance and desirable characteristics. Since then, there seems to have been an explosion of breeding conducted, improving both flower color and providing better foliage characteristics.

Today there is a vast array of ornamental foliage characteristics making coral bells a mainstay in the landscape. The lobed leaves of the various cultivars are often rounded, heart-shaped, or triangular and take on numerous hues including amber gold, green, bronze-green, purple, and deep purple. The foliage is smooth, wavy, or ruffled and may contain differing degrees of silver splotching. In most climates, coral bells remain evergreen throughout the year.

Many cultivars are very prolific bloomers, while others are grown strictly for their foliage. The bell-shaped flowers are borne on wiry stems held above the foliage and usually consist of various shades of pink, red, or white. Depending on the cultivar, blooming may be short term, continue from May till August, or rebloom throughout the growing season. The blooms may be used as cut flowers or to attract butterflies and hummingbirds to the garden.

The genus *Heuchera* was named after an eighteenth-century German professor of medicine and botanist, Johann Heinrich von Heucher (1677–1747), who specialized in medicinal plants. *Heuchera* belongs to the family Saxifragaceae, which includes many other popular perennials such as *Astilbe, Bergenia,* X *Heucherella, Rodgersia, Saxifraga,* and *Tiarella,* to name a few.

Coral bells are commonly used as accent plants, border plants, or components in mixed containers. They are heat and cold hardy in USDA Hardiness Zones 3–9 and AHS Heat Zones 8–1, but there are hardiness and heat tolerance differences between cultivars. *Heuchera* cultivars will grow more vigorously and have the best leaf coloration when grown in partial shade. In locations with full sun or full shade, coral bells will generally survive but not look as lively.

Propagation

Depending on the cultivar, growers use seed, tip cuttings, or tissue culture to propagate coral bells. Many cultivars are patented, and unlicensed propagation is prohibited. To avoid illegal activities and the consequences, always check for plant patents before attempting to propagate.

Many of the new cultivars are propagated by tissue culture and produced as plug liners. Tissue culture has allowed these cultivars to reach the market in a relatively short period of time. Most growers will purchase cultivars from tissue culture as 72-cell or larger plugs from licensed propagators. When transplanting materials from tissue culture (stage 3), provide similar conditions to those outlined below for tip cuttings.

Due to variability with the seedlings coming true to type, there are only a handful of cultivars that are propagated from seed. The seeds are small and difficult to handle. Sow seeds in the intended plug flat and do not cover the seed with germination mix or vermiculite. Light is required for germination; covering reduces the amount of light reaching the seed and may adversely affect the germination process. The seed flats should be moistened and moved to a warm environment, where the temperatures can be maintained at 65–70° F (18–21° C) for germination.

Many growers utilize germination chambers during this stage to provide uniform moisture levels and temperatures. It is important to keep the medium very moist (not saturated) during this stage, or the germination rates are likely to decrease. For germination, it is best to maintain high humidity (90% RH) of the air rather than focusing on the moisture level of the growing mix.

Seedlings will emerge fourteen to twenty-one days after sowing. Following germination, reduce the moisture levels somewhat, allowing the growing medium to dry out slightly before watering to help promote rooting. The light levels can gradually be increased as the plug develops, from 500 f.c. (5.4 klux) following germination to 2,500 f.c. (27 klux) at the final stage. Fertilizers are usually applied once the true leaves are present, applying 100 ppm nitrogen every third irrigation using a balanced water-soluble source. When plugs are grown at 65° F (18° C), they are usually ready for transplanting in ten to twelve weeks. Growers often have difficulty germinating coral bells during the summer months. It is recommended to avoid starting them during this time of year unless cool conditions can be provided until the first true leaves are present.

Tip cuttings should be taken while they are vegetative. They should measure approximately 2–3 in. (5–8 cm) and contain several nodes. Cuttings are generally stuck into large plug liners, such as 72-cell trays, for rooting. The well-drained rooting medium should be moistened prior to sticking. The base of the cuttings can be dipped into a rooting hormone, such as a solution of indolebutyric acid (IBA) at rates of 750–1,250 ppm. *Heuchera* can successfully root without rooting compounds but tend to root slightly faster and more uniformly when these treatments are provided.

The cuttings should be placed under low misting regimes for the first seven to ten days of propagation. It is usually best to propagate under high humidity levels (90% relative humidity) with minimal misting. The misting can gradually be reduced as the cuttings form calluses and root primordia. The cuttings are usually rooted in less than three weeks with soil temperatures of 68–75°F (20–24° C). For best results, the air temperature during rooting should be maintained above 60° F (16° C) and below 80° F (27° C). It is beneficial to begin constant liquid feeding with 200 ppm nitrogen at each irrigation beginning ten days from sticking. Depending on the size of the plug cell, coral bells will be fully rooted and ready to transplant in approximately six weeks from sticking.

Production

Most growers receive starting materials as finished plug liners. They are commonly planted in 1 qt. (1 L) or 1 gal. (3.8 L) containers after they have been received. Coral bells perform best when they are grown in a moist, well-drained medium with a slightly acidic pH, 5.8–6.2. Most commercially available peat- or bark-based growing mixes work well, provided there is adequate drainage.

When planting, take care to not bury the crown of the plant too deeply. Place the plug in the center of the pot and plant so that the original soil line of the plug is even with the surface of the growing medium of the new container. Planting the crown too deep will lead to crop variability and losses. After planting, I recommend drenching with a broad-spectrum fungicide such as etridiozole + thiophanate-methyl

(Banrot) or the combination of mefenoxam (Subdue Maxx) and thiophanate-methyl (Cleary's 3336).

Heuchera can be grown using moderate fertility levels. Growers using water-soluble fertilizers either apply 150–200 ppm nitrogen as needed or feed with a constant liquid fertilization program using rates of 50–100 ppm nitrogen with every irrigation. Controlled-release fertilizers are commonly used to produce containerized coral bells. Growers commonly apply time-release fertilizers as a topdressing onto the medium's surface using the medium rate or incorporated into the growing medium prior to planting at a rate equivalent to 1 lb. nitrogen per cubic yard (454 g/0.76 m^3) of growing medium.

Coral bells require an average amount of irrigation, as they do not tolerate very wet or overly dry conditions. Under high light intensities, marginal leaf burn may occur if the plants become water stressed. When irrigation is necessary, water them thoroughly, allowing the soil to dry slightly between waterings.

Insects

Aphid	Mealybug
Black vine weevil	Root weevil
Foliar nematode	Slug
Four-lined plant bug	Spider mite
Fungus gnat	Strawberry root weevil
Grasshopper	Thrips
Japanese beetle	Whitefly
Leafhopper	

There are a number of insects that feed on *Heuchera.* Occasionally, aphids, Japanese beetles, mealybugs, slugs, or whiteflies may appear, causing only a minimal amount of crop injury. In outdoor production sites or landscape situations, black vine weevils or strawberry root weevils could infest and cause significant damage to the stems and crowns. None of these insect pests require preventative control strategies. Growers should have routine scouting programs to detect the presence of insects and to determine if and when control strategies are necessary.

Diseases

Botrytis	*Rhizoctonia*
Botrytis blight	Rust
Colletotrichum (anthracnose)	*Septoria*
Corynebacterium (leaf gall)	*Sclerotium*
Phytophthora	*Sphaceloma* (anthracnose)
Powdery mildew	Stem Smut
Pseudomonas	*Xanthomonas*
Pythium	

When the proper environmental and cultural factors are provided, coral bells can be produced without the occurrence of plant pathogens. The primary disease growers should watch for is *Rhizoctonia* crown rot. Plants are most susceptible to *Rhizoctonia* when grown under cool conditions, such as going into or coming out of winter dormancy. *Botrytis* is another disease that could become problematic. *Botrytis,* like *Rhizoctonia,* often occurs around the overwintering process but is also likely to occur under dense plant canopies.

There are two bacteria, *Pseudomonas* and *Xanthomonas,* that have been observed on some *Heuchera* varieties but are not a major concern at this time. *Pseudomonas* will appear as reddish brown spots that may cause the leaf to distort, and *Xanthomonas* takes the appearance of small brown angular to circular spots with yellow halos.

To control these diseases, manage the environment by providing proper plant spacing and adequate air movement, controlling the humidity, maintaining good watering practices, and, if desired, following a preventative spray program using the appropriate chemicals.

Plant Growth Regulators

Height control is seldom necessary to control the growth of the leaves unless they are grown at high plant densities. However, reducing the height of the flowering stems or panicles is often beneficial for growers who ship plants in bloom. Apply PGRs as the panicles are just

PGRs for *Hemerocallis* Cultivars

Trade Name	Active Ingredient	Northern Rate
Bonzi, Piccolo, or Paczol	Paclobutrazol	30 ppm
Sumagic	Uniconazole-P	5 ppm

These are northern rates and need to be adjusted to fit the location where the crop is being produced. These rates are for spray applications and geared to growers using multiple applications.

beginning to elongate above the foliage. Two applications applied seven days apart will provide an adequate reduction of the flowering stalk without altering the overall appearance of the plant. When growing under high plant densities, plant height can be controlled by applying PGRs to the foliage as the plant canopy begins to enclose.

Forcing

Heuchera is easy to produce as a foliage plant. When transplanting from 72-cell plugs, coral bells grown at 65° F (18° C) typically take eight weeks to finish 1 gal. (3.8 L) pots in the summer or eleven weeks during the winter months.

When producing blooming plants is the goal, a few requirements should be met in order to produce flowering plants consistently. Research has shown that most *Heuchera* cultivars have a juvenility phase and must reach a particular size or maturity before they are capable of producing flowers. The age of the plant is often determined by counting the number of leaves that have formed. The exact number of leaves to satisfy the juvenility requirement varies with species and cultivar (usually from six to eighteen) and has not been determined for every cultivar currently being produced. Therefore, I recommend growing coral bells at natural day lengths for a minimum of six weeks to allow them to reach a mature size before providing the additional requirements necessary for flowering. During times of the year where the light levels are naturally low, providing 400–500 f.c. (4.3–5.4 klux) of supplemental lighting will dramatically increase the number of leaves formed compared to those produced under lower light levels.

Coral bells have an obligate cold requirement in order for them to flower. It is recommended to vernalize small containers of mature, bulked up coral bells for a minimum of twelve weeks at 35–44° F (2–7° C). Since they remain evergreen throughout the cold treatment, provide 100–200 f.c. (1.1–2.2 klux) of light during this period. If they are overwintered in a greenhouse or cold frame, the natural light levels should be sufficient. When coolers are used, supplemental lighting is best provided using cool-white fluorescent lamps.

After the cold requirement is achieved, coral bells can be grown at any day length, as they are day-neutral plants. The length of the photoperiod does not have any effect on the time to flower or the number of blooms produced. The time to bloom after vernalization is a function of cultivar and temperature. Depending on the cultivars being produced, coral bells will reach full bloom in approximately five to eight weeks. Producing them at cooler temperatures increases the time to flower but will improve the overall quality characteristics of the plant, such as the color of the foliage and flowers. To achieve the highest plant quality, I recommend growers produce coral bells using cool temperatures, 60–65° F (16–18° C), and allow a couple of additional weeks for crop production.

References

Greenhouse Grower Magazine and Michigan State University. *Firing Up Perennials: The 2000 Edition.* Willoughby, Ohio: Meister Publishing. 2000.

Nau, Jim. *Ball Perennial Manual: Propagation and Production.* Batavia, Ill.: Ball Publishing. 1996.

Pilon, Paul. Perennial Solutions: "*Heuchera* 'Fireworks'." *Greenhouse Product News.* September 2002.

Hibiscus moscheutos, H. x *hybrida*

Hardy Hibiscus, Rose Mallow, Southern Bells

Hibiscus is commonly referred to as hardy hibiscus, rose mallow, or southern bells. Hardy hibiscus is an old-time favorite perennial that has seen a resurgence of popularity. The cultivars available today are much improved over the varieties of the past, providing greater flower coloration, flower size, and bloom time. They also have more appealing plant habits and tolerate a wider range of environmental conditions.

Figure 12.33. *Hibiscus moscheutos* 'Lady Baltimore'

Today's varieties are cultivars of *Hibiscus moscheutos* or *Hibiscus* x *hybrida. Hibiscus moscheutos* is a marshland native of the eastern United States. Rose mallows are tough perennials and can withstand adversities such as poor soils and drought. Hibiscus are heat and cold tolerant, withstanding temperatures of -25–100° F (-32–38° C), allowing them to be grown throughout USDA Hardiness Zones 4–9 and AHS Heat Zones 8–1.

They produce numerous blooms beginning in midsummer and lasting until frost. The color of the flowers varies by cultivar and mostly consists of shades of red, pink, white, or some combination of these. The flowers range in size from 6–12 in. (15–30 cm) across, creating an awesome display, sure to be an eye-catcher and a conversation starter for those unfamiliar with this plant. In the landscape, rose mallows commonly reach 3–6 ft. (0.9–1.8 m) in height and are used as specimen plants, mass plantings, or in hedges. They prefer to be produced in moist, sunny locations with an adequate supply of irrigation. As mentioned above, they can tolerate extreme conditions, but will perform best when their preferences are provided for.

Commonly Grown *Hibiscus moscheutos* and *H.* x *hybrida* Cultivars

'Anne Arundel'
'Baboo'
'Barbara's Blush'
'Blue River II'
'Bordeaux'
'Chablis'
'Disco Belle Mix'*
'Disco Belle Pink'*
'Disco Belle Rosy'* 'Disco Belle Red'*
'Disco Belle White'*
'Everest White'
'Etna Pink'
'Fantasia'
'Fireball'
'Galaxy'*
'Grenache'
'Kopper King'
'Lady Baltimore'
'Lord Baltimore'
'Luna Blush'*
'Luna Pink Swirl'*
'Luna Red'*
'Luna White'*
'Pinot Grigio'
'Pinot Noir'
'Plum Crazy'
'Pyrenees Pink'
'Ranier Red'
'Southern Belle Mix'*
'Sweet Caroline'
'Turn of the Century'
'100 Degrees'

*Propagated from seed

Hibiscus belongs to the Malvaceae family. There are several perennials in this family being commercially produced, including *Alcea, Althea, Callirhoe, Lavatera, Malva,* and *Sidalcea.*

Propagation

Depending on the cultivar, growers use seeds or tip cuttings to commercially propagate *Hibiscus.* Many cultivars are patented, and unlicensed propagation is prohibited. Most growers associate patented cultivars as those exclusively propagated vegetatively. This is not always the case; some seed-propagated cultivars, such as the 'Luna' series, are patented and have royalties that must be paid. To avoid legal consequences, always check for plant patents before attempting to propagate rose mallow.

Starting *Hibiscus* from seed is relatively easy, as they have a high germination percentage, grow quickly, and are relatively trouble free. In the past, the most difficult aspect of starting rose mallow from seed was physically obtaining the seed, as availability was low. Today, there are more seed producers and more *Hibiscus* cultivars propagated by seed, which should alleviate some of the frustrations growers have had in the past.

Due to the size of the seed and their growing habit, it is best to propagate rose mallows in plug trays with large cell sizes. Most growers sow seeds in 128-cell or larger plug flats. Growers observe better germination and performance when using fresh seed. Sow a single *Hibiscus* seed per cell in the intended plug flat and cover heavily with germination mix or medium-grade vermiculite. The covering helps to maintain a suitable environment around the seed during this phase.

The seed flats should be moistened and moved to a warm environment, where the temperatures can be maintained at 68–78° F (20–26° C) for germination. It may take slightly longer for seeds to germinate at low temperatures (68–72° F [20–22° C]), but the uniformity of the crop will be greater. Many growers utilize germination chambers during this stage to provide uniform moisture levels and temperatures. Using germination chambers is optional, as *Hibiscus* will successfully germinate in the greenhouse. The relative humidity is not as important as keeping the germination mix uniformly moist. The seeds begin to sprout within a few days. If they are being germinated in a chamber, remove them as soon as the radicle is present to prevent the seedlings from stretching.

Following germination, continue to grow them at 68–75° F (20–24° C) for five to seven weeks until they are ready for transplanting. Fertilizers are usually applied once the true leaves are present, applying 100 ppm nitrogen every third irrigation or 50 ppm with every irrigation, using a balanced water-soluble source. As they develop, gradually increase the rate of fertilizer and/or increase the frequency of application to match the plant's needs. Keep the growing medium evenly moist, and do not allow the seedlings to wilt.

Vegetative tip cuttings can be rooted anytime they are actively growing. The best results are obtained when the cuttings are stuck in the spring. Growers often maintain their own stock plants or buy in unrooted cuttings. Tip cuttings containing several nodes and measuring approximately 3 in. (8 cm) in length are generally stuck into plug liners for rooting. The well-drained rooting medium should be moistened prior to sticking. The base of the cuttings can be dipped into a rooting hormone, such as a solution of indolebutyric acid (IBA), at rates of 1,000–1,500 ppm, prior to sticking. Hardy hibiscus can successfully root without rooting compounds but root slightly faster and more uniformly when these treatments are provided.

Cuttings should be placed under low misting regimes for about the first ten to sixteen days of propagation. It is usually best to propagate under high humidity levels (90% relative humidity) with minimal misting. The misting can gradually be reduced as the cuttings form calluses and root primordia. The cuttings are usually rooted in less than three to four weeks with soil temperatures of 68–74° F (20–23° C). For best results, the air temperature during rooting should be maintained above 65° F (18° C) and below 85° F (29° C). It is beneficial to begin constant liquid feeding with 150 ppm nitrogen at each irrigation beginning ten days from sticking. In most circumstances, they will be fully rooted and ready to be transplanted into the final container size in five to seven weeks.

Production

For container production, hardy hibiscus cultivars are most suitable for 1 gal. (3.8 L) or larger containers. Most growers receive starting materials of hibiscus as finished plug liners. The plugs are usually fresh (non-vernalized) for seed varieties and are often vernalized for cutting cultivars. Propagators producing cutting cultivars often have to propagate their crop for this spring's sales during the previous year and overwinter the plugs.

In some cases, growers may receive their plugs early in the season, before they have broken dormancy. Dormant rose mallows appear as small sticks protruding from the soil and show no signs of life or potential growth. They naturally emerge later in the spring than do most perennials. To prevent crop variability, I recommend growers wait until the *Hibiscus* liners break dormancy and begin actively growing before transplanting them into their final containers. Upon receipt of dormant materials, unpack them and place them in a warm greenhouse environment to break their dormancy. They should be kept moist but not saturated until planting.

When growing 1 or 2 gal. (3.8 or 7.6 L) pots, generally one plug per container is planted. For larger container sizes, planting multiple plug cells per container may be appropriate. Rose mallows perform best when they are grown in a moist, well-drained medium with good water-holding capacity. There are a number of peat- and bark-based growing mixes with these characteristics growers can use. They perform best when an adequate amount of moisture is present at all times. When planting,, be careful not to plant the plugs too deeply, as this could lead to poor plant establishment, crop variability, crown rot, and losses. The plugs should be planted so the original soil line of the plug is even with or just below the surface of the growing medium of the new container.

Some growers have found it beneficial to pinch *Hibiscus* prior to or shortly after they are planted in the final container. This is generally a soft pinch, only removing the growing point of the plant, and should leave four to six leaves on each stem. Pinching increases lateral branching and the total number of flowers produced on each plant. Many cultivars are said to branch freely and do not require a mechanical pinch; however, I have found this early pinch dramatically improves the quality of the finished product. As the crop develops, an additional pinch can occur to create a fuller appearing plant. Generally, it will take six weeks from the time of pinching for the plant to reach flowering. Always look ahead at the desired sales date to determine if there is adequate time to pinch them back. It may not be necessary to pinch *Hibiscus* that are being grown in small containers, as they will fill out the pot fine without a pinch.

During production the pH of the medium should be maintained at 6.0–6.5. *Hibiscus* are moderate to heavy feeders. Fertility can be delivered using water-soluble or controlled-release fertilizers. Growers using water-soluble

Figure 12.34. **Pinching *Hibiscus* prior to (shown) or shortly after planting in the final container will promote lateral branching and form more flowers per plant.**

fertilizers either apply 200–250 ppm nitrogen as needed, or feed with a constant liquid fertilization program using rates of 100–150 ppm nitrogen with every irrigation. Grower's commonly incorporate time-release fertilizers into the growing medium prior to planting at a rate equivalent to 1.25–1.50 lb. nitrogen per cubic yard (567–680 g/0.76 m^3) of growing medium or apply them as a topdressing on the medium's surface using the medium or high rates recommended on the product's label.

Due to the frequency of irrigation and the volume of water applied, I find it is better to rely on a controlled-release fertilizer program. Combining time-release with water-soluble fertilizer programs is also an effective nutrient management strategy. In these cases, growers intentionally provide most of the fertility requirements using controlled-release fertilizers and supplement as needed using water-soluble products. This allows them to grow a healthy crop without providing too many nutrients up front and gives them the ability to tone the crop prior to sales or shipping.

Good irrigation management is important throughout crop production, from the time they are planted to the day they are sold. *Hibiscus* should be kept evenly moist and never allowed to wilt the entire time they are being grown. Growing them too dry and allowing them to wilt could result in lower leaf yellowing and flower bud abortion. When irrigation is necessary, water them thoroughly and allow the growing medium to dry slightly between waterings. During container production, they are extremely susceptible to dry conditions; once they are established in the landscape, they are considered to be drought tolerant.

The best quality is achieved when *Hibiscus* are grown in full sun or in greenhouses with no shading materials. High light intensities promote better branching, more flowers per plant, and short plants. Where possible, growers should place them in an environment with no less than 4,000 f.c. (43 klux) of light. Rose mallow grown at low light levels tend to be of lower quality as they become elongated, form less branches, and do not flower as profusely as those produced at ambient light levels. To promote better branching, I recommend spacing the plants in such a manner that there is good light penetration to the basal branches. Plant quality can be improved during short days or periods where the light levels are naturally low by providing 400–500 f.c. (4.3–5.4 klux) of supplemental lighting using high-pressure sodium lamps.

Insects

Aphid	Sawfly
Caterpillar	Scale
Flea beetle	Slug
Grasshopper	Snail
Japanese beetle	Spider mite
Leafhopper	Thrips
Mealybug	Whitefly

There are a number of insects that feed on hibiscus. Aphids, Japanese beetles, spider mites, thrips, and whiteflies are often prevalent and only cause minimal injury to the crop. Growers should have routine scouting programs to determine the presence of these pests and to determine if and when control strategies are necessary. For aphid and whitefly prevention, some growers implement preventative spray programs using systemic chemicals, such as pymetrozine (Endeavor), thiamethoxam (Flagship), imidacloprid (Marathon II), dinotefuran (Safari), or acetamiprid (TriStar) to provide up to one month of control for these pests following the application.

Diseases

Alternaria leaf spot	*Nectria* canker
Cercospora leaf spot	Powdery mildew
Crown gall	*Pseudomonas*
Diaporthe canker	*Rhizoctonia*
Erwinia	Rust
Fusarium canker	*Sclerotium*
Leaf spot	*Xanthomonas*

PGRs for *Hibiscus moscheutos* Cultivars

Trade Name	Active Ingredient	Northern Rate
B-Nine or Dazide	Daminozide	3,750 ppm
B-Nine or Dazide + Cycocel	Daminozide + chlormequat chloride	2,500 ppm + 1,250 ppm
Bonzi, Piccolo, or Paczol	Paclobutrazol	45 ppm
Sumagic	Uniconazole-P	7.5 ppm
Bonzi, Piccolo, or Paczol drench	Paclobutrazol	5 ppm
Sumagic drench	Uniconazole-P	1 ppm

These are northern rates and need to be adjusted to fit the location where the crop is being produced. Unless noted otherwise, these rates are for spray applications and geared to growers using multiple applications. Drench applications are usually applied once per crop.

Under the right circumstances, certain diseases may become problematic when growing *Hibiscus* cultivars. Of the diseases they are susceptible to, growers most frequently observe leaf spots caused by *Alternaria*, *Cercospora*, and other pathogens. To control these diseases, it is best to manage the environment by providing the proper plant spacing and adequate air movement, maintaining a relative humidity below 70%, and, if desired, following a preventative spray program using the appropriate chemicals.

Plant Growth Regulators

Depending on the environmental conditions and crop spacing, container growers may need to use chemical growth regulators. Providing adequate space between the plants is the best and most effective method to control plant height during production. Depending on your geographic location, apply foliar applications of PGRs beginning with the rates listed in the above table. It is best to begin PGR applications about one week following a pinch, applying them at seven-day intervals if additional control is necessary. One-time drench applications can be applied just following a pinch to control the height of rose mallow crops.

Forcing

Until they are blooming, hibiscus remain as green leafy plants with uninteresting foliage and plant appeal. Once they begin to flower, they are very dynamic and attract great interest. To improve sales and marketability, *Hibiscus* cultivars must be forced to bloom. The two most important factors to understand are photoperiod and temperature.

Hibiscus are obligate long-day plants and will not flower unless they are grown under long-day conditions. Growers should produce them during times of the year when the photoperiods are naturally long or provide long-day photoperiods during production. When the natural photoperiods are short (less than fourteen hours), growers should provide supplemental lighting to extend the day length, ensuring a minimum of sixteen hours of light or supplying a four-hour night interruption.

After long days are provided, the time to flower largely depends on the temperatures *Hibiscus* crops are grown at. Warm production temperatures are highly recommended and will dramatically decrease the production time needed to produce this crop. For optimal development, growers should provide 70–80° F (21–27°C) day temperatures and 68–72° F (20–22° C) night temperatures. Growers should ensure that the twenty-four-hour average temperature does not fall below 68° F (20° C); low average temperatures dramatically decrease plant development and increase the overall production time. Low temperatures also cause the foliage to appear chlorotic.

The amount of time to properly schedule the crop depends on the factors mentioned above (photoperiod and temperature) and is influenced by pinching. When a 1 gal. (3.8 L) crop is not pinched, plants can bloom in as little as eight weeks. If a pinch is conducted at the time of transplant as described above, the time to finish the crop increases to ten weeks. For larger container sizes, such as 2 gal. (7.6 L) containers, which may need two pinches to produce an aesthetic product, a minimum of fourteen weeks should be anticipated.

References

Pilon, Paul. Perennial Solutions: *"Hibiscus* 'Pyrenees Pink'." *Greenhouse Product News.* July 2003.
Rusch, S. "Crop Culture Report: *Hibiscus* 'Luna'." *Greenhouse Product News.* January 2004.
Wang, S., R. Heins, W. Carlson, and A. Cameron. "Forcing Perennials: *Hibiscus moscheutos* 'Disco Bell Mixed'." *Greenhouse Grower.* February 1998.

Hosta

Plantain Lily

For the last several years, hostas have been the top selling perennial throughout North America and are a significant part of today's landscapes. Hostas are hardy herbaceous perennials grown primarily for their beautiful foliage. The leaves come in a wide range of shapes, sizes, colors, and textures. The foliage of most hostas is variegated in a variety of combinations of blue, gold, green, or white. Other varieties have leaves of solid colorations, usually various hues of blue, gold, or green. Currently, there are over three thousand different cultivars on the market, though many of these are not readily available or mass produced.

Figure 12.35. *Hosta* 'Patriot'

American Hosta Growers Association Hostas of the Year

1996: 'So Sweet'	2002: 'Guacamole'
1997: 'Patriot'	2003: 'Regal Splendor'
1998: 'Fragrant Bouquet'	2004: 'Sum and Substance'
1999: 'Paul's Glory'	2005: 'Striptease'
2000: 'Sagae'	2006: 'Stained Glass'
2001: 'June'	

Additional Commonly Produced *Hosta* Cultivars

'Albo-Marginata'	'Hyacinthina'
'August Moon'	'Krossa Regal'
'Blue Angel'	'Love Pat'
'Bressingham Blue'	'Liberty'
'Bright Lights'	'Loyalist'
'Crowned Imperial'	'Minuteman'
fortunei 'Aureomarginata'	'Night before Christmas'
'Francee'	'On Stage'
'Frances Williams'	'Pandora's Box'
'Guardian Angel'	'Paradigm'
'Great Expectations'	*sieboldiana* 'Elegans'
'Golden Tiara'	'Spilt Milk'
'Gold Standard'	'Sun Power'
'Green Gold'	'Twilight'
'Hadspen Blue'	*undulata* 'Variegata'
'Halcyon'	'Whirlwind'
'Inniswood'	'Wide Brim'

Hostas are considered shade-loving plants, although most hostas do not perform well in deep shade. Numerous varieties can tolerate some degree of direct sunlight. It is usually best to provide exposure to sun in the morning hours and shade in the afternoon. Many cultivars, especially variegated ones, will show signs of marginal leaf burn when grown in full afternoon sun. In general, gold- or yellow-leafed cultivars can tolerate more sun, while blue-leafed hostas require shade.

Hostas originated from China, Japan, and Korea and reached the United States in the middle 1800s. Today they are commonly found in landscapes across the country throughout USDA Hardiness Zones 3–9. Cultivars range in size at maturity in the landscape from only a few inches in diameter up to 8 ft. (2.4 m) across. Commercially, hostas are well suited for production in 1 qt. (1 L) up to 2 gal. (7.6 L) containers.

Sun-tolerant *Hosta*

'Allen P. McConnell'	'Miss Saigon'
'Antioch'	'Old Glory'
'August Moon'	'Patriot'
'Aureomarginata'	'Regal Splendor'
'Blue Umbrellas'	'Rising Sun'
'Fragrant Bouquet'	'Royal Standard'
'Francee'	'So Sweet'
'Frances Williams'	'Stained Glass'
'Fried Bananas'	'Striptease'
'Fried Green Tomatoes'	'Sum and Substance'
'Geisha'	'Sum of All'
'Gold Regal'	'Summer Fragrance'
'Green Gold'	'Sun Power'
'Guacamole'	'Sundance'
'Invincible'	'Super Nova'
'Krossa Regal'	'Wide Brim'
'Minuteman'	'Zippity Do Dah'

Propagation

Hostas are propagated most commonly by division. Many of the new cultivars are propagated by means of tissue culture and produced as plug liners. Growers attempting to propagate any hosta (except *Hosta ventricosa*) from seed will not produce plants that are true to type. Vegetative propagation is used to retain the desirable characteristics of the mother plant, such as leaf color, shape, and size.

Division entails dividing, or splitting, the crown into smaller sections containing at least one bud and several adjoining roots. Growers can divide hostas throughout the growing season from spring to fall. If the foliage is flushed, it is recommended to trim or mow the plants prior to lifting and dividing.

At certain times of the year, propagation of hostas or planting a bare-root propagule may lead to disappointing results. Several weeks after planting, it is not uncommon for growers to observe fully flushed plants, ready to ship or sell, only to discover that a root system does not accompany the nice foliage. This becomes problematic when consumers attempt to plant their new hosta into the landscape only to discover they have purchased a pot of soil with an unrooted plant sitting on top.

In the Midwest, I have observed that the later in the year a division is planted, fewer roots (if any) are made and the survival rate during the winter is greatly reduced. In many cases, they will overwinter without a root system and will flush top growth in the early spring but still will not generate any new roots. Rooting usually begins during mid-spring, once the days get longer and the temperatures increase. Each cultivar varies slightly, but generally it is best to schedule your hosta plantings to be done by mid-August to allow adequate time for rooting before the winter months.

The rooting phenomenon (or lack of) is also observed when growers plant hostas during the winter months from bare-root divisions. Hostas tend to root poorly up until April or May; again, they will flush top growth fine, but will not develop a root system. I would recommend growers plant their hostas from June till August of the current year for next spring's sales to avoid any possibility of providing unrooted hostas into the marketplace.

If plantain lilies must be planted in the late winter, it is usually best to transplant large plugs or small containers with established root systems into the larger finished containers. A hosta with a root system will continue to flush new roots and will provide a better finished product when planted during this time of year. If bare-root divisions must be planted before April, I recommend growers provide long days through photoperiodic lighting and grow at temperatures averaging 70° F (21° C) to prevent the tops from going dormant. It will still take an extended period of time for them to root, but they root faster than if long days and warm temperatures were not provided.

Production

Most hosta cultivars are relatively easy to grow, provided a few simple requirements are delivered. Bare-root divisions or large plugs should be planted in a well-drained potting medium. Most growers use bark-based mixes that provide adequate drainage for long periods of time. The divisions should be planted so that the eyes, or the growing points, are at or just below the soil surface. It is not uncommon for bare-root divisions to develop a surface mold on them during storage before they are planted. This mold is rarely a concern and should not be detrimental to crop production. Whether a mold is present or not, it is a good practice to apply a preventative fungicide drench after planting using thiophanate-methyl (Cleary's 3336, OHP 6672, FungoFlo).

Marginal leaf necrosis is often observed on newly transplanted hostas or with plants lacking an adequate root system, particularly if the roots are exposed to insufficient moisture levels. This leaf necrosis is also commonly observed on bare-root plants potted in the fall or late winter months that do not have an established root system going into the winter months and begin flushing leaves in the early spring. Hostas with an adequate root system usually only show marginal leaf necrosis when they are produced under high light levels or drought conditions. Once established, hostas can be allowed to dry out between each watering. It is usually best to water hostas early in the day to allow the foliage to dry before light levels get high and to ensure there is adequate moisture available to the roots.

Hostas are light to moderate feeders, requiring only modest amounts of fertilizer. Although liquid fertilization programs providing 100 ppm nitrogen are adequate to produce hostas, due to the longevity of this crop, it is generally better to use controlled-release fertilizers. For plants potted in the late summer for the following spring's sales, I recommend using a controlled-release fertilizer with at least an eight-month release pattern. If possible, incorporate the time-release fertilizer into the growing medium prior to planting at a rate equivalent to 1 lb. elemental nitrogen per cubic yard (454 g/0.76 m^3). Otherwise, I would recommend growers apply a topdressing of these materials to the top of the growing medium after the hostas have been potted. When top-dressing, growers should use the medium rate recommendations provided on the fertilizer label. Although generally not a concern, the optimal pH of the growing medium is 5.8–6.5.

When hostas are being produced for container sales, it is best to produce them in an area that receives at least partial shade. Many growers use a 30% shade cloth over the production area to reduce the exposure to direct sunlight. Many hosta cultivars, such as 'Sum and Substance', can be produced under full sun when they are grown in the landscape. However, when being produced in containers, these cultivars tend to prefer some exposure to shady conditions. Container production provides more stressful conditions, such as higher root zone temperatures, and more dramatic swings in the moisture level surrounding the roots than when plants are grown in the ground. These stresses make even sun-loving cultivars more sensitive to exposure to high light levels and more prone to sunscald on their leaves. Generally, keeping the light levels below 4,000 f.c. (43 klux) will keep most varieties from getting sunscald.

Insects

Aphid	Leafhopper
Beet armyworm	Mealybug
Black vine weevil	Slug
Bulb mite	Snail
Caterpillar	Soft scale
Foliar nematode	Spider mite
Fungus gnat	Thrips
Grasshopper	Whitefly

Hostas are often considered trouble-free perennials without many insect and disease problems. The most prevalent pests to hostas are slugs, which create "shot holes" in the foliage. For controlling slugs, I would recommend using slug baits such as metaldehyde (Deadline), iron phosphate (Sluggo), or methiocarb (Mesurol) around the production area on a monthly basis. Aphids, spider mites, and thrips may also be observed feeding on hostas, but rarely do they become problematic. During the overwintering period, mice and voles commonly feed on the crowns and roots below the soil surface. Apply mouse baits such as diphacinone (Ramik) throughout the production site during the winter months.

Diseases

Alternaria	*Pellicularia* crown rot
Botryotinia rot	*Pythium*
Botrytis	*Rhizoctonia*
Botrytis blight	*Sclerotinia* crown rot
Colletotrichum (anthracnose)	*Sclerotium* petiole rot
Erwinia	*Sphaceloma* (anthracnose)
Mycoleptodiscus crown rot	*Stromatinia* rot

Hostas generally are disease free and easy to produce. Of the diseases, *Sclerotium rolfsii* var. *delphinii,* or petiole rot, often catches growers by surprise and can be very devastating to a crop. Petiole rot causes soft brown areas at the base of the petioles, followed by withering and a collapse of the leaves. It can be identified by the presence of little inconspicuous spheres called sclerotia near the site of the infection. Sclerotia contain the fungus that will attack other plants when the opportunity arises. They are the size of mustard seeds and range from white to brick red in color. It is best to carefully remove and discard infected plants from the production area, as no fungicide has been effective at eradicating this disease. Metribuzin (Contrast) and PCNB (Terraclor 400) have been shown to suppress the activity of *Sclerotium* when used preventively or when detected and applied at the early stages of this disease.

The last couple of years have brought a lot of concern and attention to a fairly new virus, hosta virus X (HVX), which has infected hostas at an alarming rate. Most commonly, the virus causes mottling, stunting, twisting, and/or puckering of the leaves. It can take three years or longer for infected plants to become symptomatic. The virus is spread by contact of the sap of an infected plant. The most common means of spreading the virus is through routine maintenance and propagation practices such as trimming, division, and tissue culture. There is no treatment available for virus-infected plants; symptomatic plants should be tested and destroyed if the results are positive. Tobacco rattle virus (TRV), tobacco mosaic virus (TMV), and impatiens necrotic spot virus (INSV) have also been observed infecting hostas.

Figure 12.36. Hosta virus X symptoms on *Hosta* 'Peedee Gold'

Growers can refer to the Hosta Library (www.hostalibrary.org/firstlook/HVX.htm) and the American Hosta Society (http://www.hosta.org/About_Hosta/pests_diseases.htm) for further information regarding hosta viruses and a listing of cultivars that have been found to have infections.

Plant Growth Regulators

Depending on the environmental conditions and crop spacing, container growers may need to use PGRs to control plant height. Providing adequate space between the plants is the best and most effective method growers can practice to control plant height during production. At high plant densities, the petioles of hosta leaves tend to stretch excessively, reducing the appearance and quality characteristics of the plant. I have had the most success applying PGRs to hostas using foliar applications of the tank mix of daminozide at 2,500 ppm and uniconazole-P at 5 ppm. The first application usually is applied when the hosta leaves begin to unfold or expand, and the second is applied seven days later. A third application may be needed and should be applied seven days after the second. One-time drench applications of ancymidol at 5ppm or uniconazole-P at 1 ppm are also highly effective at reducing petiole stretch of hostas. Drenches should be applied after the first few leaves are fully expanded, as earlier applications may provide too much control.

PGRs for *Hosta* Cultivars

Trade Name	Active Ingredient	Northern Rate
A-Rest	Ancymidol	25–50 ppm
Sumagic	Uniconazole-P	10 ppm
B-Nine or Dazide + Sumagic	Daminozide + uniconazole-P	2,500 ppm + 5 ppm
A-Rest drench	Ancymidol	5 ppm
Sumagic drench	Uniconazole-P	1 ppm

These are northern rates and need to be adjusted to fit the location where the crop is being produced. Unless noted otherwise, these rates are for spray applications and geared to growers using multiple applications. Drench applications are usually applied once per crop.

Forcing

Hostas will naturally become dormant as the day length shortens in the fall. A cold period of at least six weeks at 40° F (4° C) is recommended to break dormancy and reinitiate growth. An established plant usually overwinters well when a minimum amount of protection is provided. It is usually best to place plants in a covered structure, such as a cold frame or an unheated greenhouse, during this period. After they have gone completely dormant and the temperatures are routinely at or near freezing, they should be covered with insulating materials, such as insulation blankets, to provide sufficient winter protection. For plants that are not established, potted after mid August, or considered expensive varieties, it is recommended to overwinter them in a facility where minimum heat can be provided, keeping them below 40° F (4° C) and above freezing.

After the cold period, regardless of day length, hostas will flush leaves when temperatures are adequate (58–74° F [14–23° C]). Many growers forcing them under short days (natural day lengths less than fourteen hours) often observe an initial flush of leaves followed by a vegetatively dormant state. Hostas have an obligate long-day requirement for continued leaf formation. When producing hostas for early sales, assuming the natural day length is less than fourteen hours, I recommend providing long-day photoperiods by extending the day length or providing night-interruption lighting for four hours between 10 P.M. and 2 A.M., to prevent the dormant phase.

Hostas are commonly flushed at production temperatures of 60–70° F (16–21° C). At 65° F (18° C), it will take approximately eight weeks to force an overwintered crop of 1 gal. (3.8 L) hostas to shippable size. Hosta cultivars will develop at various rates when grown at 65° F (18° C); use this temperature recommendation only as a guideline. Growing at cool temperatures (58–65° F [14–18° C]) may take longer to produce a finished crop, but it will enhance or intensify the leaf coloration and overall appearance of the crop. Conversely, high production temperatures (over 80° F [27° C]) will cause the leaves to lose many of their cultivar-specific characteristics such as leaf size, shape, and color.

Spring-planted hostas typically take longer to finish, from twelve to sixteen weeks, depending on the cultivar and size of the starting materials. When planting hostas propagated via tissue culture, it is usually best to plant them in the finished container during the summer before they are to be sold, allowing them to bulk up and become established.

References

Edmonds, Brooke, Mark Gleason, and Ursula Schuch. "Hosta Takeover." *American Nurseryman*. October 1, 1999.

Elenbaas, Evan, and Jeff Westendorp. "From Start to Finish—Hosta." *Greenhouse Grower*. August 1996.

Fausey, Beth, Arthur Cameron, Royal Hiens, and Will Carlson. "Forcing Perennials: Hosta. *Greenhouse Grower*. November 1999.

Heinke, Gretchen, and Jane Martin. *Growing Hostas.* Ohio State University Extension FactSheet HYG-1239-02.

Pilon, Paul. Perennial Solutions: "Hosta 'Sum and Substance'." *Greenhouse Product News*. July 2004.

Lamium maculatum

Spotted Dead Nettle

Lamium maculatum cultivars, also known as dead nettles or spotted dead nettles, are versatile, mat-forming perennials. They can be utilized in a number of commercial and garden situations. They look great in containers alone or mixed with other plants in patio pots, window boxes, or hanging baskets. In the landscape, they work well as a groundcover in mass plantings or to add foliage color and texture to mixed annual or perennial beds. They bear clusters of lavender, pink, or white flowers over the variegated foliage (silver-green or chartreuse-green) during both the spring and fall seasons. *Lamium* will flower sporadically during the summer months under certain growing conditions.

Figure 12.37. *Lamium maculatum* 'Golden Anniversary'

Lamium thrive in partial sun to partial shade and grow rapidly, spreading 16–24 in. (41–61 cm) wide, while maintaining a height of 4–10 in. (10–25 cm) tall. Most dead nettles are evergreen in southern climates or regions that experience mild winters. They are both cold and heat hardy in USDA Hardiness Zones 3–9 and AHS Heat Zones 8–1. Each cultivar may vary slightly in its ability to tolerate cold, heat and light levels. The foliage of many cultivars becomes scorched under the heat and high light levels during the summer. They can be cut back to 2–3 in. (5–8 cm) during the late summer to encourage a new flush of growth and blooms during the fall.

Commonly Grown *Lamium maculatum* Cultivars

'Anne Greenaway'	'Orchid Frost'
'Aureum'	'Pink Pewter'
'Beacon Silver'	'Purple Dragon'
'Beedham's White'	'Red Nancy'
'Chequers'	'Shell Pink'
'Golden Anniversary'	'White Nancy'

Lamium belongs to the Lamiaceae (Labiatae) family, commonly called the mint family. There are numerous commercially produced perennials in this family, including *Agastache, Ajuga, Calamintha, Lamiastrum, Lavandula, Monarda, Nepeta, Salvia, Stachys,* and *Thymus. Lamium* is derived from the Greek word *laimos,* which means "throat," describing the throat-like appearance of the lower portion of the blossoms. *Maculatum* translates as spotted, describing the silvery markings on the leaves.

Propagation

Lamium maculatum is vegetatively propagated by tip cuttings or bare-root divisions. Commercially, they are primarily vegetatively propagated by cuttings. Many cultivars are patented, and unlicensed propagation is prohibited. To avoid illegal activities and the consequences, always check for plant patents before attempting to propagate.

Tip cuttings should be taken while they are vegetative and non-flowering. Cuttings that are already in bloom will root successfully, but they will take longer to root and have a lower survival rate than purely vegetative starting materials. The cuttings should measure approximately 2 in. (5 cm) in length and contain several nodes. They can be stuck in various sized plug liners (128-cell size to 3 in. [8 cm] containers) for rooting. The well-drained rooting medium should be moistened prior to sticking. The base of the cuttings can be dipped in a rooting hormone, such as a solution of indolebutyric acid (IBA) at rates of 750–1,250 ppm prior to sticking. Dead nettles root easily without rooting compounds but tend to root slightly faster and more uniformly when these treatments are provided.

Cuttings should be placed under low misting regimes for about the first ten days of propagation. It is usually best to propagate under high humidity levels (90% relative humidity) with minimal misting. Misting can gradually be reduced as the cuttings form calluses and root primordia. They are usually rooted in approximately three to four weeks with soil temperatures of 70–75° F (21–24° C). For best results, the air temperature during rooting should be maintained above 65° F (18° C) and below 80° F (27° C). It is beneficial to begin constant liquid feeding with 150 ppm nitrogen at each irrigation beginning ten days from sticking.

Production

Lamium cultivars are commonly produced in 1 qt. (1 L) to 1 gal. (3.8 L) containers. They perform best when grown in a moist, well-drained growing medium with the pH maintained at 5.5–6.3. Many commercially available peat- or bark-based growing mixes work well, provided there are good water-holding ability and adequate drainage. When planting, the plugs should be planted so the original soil line of the plug is even with the surface of the growing medium of the new container.

For plant establishment, maintain average temperatures of at least 65° F (18° C). Once they are established, they can be grown at 55–70° F (13–21° C) day temperatures and night temperatures between 55-60° F (13–16° C). When they are grown under naturally low light levels, warmer production temperatures will often cause them to become straggly and have reduced quality characteristics.

Dead nettle requires an average amount of irrigation, preferring to be kept moist but not consistently wet. Water as needed when the plants are young and becoming established. Once the plants are large, they will require more frequent irrigations, as they dry out rather quickly. Plants grown under drier conditions may struggle but will persist. When irrigation is necessary, water them thoroughly, allowing the soil to dry slightly between waterings.

Lamium maculatum is a moderate feeder; fertility can be delivered using water-soluble or controlled-release fertilizers. Growers using water-soluble fertilizers either apply 150–200 ppm nitrogen as needed or feed with a constant liquid fertilization program using rates of 75–100 ppm nitrogen with every irrigation. Controlled-release fertilizers are commonly applied as a topdressing on the medium's surface using the medium recommended rate on the fertilizer label or incorporated into the growing medium prior to planting at a rate equivalent to 1.0–1.25 lb. nitrogen per cubic yard (454–567 g/0.76 m^3) of growing medium. Plants grown under high fertility regimes generally become very lush and may take longer to flower.

It is usually best to produce *Lamium maculatum* under moderate light levels, 3,000–4,000 f.c. (32–43 klux). Some cultivars do not perform as well at high light levels, often getting scorched leaves as a result, and should be produced under slightly lower light levels. Dead nettles can be produced at high plant densities. However, the highest quality plant is produced when they are grown at minimum plant spacing, i.e., 1 qt. (1 L) pots grown at 5 in. (13 cm) centers or 1 gal. (3.8 L) pots grown at 10–12 in. (25–30 cm) centers. They grow very quickly and should be spaced before the foliage from adjacent plants touch each other.

Pinching is often recommended but may not be necessary, as most cultivars branch freely. Growers who prefer to pinch dead nettles either pinch the plugs prior to planting or two weeks after they have been planted. *Lamium* may be trimmed to shape growth or cut back if desired. Many growers trim or mow the foliage back after plants have flowered to keep the foliage free of leaf spots and to rejuvenate plant growth and flowering.

Insects

Aphid	Southern root knot nematode
Caterpillar	Sow bug
Leaf miner	Spider mite
Plant Bug	Thrips
Slug	Whitefly
Snail	

PGRs for *Lamium maculatum* Cultivars

Trade Name	Active Ingredient	Northern Rate
B-Nine or Dazide	Daminozide	2,500 ppm
Bonzi, Piccolo, or Paczol	Paclobutrazol	30 ppm
Cycocel	Chlormequat chloride	1,250 ppm
Sumagic	Uniconazole-P	5 ppm

These are northern rates and need to be adjusted to fit the location where the crop is being produced. These rates are for spray applications and geared to growers using multiple applications.

The occurrence of insects is not uncommon, but rarely do they become problematic. Whiteflies are the most troublesome insect pests of *Lamium,* but occasional outbreaks of aphids, slugs, and snails may occur, causing only a minimal amount of crop injury. None of these insect pests require preventative control strategies. Growers should have routine scouting programs to detect the presence of insects early and to determine if and when control strategies are necessary.

Diseases

Botrytis	*Myrothecium* (blight)
Downy mildew	*Pythium*

Under the right circumstances, certain diseases may become prevalent when growing *Lamium maculatum. Botrytis* is occasionally a problem on the lower foliage where air movement is limited and the foliage often stays wet after irrigation for extended periods of time. Downy mildew can usually be detected by the presence of purple spots on the upper leaf surfaces. Directly beneath these spots on the undersides of the leaf, growers may observe a mass of white or gray spores. To control these diseases, it is best to manage the environment by providing proper plant spacing and adequate air movement, maintaining a relative humidity below 70%, and, if desired, following a preventative spray program (especially for downy mildew) using the appropriate chemicals. Excessive irrigation, poorly drained growing medium, and excessive periods of heavy rainfall could lead to crown or stem rots and ultimately plant loss.

Plant Growth Regulators

During the winter months, periods of low light levels, or when grown at high plant densities, excessive plant stretching might occur, requiring some type of height management strategy. When produced under high light intensities and adequate spacing, height control is seldom an issue. If controlling plant height is necessary, the PGRs in the above table can be used. Begin the PGR applications as the plant canopy is beginning to enclose. It will usually require two or three applications at seven-day intervals to provide adequate height control.

Forcing

For container production, *Lamium maculatum* cultivars are well suited for forcing into bloom throughout the year. They do not have juvenility or cold requirements that must be met

Figure 12.38. **Flowers of *L. maculatum* 'Anne Greenway'**

prior to forcing. Plants will easily bloom when propagated from cuttings the first growing season, sometimes before they are fully rooted in the plug liner. During times of the year where the light levels are naturally low, providing 400–500 f.c. (4.3–5.4 klux) of supplemental lighting will dramatically improve the quality characteristics compared to those produced under lower light levels.

Dead nettles can be grown at any day length, as they are day-neutral plants. The length of the photoperiod does not have any effect on the time to flower or the number of blooms produced. The time to bloom is a function of cultivar and temperature. Depending on the cultivars being produced, dead nettles will reach full bloom in approximately eight to ten weeks when they are grown with temperatures of 60–70° F (16–21° C). Producing them at cooler temperatures increases the time to flower but will improve the color intensity of the foliage and flowers. To achieve the highest plant quality, I recommend growers produce *Lamium maculatum* using cool temperatures, 60–65° F (16–18° C), and allow a couple of additional weeks for crop production.

Growers producing *Lamium* in cold frame structures can produce flowering plants for spring sales without providing heat to the crop. When producing plants in this manner, growers often plant plugs in the final containers during the late summer. This allows them to become established and fill out the pots prior to winter. Of course, every spring and every location is different, but flowering plants are often ready for shipping by mid-April.

References

Pilon, Paul. Perennial Solutions: *"Lamium* 'Orchid Frost'." *Greenhouse Product News.* October 2002.

Lavandula angustifolia

Lavender

Lavenders are very popular perennials in today's gardens. With their distinguishing characteristics, *Lavandula angustifolia* cultivars are widely used for a variety of landscape applications, including use as accent plants, border plants, foliage plants, cut flowers, and in mass and container plantings. Besides being a useful plant in the landscape, lavenders have been used throughout the years for aromatherapy in perfumes and potpourri, as well as a useful herb. In fact, the genus name is derived from the Latin word *lavo,* which translates to "I wash," referring to an aromatic wash of lavender water once used.

Figure 12.39. ***Lavandula angustifolia* 'Hidcote Superior'**
Jelitto Perennial Seeds

These native Mediterranean perennials are commonly grown across the United States in USDA Hardiness Zones 5–9 and AHS Heat Zones 9–1. They naturally bloom from June till September, bearing numerous blue, pink, purple, or white flowers. Like most silver-leafed perennials or Mediterranean plants, lavenders grow best in sunny locations.

There are numerous marketing opportunities available for producers and retailers who offer lavender cultivars. They can be easily

Commonly Grown
***Lavandula angustifolia* Cultivars**

'Alba'	'Lavender Lady'
'Dwarf Blue'	'Munstead Strain'
'Hidcote Blue'	'Rosea'
'Hidcote Superior'	'Silver Edge'
'Jean Davis'	

marketed as an herb or a perennial. An additional marketing strategy is to sell lavenders as potted flowering houseplants that customers can plant outdoors after they have enjoyed them indoors for several weeks, providing years of enjoyment following the initial purchase. Flowering plants sell well at the retail site; however, with its silvery gray foliage, lavenders usually sell well even when they are not in bloom.

Propagation

Although most *Lavandula angustifolia* cultivars are propagated by seed, many can be vegetatively propagated by tip cuttings. Tip cuttings should be taken while they are vegetative. Many growers have found the best time to propagate lavenders is in early to mid-spring or late summer to early fall, when they are actively growing and vegetative. Acquiring unrooted cuttings from overseas is becoming a viable option for year-round propagation of lavenders.

Figure 12.40. **Ideal lavender tip cuttings**

Tip cuttings should measure approximately 1.5–2 in. (4–5 cm) in length and contain several nodes. The well-drained rooting medium should be moistened prior to sticking. The base of the cuttings can be dipped into a rooting hormone, such as a solution of indolebutyric acid (IBA) at rates of 750–1,250 ppm, prior to sticking. Lavenders can successfully root without rooting compounds but tend to root slightly faster and more uniformly when these treatments are provided. They are generally stuck into plug liners for rooting.

It is best to propagate lavenders under high humidity levels with minimal or no misting. Too much misting will cause the leaves to rot. Certain times of the year (mid-fall to mid spring), misting is not required to successfully root lavenders. If misting is required, it should only be used during the hottest portion of the day and kept to a minimum. Usually only the first seven days of propagation require any type of misting. Once the cuttings appear to tolerate the stress caused from the light levels each day, the misting is no longer required. Again, prolonged exposure to misting may cause the leaves to rot during propagation; only use misting when needed.

To reduce the incidence of *Botrytis*, it is beneficial to provide sufficient air circulation during propagation. I recommend beginning constant liquid feeding with 150 ppm nitrogen at each irrigation beginning seven days from sticking. During certain times of the year, the cuttings will benefit from bottom heat to maintain the soil temperature at 65–75° F (18–24° C). At these temperatures, lavenders should be rooted in three to four weeks.

Most growers propagate *Lavandula angustifolia* cultivars by seed. To increase the germination rate and decrease the germination time, it is beneficial to stratify the seed prior to germination. Stratification is useful to overcome seed dormancies and will decrease the germination time and increase the uniformity and rate of germination. The seeds should be sown in the intended plug flat. Do not cover the trays with germination mix or vermiculite. Light is required for germination, and covering reduces the amount of light reaching the seed and may adversely affect the germination process. The seed flats should be moistened and moved to an environment, such as a cooler, where the temperatures can be maintained at 38–40° F (3–4° C) for four weeks.

After the cold treatment, move the plug trays into a warm environment (65–75° F [18–24° C]) for germination. Many growers utilize germination chambers during this stage to provide uniform moisture levels and temperatures. The medium should be kept moist but not saturated. Germination begins within a few days and takes up to ten days to be completed. Following germination, reduce the moisture levels somewhat, allowing the growing medium to dry out slightly before watering to help promote rooting. Fertilizers are usually applied once the true leaves are present, applying 100 ppm nitrogen every third irrigation or 50 ppm with every irrigation, using a balanced water-soluble source. The light levels can gradually be increased as the plug develops, from 500 f.c. (5.4 klux) following germination to 2,500 f.c. (27 klux) at the final stage. When produced at temperatures averaging 70° F (21° C), it will take eight to ten weeks until they are ready for transplanting.

Production

For container production, lavender cultivars are suitable for 1 qt. (1 L) to 1 gal. (3.8 L) containers. Most growers receive starting materials as finished plug liners. They can be started using bare-root materials; however, it is critical that bare-root starter materials are planted immediately after harvesting and that they do not dry out prior to potting. It is not uncommon for growers to lose some plants when bare-root materials are used.

When planting large containers, such as 1 gal. (3.8 L) pots or larger, I recommend planting two plug cells per container to properly fill out the pot and reduce the production time. Lavenders perform best when they are grown in a well-drained, porous growing medium with a slightly acidic pH, 5.8–6.2. Most commercially available peat- or bark-based growing mixes work well, provided there is adequate drainage.

When planting, the plugs should be planted so the original soil line of the plug is even with the surface of the growing medium of the new container. Planting the crown too deeply will lead to crop variability and losses. After planting, particularly if bare-root starting materials are used, I recommend drenching with a broad-spectrum fungicide such as etridiozole + thiophanate-methyl (Banrot) or the combination of mefenoxam (Subdue Maxx) and fludioxonil (Medallion). The best quality is achieved when plants are grown in full sun or in greenhouses with high light intensities—2,500–3,500 f.c. (27–38 klux) is sufficient. During the winter months, crop quality can be greatly improved when 400–500 f.c. (4.3–5.4 klux) of supplemental lighting is provided.

Lavenders require an average amount of irrigation, as they do not tolerate real wet or overly dry conditions. When moist or wet conditions occur, they are very susceptible to root rots. When irrigation is necessary, water them thoroughly, allowing the soil to dry slightly between waterings.

They are light feeders; fertility can be delivered using water-soluble or controlled-release fertilizers. Growers using water-soluble fertilizers either apply 100–150 ppm nitrogen as needed or feed with a constant liquid fertilization program using rates of 50–75 ppm nitrogen with every irrigation. Growers commonly apply time-release fertilizers as a topdressing onto the medium's surface using the medium rate or incorporated into the growing medium prior to planting at a rate equivalent to 0.75 lb. nitrogen per cubic yard (340 g/0.76 m^3) of growing medium.

Insects

Aphid	Slug
Caterpillar	Spider mite
Four-lined plant bug	Spittlebug
Leafhopper	Whitefly
Root knot nematode	

There are not many insects that cause significant amounts of damage to lavenders. Occasional outbreaks of aphids and whiteflies may appear, causing only a minimal amount of crop injury. None of these insect pests require preventative control strategies. Growers should have routine scouting programs to detect the presence of insects early and to determine if and when control strategies are necessary.

Diseases

Armillaria root rot	*Rhizoctonia*
Botrytis	*Septoria*
Phytophthora	*Xanthomonas*
Pythium	

Lavandula angustifolia can generally be grown free of plant diseases. *Phytophthora* and *Rhizoctonia* are the most common diseases of lavenders and may occur at anytime throughout the production cycle. Reducing the humidity and practicing sound irrigation methods are the best methods of controlling the occurrence of these pathogens. If chemical control is necessary, there are several chemicals on the market, such as fludioxonil (Medallion 50W), PCNB (Terraclor 75WP), azoxystrobin (Heritage), and flutolanil (Contrast 70WSP), which provide excellent control of these diseases.

To reduce the occurrence of lavender diseases, it is best to manage the environment by providing proper plant spacing and adequate air movement and controlling the humidity. Growers should carefully watch the moisture levels during adverse times of the year and avoid overwatering their plants. Routine scouting is useful and recommended to detect plant diseases early, allowing the appropriate control strategies to be implemented before significant crop injury or mortality occurs.

Plant Growth Regulators

Most *Lavandula angustifolia* cultivars are fairly compact and usually do not require height control. Controlling stem elongation may be necessary when they are grown in a greenhouse during the early spring. When necessary, I recommend applying uniconazole-P at a minimum rate of 5 ppm (the rate will vary with location; this is a northern rate). Two applications seven days apart should provide adequate control. Before applying these chemicals, the height can often be effectively controlled by providing adequate spacing between the plants and by withholding water and nutrients.

Forcing

Lavandula angustifolia is easy to produce as a foliage plant. When transplanting from 72-cell plugs, lavenders grown at 65° F (18° C) typically take eight weeks to finish 1 gal. (3.8 L) pots in the summer or eleven weeks during the winter months.

To improve marketability, lavender cultivars can be forced to bloom throughout the year. Growers wishing to produce blooming plants need to understand and follow the guidelines discussed below.

Research has shown that most lavender cultivars must reach a particular size, or maturity, before they are capable of flowering uniformly and consistently. I recommend beginning the forcing process using fairly large plant materials, consisting of at least forty leaves. Smaller plant materials will most often result in less uniformity, less flowers per plant,

PGRs for *Lavandula angustifolia* Cultivars

Trade Name	Active Ingredient	Northern Rate
A-Rest	Ancymidol	25 ppm
B-Nine or Dazide	Daminozide	2,500 ppm
Bonzi, Piccolo, or Paczol	Paclobutrazol	30 ppm
Sumagic	Uniconazole-P	5 ppm

These are northern rates and need to be adjusted to fit the location where the crop is being produced. These rates are for spray applications and geared to growers using multiple applications.

and a decreased percentage of flowering plants. Therefore, it is best to grow lavenders at natural day lengths for a minimum of six to eight weeks to allow them to reach a mature size before providing the additional requirements necessary for flowering.

For uniform and rapid flowering, it is best to vernalize lavenders prior to forcing. The best results are achieved when they are cooled for at least ten weeks at 35–44° F (2–7° C). Uncooled plants will still flower, but they will be more inconsistent and it will take considerably longer for them to reach flowering.

Once the cold period is achieved, lavenders can be forced to bloom under any photoperiod. The time to bloom, uniformity of flowering, and the flowering percentage are greatly influenced by the amount of cold the crop has received and determine to some extent the type of photoperiod growers should provide. For example, uniform flowering and consistency can be achieved forcing lavenders that have received fifteen weeks of cold under natural photoperiods. There is no additional benefit from providing long days to plants that have received this amount of cold. Crops receiving less than fifteen weeks of cold will flower under natural day lengths, but the flowering percentages and uniformity will be reduced. Long days can be used to some extent as a substitute for cold, for situations where either imperfect cooling temperatures or insufficient durations occur.

Regardless of the vernalization temperatures or durations, I recommend providing photoperiods of sixteen hours by extending the day if necessary, or using a four-hour night interruption during the middle of the night, providing a minimum of 10 f.c. (108 lux) of light at plant level. Long-day photoperiods help growers achieve rapid flowering, overcome imperfect vernalization, and increases the uniformity of flowering and flowering percentages. Photoperiodic lighting should be provided from the start of forcing until flower buds are visible.

After the cold treatment and long days are provided, the time to flower depends on the temperature the plants are grown at. The following temperature recommendations fit most varieties, but each cultivar will have a slightly different finishing time at each of the temperatures illustrated. The time to flowering also varies with the amount of vernalization the crop has received. Lavenders grown at 60° F (16° C) will flower in about ten weeks, while plants grown at 68° F (20° C) will flower in approximately eight weeks.

Lavandula angustifolia grown at temperatures less than 65° F (18° C) tend to have higher quality characteristics, such as a more compact appearance, an upright rather than sprawling habit, and more flowers per plant than those grown at warmer temperatures. With temperatures above 73° F (23° C), plant quality decreases; lavenders will have a weak, floppy growth habit and reduced flowering percentages. To obtain the best plant quality, I recommend producing lavender cultivars at 63–68° F (17–20° C).

References

Greenhouse Grower Magazine and Michigan State University. *Firing Up Perennials: The 2000 Edition.* Willoughby, Ohio: Meister Publishing. 2000.

Levy, Miriam. "Lavender Deciphered." *GrowerTalks.* May 1996.

Pilon, Paul. Perennial Solutions: *"Lavandula angustifolia* 'Hidcote Superior'." *Greenhouse Product News.* May 2003.

Leucanthemum x superbum

Shasta Daisy

Shasta daisies are very popular perennials in today's gardens. With the differing height characteristics of the various cultivars, ranging from 8–36 in. (15–91 cm), they can be used for numerous landscape applications. Gardeners and landscapers commonly use them as accent plants, border plants, cut flowers, and in mass and container plantings. Most cultivars have classic daisy-like flowers with white ray petals encircling a raised center of tiny yellow disc flowers. Several varieties have semi-double or fully double inflorescences with various sized yellow centers, making them appear fringed

Figure 12.41. ***Leucanthemum* x *superbum* 'Snow Cap'**

and frilly. Shasta daisies have become slightly more popular since the Perennial Plant Association selected *Leucanthemum* 'Becky' as the Plant of the Year in 2003.

The American plant breeder Luther Burbank (1849–1926) created the Shasta daisy in 1890. Burbank is best known for developing the Burbank potato, which was resistant to blight and helped relieve the great Irish potato famine. *Leucanthemum* belongs to the Asteraceae family (formerly Compositae). Among the numerous perennials in the Asteraceae family being commercially produced are *Achillea, Aster, Bellis, Centaurea, Coreopsis, Doronicum, Echinops, Echinacea, Erigeron, Gaillardia, Helenium, Heliopsis, Ligularia, Rudbeckia,* and *Stokesia.*

Commonly Grown *Leucanthemum* x *superbum* Cultivars

'Aglaia'	'Polaris'*
'Alaska'*	'Silver Princess'*
'Becky'	'Snowcap'
'Crazy Daisy'*	'Snow Lady'*
'Ester Read'	'Sonnenschein'
'Exhibition'*	'Summer Snowball'
'Highland White Dream'	'Sunny Side Up'
'Little Princess'	'Thomas Killien'
'Marconi'*	'White Knight'*

*Propagated from seed.

Leucanthemum cultivars flower from June till August. When growers cut back or deadhead the plants during the growing season, I have seen flowering into the month of October. The foliage is always attractive and will remain evergreen in southern states. Many cultivars are suitable for production in USDA Hardiness Zones 3–9 and AHS Heat Zones 9–1, with hardiness and heat tolerance differences between cultivars. They prefer to be grown in full sun, although locations receiving partial sun are often acceptable. Shasta daisies perform well as cut flowers and attract butterflies to the garden, which can be used as an added selling point for consumers.

In recent years, there have been changes to the genus names *Leucanthemum* and *Chrysanthemum,* which has brought about some confusion as to how Shasta daisies should be classified. They are currently considered to be members of the genus *Leucanthemum.* Previously, Shasta daisies were classified as *Chrysanthemum* x *superbum* and *Chrysanthemum maximum,* and today these classifications are often used synonymously. Cultivars of *Leucanthemum* x *superbum* are crosses between *L. lacustre* and *L. maximum* (*C. lacustre* and *C. maximum*).

Propagation

Depending on the cultivar, Shasta daisies are started from seed or vegetatively propagated by tip cuttings or division. The method of propagation is often dependent on cultivar, availability, and cost. Although using bare-root divisions provides for a quicker finishing time, it often leads to crop variability and lower survival rates. Most commonly, growers propagate them from seed or tip cuttings.

Sow *Leucanthemum* seeds in the intended plug flat and cover them lightly with germination mix or medium-grade vermiculite. The covering helps to maintain a suitable environment around the seed during this phase. The seed flats should be moistened and moved to a warm environment, where the temperatures can be maintained at 65–70° F (18–21° C) for germination. Many growers utilize germination chambers during this stage to provide uniform moisture levels and temperatures. Using germination chambers is optional, as Shasta daisies will successfully germinate in the greenhouse.

The seeds should be germinated in seven to ten days. Following germination, reduce the moisture levels somewhat, allowing the growing medium to dry out slightly before watering; this will help promote rooting. The light levels can gradually be increased as the plug develops, from 500 f.c. (5.4 klux) following germination to 2,500 f.c. (27 klux) at the final stage. Fertilizers are usually applied once the true leaves are present, applying 100–150 ppm nitrogen every third irrigation or 75 ppm with every irrigation, using a balanced water-soluble source. When plugs are grown at 65° F (18° C), they are usually ready for transplanting in five to seven weeks.

Tip cuttings should be taken while they are vegetative. Cuttings that are already in bloom will take longer to root and have a lower survival rate than purely vegetative starting materials. Tip cuttings should measure approximately 2–3 in. (5–8 cm) and contain several nodes. The well-drained rooting medium should be moistened prior to sticking. The base of the cuttings can be dipped into a rooting hormone, such as a solution of indolebutyric acid (IBA), at rates between 1,000–1,500 ppm prior to sticking. *Leucanthemum* can successfully root without rooting compounds, but they tend to root slightly faster and more uniformly when these treatments are provided. They are generally stuck into plug liners for rooting, although they can be direct stuck into the final container for rooting.

Cuttings should be placed under low misting regimes for about the first two weeks of propagation. It is very common for them to appear wilted on the misting bench during the brightest, warmest times of the day. I usually do not like to see flaccid plants during the propagation stage, but have not experienced significant plant losses due to this wilt. Sticking cuttings with a smaller leaf size does seem to reduce the wilting somewhat. Once the cuttings form roots, usually ten to fourteen days after sticking, the daily wilting will no longer occur. It is usually best to propagate under high humidity levels (90% relative humidity) with minimal misting. Prolonged exposure to mist may cause the leaves to rot during propagation. The misting can gradually be reduced as the cuttings form calluses and root primordia. It is beneficial to begin constant liquid feeding with 200 ppm nitrogen at each irrigation beginning ten days from sticking.

The cuttings are usually rooted in less than three weeks with soil temperatures at 70–75° F (21–24° C). For best results, the air temperature during rooting should be maintained above 60° F (16° C) and below 80° F (27° C). To promote ideal rooting, provide photoperiods of less than thirteen hours per day during propagation.

Production

Depending on the cultivar, Shasta daisies are commonly produced in 1 qt. (1 L) to 2 gal. (7.6 L) containers. They perform best when they are grown in a moist, well-drained medium with a slightly acidic to neutral pH of 5.8–6.7. Many commercially available peat- or bark-based growing mixes work well, provided there is good water-holding ability and adequate drainage.

Leucanthemum are moderate feeders. Fertility can be delivered using water-soluble or controlled-release fertilizers. Growers using water-soluble fertilizers either apply 150 ppm nitrogen as needed or feed with a constant liquid fertilization program using rates of 75–100 ppm nitrogen with every irrigation. Controlled-release fertilizers are commonly applied as a topdressing onto the medium's surface using the medium recommended rate on the fertilizer label or incorporated into the growing medium prior to planting at a rate equivalent to 1.0–1.25 lb. nitrogen per cubic yard (454–567 g/0.76 m^3) of growing medium. Plants grown under lower fertility regimes generally remain shorter than those with ambient fertilizer.

Shasta daisies prefer to be grown in a moist growing medium. Water as needed when the plants are young and becoming established. Once they are large, they will require more frequent irrigations, as they will dry out rather quickly. When irrigation is needed, water them thoroughly, ensuring the entire growing medium is wet or nearly saturated. They can tolerate and fully recover from short periods of drought stress; however, prolonged periods of drought stress may cause the leaf margins to turn necrotic. Plants grown too dry will often be shorter, exhibit a delay of flowering, and produce fewer flowers.

Insects

Aphid	Leaf miner
Caterpillar	Slug
Foliar nematode	Spider mite
Four-lined plant bug	Thrips
Leafhopper	Root knot nematode

Aphids, whiteflies, and thrips occasionally will become problematic. Of these insect pests, aphids are the most prevalent. All of these pests can be controlled after they are detected, or they can be prevented using proactive strategies. Most insects can be detected with routine crop monitoring. Control strategies may not be necessary unless the scouting activities indicate actions should be taken.

Many cultivars are sensitive to pesticide applications and may show phytotoxicity to certain chemicals. Symptoms of chemical sensitivity include leaf burn, browning of the foliage, chlorosis, and plant death with some cultivars. The insecticides abamectin (Avid) and acephate (Orthene) in particular cause injury to Shasta daisies. Cultivars propagated from seed are usually more sensitive to pesticides than are the vegetative varieties. Always use caution when applying pesticides. If applying chemicals for the first time to *Leucanthemum,* trial them on a small scale before making applications to the entire crop.

Diseases

Acremonium (wilt)	Crown gall
Agrobacterium tumefaciens	*Erwinia*
Alternaria	Powdery mildew
Botrytis	*Pythium*
Corynebacterium fascians	*Septoria*

Generally, *Leucanthemum* are relatively disease free, and under normal growing conditions do not usually require the use of fungicides. Occasionally, galls or very numerous small shoots on the crown near the soil line are observed. This shoot proliferation and formation of galls are caused by the bacteria *Agrobacterium tumefaciens* and *Corynebacterium fascians.* These bacterium are most likely spread from plant to plant through vegetative propagation or maintenance activities, such as trimming. Since there is no known treatment for plants with these symptoms, the infected plants should immediately be discarded and the area sanitized to prevent spreading to uninfected areas. To help eliminate galls and shoot proliferation from occurring, I recommend sterilizing all propagation and trimming tools with a disinfectant, such as a 10% bleach solution, between each use.

Plant Growth Regulators

Height control is often necessary to produce a high-quality product under greenhouse conditions. Many cultivars are naturally short and may not require height control. Providing adequate spacing between the plants will reduce plant stretch caused by competition. To a certain extent, the height can also be reduced by withholding water and nutrients. Under certain growing conditions or under high plant densities, it may be necessary to use chemical PGRs. Several of the commercially available PGRs are effective at controlling plant height when they are applied using the appropriate rates, frequency, and timing. Depending on your geographic location, apply foliar applications of uniconazole-P at 5 ppm or paclobutrazol at 30 ppm. It will usually require one or two applications at seven-day intervals to

PGRs for *Leucanthemum* x *superbum* Cultivars

Trade Name	Active Ingredient	Northern Rate
A-Rest	Ancymidol	25 ppm
B-Nine or Dazide	Daminozide	2,500 ppm
Bonzi, Piccolo, or Paczol	Paclobutrazol	30 ppm
Sumagic	Uniconazole-P	5 ppm
B-Nine or Dazide + Bonzi	Daminozide + paclobutrazol	2,500 ppm + 15 ppm
B-Nine or Dazide + Sumagic	Daminozide + uniconazole-P	2,000 ppm + 3 ppm

These are northern rates and need to be adjusted to fit the location where the crop is being produced. These rates are for spray applications and geared to growers using multiple applications.

provide adequate height control. To achieve the best results, these applications should begin as the stems are rapidly elongating and before the flower buds appear.

Forcing

Leucanthemum can be forced into bloom out of season by following a few guidelines. Most cultivars are considered to be cold beneficial, although some cultivars, such as 'Snow Lady', will flower reliably without a cold treatment. Several cultivars flower more quickly, more uniformly, and produce more flowers when cold treatments are provided. Other varieties may flower without a cold treatment but the entire population may not flower, or flowering may be sporadic and unpredictable. I recommend vernalizing plugs or small containers of Shasta daisies for a minimum of six weeks at 35–45° F (2–7° C).

Most *Leucanthemum* x *superbum* cultivars are considered to be long-day beneficial plants, meaning they flower best under long day conditions. Following the cold treatment, most Shasta daisies will flower under any photoperiod but will flower faster and produce more blooms when grown under long days. Some cultivars, such as 'Snow Cap', have an obligate long-day requirement for flowering when no cold treatment is provided. Providing more than six weeks of cold is recommended when long days cannot be provided, but increasing the cold duration has no effect on flower times when forcing with long day photoperiods. For all *L.* x *superbum* cultivars, growers should provide photoperiods of sixteen hours by extending the day if necessary or use a four-hour night interruption during the middle of the night, providing a minimum of 10 f.c. (108 lux) of light at plant level.

After the cold treatment and long days are provided, the time to flower depends on the temperature the plants are grown at. The following temperature recommendations fit most varieties, but each cultivar will have a slightly different finishing time at each of the temperatures illustrated. Shasta daisies grown at 60° F (16° C) will flower in about ten weeks, while plants grown at 72° F (22° C) will flower in as little as six weeks. For the largest flower size, most flowers, and shortest plants, grow them at cooler temperatures. To obtain the best plant quality, I recommend producing them at 65–68° F (18–20° C).

References

Greenhouse Grower Magazine and Michigan State University. *Firing Up Perennials: The 2000 Edition.* Willoughby, Ohio: Meister Publishing. 2000.

Nau, Jim. *Ball Perennial Manual: Propagation and Production.* Batavia, Ill.: Ball Publishing. 1996.

Pilon, Paul. Perennial Solutions: "*Leucanthemum* x *superbum* 'Becky'." *Greenhouse Product News.* December 2002.

Lilium, Asiatic Hybrids

Asiatic Lily

Asiatic lilies are becoming more widely produced and marketed as landscape perennials. They are easy to grow, provide an impressive display of color, and fit the production plans of numerous perennial operations. Asiatic lilies are extensively grown throughout the United States and Canada for cut flower, potted plant, and perennial production.

Unlike most perennials that originate from seed, cuttings, or bare-root divisions, Asiatic lilies are started from bulbs. The bulbs have a solid basil plate that produces roots from its bottom and a concentric series of fleshy, overlapping scales. Asiatic lilies are available to the industry as 10/12 cm, 12/14 cm, 14/16 cm, and 16/18 cm bulbs. The majority of the varieties used for commercial production have high bud counts with small bulb sizes. Most lily bulbs are grown in the Netherlands, New Zealand, or the Northwestern United States and distributed to growers throughout the world.

Lilies produce a single, unbranched stem bearing linear leaves in a whorled pattern. By the time Asiatic lilies bloom, the plant will reach 12–60 in. (0.3–1.5 m) in height. The final height of the plant is determined partially by genetics and partly by environmental factors. Asiatic lilies bear cream, orange, peach, pink, red, yellow, or white flowers in umbels. The 6–8 in. (15–20 cm) diameter flowers are "lily-like" and best described as being funnel or bell shaped. Many cultivars have speckled flowers or blooms with secondary colorations. They may be borne erect, horizontal, or drooping. The flowers open individually and create an impressive display of color and flower power when numerous blooms are open simultaneously.

Figure 12.42. Asiatic lily 'Petit Bridgette'

Depending on cultivar, geographic location, and average temperatures, lilies naturally bloom from May till September in USDA Hardiness Zones 3–9 and AHS Heat Zones 12–1. Gardeners can manipulate the flowering times by selecting different cultivars and staggering the springtime planting dates of pre-cooled, stored bulbs.

Propagation

Commercial perennial growers obtain fully mature, pre-cooled, and ready-to-plant bulbs from various bulb suppliers. Although Asiatic lilies can be propagated from seed and vegetatively, it is not economical for the average grower to take the steps and time necessary to propagate and produce a lily from the starter material, grow it to maturity, and produce a finished crop.

Production

Growers most commonly receive Asiatic lily bulbs packed in peat moss inside black plastic bulb crates or wood/cardboard boxes. The peat moss is used to help retain some moisture around the bulbs, preventing them from drying out during storage and shipping. In order for them to flower, the bulbs have to be precooled, or vernalized. Most cultivars require a minimum of six weeks at 34–36° F (1–2° C) for proper vernalization. Once the bulbs have been vernalized, they should be planted or kept frozen. Keeping the bulbs at the temperatures used for vernalization for extended periods of time will cause the bulbs to sprout.

Most growers receive lilies that have been precooled prior to receipt. In many cases, they are both vernalized and frozen when growers

Commonly Grown Asiatic Lily Cultivars

'Admiration'	'Fancy'	'Pixie Buff'
'Alaska'	'Foxtrot'	'Pixie Butter'
'Aristo'	'Fullspeed'	'Pixie Crimson'
'Blazing' 'Dwarf'	'Gironde'	'Pixie Ivory'
'Black Bird'	'Golden Dwarf'	'Pixie Lemon'
'Brunello'	'Horizon'	'Pixie Orange'
'Cameleon'	'Kansas'	'Pixie Peach'
'Cancun'	'Lativa'	'Pixie Pink'
'Cannes'	'Lolly Pop'	'Polka'
'Charisma	'London'	'Puccini'
'Cote D'Azur'	'Mirbella'	'Reinesse'
'Crimson Sun'	'Monte Negro'	'Scarlet Dwarf'
'Dandy'	'Navona'	'Shiraz'
'Denia'	'Nerone'	'Sphinx'
'Detroit'	'New Wave'	'Sunray'
'Disco'	'Orange Delight'	'Symphony'
'Dominator'	'Partner'	'Vermeer'
'Double Sphinx'	'Petit Brigitte'	'White Baby'

obtain them. Occasionally growers will need to provide the vernalization treatment prior to forcing. The precooling can occur inside the peat filled boxes the bulbs were received in or after they have been potted into the final containers.

If precooled bulbs must be kept in storage for extended periods of time, they should be frozen and held at 28° F (-2° C) until needed. The primary reason for freezing the bulbs is to keep them from sprouting. Asiatic lily bulbs should only be frozen after they have been vernalized. When frozen bulbs have been received from the supplier, they have already been precooled and should be kept frozen until just prior to planting. To prevent frost damage, which could lead to flower and leaf disorders, growers should keep them frozen and not allow the bulbs to defrost and thaw and then freeze again.

If you cannot store frozen bulbs in a freezer after receiving them, store them in a cooler at 34–36° F (1–2° C). If they are frozen upon receipt, they can be held at these temperatures for up to two weeks. Warmer storage temperatures and storing them for extended time periods may cause the bulbs to sprout prior to planting. When the bulbs have sprouted in the peat moss they are packed in, the young shoots can become severed from the bulb when they are removed from the packing materials. Once the tender shoot is broken, the bulb is useless and should be discarded.

When you are ready to pot the bulbs, it is important to gradually thaw them out for several days. Remove only the necessary trays needed from the freezer and place them in a warm environment (55–60° F [13–16° C]) for several days. To speed up the thawing process, lay the trays in a single layer (not stacked), leaving several inches between each tray. They should be checked daily to determine if they have thawed out enough for planting. If there is still a solid, frozen core in the center of the tray, they need more time to thaw out. Plant the bulbs as soon as all have thawed.

Prior to planting, the bulbs need to be carefully separated from the peat moss they are packaged in. Some growers dump the tray of bulbs out on a table and sift through the peat moss and remove the bulbs; other growers use specialized equipment that vibrates or gently shakes and separates the bulbs from the peat moss.

It is beneficial to dip the bulbs in water for at least ten minutes before planting them. Bulb dipping helps to rehydrate the bulbs, bringing them to a uniform moisture level. If the bulbs are not at a uniform moisture level, they will emerge unevenly, thus making crop management more difficult.

Once the bulbs are thawed out and have been rehydrated, they should be planted immediately. If planting cannot be done right away, the bulbs should be stored at 34–38° F (1–3° C) up until the time of planting, which should be within two days after complete thawing. It is recommended to cover the bulbs with plastic during this cool storage to prevent them from drying out. When they are removed from storage, it is not necessary to re-dip them, but it is beneficial to run water over them using a garden hose to rewet any surfaces that have dried out, particularly along the edges. Remove the amount of pre-soaked bulbs from storage that will realistically get potted in a two-hour time period. Apply water over the top of the bulbs prior to potting. Do not allow the bulbs to remain in the potting area overnight, and always cover them with plastic and place them back in cool storage. Planting dehydrated bulbs will lead to erratic emergence and poor crop uniformity.

For container production, Asiatic lilies are suitable for 1 qt. (1 L) to 1 gal. (3.8 L) containers. Small container sizes, such as 4 in. (10 cm) pots, usually contain one bulb each, whereas larger containers, such as 1 gal. (3.8 L) pots, often have up to three bulbs per container. The number of bulbs used depends on the size of the container, the specifications of the customer, and the price point of the product.

Lilies perform best when they are grown in a moist, well-drained, porous growing medium with a slightly acidic pH, 5.8–6.4. Most commercially available peat- or bark-based growing mixes work well, provided there is adequate drainage.

Although some roots grow from the bulb, a majority of the root system develops on the stem between the bulb and the top of the soil surface. These roots are referred to as stem roots. Adequate stem root development is crucial to the production of high-quality lilies. Stem roots serve as the primary source of nutrition and moisture uptake, as well as supporting and anchoring the plant.

Asiatic lilies must be planted deeply. The bulbs should be planted with a minimum of 2 in. (5 cm) of growing mix above the bulb. It is not uncommon to plant them with 3–5 in. (8–13 cm) of growing mix above the bulb. There should also be a minimum of 1 in. (3 cm) of growing mix under the bulb. When planting, place 1 in. (3 cm) of soil in the bottom of the pot. Then place the bulb or bulbs near the center of the pot. The bulbs should be pointed upward, not sideways, to help the stems come up near the center of the pot; this will reduce the labor of straightening stems later. Finally, fill the medium around the bulbs, being careful to not let the bulbs fall sideways while filling. Thoroughly water them in after planting.

During emergence, uniform moisture is important. The growing medium should be watered well after planting and kept uniformly moist thereafter. The growing mix should not be allowed to dry out, but neither should it remain

Figure 12.43. **As seen here, Asiatic lilies do not emerge uniformly and often grow away from the center of the pot. To improve the appearance of the product, many growers straighten the stems by moving them gently toward the center of the pot.**

excessively wet. Watering must be carried out sparingly until they have emerged. At this stage, it is best to water them as needed by hand rather than using overhead irrigation systems.

In many cases, the emerging stems appear to be growing toward the edges of the containers they are being grown in. In most cases, they grow in this manner because the bulb was not centered in the pot or while the container was filled or the bulb shifted slightly, resulting in the new growth appearing around the container's perimeter. As the young stems emerge, it is often desirable—for aesthetic reasons—to straighten them, or move them slightly, so they are positioned more to the center of the pot.

To straighten the stems, gently shift them toward the center of the pot while tucking growing mix behind the stems to help hold them in place. The emerging stems are tender, and moving them too briskly or with too much pressure could break them. If the stems break, the entire pot should be discarded, as the bulb will not produce another flowering stem. Since emergence occurs over a period of time, it is normally necessary to check and straighten the stems two times. This activity should occur as quickly as possible following emergence, before the stems become lignified or begin to form adventitious roots. Waiting for too long to straighten the stems may lead to more broken stems or plants that have already set down stem roots, making them difficult or impossible to straighten.

Asiatic lilies are light feeders, as there is a great amount of stored energy within the bulb. Early in production, fertility should not be an issue because the bulbs contain enough nutrients to begin shoot growth. As the lilies reach 2–3 in. (5–8 cm) tall, begin applying fertilizers to the crop. Fertility can be delivered using water-soluble fertilizers. The rate and type of fertilizer to apply is best determined by conducting routine soil tests. Growers using water-soluble fertilizers either apply 100–150 ppm nitrogen as needed or feed with a constant liquid fertilization program using rates of 50–75 ppm nitrogen with every irrigation.

Controlled-release fertilizers are frequently used to produce containerized Asiatic lilies. Growers commonly incorporate time-release fertilizers into the growing medium prior to planting at a rate equivalent to 1 lb. nitrogen per cubic yard (454 g/0.76 m^3) of growing medium. Time-release fertilizers work rather well for lily crops, as it often takes a couple of weeks for the fertilizer to start releasing. This coincides with the period of time the plant naturally derives its food from the nutrient reservoirs of the bulb.

Good irrigation management is important throughout crop production, from the time they are planted to the day they are sold. The amount of water applied can safely be increased when the shoots are 3–6 in. (8–15 cm) tall, as the stem roots are usually well formed at this time. When providing irrigation, water thoroughly, allowing them dry out slightly between waterings. Growers should avoid overwatering, as excessive moisture can injure the root systems and increase the chance of root rots. Once the flower buds are visible, the growing medium should never dry out. Overly dry conditions at this stage could cause bud abortion, also referred to as bud blasting.

Insects

Aphid	Mite
Beetle	Shore fly
Bulb mite	Stalk borer
Foliar nematode	Thrips
Fungus gnat	Weevil
Scale	

There are relatively few insects affecting the production of Asiatic lilies, and seldom does significant plant injury or loss occur. Aphids, fungus gnats, and shore flies are the insects observed most frequently. Fungus gnats and shore flies are best controlled preventatively using the appropriate larvicides and/or insect growth regulators, and aphids can be controlled as needed using contact insecticides. Growers should have routine scouting programs to determine the presence of these pests and to determine if and when control strategies are necessary.

Diseases

Alternaria blight	*Phoma* stem canker
Ascochyta blight	*Pythium*
Botrytis	*Rhizoctonia*
Bulb rot	Rust
Colletotrichum	*Sclerotinia* blight
Erwinia	*Sclerotium*
Fusarium	*Streptotinia* blight
Pectobacterium soft rot	Virus
Pellicularia blight	

Botrytis often occurs on dead flower petals and can quickly attack the entire plant. *Botrytis* can be prevented by providing adequate air movement, avoiding overhead irrigation while the plants are blooming, and watering early in the day. Once the plants are blooming, it is beneficial to apply preventative fungicide applications using chemicals such as trifloxystrobin (Compass) or azoxystrobin (Heritage), which are effective at controlling this disease but do not leave any unsightly chemical residues on the foliage and flowers.

Root rots caused by the pathogens *Pythium* and *Rhizoctonia* are likely to occur when lilies are grown under wet conditions. Root roots can be prevented by practicing sound irrigation practices and/or using preventative fungicide drench applications. Growers who observe the highest amount of disease pressure from the various pathogens are those who carry over Asiatic lily crops from year to year. When fresh bulbs are used, there is little incidence of disease. Take aggressive actions to control any diseases on carried-over plant materials. Try to keep the foliage clean and healthy. Overwintering diseased materials only perpetuates the problem during the next growing season. Whenever possible, do not carry over Asiatic lily crops.

Leaf Yellowing

Shortly after visible bud, lilies are infamous for developing yellow leaves at the base of the plant. Usually, the higher the plant density or tighter the crop spacing, the greater the occurrence of yellow leaves. To help prevent yellow leaves, growers can spray the plants with the growth regulator Fascination (benzyladenine + gibberellins) using 36 oz. per 100 gal. (1.1 L/ 379 L). Adding a surfactant to the spray solution will help growers achieve better coverage. It is important to get complete coverage of the lower leaves. This application is only effective on the leaves it comes in contact with. In fact, only the portion of a leaf that receives coverage is protected. For example, when only the tip of the leaf receives spray coverage of Fascination and the basal half of the leaf does not come in contact with the chemical, then the tip of the leaf will remain green while the basal half of the leaf may turn yellow.

Fascination can be applied over the entire plant, including the flower buds. The application should be made around the time the buds are becoming visible but before the plant canopy closes in, making coverage difficult. This application will cause a small amount of stem elongation in some circumstances, particularly when no height control products have been previously applied to the crop. This increase in plant height is usually insignificant and does not reduce the appearance or quality of the plant. Not all lilies respond in a similar manner; *Lilium longiflorum* (Easter lily) elongates excessively when Fascination is applied over the entire plant (particularly the growing point). For Easter lilies, it is important to keep the Fascination spray only on the lower leaves.

If increasing the height is a concern, growers can tank mix Fascination with Sumagic. This seems to counteract the potential for stem elongation and still provides the protection of the leaves from yellowing. Sumagic should be tank mixed with Fascination (at the labeled rate) at about half of the normal rate applied to achieve height control.

Plant Growth Regulators

Controlling the height of Asiatic lilies is often necessary to maintain an adequate plant height for both shipping and aesthetic purposes. Plant height can be reduced by manipulating temperatures using DIF (DIF = day temperature – night temperature). Stem elongation decreases under negative DIF conditions (when day temperatures are cooler than night temperatures). For more information on implementing DIF and using temperature to control plant height, refer to the temperature section of chapter 8.

Depending on the cultivar, environmental conditions, and crop spacing, container growers may need to use chemical growth regulators to control plant height. A few of the commercially available plant growth regulators (PGRs) are effective at controlling plant height when they are applied using the appropriate rates, frequency, and timing. Depending on your geographic location, apply foliar applications of PGRs beginning with the rates listed in the table below. Asiatic lilies are particularly sensitive to these plant growth regulators, and care should be taken to not overapply these products. Growth regulators are especially effective when they are applied to plants that are also being exposed to negative DIF conditions.

It is best to begin PGR applications after the plants have enough leaf and stem tissues developed to absorb these chemicals (usually by the time they are 2–3 in. [5–8 cm] tall). If additional applications need to be made, allow seven days between them, otherwise too much height control could occur. It may take two or three applications to provide sufficient levels of height control.

Temperature and Scheduling

The main factor influencing the proper timing of an Asiatic lily crop is temperature. Altering the temperature is the best tool growers can use to speed up or slow down crop development. To ensure the strongest stems and plants of the highest quality, it is best to force Asiatic lilies at cool temperatures.

During emergence, it is important to maintain soil temperatures at 55° F (13° C); the air temperature is not important during this stage of growth. To achieve soil temperatures of 55° F (13° C), growers may have to run the air temperatures warmer. The air temperature might have to be 60–65° F (16–18° C) to maintain this average soil temperature. It is important to monitor the soil temperature and adjust the air temperatures accordingly. Some operations have the ability to provide bottom heat to maintain proper soil temperatures while running cooler air temperatures. Bottom heat allows the grower to be energy efficient and gives them the ability to produce uniform crops.

After the majority of the lilies have emerged, usually within two to three weeks, the air temperature becomes more important to control than does the soil temperature. Following emergence, the air temperatures can be maintained at not higher than 68° F (20° C). When possible, the average twenty-four-hour temperatures should be maintained at around 60° F (16° C). Cooler temperatures dramatically delay crop development, and warmer temperatures hasten development and most often reduce the quality of the crop. Typical production regimes maintain 50–55° F (10–13° C) night temperatures and 65–70° F (18–21° C) day temperatures. Asiatic lilies do

PGRs for Asiatic Lily Cultivars

Trade Name	Active Ingredient	Northern Rate
A-Rest	Ancymidol	25 ppm
Bonzi, Piccolo, or Paczol	Paclobutrazol	15 ppm
Sumagic	Uniconazole-P	2.5 ppm

These are northern rates and need to be adjusted to fit the location where the crop is being produced. Unless noted otherwise, these rates are for spray applications and geared to growers using multiple applications.

Figure 12.44. Asiatic lily 'Pixie Crimson', ready for shipping just as the flower buds begin to show color.

not perform well at high temperatures; daytime temperatures above 85° F (29° C) should be avoided.

Bright, sunny conditions generally promote rapid development due to increased leaf temperatures. Conversely, dark, overcast periods delay crop development. Growers should watch the weather each day and adjust the temperature settings accordingly. For example, on sunny days the daytime settings can be reduced 5–10° F (3–5° C) to compensate for the increased leaf temperatures. Growers who do not make these adjustments will most likely observe blooming crops much earlier than expected during sunny springs or much later than anticipated during cloudy growing seasons.

Different cultivars require differing lengths of time to produce. Your bulb supplier should be able to furnish you with the number of days necessary from planting to open flower and the temperatures that are necessary to produce them. Most cultivars take between sixty to eighty days to finish. Remember, these are just guidelines and may vary dramatically from location to location and from year to year. The forcing time tends to decrease as the season progresses. Lilies take longer to force in February than they do in May. The time necessary to produce Asiatic lilies is also affected by the planting date, growing environment, temperature, light level, and crop density.

With most cultivars, it takes about thirty days from visible bud to open bloom. Growers can use this rule of thumb as a guideline to help determine if the temperatures need to be increased or decreased in order to meet the anticipated sales date.

Postharvest

Asiatic lilies are beautiful and draw a lot of attention at the retail site. They practically sell themselves. The most common drawback of producing this crop is the narrow marketing window. Lilies are difficult to sell when they are green—even budded plants are difficult to sell unless they have some color. Once the flower buds swell and show signs of color, their market appeal goes up dramatically. However, each flower typically only lasts a few days, each plant only produces a certain number of flowers, and the plant does not rebloom. There is about a seven-to-ten-day time period where these plants have to get from the greenhouse to the retail site and then to the customer's home. When the flowering is done, the potential to sell them at a premium price is gone.

In the greenhouse, growers can reduce the production temperatures to reduce the rate of flower development and try to hold the crop until the desired ship date. This is sometimes easier said than done, as it is difficult to hold them back during warm, sunny days. Some growers have had success removing lilies in the puffy bud stage (just showing signs of color) from the greenhouse and placing them in a cooler at 38° F (3° C). Placing them in the cooler adds to production costs and reduces the profitability, but it allows growers to ship Asiatic lilies during their desired marketing window.

A few years ago, researchers and growers found a method of treating the plants while they are in the greenhouse that would considerably extend the shelf life of each bloom and the sales window for the crop. By applying the plant growth regulator Fascination (benzyladenine + gibberellins) to the flower buds, growers can increase the shelf life from seven up to twenty-one days. This is the same chemical used to prevent lower leaf yellowing, only it is applied at a different stage of development and at different rates. Growers should spray Fascination at 71.2 oz. per 100 gal. (2.1 L/379 L) to the lily crop just as the first flowers are beginning to show color. This is the labeled rate for this type of application; however, I have observed excellent results using rates from 35.6–71.2 oz. per 100 gal. (1.1–2.1 L/379 L).

References

Balge, R., S. Gill, E. Dutky, W. Maclachlan, and S. Klick. "Production of Asiatic and Oriental Lilies as Cut Flowers." University of Maryland. College of Agriculture and Natural Resources. Bulletin FS-687. 1996.

De Hertogh, G. "Growing Hybrid Lilies." *GrowerTalks*. January 2000.

"Hybrid Lily Program." Fred C. Gloeckner & Company, Inc.

Phlox paniculata

Tall Phlox, Garden Phlox

Phlox paniculata, commonly known as garden phlox or tall phlox, is one of the best known, old-fashioned garden perennials. Phlox is widely grown and used for its colorful, bright, showy flowers during midsummer. The long-lasting, intensely colored 8–10 in. (20–25 cm) flower clusters contain numerous, fragrant, 1–1.5 in. (3–4 cm) five-lobed florets. The flowers consist of various shades and color combinations of blue, lavender, pink, purple, red, salmon, and white.

'David' is the most recognized cultivar of garden phlox and was selected by the Perennial Plant Association as the Plant of the Year in 2002 for its landscape performance and desirable characteristics. Since then, there has been a revived interest from gardeners and landscapers alike in using *Phlox paniculata* as a garden staple. Plant breeders are always selecting cultivars for better garden performance, disease resistance, shorter plant habits, and improved blooming characteristics. The cultivars available today are by far superior to the cultivars grown in decades past.

Figure 12.45. ***Phlox paniculata* 'David'**

Most *Phlox* reach 30–40 in. (76–102 cm) in height. Many cultivars, such as the 'Flame' series have been selected with shorter growth habits. These varieties often reach 12–18 in. (30–46 cm) while they are in bloom.

Phlox paniculata is widely grown throughout USDA Hardiness Zones 3–8 and AHS Heat Zones 8–1. There are slight hardiness and heat tolerance differences between cultivars. Many cultivars perform better in the North than in the South, as they do not tolerate extreme heat that well. Garden phlox prefers full sun, although in the South they perform best when partial shade is provided.

This native American perennial belongs to the Polemoniaceae family. The only other perennial in this family being commercially produced is *Polemonium* (Jacob's ladder). *Phlox* are used as aromatic border plants to attract hummingbirds and butterflies into the garden, for accent plantings, and as cut flowers.

Propagation

Growers use tip cuttings, stem cuttings, root cuttings, or divisions to propagate *Phlox paniculata.* Tip and stem cuttings are the most common means of propagating garden phlox. Although using bare-root divisions provides for a quicker finishing time, it often leads to crop variability and lower survival rates. Propagating *Phlox* with seed is very difficult and does not produce named cultivars. Commercially, seed of 'Beltsville Beauty' and 'New Hybrids Mixed' are offered to the industry. The seed lines are a mix of various colors. To maintain the desirable characteristics of the named cultivars, vegetative propagation is a must.

Commonly Grown *Phlox paniculata* Cultivars

'Amethyst'	'Harlequin'	'Orange Perfection'
'Andre'	'Hesperis'	'Prime Minister'
'Becky Towe'	'Jubilee'	'Purple Eyes'
'Blue Boy'	'Laura''	'Rainbow'
'Blue Paradise'	'Little Boy'	'Red Magic'
'Bright Eyes'	'Little Princess'	'Red Riding Hood'
'Darwin's Joyce'	'Magic Blue'	'Red Super'
'David'	'Mini Star'	'Robert Poore'
'Eden's Crush'	'Miss Elie'	'Salmon Glow'
'Eden's Glory'	'Miss Holland'	'Sandra'
'Eden's Smile'	'Miss Jessica'	'Shortwood'
'Elizabeth'	'Miss Jill'	'Shorty White'
'Ending Blue'	'Miss Kelly'	'Snow White'
'Eva Cullum'	'Miss Mary'	'Starfire'
'Flame' series	'Miss Universe'	'Tenor'
'Flamingo'	'Nicky'	'The King'
'Franz Schubert'	'Nora Leigh'	

Garden phlox can be propagated by vegetative tip or stem cuttings in the spring or early summer. Stem cuttings are similar to tip cuttings, as they consist of one or more nodes but do not contain a growing point. Both stem and tip cuttings should be harvested from stock plants before they are flowering. The survival rate of cuttings taken from stock plants that have initiated flowering is greatly lower than those from vegetative stock plants.

To maintain vegetative growth, the stock plants should be maintained at intermediate day lengths of twelve to thirteen hours. At longer day lengths they are likely to go reproductive, and at shorter photoperiods they could go dormant.

Tip cuttings containing several nodes or stem cuttings containing at least one node and measuring approximately 2 in. (5 cm) in length are generally stuck into plug liners for rooting. The well-drained rooting medium should be moistened prior to sticking. The cuttings are usually dipped in 1,000 ppm IBA to promote root development, increase the rooting percentage, provide uniformity, and decrease the rooting time.

Provide twelve- to thirteen-hour photoperiods during propagation to maintain vegetative growth. The amount of light the cuttings receive during the rooting process greatly affects the time to root and rooting uniformity. *Phlox* cuttings under low light levels (constantly less than 500 f.c. [5.4 klux]) require more time for rooting and root less uniformly. Conversely, rapid, uniform rooting occurs with higher light levels (600–900 f.c. [6.5–9.7 klux] for twelve to thirteen hours each day).

Cuttings should be placed under low misting regimes for about the first seven to ten days of propagation. When possible, it is usually best to propagate under high humidity levels (90% relative humidity) with minimal misting. The misting and the humidity levels can gradually be reduced as the cuttings form calluses and root primordia.

The cuttings are usually rooted in three to four weeks with soil temperatures ranging from 72–76° F (22–24° C). For best results, the air temperature during rooting should be maintained above 65° F (18° C) and below 80° F (27° C). It is beneficial to begin constant liquid feeding with 150 ppm nitrogen at each irrigation as the cuttings are forming roots (usually between fourteen and twenty-one days).

During propagation, *Phlox* are sensitive to many of the chemicals commonly used to control insects and diseases. Unless you have specific experience using these products on *Phlox paniculata* during propagation, it is best to avoid using any chemicals until plants become established. Once rooted, they seem to tolerate normal chemical applications just fine.

Using root cuttings is another common method of propagating garden phlox cultivars. Root cuttings are taken from dormant plants during the late fall or early winter. The roots measuring 0.125–0.25 in. (3–6 mm) thick are cut up into 1–3 in. (3–8 cm) long pieces and planted vertically in open flats. It is important to place the roots in the proper orientation, with the top of the root is facing up and the bottom of the root down. To help identify the root orientation, it is helpful to mark the roots by the way they are cut. For example, cut the top of the roots straight and the bottom of the roots with an angled cut.

The root cuttings can be left at cool temperatures (35–44° F [2–7° C]) for the duration of the winter. Keep the cuttings moist but not wet during this period, as they rot easily under saturated conditions. After the roots have gone through at least six weeks of cold, the temperatures can gradually be increased up to 65° F (18° C) to promote root and shoot development. Following the cold period, it takes eight to ten weeks for rooting to occur.

Production

Phlox paniculata are most commonly planted from rooted liners or bare-root divisions. Most growers prefer to use plug liners, as bare-root divisions often have more problems associated with carryover diseases. They perform best when grown in a moist, well-drained medium with a slightly acidic pH, 6.0–6.5. Many commercially available peat- or bark-based growing mixes work well, provided there is adequate drainage.

Due to their large size when flowering, it is best to produce tall phlox in 1 gal. (3.8 L) or larger containers. When planting, use two 72-cell plugs per 1 gal. (3.8 L) container. Using multiple plugs provides more shoots per pot, creating a fuller product and produces more blooms. If larger sized plug cells are used, planting one plug per pot is often appropriate. Some growers obtain more branches and flowers per pot by pinching plants just prior to or shortly after they are planted in the final container. When planting, be careful to not plant them too deeply, as this could lead to poor plant establishment, crop variability, crown rot, and losses.

Garden phlox performs best under "average" watering regimes. They can be grown slightly drier until they become established, when more normal watering practices usually begin. Unless the growing mix is excessively porous or holds onto too much moisture, normal irrigation practices should suffice. When irrigation is needed, water thoroughly, allowing the medium to dry between waterings.

They can be grown using moderate fertility levels. Fertilizers can be delivered using water-soluble or controlled-release sources. Growers using water-soluble fertilizers either apply high rates (150–200 ppm) of nitrogen as needed or feed with a constant liquid fertilization program using rates of 75–125 ppm nitrogen with every irrigation. Controlled-release fertilizers are commonly applied as a topdressing onto the medium's surface using the medium rate listed on the product's label or incorporated into the growing medium prior to planting at a rate equivalent to 1–1.25 lb. nitrogen per cubic yard (454–567 g/0.76 m^3) of growing medium.

Insects

Aphid	Scale
Beetle	Slug
Caterpillar	Spider mite
Foliar nematode	Spittlebug
Fungus gnat	Stem nematode
Grasshopper	Thrips
Leafhopper	Whitefly
Plant bug	Wireworm

Aphids, spider mites, thrips, and whiteflies are the most common insects that are observed feeding on *Phlox paniculata*. In most cases, these pests can be controlled after they are detected or prevented using proactive strategies. Other insect pests listed in the table are often observed feeding on garden phlox but rarely become problematic. Most insects can be detected with routine crop monitoring. Control strategies may not be necessary unless the scouting activities indicate otherwise.

Diseases

Alternaria	*Pyrenochaeta* stem blight
Anthracnose	*Pythium*
Aster yellows	Powdery mildew
Botrytis	*Rhizoctonia*
Cercospora leaf spot	Rust
Colletotrichum	*Sclerotinia*
Corynebacterium leaf gall	*Sclerotinum*
Downy mildew	*Septoria* leaf spot
Fusarium	Stem canker
Leaf blight	*Verticillium* wilt virus
Mycoleptodiscus crown rot	*Xanthomonas*
Pellicularia crown rot	

Of the plant pathogens *Phlox* are the most susceptible to, powdery mildew is the most problematic to growers. At first, powdery mildew appears as small white, talcum-like colonies on the upper leaf surfaces, but under the right conditions they may engulf the plant with a powdery appearance. To control this disease and most foliar diseases, it is best to manage the environment by providing the proper plant spacing and adequate air movement, controlling the humidity, and, if desired, following a preventative spray program using the appropriate chemicals. In many cases, the incidence of powdery mildew can be greatly reduced by cultivar selection.

Alternaria is another disease that is observed frequently by growers. *Alternaria* can be identified at various stages of infection. The early stages often go undetected, and growers often do not react until the pathogen has seriously infected the plant. At first, this pathogen can be identified as small white spots on the leaves. These spots turn brown, enlarge, and take on a dry appearance, eventually killing the leaf. On some cultivars the leaves turn yellow in the proximity of the infection. This disease usually attacks the middle to lower leaves of the plant. Severe infections often kill the leaves on the lower third of the plant. Growers consider *Alternaria* to be equally as troublesome as powdery mildew. It is best to prevent this disease by providing good air circulation, reducing the free moisture on the leaves, providing proper plant spacing, and ensuring sufficient light levels. Many chemicals work well on a preventative basis, but none will clean up an established infestation. It is highly recommended to implement preventative spray programs, making applications every fourteen days, and rotating products effective at controlling this disease including iprodione (Chipco 26019), chlorothalonil (Daconil Ultrex), and mancozeb (Protect T/O).

When using bare-root starter materials, I recommend drenching the containers after planting with a broad-spectrum fungicide, such as thiophanate-methyl (Cleary's 3336), to reduce pressure from pathogens and to hasten plant establishment. Growers usually observe a higher incidence of leaf spots and other foliar diseases when bare-root starting materials are used. These growers can greatly reduce the occurrence of these pathogens when implementing monthly preventative fungicide spray programs.

Plant Growth Regulators

Phlox may require height control when they are grown under greenhouse and nursery conditions. Providing adequate spacing between the plants will reduce plant stretch caused by competition. The height can also be reduced by withholding water and nutrients. Pinching is a great mechanical method of reducing the overall height of the crop and improving plant quality by producing more lateral branches per plant.

Under certain growing conditions or at high plant densities, it may be necessary to use chemical PGRs. Compared to many perennials, the PGR rates needed to achieve sufficient control on phlox cultivars are relatively high (see chart). For each of these products

PGRs for *Phlox paniculata* Cultivars

Trade Name	Active Ingredient	Northern Rate
B-Nine or Dazide	Daminozide	3,750 ppm
Bonzi, Piccolo, or Paczol	Paclobutrazol	45 ppm
Sumagic	Uniconazole-P	10 ppm
B-Nine or Dazide + Sumagic	Daminozide + uniconazole-P	2,500 ppm + 5 ppm

These are northern rates and need to be adjusted to fit the location where the crop is being produced. These rates are for spray applications and geared to growers using multiple applications.

applied as foliar sprays, it usually requires two or three applications at seven-day intervals to provide adequate height control. Begin the applications early in the crop, as the stems are rapidly elongating.

Forcing

Phlox paniculata naturally bloom during midsummer. To improve sales and marketability, it is often beneficial to force phlox into bloom earlier to capture additional spring sales of this popular perennial.

To promote branching and produce full pots, garden phlox should be bulked up prior to forcing. After the plugs are planted into the final container, they should grown at twelve- to thirteen-hour photoperiods with temperatures of 64–70° F (18–21° C) to keep them actively growing. The length of the bulking period depends on the size of the pot; 1 qt. (1 L) pots may require three weeks of bulking, while 1 gal. (3.8 L) containers might require four weeks, and so on. To promote branching, pinch the plants, leaving six nodes, just prior to or at the time of planting.

Figure 12.46. *Phlox paniculata* 'Laura' in flower and ready for sale

The requirement and benefits of providing a cold treatment varies widely with the cultivar and the environmental conditions in which the stock plants and the rooted liners were grown. Some cultivars do not require vernalization for flowering, while others will not flower without it. For the varieties that have a cold requirement, they will flower if the stock plant they originated from received vernalization in the same year (the effects of receiving cold the previous year does not carryover). It is also acceptable to provide cold to the rooted liners prior to forcing.

All *Phlox*, cold requiring or not, show great benefits from exposure to cold. Following vernalization, they grow more vigorously and tend to flower more rapidly and uniformly than plants without this treatment. I recommend providing at least five weeks of vernalization at temperatures of 35–44° F (2–7° C). During the cold treatment, it is beneficial to provide 50 f.c. (538 lux) of light for twelve hours each day.

Garden phlox cultivars are considered long-day plants, requiring at least fourteen hours of light for flowering. Natural photoperiods of less than fourteen hours will cause them to flower poorly or not at all. Provide fourteen-hour photoperiods or night interruption lighting when the natural photoperiod is less than fourteen hours.

Phlox thrive and will produce the highest quality plants when they are produced in high light environments (minimum 3,000 f.c. [32 klux]). Under low light levels, the size and quantity of flowers per plant are often reduced, and the stems are often weak and cannot support the weight of the flower heads. To improve the quality when the light levels are naturally low, 300–400 f.c. (3.2–4.3 klux) of supplemental lighting should be provided.

The time to bloom after vernalization and the proper photoperiod is provided is a function of temperature. *Phlox paniculata* cultivars grown at 68° F (20° C) will take eleven to twelve weeks to reach flowering, while plants grown at 62° F (17° C) will flower in approximately fourteen weeks. Plant quality and the flowering characteristics are greatly reduced at high temperatures (greater than 74° F [23° C]). To obtain the best plant quality, I recommend producing phlox cultivars at 64–68° F (18–20° C). Plants that have received a cold treatment typically flower one to two weeks earlier than non-cooled plants.

References

Armitage, Allan M. *Herbaceous Perennial Plants: A Treatise on Their Identification, Culture, and Garden Attributes*. Second edition. Champaign, Ill.: Stipes Publishing. 1997.

Enfield, A., E. Runkle, R. Heins, and A. Cameron. "Herbaceous Perennials: *Phlox paniculata.*" *Greenhouse Grower.* June 2003.

Nau, Jim. *Ball Perennial Manual: Propagation and Production*. Batavia, Ill.: Ball Publishing. 1996.

Platycodon grandiflorus

Balloon Flower

Platycodon grandiflorus is a popular but underutilized perennial in the American landscape. The common name, balloon flower, refers to the puffy balloon-like flower buds of this plant before they open. Most cultivars are suitable for production in USDA Hardiness Zones 3–8 and AHS Heat Zones 8–1. There may be slight hardiness and heat tolerance differences between cultivars. Balloon flowers prefer to be grown in full sun, although in the South they perform best when grown under partial shade.

Certain cultivars have a compact habit, forming small floriferous mounds that reach 8–10 in. (20–25 cm) tall, while others grow to 20 in. (41 cm) in height. Balloon flower cultivars are commonly used as accent plants, border plants, in rock gardens, container plantings, and as potted houseplants. The taller cultivars, such as the 'Fuji' series and 'Mariesii', are commonly used as cut flowers. They naturally bloom during the midsummer. The balloon-like flower buds open into single or fully double bell-shaped flowers with blue, pink, or white colorations.

Figure 12.47. ***Platycodon grandiflorus* 'Fairy Snow'**

Jelitto Perennial Seeds

Platycodon species are native to eastern Asia and belong to the Campanulaceae family. There are several perennials in this family being commercially produced, including *Adenophora, Campanula, Codonopsis, Jasione,* and *Lobelia.*

Commonly Grown *Platycodon grandiflorus* Cultivars

'Astra' series	'Hakone White'
'Double Blue'	'Mariesii'
'Double White'	'Misato Purple'
'Fairy Snow'	'Miss Tilly'
'Fuji' series	'Sentimental Blue'
'Hakone Blue'	'Shell Pink'

Propagation

Platycodon grandiflorus cultivars are propagated from seed. Sow seeds in the intended plug flat and cover them lightly with germination mix or medium-grade vermiculite. The covering helps to maintain a suitable environment around the seed during this phase. Light is necessary for germination. As long as the covering is put on the flats lightly, there is generally enough light available to satisfy this requirement. It is recommended to provide 10–100 f.c. (0.1–1.1 klux) of light until the seeds have germinated. The seed flats should be moistened and moved to a warm environment, where the temperatures can be maintained at 68–72° F (20–22° C) for germination. Many growers utilize germination chambers during this stage to provide uniform moisture levels and temperatures. It is important to keep the medium very moist but not saturated during this stage or the germination rates are likely to decrease. Using germination chambers is optional since *Platycodon* will successfully germinate in the greenhouse when the proper temperatures and moisture levels are provided.

The seeds should be germinated in seven to fifteen days. Some growers have found it beneficial to cover the emerging seedlings with medium-grade vermiculite after germination to help maintain moisture levels. Following germination, reduce the moisture levels somewhat, allowing the growing medium to dry out slightly before watering to help promote rooting. Fertilizers are usually applied once the true leaves are present, applying 100 ppm nitrogen every third irrigation or 50 ppm with every irrigation, using a balanced water-soluble source. When plugs are grown at 65° F (18° C), they are usually ready for transplanting in eight to ten weeks.

The light levels can gradually be increased as the plug develops, from 500 f.c. (5.4 klux) following germination to 2,500 f.c. (27 klux) at the final stage. When producing plugs during naturally short days, it is best to provide long days using day extension or night interruption lighting. Balloon flowers are very sensitive to short days and will naturally begin to go dormant when the day length is below nine hours.

Production

For container production, *Platycodon grandiflorus* is suitable for 1 qt. (1 L) to 1 gal. (3.8 L) containers. Most growers receive starting materials of balloon flowers as finished plug liners. When planting large containers, such as 1 gal. (3.8 L) pots, plant at least two plug cells per container to properly fill out the pot. Balloon flowers perform best when they are grown in a moist, well-drained medium with a slightly acidic pH, 5.5–6.0. Most commercially available peat- or bark-based growing mixes work well, provided there is adequate drainage.

When planting, the plugs should be planted so the original soil line of the plug is even with or just below the surface of the growing medium of the new container. The fleshy tap root is susceptible to root rots, particularly after transplanting or when overly wet conditions occur.

After planting, I recommend drenching with a broad-spectrum fungicide such as etridiozole + thiophanate-methyl (Banrot) or the combination of mefenoxam (Subdue Maxx) with thiophanate-methyl (Cleary's 3336). The best quality is achieved when plants are grown in full sun or in greenhouses with high light intensities.

It is beneficial to pinch balloon flowers prior to or shortly after they are planted in the final container. Pinching increases lateral branching and the total number of flowers produced on each plant. Flowering will be delayed three to four weeks when *Platycodon grandiflorus* are pinched; therefore, growers should modify their crop schedules or ready dates accordingly.

Platycodon can be grown using light to moderate fertility levels. Fertility can be delivered using water-soluble or controlled-release fertilizers. Growers using water-soluble fertilizers either apply 100–150 ppm nitrogen as needed or feed with a constant liquid fertilization program using rates of 50–75 ppm nitrogen with every irrigation. Growers commonly apply time-release fertilizers as a topdressing on the medium's surface using the medium rate or incorporated into the growing medium prior to planting at a rate equivalent to 1 lb. nitrogen per cubic yard (454 g/0.76 m^3) of growing medium.

Balloon flowers require an average amount of irrigation, as they do not tolerate very wet or overly dry conditions. Root zones that remain waterlogged tend to get root rot pathogens and can quickly lead to crop losses. Overly dry growing conditions greatly reduce crop quality and delay flowering. Growers will usually observe yellowing of the leaves if the plants have recently wilted due to moisture stress. When irrigation is necessary, water them thoroughly, allowing the soil to dry slightly between waterings.

Insects

Aphid	Thrips
Leaf miner	Whitefly
Spider mite	

Balloon flowers are relatively free of serious problems associated with insects. Occasionally, aphids, spider mites, and whiteflies may appear, causing only a minimal amount of crop injury. Of these insect pests, aphids are the most prevalent. None of these insect pests require preventative control strategies, and all can be controlled after they are detected or can be prevented using proactive strategies. Most insects can be detected with routine crop monitoring. Control strategies may not be necessary unless the scouting activities indicate actions should be taken.

Diseases

Alternaria	*Pyrenochaeta* stem blight
Botrytis	*Rhizoctonia*
Heterosporium stem canker	*Sclerotinia*
Phoma	*Sclerotium*

Balloon flowers can generally be grown free of plant pathogens. The primary diseases growers should watch for are *Rhizoctonia* crown rot and *Heterosporium* stem canker. They are most susceptible to these diseases when they are grown under cool, wet conditions, such as during the fall just prior to entering the winter dormancy. *Botrytis* is another disease that could become problematic. *Botrytis*, like *Heterosporium* and *Rhizoctonia*, often occurs around the overwintering process but is also likely to occur under dense plant canopies while the plants are actively growing.

To control these diseases, it is best to manage the environment by providing proper plant spacing and adequate air movement and controlling the humidity. Growers should carefully watch the moisture levels during adverse times of the year and avoid overwatering their plants. Many growers apply a broad-spectrum fungicide drench before and just after the overwintering period to prevent *Rhizoctonia*. Stem cankers can be controlled by applying preventative sprays of trifloxystrobin (Compass) or thiophanate-methyl (Cleary's 3336) when the conditions for the disease are optimal.

Plant Growth Regulators

Platycodon grandiflorus cultivars have various growth habits. Cultivars such as 'Sentimental Blue' have naturally compact growth habits and will usually not require plant growth regulators to control plant height. Other varieties, such as 'Double Blue', have taller growth habits and may need height control while they are being commercially produced. During the winter months, periods of low light levels, or when grown at high plant densities, excessive plant stretching might occur, requiring some type of height management strategy. Most commercially available growth regulators are effective at controlling plant height of balloon flowers. Before applying these chemicals, the height can often be effectively controlled by providing adequate spacing between the plants and by withholding water and nutrients. For toning and shaping purposes, one application is often adequate. When additional height control is necessary, make a second application seven to ten days following the first.

Forcing

For container production, *Platycodon grandiflorus* cultivars are well suited for forcing into bloom throughout the year. When following a few guidelines, growers can successfully produce uniform, consistent, high-quality flowering plants.

PGRs for *Platycodon grandiflorus* Cultivars

Trade Name	Active Ingredient	Northern Rate
A-Rest	Ancymidol	25 ppm
B-Nine or Dazide	Daminozide	2,500 ppm
Bonzi, Piccolo, or Paczol	Paclobutrazol	30 ppm
Sumagic	Uniconazole-P	5 ppm
B-Nine or Dazide + Sumagic	Daminozide + uniconazole-P	2,000 ppm + 3 ppm

These are northern rates and need to be adjusted to fit the location where the crop is being produced. These rates are for spray applications and geared to growers using multiple applications.

Balloon flowers do not have a cold requirement that must be met prior to forcing. They will easily bloom from plants started by seed during the first growing season. However, crops that have received a cold treatment will often generate numerous shoots from the swollen taproot, which ultimately produces more flowers on each plant than when uncooled plugs are used. Established containers of balloon flowers can successfully be overwintered. Growers have had difficulties overwintering plug trays and, in many cases, have experienced significant crop losses when attempting to do so. Regardless of the size of container being overwintered, it is important to avoid moisture extremes and not allow them to become overly wet or dry.

Platycodon are very sensitive to cold temperatures, which causes the foliage to die back quickly. Because they naturally emerge late in the spring, similar to hibiscus, many growers get nervous about losing the crop. Eventually the shoots begin to sprout, often producing more shoots than anticipated, creating full pots and numerous blooms. Once they are provided warm temperatures, such as those recommended below, balloon flowers will readily break dormancy and growth will commence.

Figure 12.48. *Platycodon grandiflorus* 'Sentimental Blue' grown at an outdoor production facility are ready for shipping.

Balloon flowers do not have a juvenility period and have been observed flowering with as few as thirteen leaves present. They can be forced into flower under any photoperiod, though they are long-day beneficial plants and will flower seven to ten days faster under long days as compared to plants grown under short days.

The primary factor for flowering is temperature. Balloon flowers grow relatively slowly compared to other perennials. When warm production temperatures (greater than 70° F [21° C]) are provided, plant development is hastened, reducing the production time. Platycodon cultivars grown at 65°F (18° C) will take twelve to thirteen weeks to reach flowering, while plants grown at 75° F (24° C) will flower in nine to ten weeks.

Platycodon grandiflorus prefer to be grown at high light levels. Where possible, growers should place them in an environment with no less than 4,000 f.c. (43 klux) of light. Plants grown at low light levels tend to be of lower quality, becoming elongated and not flowering as profusely as those produced at ambient levels. During the winter months, providing 400–500 f.c. (4.3–5.4 klux) of supplemental lighting helps to improve the quality of the crop.

References

Greenhouse Grower Magazine and Michigan State University. *Firing Up Perennials: The 2000 Edition.* Willoughby, Ohio: Meister Publishing. 2000.

Nau, Jim. *Ball Perennial Manual: Propagation and Production.* Batavia, Ill.: Ball Publishing. 1996.

Rudbeckia fulgida

Black-eyed Susan

Black-eyed Susan is a popular perennial garden plant. The long-lasting display of golden, daisy-like flowers with a distinctive dark brown central cone delivers an impressive display to the landscape. 'Goldsturm' is the most widely grown cultivar of *Rudbeckia fulgida* and was selected by the Perennial Plant Association as the Plant of the Year in 1999 for its landscape performance and desirable characteristics. Since then, the popularity of *Rudbeckia fulgida* and other species, namely *R. hirta,* has skyrocketed, making black-eyed Susans a mainstay in sunny perennial gardens across the country.

Figure 12.49. ***Rudbeckia fulgida* 'Goldsturm** Goris Passchier

Rudbeckia belongs to the Asteraceae family (formerly Compositae). There are numerous perennials in the Asteraceae family being commercially produced, including *Achillea, Aster, Bellis, Centaurea, Coreopsis, Doronicum, Echinops, Echinacea, Erigeron, Gaillardia, Helenium, Heliopsis, Leucanthemum, Ligularia,* and *Stokesia.*

Black-eyed Susans perform well across a wide portion of the United States, throughout USDA Hardiness Zones 4–10 and AHS Heat Zones 9–2. They prefer full sun, although in the South they perform best when partial shade is provided. In the landscape, *Rudbeckia fulgida* reaches 24–36 in. (61–91 cm) in height. This American native is used as an accent plant, border plant, in mass plantings, and makes an excellent cut flower.

Propagation

Growers use seed and divisions to propagate *Rudbeckia fulgida.* Although using bare-root divisions provides for a quicker finishing time, it often leads to crop variability and lower survival rates. Propagation from seed is the most common method used by commercial growers since it is less expensive than bare-root divisions.

When propagating *Rudbeckia* from seed, it is beneficial to stratify the seed prior to germination. Stratification is useful to overcome seed dormancies and will increase the uniformity and rate of germination. The seeds should be sown in the intended plug flat and covered lightly with germination mix or medium-grade vermiculite. The covering helps to maintain a suitable environment around the seed during this phase. The seed flats should be moistened and moved to an environment, such as a cooler, where the temperatures can be maintained at 40° F (4° C) for four weeks. After the cold treatment, move the plug trays into a warm environment (75–85° [24–29° C]) for germination. Many growers utilize germination chambers during this stage to provide uniform moisture levels and temperatures. Germination begins within a few days, but it often takes a couple of weeks for all of the seeds to germinate. Following germination, they can be grown at 70° F (21° C) for eight to nine weeks, until they are ready for transplanting.

Jelitto Perennial Seeds has developed a procedure for eliminating the stratification requirement of many perennials, allowing for quick and uniform germination without the need to provide a cold treatment. Their patented process is offered to the industry as Jelitto Gold Nugget Seed, and at the time of writing they offered over a hundred varieties of perennials. including *Rudbeckia* 'Goldsturm'. This pretreated seed eliminates the four weeks of cold recommended above and reduces the finish time to seven weeks, nearly cutting in half the total time necessary to produce 'Goldsturm' from traditional, untreated seeds.

Production

Rudbeckia fulgida perform best when grown in a moist, well-drained medium with a pH at 5.8–6.4. They perform well in many commercially available peat- or bark-based growing mixes, provided there is good water-holding ability and adequate drainage. Due to their large size when flowering, it is best to produce them in 1 gal. (3.8 L) or larger containers.

Black-eyed Susans prefer to be grown in a moist growing medium. Water as needed when the plants are young and becoming established. Once they are large, they will require more frequent irrigations, as they will dry out rather quickly. Under stressful growing conditions, such as warm temperatures and high light levels, they wilt very easily. Generally, if they are watered within a reasonable amount of time after they have begun to wilt and if the water stress was not severe, they will recover quickly. In extreme cases, leaf necrosis or tip burn may occur. When irrigation is needed, water them thoroughly, ensuring the entire growing medium is wet or nearly saturated. It is best to allow the growing medium to dry slightly between irrigations.

They are moderate feeders. Fertility can be delivered using water-soluble or controlled-release fertilizers. Growers using water-soluble fertilizers either apply at high rates (200–300 ppm) of nitrogen as needed or feed with a constant liquid fertilization program at rates of 75–150 ppm nitrogen with every irrigation. Controlled-release fertilizers are commonly applied as a topdressing onto the medium's surface using the medium recommended rate on the fertilizer label or incorporated into the growing medium prior to planting at a rate equivalent to 1–1.25 lb. nitrogen per cubic yard (454–567 g/0.76 m^3) of growing medium.

Insects

Aphid	Mealybug
Caterpillar	Slug
Four-lined plant bug	Spider mite
Grasshopper	Spittlebug
Japanese beetle	Thrips
Leafhopper	Whitefly

Generally, *Rudbeckia fulgida* can be produced relatively insect free. Aphids and whiteflies occasionally will become problematic. These pests can be controlled after they are detected or prevented using a proactive strategy. The other insect pests listed in the table are often observed feeding on black-eyed Susans but rarely become problematic. Most insects can be detected with routine crop monitoring. Control strategies may not be necessary unless the scouting activities indicate actions should be taken.

Diseases

Aster yellows	*Pseudomonas*
Botrytis	*Pythium*
Botrytis blight	*Rhizoctonia*
Cercospora leaf spot	Rust
Colletotrichum (anthracnose)	*Sclerotinia*
Downy mildew	*Sclerotium*
Entyloma (smut)	*Septoria* leaf spot
Macrophomina (charcoal rot)	*Synchytrium* (leaf gall)
Pellicularia (crown rot)	*Verticillium*
Powdery mildew	*Xanthomonas*

Plant diseases may be observed when environmental conditions are favorable for their development. The most common diseases observed attacking *Rudbeckia fulgida* crops are downy mildew and *Septoria* leaf spot. As with many perennials, the occurrence of plant diseases can be negated or greatly reduced when the proper cultural practices are followed. To control foliar diseases, it is best to manage the environment by providing proper plant spacing and adequate air movement, controlling the humidity, and, if desired, following a preventative spray program using the appropriate chemicals.

In some parts of the country, downy mildew is becoming a major disease of black-eyed Susans. Downy mildew usually appears first on the undersides of the leaves as a mass of white or gray spores, and often the upper leaf surface (directly above where the spores are observed) will appear mottled, discolored, or blistered. To control this disease, it is best to manage the environment by providing proper plant spacing and adequate air movement, controlling the humidity, and watering early in the day, which allows the foliage to be dry before night. Many chemicals work well on a preventative basis, but none will clean up an established infestation. Preventative fungicide applications act as a barrier, not allowing the disease to infect the plant. The best strategy is to rotate products, such as fosetyl-aluminum (Aliette), trifloxystrobin (Compass), azoxystrobin (Heritage), copper-hydroxide + mancozeb (Junction), mancozeb (Protect T/O), dimethomorph (Stature DM), and mefenoxam (Subdue), making applications every seven to ten days beginning at the onset of favorable conditions for this disease.

Some growers have observed plants with cupped leaves or deformed terminal buds, leaves, and inflorescences. These abnormalities are the result of a calcium deficiency. In most cases, the symptoms occur following cloudy conditions or after periods of high relative humidity, which greatly decreases the translocation of calcium up the plant. Many growers have been successful applying foliar sprays of calcium chloride at 100 ppm every seven to ten days during the growing season or during periods were the translocation of calcium is greatly reduced.

Plant Growth Regulators

Black-eyed Susans may require height control when they are grown under greenhouse and nursery conditions. Providing adequate spacing between the plants will reduce plant stretch caused by competition. Under certain growing conditions or at high plant densities, it may be necessary to use chemical plant growth regulators. Compared to many perennials, the PGR rates needed to achieve sufficient control are relatively high (see chart). For foliar applications, it usually requires two or three applications at seven-day intervals to provide adequate height control. Begin the applications when the flower stalks are near the leaf canopy, as they are beginning to elongate or bolt.

Forcing

Rudbeckia fulgida naturally bloom in mid to late summer. Unfortunately, most perennial sales occur in the late spring. To extend the shipping season of flowering black-eyed Susans, they can be forced to bloom throughout the year.

They have a juvenile period where they will not flower, even if all of the other requirements for flowering are met, until they are mature. Plants that have at least ten leaves will flower successfully, while those with less will remain vegetative, flower sporadically, or take an extended period to reach bloom. Grow plants to maturity using short days, or photoperiods no longer than twelve hours, until the plants have an average of at least ten leaves. Temperatures of 70–75° F (21–24° C) will promote rapid development during this growth phase. Once they are mature and have been provided the proper photoperiod for flowering, they will develop an additional twelve to fifteen leaves before the first flower bud.

Rudbeckia do not require a cold treatment for flowering. However, they are considered to be cold beneficial, as flowering will occur two to three weeks earlier following a cold period. I

PGRs for *Rudbeckia fulgida*

Trade Name	Active Ingredient	Northern Rate
A-Rest	Ancymidol	50 ppm
B-Nine or Dazide	Daminozide	3,750 ppm
Bonzi, Piccolo, or Paczol	Paclobutrazol	45 ppm
Sumagic	Uniconazole-P	10 ppm
B-Nine or Dazide + Cycocel	Daminozide + chlormequat chloride	2,500 ppm + 1,250 ppm
B-Nine or Dazide + Sumagic	Daminozide + uniconazole-P	2,500 ppm + 5 ppm

These are northern rates and need to be adjusted to fit the location where the crop is being produced. These rates are for spray applications and geared to growers using multiple applications.

recommend vernalizing black-eyed Susans for a minimum of ten weeks at 35–41° F (2–5° C). They can be vernalized as a plug or in the final container. Regardless of the container size, make sure they are fully rooted and past the juvenile stage prior to exposing them to cold temperatures.

They are considered to be obligate long day plants, absolutely requiring long days for them to flower. With photoperiods of less than thirteen hours, plants not receiving a cold treatment will remain vegetative. If plants have undergone a cold treatment, flowering occurs when the photoperiod is greater than thirteen hours. I recommend providing at least fourteen-hour photoperiods or night interruption lighting when the natural photoperiod is less than fourteen hours. During naturally short days, using high-pressure sodium lights to deliver a four-hour night interruption between 10 P.M. and 2 A.M. is effective in promoting flowering. Using incandescent light sources will promote flowering but will cause the internodes to stretch excessively, reducing the overall quality of the crop. Research has shown that extending the duration of the night interruption to more than four hours will provide more rapid and uniform flowering.

The time to bloom after vernalization and the proper photoperiod is provided is a function of temperature. *Rudbeckia fulgida* grown at 68° F (20° C) will take twelve to thirteen weeks to reach flowering, while plants grown at 60° F (16° C) will flower in sixteen to eighteen weeks. The size and number of flowers produced decreases with increasing temperatures. To obtain the best plant quality, I recommend producing them at 65–68° F (18–20° C). Black-eyed Susans not receiving a cold treatment typically take two to four weeks longer (depending on temperature) than the durations listed above.

To obtain the highest quality finished plants, particularly for large container sizes, growers should consider planting rooted liners in the desired pot during the late summer of the year prior to the intended date of sale. Established containers should be overwintered in a protected area prior to spring forcing. Planting liners of *Rudbeckia fulgida* in this manner will allow them to bulk up, produce more flowers per plant, and bloom earlier than when they are planted and grown only in the spring. When small container sizes, such as 1 qt. (1 L) pots, are desired, it is usually best to transplant vernalized plugs into the final container and immediately provide the proper photoperiod.

References

Greenhouse Grower Magazine and Michigan State University. *Firing Up Perennials: The 2000 Edition.* Willoughby, Ohio: Meister Publishing. 2000.

Salvia x *sylvestris*

Meadow Sage

There is often great confusion between the parentage of several commercially produced cultivars of *Salvia*, namely those originally associated with *Salvia nemorosa, S. superba, and S.* x *sylvestris.* Although there are often distinguishing characteristics between these species, the cultivars associated with them are often

Figure 12.50. ***Salvia* x *sylvestris* 'Caradonna'** Walter's Gardens

classified as *Salvia* x *sylvestris*. *Salvia* belongs to the Lamiaceae, or mint, family, which contains numerous commercially produced perennials, including *Agastache, Ajuga, Lamium, Lamiastrum, Lavandula, Monarda, Nepeta, Stachys,* and *Thymus.*

Commonly Grown of *Salvia* x *sylvestris* Cultivars

'Blue Hill'	'May Night'
'Blue Queen'*	'Plumosa'
'Caradonna'	'Rose Queen'*
'East Friesland'	'Snow Hill'
'Lubeca'	'Viola Klose'
'Marcus'	'Violet Queen'*

*Started by seed

Salvia, commonly called meadow sage, is best recognized by the distinctive, upright spikelike inflorescences. The coloration of the blooms is mostly various shades of blue and violet purple, though cultivars with rose or white flowers are also available. 'May Night' is the most recognized cultivar and was selected by the Perennial Plant Association as the Plant of the Year in 1997 for its landscape performance and desirable characteristics. Since then, 'May Night' and other cultivars have become increasingly popular in American landscapes.

Meadow sages can be used as accents plants, in borders, mass plantings, containers, and as cut flowers. In addition to the attractive, often scented foliage and colorful flowers, one of the main benefits of growing *Salvia* is the hummingbirds and butterflies they attract. They naturally bloom during the late spring and early summer. Removing the flower spikes after they have faded will encourage continuous flowering throughout the summer.

Meadow sage performs well across a wide portion of the United States, throughout USDA Hardiness Zones 4–9 and AHS Heat Zones 12–1. There are hardiness and heat tolerance differences between cultivars. For best flowering, *Salvia* generally prefer direct sun, but in the southern United States filtered shade is desirable. Once established, most cultivars are quite drought tolerant.

Propagation

Depending on the cultivar, *Salvia* is propagated by vegetative tip cuttings or bare-root divisions in the spring and early summer or by seed any time of the year. Tip cuttings should be harvested from stock plants before they are flowering. Cuttings that are already in bloom will take longer to root and will have a lower survival rate than purely vegetative starting materials. During the summer heat, it is difficult to propagate meadow sages; they will often sit in the production flat for a long time without initiating roots, turn yellow, and may eventually die.

Tip cuttings containing several nodes and measuring approximately 2 in. (5 cm) in length are generally stuck into plug liners for rooting. The well-drained rooting medium should be moistened prior to sticking. The base of the cuttings can be dipped into a rooting hormone, such as a solution of indolebutyric acid (IBA), at rates of 500–1,000 ppm prior to sticking. *Salvia* can successfully root without rooting compounds but tend to root slightly faster and more uniformly when these treatments are provided.

Cuttings should be placed under low misting regimes for about the first seven to ten days of propagation. It is usually best to propagate under high humidity levels (90% relative humidity) with minimal misting. The misting can gradually be reduced as the cuttings form calluses and root primordia. The cuttings are usually rooted in less than three to four weeks with soil temperatures ranging from 68–74° F (20–23° C). For best results, the air temperature during rooting should be maintained above 60° F (16° C) and below 80° F (27° C). It is beneficial to begin constant liquid feeding with 150 ppm nitrogen at each irrigation beginning ten days from sticking.

For the most successful vegetative propagation, harvest the cuttings before flowering occurs or produce stock plants under conditions that do not promote flowering. Stock plants should be produced under short days or with ten to twelve hours of light per day. To maintain short-day conditions, it is often necessary to pull black cloth over the crop daily, providing a dark period for a minimum of twelve hours.

Meadow sages are relatively easy to propagate from seed. Sow the seeds in the intended plug flat and cover them lightly with germination mix or vermiculite. The covering helps to maintain a suitable environment around the seeds during this phase, but they will also germinate successfully without the covering. The seed flats should be moistened and moved to a warm environment, where the temperatures can be maintained at 68–72° F (20–22° C) for germination. Placing the plug flats in a germination chamber will improve the germination rate and decrease the time to germinate, but is not necessary to successfully produce *Salvia* from seed. At these temperatures, germination will occur in about seven to ten days.

Following germination, reduce the moisture levels somewhat, allowing the growing medium to dry out slightly before watering to help promote rooting. The light levels can gradually be increased as the plug develops, from 500 f.c. (5.4 klux) following germination to 2,500 f.c. (27 klux) at the final stage. Fertilizers are usually applied once the true leaves are present, applying 100–150 ppm nitrogen every third irrigation using a balanced water-soluble source. When plugs are grown at 65° F (18° C), they are usually ready for transplanting in five to seven weeks.

Production

Salvia is planted from plugs, rooted liners, or bare-root divisions. They are commonly produced in 1 qt. (1 L) or 1 gal. (3.8 L) containers. The type of starter material used varies largely depending on the cultivar, availability, and grower preference. Growers may also utilize different starter materials depending on the anticipated ready or ship date. For example, a grower expecting to achieve flowering by the April 1 may opt to start the finished container in the late summer of the previous year using plugs or liners as the starting materials, bulk the container up, and provide the cold treatment prior to spring forcing. The same grower might consider using bare-root materials planted in late winter to achieve flowering in the finished container by May 1. And for flowering on June 1, this grower could use vernalized plugs or liners. Fortunately, there are a lot of options available, providing growers with some flexibility when laying out the crop schedules.

Salvia perform best in a moist, well-drained medium with a pH at 5.6–6.2. Most peat- or bark-based mixes are acceptable, provided there is adequate drainage. They do not tolerate excessive amounts of water; when overly moist or wet conditions occur, they are very susceptible to crown and root rots. Unless the growing mix is excessively porous or holds onto too much moisture, normal irrigation practices should suffice. When providing irrigation, water thoroughly and let them dry out between waterings.

They are light to moderate feeders, requiring only modest amounts of fertility. Generally when planting *Salvia,* I incorporate a controlled-release fertilizer into the growing medium at a rate equivalent to 0.75 lb. nitrogen per cubic yard (340 g/0.76 m^3) of growing medium. Another method to deliver fertility to this crop would be using a constant liquid fertilizer program, delivering 75–100 ppm nitrogen to the crop at every irrigation.

Meadow sages prefer to be grown at high light levels. Where possible, growers should place them in an environment with no less than 4,000 f.c. (43 klux) of light. Plants grown at low light levels tend to be of lower quality, as they become elongated and do not flower as profusely as those produced at ambient levels. During the winter months, providing 400–500 f.c. (4.3–5.4 klux) of supplemental lighting helps to improve the quality.

Insects

Aphid	Scale
Caterpillar	Slug
Foliar nematode	Spider mite
Grasshopper	Spittlebug
Leafhopper	Whitefly
Root knot nematode	

Of the insects listed above, the two-spotted spider mite is usually the most cumbersome and difficult to control. Unless regular scouting occurs, the presence of spider mites often goes undetected until they have caused significant amounts of plant injury. At first glance the injury to the leaves might be confused with a nutritional deficiency, as from a distance the leaves appear to be turning yellow. Looking more closely, you will observe the damaged leaves are stippled with small yellowish to silver-gray speckles. Spider mites are usually found feeding on the undersides of the leaves. Controlling spider mites is not easy because it is difficult to deliver the chemicals to the lower leaf surfaces where they are feeding. Several mite life stages present at any time, and they build resistance quickly to pesticides. I have found success by rotating chemical classes at every application, tank-mixing ovicides such as hexythiazox (Hexygon) and clofentezine (Ovation) with adulticides such as abamectin (Avid) and bifenazate (Floramite), and by ensuring good coverage of the sprays to the crop. Always follow the chemical labels for rates, application frequency, and the total number of applications allowed per crop.

Diseases

Alternaria	*Pseudomonas*
Botrytis	*Pythium*
Corynespora (leaf spot)	*Rhizoctonia*
Downy mildew	Rust
Macrophomina (black canker)	*Sclerotium*
Powdery mildew	*Sphaeropsis* (stem rot)
Phytophthora	*Verticillium*

There are relatively few diseases affecting the production of *Salvia,* and seldom does significant plant injury or loss occur. As mentioned earlier, crown and root rots are likely to occur, especially when grown under wet conditions. *Botrytis* outbreaks in the foliage are often observed in situations where there is a dense plant canopy, little air movement, and when water remains on the leaves for long durations.

Plant Growth Regulators

Controlling the plant height may be required while producing containerized *Salvia* under greenhouse conditions. Before using chemicals to reduce stem elongation, it is usually beneficial to provide adequate space between each plant, which will reduce the competition between plants for light and prevent the plants from growing taller. Under certain growing conditions or under high plant densities, it may be necessary to use chemical plant growth regulators to reduce or control plant height. If chemical PGRs are required, daminozide has shown the most effectiveness. In the northern parts of the country, I would recommend beginning with an application rate of 2,500 ppm, applying it two or three times at weekly intervals. In other locations, it might be necessary to apply the weekly applications beginning with a higher rate.

Forcing

Producing flowering *Salvia* x *sylvestris* out of season is relatively easy, provided a few guidelines are followed. Generally, I recommended vernalizing plugs or small containers of *Salvia*

Figure 12.51. *Salvia* x *sylvestris* 'May Night' in full bloom, produced under long days inside a Quonset structure

PGRs for *Salvia* x *sylvestris* Cultivars

Trade Name	Active Ingredient	Northern Rate
A-Rest	Ancymidol	25–50 ppm
B-Nine or Dazide	Daminozide	2,500 ppm
Bonzi, Piccolo, or Paczol	Paclobutrazol	30 ppm
Cycocel	Chlormequat chloride	1,250 ppm
Sumagic	Uniconazole-P	5 ppm
B-Nine or Dazide + Sumagic	Daminozide + uniconazole-P	2,000 ppm + 3 ppm

These are northern rates and need to be adjusted to fit the location where the crop is being produced. These rates are for spray applications and geared to growers using multiple applications.

for six to nine weeks at 35–41° F (2–5° C). Meadow sages are cold beneficial; the cooling period enhances uniformity, reduces the time it takes to reach flowering, and improves the overall quality of the crop. If you provide the cold treatment to the finished containers, bulk them under natural short days for four weeks for 1 qt. (1 L) containers or six weeks when 1 gal. (3.8 L) pots are produced. Seed propagated varieties, such as 'Blue Queen', will flower the first year without receiving a cold treatment. Shorter plants, greater uniformity of bloom, and decreased crop forcing times are observed when cold treatments are provided. Regardless, there is no minimum size requirement for *Salvia* to perceive the cold treatment.

After the cooling is achieved, provide photoperiods of sixteen hours by extending the day, if necessary, or using a four-hour night interruption during the middle of the night, providing a minimum of 10 f.c. (108 lux) of light at plant level. Although flowering will occur under any photoperiod, they are considered to be long-day beneficial plants and will flower quicker with better flowering characteristics under long days, regardless of the vernalization time. The time it takes to reach flowering depends on the growing temperature after the plants are placed under long-day conditions. Plants grown at 64° F (18° C) will flower in about eight weeks, while plants grown at 68° F (20° C) will flower in as little as six weeks.

References

Nau, Jim. *Ball Perennial Manual: Propagation and Production*. Batavia, Ill.: Ball Publishing. 1996.

Pilon, Paul. Perennial Solutions: *"Salvia nemorosa* 'Caradonna'." *Greenhouse Product News*. March 2003.

Scabiosa columbaria

Pincushion Flower

Scabiosa columbaria, commonly known as pincushion flower, is a popular, compact, and prolific flowering perennial that draws interest in garden centers and landscapes everywhere. The common name describes the stamens, which stand above the petals and resemble pins stuck in a pincushion. These clump-forming perennials are often utilized in landscape beds, combination planters, or as specimen plants.

Figure 12.52. *Scabiosa columbaria* 'Misty Butterflies'

Jelitto Perennial Seeds

The compact, finely divided gray-green foliage form rosettes, often reaching 6–8 in. (15–20 cm) in height. Beginning in late spring and continuing throughout the summer, *Scabiosa* produce delicate lavender-blue or pink flowers on slender 12–16 in. (30–41 cm) stems. Flowering is especially extended when the spent blooms are removed. Usually the clumps maintain a nice habit throughout the year, but under shady conditions the plants have a tendency to sprawl and appear unruly. The blooms may be used as cut flowers or to attract both butterflies and hummingbirds to the garden.

Scabiosa prefer full sun, although in the South they perform best when grown under partial shade. They are both cold and heat hardy in USDA Hardiness Zones 3–10 and AHS Heat Zones 9–1, though there are hardiness and heat tolerance differences between cultivars. In general, pincushion flowers are rather short lived in the garden, often only surviving a couple of years.

Commonly Grown
***Scabiosa columbaria* Cultivars**

'Baby Blue'*	'Misty Butterflies'*
'Baby Pink'*	'Nana'*
'Blue Buttons'	'Pink Buttons'
'Blue Diamonds'*	'Pink Mist'
'Butterfly Blue'	

*Propagated from seed

Scabiosa belongs to the Dipsacaceae family, which does not contain many perennial species being commercially produced. After *Scabiosa, Knautia* is the next most popular perennial being grown from this family.

'Butterfly Blue' is the most recognized cultivar of *Scabiosa columbaria* and was selected by the Perennial Plant Association as the Plant of the Year in 2000 for its landscape performance and desirable characteristics. Since then, 'Butterfly Blue' and other cultivars have become increasingly popular and widely used by gardeners and landscapers across the country.

Propagation

Scabiosa cultivars are commercially propagated by seed or vegetatively by tip cuttings. It is also possible to propagate pincushion flowers by division. Bare-root divisions often provide growers with quicker finishing times but often lead to crop variability, carryover diseases, and lower survival rates. The propagation method is dependent on the cultivar, availability, cost, and grower preference.

Starting *Scabiosa columbaria* from seed can be challenging. Growers often observe germination over a long period of time, and it is not uncommon for the germination rates to be less than 50%. Providing a pre-chilling, or stratification treatment, will help to improve germination success.

The seeds should be sown in the intended plug flat and covered lightly with germination mix or medium-grade vermiculite. The covering helps to maintain a suitable environment around the seed during this phase. Growers commonly sow a minimum of two seeds per cell. The seed flats should be moistened and moved to an environment, such as a cooler, where the temperatures can be maintained at 40° F (4° C) for three to four weeks.

After the cold treatment, move the plug trays into a warm environment (70–75° F [21–24° C]) for germination. Many growers utilize germination chambers during this stage to provide uniform moisture levels and temperatures. Using germination chambers is optional, as *Scabiosa columbaria* will successfully germinate in the greenhouse. The seeds begin to sprout within a few days and, depending on the cultivar, may take up to two weeks for germination to be complete.

Following germination, they can be grown at 65–68° F (18–20° C) for six to eight weeks until they are ready for transplanting. Fertilizers are usually applied once the true leaves are present, applying 100 ppm nitrogen every third irrigation or 50 ppm with every irrigation, using a balanced water-soluble source.

Tip cuttings are the most common means of propagating *Scabiosa columbaria* cultivars. Due to the free-flowering nature of the plant, harvesting vegetative cuttings can often be challenging. To maintain stock plants, it is best to produce them at cool temperatures, high light levels, and under short days. For the most part, these conditions will delay flowering and keep plants vegetative longer. As a result, the stock plants remain as compact rosettes with numerous vegetative shoots that can be harvested for propagation. Growers commonly obtain unrooted cuttings from suppliers rather than producing their own.

I recommend providing short day conditions during propagation to maintain vegetative growth, promote branching, and to reduce elongation. To maintain short-day conditions, it is often necessary to pull black cloth over the crop daily, providing a dark period for a minimum of twelve hours. Providing short days is not practical for most growers and is almost never practiced commercially. Plants will root successfully under longer photoperiods, but the rooting process will take longer.

Scabiosa columbaria cuttings are best taken prior to flower bud initiation and while the foliage is somewhat smooth, prior to the formation of the finely divided leaves. Tip cuttings containing several nodes and measuring approximately 2 in. (5 cm) in length are generally stuck into plug liners for rooting. The well-drained rooting medium should be moistened prior to sticking. The base of the cuttings can be dipped into a rooting hormone, such as a solution of indolebutyric acid (IBA), at rates of 500–1,000 ppm prior to sticking. Pincushion flowers can successfully root without rooting compounds but tend to root slightly faster and more uniformly when these treatments are provided.

Cuttings should be placed under low misting regimes for about the first seven to ten days of propagation. When possible, it is usually best to propagate under high humidity levels (90% relative humidity) with minimal misting. The misting can gradually be reduced as the cuttings form calluses and root primordia. The cuttings are usually rooted in less than three to four weeks with soil temperatures ranging of 68–74° F (20–23° C). For best results, the air temperature during rooting should be maintained above 60° F (16° C) and below 80° F (27° C). It is beneficial to begin constant liquid feeding with 150 ppm nitrogen at each irrigation beginning ten days from sticking.

Production

For container production, *Scabiosa columbaria* are suitable for 1 qt. (1 L) to 1 gal. (3.8 L) containers. Most growers receive starting materials as finished plug liners. When planting large containers, such as 1 gal. (3.8 L) pots, I recommend planting two 72-cell or smaller sized plugs. If larger sized plug cells are used, planting one plug per container is often appropriate.

Pincushion flowers perform best when they are grown in a moist, well-drained medium with good aeration and water-holding capacity. Most commercially available peat- or bark-based growing mixes work well, provided there is adequate drainage. If they are going to be overwintered, it is best to plant them into a porous growing medium, preferably a nursery-type mix (bark based) rather than a traditional greenhouse (peat-vermiculite) mix. Not only is the porosity (drainage) of the medium beneficial during crop production, it will help ensure the roots are not exposed to wet conditions for long durations, reducing losses that may occur during the winter months.

When planting, be careful not to plant the plugs too deeply, as this could lead to poor plant establishment, crop variability, crown rot, and losses. The plugs should be planted so the original soil line of the plug is even with or just below the surface of the growing medium of the new container. After planting, I suggest applying a fungicide drench, using a broad-spectrum fungicide, such as etridiozole + thiophanate-methyl (Banrot) or the combination of mefenoxam (Subdue Maxx) and thiophanate-methyl (Cleary's 3336).

Some growers have found it beneficial to pinch *Scabiosa columbaria* prior to or shortly after they are planted in the final container. Pinching increases lateral branching and the total number of flowers produced on each plant. Flowering will be delayed approximately two to three weeks when plants are pinched; therefore, growers should modify their crop schedules or ready dates accordingly. It is not necessary to pinch plants that are being grown in small containers, such as 1 qt. (1 L) pots, as they will fill out fine without pinching.

During production, the pH of the medium should be maintained at 5.8–6.5. Pincushion flowers are moderate feeders. Growers using water-soluble fertilizers either apply 125–150 ppm nitrogen as needed or feed with a constant liquid fertilization program using rates of 75–100 ppm nitrogen with every irrigation. Controlled-release fertilizers are frequently used to produce containerized pincushion flowers. Growers commonly incorporate time-release fertilizers into the growing medium prior to planting at a rate equivalent to 1–1.25 lb. nitrogen per cubic yard (464–567 g/0.76 m^3) of growing medium or apply them as a topdressing on the medium's surface using the medium rate listed on the product's label.

Where water-soluble fertilizers are the primary source of nutrition, it is best to use high nitrate fertilizers. Ammonium-based fertilizers may promote excessive internode elongation, especially during cool, dark weather conditions. This additional plant stretch causes these plants to sprawl and become floppy, reducing the appearance and quality of the crop.

Scabiosa perform well under average watering regimes. To avoid overly wet conditions that could lead to crown rot, keep plants that are not fully established slightly drier than fully rooted containers. When irrigation is necessary, water them thoroughly, allowing the soil to dry slightly between waterings.

The best quality is achieved when plants are grown in full sun or in greenhouses with high light intensities. High light intensities promote more, larger flowers per plant and produces shorter plants. Where possible, growers should place them in an environment with no less than 4,000 f.c. (43 klux) of light. *Scabiosa columbaria* grown at low light levels tend to become elongated and do not flower as profusely as those produced at ambient levels. Plant quality can be improved during short days or periods where the light levels are naturally low by providing 400–500 f.c. (4.3–5.4 klux) of supplemental lighting using high-pressure sodium lamps.

Insects

Aphid	Snail
Caterpillar	Spider mite
Leafhopper	Thrips
Slug	Whitefly

Aphids, thrips, two-spotted spider mites, and whiteflies are the most common insects that are observed feeding on *Scabiosa columbaria.* Growers should have scouting programs in place to detect these pests early before their populations build up to excessive levels. Some growers implement preventative spray programs using systemic chemicals, such as pymetrozine (Endeavor), thiamethoxam (Flagship), imidacloprid (Marathon II), dinotefuran (Safari), or acetamiprid (TriStar). These should keep crops free of aphids and whiteflies for up to one month following application.

The presence of spider mites often goes undetected (since they are usually found feeding on the undersides of the leaves) until their populations get high and significant plant injury has occurred. Damaged leaves are stippled with small yellowish to silver-gray speckles. Controlling spider mites is not easy because it is difficult to deliver the chemicals to the lower leaf surfaces where they are feeding, several life stages are present at any time, and they build resistance quickly to pesticides. Spider mite control is more successful when growers rotate chemical classes at every application; tank-mix ovicides, such as hexythiazox (Hexygon) and clofentezine (Ovation), with adulticides, such as abamectin (Avid) and bifenazate (Floramite); and ensure good coverage of the sprays to the crop.

Diseases

Aster yellows	*Sclerotinia*
Pellicularia blight	*Sclerotium*
Powdery mildew	*Thielaviopsis*
Pythium	

Under the right circumstances, certain diseases may become prevalent when growing *Scabiosa columbaria* cultivars. As mentioned earlier, root rots are likely to occur, especially when grown under wet conditions. Of the plant pathogens they are the most susceptible to, powdery mildew is the disease most problematic to growers. At first, powdery mildew appears as small white, talcum-like colonies on the upper leaf surfaces, but under the right conditions they may engulf the plant with a powdery appearance. To control this disease, it is best to manage the environment by providing the proper plant spacing and adequate air movement, controlling the humidity, and, if desired, following a preventative spray program using the appropriate chemicals.

Plant Growth Regulators

When grown in greenhouses, the flower stalks tend to elongate and sprawl compared to plants grown under natural, high light environments. Height control is often necessary to produce a high-quality product under these conditions. Many cultivars, such as 'Blue Diamonds', are naturally short and may not require height control. Before using chemicals to reduce plant height, it is usually beneficial to provide adequate space between each plant, which will reduce the competition between plants for light, prevent the plants from growing taller, and reduce the conditions conducive for powdery mildew development.

Depending on your geographic location, apply foliar applications of PGRs beginning with the rates listed in the table below. *Scabiosa columbaria* will remain as compact rosettes unless the conditions for excessive elongation are present. Apply growth regulators as the flower stalks are just beginning to elongate and if the rosette appears to be elongating with the flowers. If the plant remains compact and appears to be a tight rosette, then PGRs are probably not necessary. When plant stretch is occurring, it will usually require two or three applications at seven-day intervals to provide adequate height control.

Forcing

To improve marketability, *Scabiosa columbaria* cultivars can be forced to bloom throughout the year. Pincushion flowers are prolific, free flowering perennials and can easily be forced into bloom. They bloom so prolifically that in many cases they flower before the plant is bulked up or has filled out the container. Growers producing large containers, such as 1 gal. (3.8 L) pots, often achieve flowering before the plants are marketable. To overcome early flowering or to allow the plants to bulk up, growers commonly use larger starting materials, plant multiple plugs per container, or supply short day lengths.

As mentioned previously, growers often provide short days to stock plants and throughout propagation when the photoperiods are naturally long to delay flower development and promote lateral branching. Short-day conditions are created by pulling black plastic over the production site, ensuring the plants only receive twelve hours of light each day. The blackout area should be completely dark, as *Scabiosa columbaria* can perceive only a few foot-candles of light, which could delay or negate the results of providing short days. Ideally, short-day photoperiods should begin throughout propagation and/or the beginning of crop production and should last until the foliage covers more than half of the surface area of the final container. Depending on the size of the starting materials, this may take three to four weeks.

PGRs for *Scabiosa columbaria* Cultivars

Trade Name	Active Ingredient	Northern Rate
B-Nine or Dazide	Daminozide	2,500 ppm
Bonzi, Piccolo, or Paczol	Paclobutrazol	30 ppm
Sumagic	Uniconazole-P	5 ppm

These are northern rates and need to be adjusted to fit the location where the crop is being produced. These rates are for spray applications and geared to growers using multiple applications.

Most growers have the best success when planting large plug liners into the final container during the late summer or early fall of the year prior to the desired spring sales and overwintering them as a potted plant. Although cold is not required for flowering, it does accelerate flowering of the final product. *Scabiosa* still develop leaves while exposed to above freezing temperatures. They may develop seven to nine leaves during a ten-week cold treatment. In fact, it is not uncommon to have flower buds develop and open during this period. Since they are actively growing, it is best to provide a minimum amount of light, 25–50 f.c. (269–538 lux), when in coolers; natural light levels should suffice when overwintering inside greenhouses or cold frames. The cold treatment is optional; if vernalization is provided, a minimum of ten weeks of exposure to temperatures ranging from 35–44° F (2–7° C) is recommended.

Pincushion flowers are day-neutral plants and will flower under either short or long days. Plants forced under long days tend to grow taller than those grown under short days and may require additional applications of PGRs. I recommend forcing *Scabiosa columbaria* under natural photoperiods. If compact plants are desired, it is best to force them during periods when the day lengths are naturally short.

Figure 12.53. ***Scabiosa columbaria* 'Butterfly Blue' potted in the late summer remains evergreen and still develops leaves and flowers during the overwintering period. In many cases, it can make early spring shipments while being produced in unheated cold frames.**

The time to bloom varies widely with cultivar, size of the starting materials, cold treatment, and the production temperatures. Precisely determining the time to bloom can be challenging since they flower so freely and have no cold or photoperiod requirements. In many cases, the starting materials already have visible flower buds prior to planting. Interestingly, providing a ten-week cold treatment can reduce the forcing time by three to five weeks when compared with crops that did not receive vernalization.

The time to flower largely depends on the temperatures the plants are grown at. The following temperature recommendations fit most varieties, but each cultivar will have a slightly different finishing time at each of the temperatures listed. Pincushion flowers grown at 62° F (17° C) will flower in about six weeks if they have been vernalized and approximately ten weeks when no cold treatment has been provided. Plants grown at 68° F (20° C) will flower in approximately four weeks when they are vernalized and will take nearly four additional weeks to bloom when no cold is provided. For the largest flower size, most flowers, and shortest plants, I recommend vernalizing for a minimum of ten weeks and to grow plants at cooler production temperatures, 60–67° F 16–19° C).

References

"Growers' Notebook: Scabiosa." *GMPro.* August 1999.

Runkle, E., R. Heins, A. Cameron, and W. Carlson. "Herbaceous Perennials: *Scabiosa columbaria.*" *Greenhouse Grower.* September 2001.

Thomas, M., and S. Still. "Perennial Plant of the Year 2000: *Scabiosa columbaria* 'Butterfly Blue'." *Perennial Plants.* Autumn 1999.

*Sedum spectabile**

Stonecrop

Stonecrop is a popular perennial garden plant. The succulent leaves, sturdy stems, and massive flower heads have earned them a spot on many landscapers' "must have" list. Besides their physical attributes, *Sedum* are drought tolerant and can withstand many adverse growing conditions, which also make this an ideal landscape perennial. *Sedum* are often used as specimen plants, border plants, or in mass plantings. Stonecrops also make great additions to rock gardens and "living roofs." Commercial growers find *Sedum* to be an easy to grow and popular perennial that performs well in containers and are always in demand.

Figure 12.54. *Sedum hybridum* 'Black Jack' (left) and *S. spectabile* 'Matrona' (right)
Walter's Gardens

Sedum thrive in sunny locations throughout USDA Hardiness Zones 3–10 and AHS Heat Zones 12–3, although locations receiving partial sun are often acceptable. There are hardiness and heat tolerance differences between cultivars. During the midsummer, large broccoli-like flower heads form that contain numerous green flower buds. By late summer, the flower buds swell and begin to show color; as the buds open, the color intensifies, creating an impressive color display in the landscape. The flower color varies with the cultivar, but mostly consists of different hues of pink or red.

Sedum belongs to the Crassulaceae family. There are not many perennials in this family that are being commercially produced. After *Sedum, Sempervivum* is the next most popular perennial being grown from this family. Other less commonly produced perennials from the Crassulaceae family are *Chiastophyllum, Crassula, Jovibarba, Rhodiola,* and *Rosularia.*

Propagation

Sedum cultivars are vegetatively propagated by tip cuttings or division. Many cultivars produce sterile seed, which does not germinate. Some varieties do produce viable seed, but the seed of many of these hybrids does not come true to type and should not be used for propagation. Tip cuttings are the fastest, most cost effective method of propagating *Sedum.* In fact, most bare-root producers generally plant rooted plug liners into the field as the starting materials for their bare-root crops rather than using bare-root divisions. For container production, the advantages and disadvantages of using rooted liners or bare-root divisions will be discussed later.

Tip cuttings should be taken while plants are vegetative. Many growers have found the best time to propagate *Sedum* is during the early spring when they are actively growing and vegetative, although acquiring unrooted cuttings from overseas is becoming a viable option for

Commonly Grown *Sedum spectabile* Cultivars

'Autumn Fire'	'Matrona'
'Autumn Joy'	'Mini Joy'
'Black Jack'	'Neon'
'Brilliant'	'Pink Chablis'
'Frosty Morn'	'Purple Emperor'
'Hot Stuff'	'Rosy Glow'

**The parentage of many commercially produced stonecrops is often confused with one another. The distinction between* Sedum spectabile *and* Sedum hybridum *is not clearly defined and they are often used interchangeably.*

year-round propagation of *Sedum*. Cuttings that have already initiated flower buds will root more slowly and have a lower rooting percentage than cuttings without flowers.

Tip cuttings should measure approximately 1.5–2 in. (4–5 cm) in length and contain several nodes. The well-drained rooting medium should be moistened prior to sticking. The base of the cuttings can be dipped into a rooting hormone, such as a solution of indolebutyric acid (IBA), at rates of 750–1,250 ppm prior to sticking. *Sedum* can successfully root without rooting compounds but tend to root slightly faster and more uniformly when these treatments are provided. They are generally stuck into 128-cell or larger sized plug liners for rooting.

It is best to propagate *Sedum* under high humidity levels with minimal or no misting. Too much misting will cause the leaves to rot. During certain times of the year (mid-fall to mid-spring), misting is not required to successfully root *Sedum*. If misting is required, it should only be used during the hottest portion of the day and kept to a minimum (once or twice per day). Usually only the first seven days of propagation requires any type of misting. Once the cuttings appear to tolerate the stress caused from the light levels each day, the misting is no longer required. Again, prolonged exposure to misting may cause the leaves to rot during propagation; only use misting when needed.

Begin constant liquid feeding with 150 ppm nitrogen at each irrigation beginning seven days from sticking. During certain times of the year, the cuttings will benefit from bottom heat, with soil temperatures maintained at 75–80° F (24–27° C). At these temperatures, *Sedum* should be rooted in three to four weeks.

Division entails splitting the crown into smaller sections, with each section containing at least one bud and several adjoining roots. Growers can divide *Sedum* throughout the growing season, but they are most commonly divided in the early spring or fall. When making divisions, the best results are achieved when the foliage is cut back to 4–6 in. (10–15 cm) prior to lifting and dividing. The size of the division usually depends on the size of the container in which it is being potted. For 1 qt. (1 L) production, a crown consisting of three to five eyes is commonly used. For larger containers, such as 1 gal. (3.8 L) pots, divisions of five to seven eyes are often used.

Production

For container production, *Sedum* cultivars are suitable for 1 qt. (1 L) up to 2 gal. (7.6 L) containers. Most growers receive starting materials as bare-root divisions or finished plug liners. The type of starting materials growers use often depends on availability, grower preference, production schedule, and cost. Plants started from bare-root are almost always larger, produce fuller finished containers, and require less production time than when rooted liners are used. When using plugs as starting materials for planting large containers, such as 1 gal. (3.8 l) or larger pots, I recommend planting two or more plug cells per container to properly fill out the pot and reduce the production time.

When planting in the spring, growers find it more advantageous to use bare-root starting materials, which allows them to produce a finished container faster than when starting from plugs. The size of the container will also determine which starter material a grower chooses. For example, when planting small containers, such as 1 qt. (1 L) pots, in spring, growers may decide to use plug liners, but when producing 2 gal. (7.6 L) containers at this time, the quickest method to produce a finished container and maintain the desired quality attributes would be to plant *Sedum* from bare-root divisions. Crops started in the summer a year before the desired sales date can be started from plugs or bare-root materials, regardless of the container size.

For the best performance, plant *Sedum* in a well-drained medium, preferably a nursery-type mix (bark based) rather than a traditional greenhouse (peat-vermiculite) medium. Not only is good porosity (drainage) of the medium beneficial during crop production, it is essential to successfully overwinter *Sedum*, as they do not tolerate wet soils for long durations. When planting bare-root divisions, the crown should be placed in the center of the pot, with the crown slightly covered (no more than 0.5 in. [13 mm]) or just exposed after they are thoroughly watered in. Plug liners should be planted so the original soil line of the plug is even with or just below the surface of the growing medium of the new container.

Stonecrops are light to moderate feeders. Fertility can be delivered using water-soluble or controlled-release fertilizers. Growers using water-soluble fertilizers either apply 150–200 ppm nitrogen as needed or feed with a constant liquid fertilization program using rates of 75–100 ppm nitrogen with every irrigation. Growers commonly apply time-release fertilizers as a topdressing onto the medium's surface using the medium rate, or incorporated into the growing medium prior to planting at a rate equivalent to 1 lb. nitrogen per cubic yard (454 g/0.76 m^3) of growing medium. The pH of the medium should be maintained at 6.0–6.5.

Sedum requires a below average amount of irrigation. They are succulents and can tolerate average watering regimes, but generally perform best under slightly dry conditions. I recommend, when watering is required, to water them thoroughly and allow the substrate to dry between waterings. Keeping them on the dry side will help to reduce plant stretch and intensifies the color of the leaves and stems.

Sedum prefer to be grown at high light levels. Where possible, growers should place them in an environment with no less than 4,000 f.c. (43 klux) of light. Plants grown at low light levels tend to become elongated and do not flower as profusely as those produced at ambient levels. During the winter months, providing 400–500 f.c. (4.3–5.4 klux) of supplemental lighting helps to improve the quality characteristics of *Sedum* crops.

Insects

Aphid	Root knot nematode
Black vine weevil	Slug
Caterpillar	Spittlebug
Fungus gnat larva	Thrips
Leafhopper	Whitefly
Mealybug	

Sedum are not susceptible to many insect pests. Aphids are the most prevalent insect pest observed feeding on *Sedum*. During propagation, fungus gnat larvae are frequently seen burrowing into the base of the unrooted cuttings. None of these insect pests require preventative control strategies. Growers should have routine scouting programs to detect their presence early and to determine if and when control strategies are necessary.

Diseases

Colletotrichum (anthracnose)	Powdery mildew
Erwinia	*Rhizoctonia*
Fusarium	Rust
Pellicularia (crown rot)	*Sclerotium*
Phytophthora	*Septoria*

Sedum cultivars are not susceptible to many plant diseases. Crown rot caused by *Fusarium* or *Rhizoctonia* is likely to occur when crops are overwatered or if the growing medium has insufficient drainage. None of these diseases require preventative control strategies. Growers should have routine scouting programs to detect their presence early and to determine if and when control strategies are necessary.

Height control is often necessary to produce a high-quality product under greenhouse conditions. Providing adequate spacing between the plants will reduce plant stretch caused by competition. To a certain extent, the height can also be reduced by withholding water and nutrients. Under certain growing conditions or under high plant densities, it may be necessary to use chemical plant growth regulators (see chart). Depending on your geographic location, apply foliar applications of uniconazole-P at 5 ppm or paclobutrazol at 30 ppm. It will usually require two or three applications at seven-day intervals to provide adequate height control. To achieve the greatest results, these applications should begin just as the leaves from adjacent plants are beginning to touch one another.

PGRs for *Sedum spectabile* Cultivars

Trade Name	Active Ingredient	Northern Rate
B-Nine or Dazide	Daminozide	2,500 ppm
Bonzi, Piccolo, or Paczol	Paclobutrazol	30 ppm
Sumagic	Uniconazole-P	5 ppm
B-Nine or Dazide + Bonzi	Daminozide + paclobutrazol	2,000 ppm + 15 ppm
B-Nine or Dazide + Cycocel	Daminozide + chlormequat chloride	2,000 ppm + 1,000 ppm

These are northern rates and need to be adjusted to fit the location where the crop is being produced. These rates are for spray applications and geared to growers using multiple applications.

Forcing

Sedum are often shipped when green, budded, or in full flower. Non-flowering stonecrops can be sold throughout the growing season from spring to fall. Naturally flowering stonecrops are usually available from late summer to early fall. To improve marketability, *Sedum* cultivars can be forced to bloom throughout the year.

Sedum do not have a juvenility period and may set flowers with only a few leaves being present. Under day lengths over fourteen hours, flower induction and formation begins. Flower buds may be observed with as little as three weeks of exposure to long photoperiods. Propagators often observe bud set and flowering during propagation unless short days are maintained during stock plant production and throughout the propagation cycle.

Although vernalization is not required, providing a cold treatment may be beneficial, at least from the standpoint of bulking up plants started from plug liners. Containers started from plugs during the late summer for the following spring's sales will typically be larger and have more shoots, both for filling out the container and for increasing the number of flowers per pot, as compared with plugs planted in the spring of the same year's sales. Bare-root materials do not need to be bulked up and can be successfully planted and forced into bloom during the spring.

Figure 12.55. **Stonecrops are very popular and move well when flowering at retail.**

As indicated above, *Sedum* are obligate long-day plants and will not flower under short days. As long as the photoperiod is less than thirteen hours, *Sedum* will remain as compact, non-flowering rosettes. Growers should provide a minimum of fourteen-hour photoperiods or four-hour night interruptions until the flower buds are initiated (after about five weeks) or for the entire forcing duration.

Once flower buds are initiated, growers can remove the lighting and produce them under natural day lengths to finish the remainder of the forcing. Moving the plants from long days to short days is advantageous to the grower as it reduces plant height by a couple of inches (five centimeters) and reduces the time to flower by eight to ten days.

The time to bloom after the proper photoperiod is provided is a function of temperature. *Sedum spectabile* cultivars grown at 68° F (20° C) will take twelve to fourteen weeks to reach flowering, while plants grown at 60° F (16° C) will flower in approximately sixteen weeks.

References

Greenhouse Grower Magazine and Michigan State University. *Firing Up Perennials: The 2000 Edition.* Willoughby, Ohio: Meister Publishing. 2000.

Tiarella

Foamflower

Foamflowers are very popular, yet still underutilized, perennials in shade gardens across the country. They provide interesting, deeply cut variegated foliage with contrasting leaf patterns and extended bloom times. The evergreen foliage supplies year-round interest as it transforms to an attractive reddish-bronze display in the autumn and lasts throughout the winter months. With an explosion of breeding efforts in recent years, there have been numerous new cultivars introduced with improvements of both flower color and foliage characteristics.

Figure 12.56. ***Tiarella* x *hybrida* 'Spanish Cross'**

Today a vast array of ornamental foliage characteristics makes the native American *Tiarella* a must-have in the shady landscape. Most cultivars in production today are hybrids with improved foliage or flowering characteristics. The leaves of the various cultivars range from maple-shaped to very deeply lobed starfish shapes or somewhere in between. Most cultivars have some variegation, usually with various degrees of purple pigmentation along the mid-rib, although, a few cultivars, such as 'Heronswood Mist', have more unique variegation, appearing as layers of cream, green, and pink speckled over the leaf surfaces.

Many cultivars are very prolific bloomers, while others are grown strictly for the ornamental value of the foliage. The bottlebrush-shaped flowers are held above the foliage and usually consist of various shades of pink and white. Depending on the cultivar, blooming occurs in the early spring, with a certain degree of reblooming throughout the growing season. The blooms are fragrant and often used as cut flowers.

Foamflowers are both heat and cold hardy in USDA Hardiness Zones 3–9 and AHS Heat Zones 8–1, with hardiness and heat tolerance differences between cultivars. Foamflowers will grow more vigorously and have the best leaf coloration when grown in partial to full shade.

Tiarella belongs to the family Saxifragaceae, which includes many other popular perennials such as *Astilbe, Bergenia, Heuchera, Heucherella, Rodgersia, Saxifraga,* and *Tellima,* to name a few. In the landscape, the planting spreads slowly by underground stolons, but they always maintain their clumplike appearance, often reaching 1–2 ft. (30–61 cm) across. They are commonly used as accent plants, border plants, in rock gardens, mass plantings, containers, or as components in mixed containers.

Propagation

Depending on the cultivar, growers use divisions, seeds, or tissue culture to propagate *Tiarella.* To retain the desirable characteristics and remain true to name, *Tiarella* must be propagated vegetatively. Most new cultivars are patented, and unlicensed propagation is prohibited. To avoid illegal activities and the consequences, always check for plant patents before attempting to propagate.

Many of the new cultivars are propagated by means of tissue culture and produced as plug liners. Tissue culture has allowed these cultivars to reach the market in a relatively short period of time. Most growers purchase cultivars from tissue culture as 72-cell or larger sized plugs from licensed propagators. Some

Commonly Grown
***Tiarella* Hybrids and Cultivars**

'Black Snowflake'	'Pink Sky Rocket'
'Black Velvet'	'Pirate's Patch'
'Candy Striper'	'Sea Foam'
'Crow Feather'	'Slick Rock'
'Dark Eyes'	'Spanish Cross'
'Inkblot'	'Spring Symphony'
'Iron Butterfly'	'Starfish'
'Jeepers Creepers'	'Stargazer Mercury'
'Mint Chocolate'	'Sugar and Spice'
'Neon Lights'	*T. wherryi**
'Pink Bouquet'	*T. wherryi* 'Heronswood Mist'
'Pink Brushes'	

*Propagated by seed

perennial propagators receive stage 3 tissue culture materials. These are small plantlets without an established root system and should be handled in a similar manner as unrooted cuttings. They are generally stuck into large plug liners, such as 72-cell trays, for rooting. The well-drained rooting medium should be moistened prior to sticking.

Stage 3 materials should be placed under low misting regimes for the first seven to ten days of propagation. It is usually best to propagate under high humidity levels (90% relative humidity) with minimal misting. The misting can gradually be reduced as they form calluses and root primordia. *Tiarella* are usually rooted in three to four weeks with soil temperatures at 68–75°F (20–24° C). For best results, the air temperature during rooting should be maintained above 60° F (16° C) and below 80° F (27° C). It is beneficial to begin constant liquid feeding with 150 ppm nitrogen at each irrigation beginning ten days from sticking. Depending on the size of the plug cell, *Tiarella* will be fully rooted and ready for transplanting approximately six weeks from sticking.

Due to crop variability because most seedlings do not grow true to type, there are only a handful of cultivars that are propagated from seed. The most common seed-propagated foamflower is *Tiarella wherryi*. The seeds are small and difficult to handle. Sow *T. wherryi* seeds in the intended plug flat and do not cover them with germination mix or vermiculite. Light is required for germination; covering reduces the amount of light reaching the seed and may adversely affect the germination process. The seed flats should be moistened and moved to a warm environment, where the temperatures can be maintained at 65–70° F (18–21° C) for germination. Many growers utilize germination chambers during this stage to provide uniform moisture levels and temperatures. It is important to keep the medium very moist but not saturated during this stage or the germination rates are likely to decrease.

Seedlings will emerge over a period of time, ranging from fourteen to twenty-one days after sowing. Following germination, reduce the moisture levels somewhat, allowing the growing medium to dry out slightly before watering to help promote rooting. The light levels can gradually be increased as the plugs develop, from 500 f.c. (5.4 klux) following germination to 2,500 f.c. (27 klux) at the final stage. Fertilizers are usually applied once the true leaves are present, applying 100 ppm nitrogen every third irrigation using a balanced water-soluble source. When plugs are grown at 65° F (18° C), they are usually ready for transplanting in ten to twelve weeks. Growers often have difficulty germinating *Tiarella* during the summer months. It is recommended to avoid starting them during this time of year, unless cool conditions (from a cool germination chamber) can be provided until the first true leaves are present.

Production

Most growers receive finished plug liners as starting materials. They are commonly planted in 1 qt. (1 L) or 1 gal. (3.8 L) containers after they have been received. Foamflowers perform best in a porous growing medium with both good water retention and aeration characteristics. Many commercially available peat- or bark-based growing mixes work well, provided there is adequate drainage.

When planting, take care to not bury the crown of the plant too deeply. Place the plug in the center of the pot and plant so the original soil line of the plug is even with the surface of the growing medium of the new container. Planting the crown too deep will lead to poor plant establishment, crop variability, crown rots, and crop losses. After planting, I recommend drenching with a broad-spectrum fungicide such as etridiozole + thiophanate-methyl (Banrot) or the combination of mefenoxam (Subdue Maxx) and thiophanate-methyl (Cleary's 3336).

The root zone should be kept uniformly moist until the roots reach the outside of the growing medium. Once rooted, they can dry partially between waterings. *Tiarella* require an average amount of irrigation, as they do not tolerate very wet conditions or overly dry conditions. When overly wet conditions persist, foamflowers are particularly susceptible to crown and root rot pathogens. Additionally, when they are kept too moist over extended periods of time, the foliage appears slightly chlorotic. When irrigation is necessary, water them thoroughly, allowing the soil to dry slightly between waterings.

During the winter and early spring, foamflowers can be produced under natural light levels. As the spring progresses and the outdoor light levels increase, the production sites should have some type of shading materials applied, ultimately reducing the light levels by 35–50%. Under high light intensities, marginal leaf burn may occur if the plants become water stressed. Producing foamflowers under shady conditions will ensure more active growth and will reduce the potential for the leaves to become scorched.

Foamflowers can be grown using light to moderate fertility levels. The pH of the growing medium should be maintained at 5.5–6.5 throughout crop production. Fertility can be delivered using water-soluble or controlled-release fertilizers. Growers using water-soluble fertilizers either apply 150–200 ppm nitrogen as needed or feed with a constant liquid fertilization program using rates of 50–100 ppm nitrogen with every irrigation. Growers commonly apply time-release fertilizers as a topdressing onto the medium's surface using the medium rate listed on the product's label or incorporate them into the potting substrate prior to planting at a rate equivalent to 1 lb. nitrogen per cubic yard (454 g/0.76 m^3) of growing medium.

Insects

Aphid	Slug
Black vine weevil	Whitefly

Generally, *Tiarella* can be commercially produced in containers relatively free of insects. Aphids, black vine weevils, slugs, and whiteflies may occasionally be observed feeding on foamflowers, but they rarely become problematic. None of these insect pests require preventative control strategies. Growers should have routine scouting programs to detect the presence of insects early and to determine if and when control strategies are necessary.

Diseases

Botrytis	Powdery mildew
Colletotrichum (anthracnose)	*Pythium*
Phytophthora	*Rhizoctonia*
	Rust

When the proper environmental and cultural factors are provided, foamflowers generally can be produced without the occurrence of plant pathogens. The primary diseases growers should watch for are crown and root rots caused by the pathogens *Pythium*, *Phytophthora*, and *Rhizoctonia*. These diseases are most often observed shortly after planting if the crown was planted too deeply or when wet conditions persist for long durations. They are most susceptible to these diseases when they are grown under cool, wet conditions, such as going into or coming out of winter dormancy.

Growers should carefully monitor the moisture levels during adverse times of the year and avoid overwatering their plants. Many growers apply a broad-spectrum fungicide drench before and just after the overwintering period to reduce the population of these pathogens. Routine scouting is useful and recommended to detect plant diseases early, allowing the appropriate control strategies to be implemented before significant crop injury or mortality occurs.

Plant Growth Regulators

Height control is seldom necessary to control the growth of foamflowers unless they are grown at high plant densities. Providing adequate space between the plants is the best and most effective method growers can practice to control plant height during production. Under certain growing conditions, it may be necessary, although not common, to use chemical plant growth regulators. If height control is necessary, I recommend applying uniconazole-P at 5 ppm. This is a northern rate and will need to be adjusted to match the location where the crop is being grown. Apply growth regulators to the foliage as the plant canopy is beginning to enclose.

Forcing

Tiarella are easy to produce as either foliage or flowering plants. A foliage plant or one with sporadic flowering can easily be produced by simply transplanting plugs into larger container sizes. When transplanting from 72-cell plugs, foamflowers grown at 65° F (18° C) typically take seven weeks to finish 1 gal. (3.8 L) pots in the summer or ten weeks during the winter months.

Growers wishing to produce blooming plants need to understand and follow the guidelines discussed below.

They do not have a known juvenility period; plants with as few as ten leaves have flowered. Although juvenility is not a concern, I recommend bulking them up prior to forcing them into bloom. To obtain full, flowering foamflowers for spring sales, it is beneficial to plant them during the late summer of the previous season. Therefore, it is best to grow foamflowers at natural day lengths for a minimum of six weeks to allow them to reach a mature size before providing the additional requirements necessary for flowering.

Tiarella cultivars are considered to be cold-beneficial plants. Some cultivars will flower without a cold treatment regardless of the photoperiod, while others are capable of flowering without a cold treatment only when they are grown under long day photoperiods. All *Tiarella* cultivars will flower faster and will bloom under any photoperiod when vernalization is provided. Vernalize small bulked-up containers of foamflowers for a minimum of twelve weeks at 35–44° F (2–7° C). Since they remain evergreen throughout the cold treatment, provide 100–200 f.c. (1.1–2.2 klux) of light during this period. If they are overwintered in a greenhouse or cold frame, the natural light levels should be sufficient. When coolers are used, supplemental lighting is best provided using cool-white fluorescent lamps.

After the cold requirement is achieved, foamflowers can be grown at any day length, as they are day-neutral plants. Plants that do not receive a cold treatment are considered to be facultative long-day plants (long-day beneficial) and will flower after several weeks of exposure to long-day photoperiods. Following a cold treatment, the length of the photoperiod does not have any effect on the time to flower or the number of blooms produced.

The time to bloom after vernalization is a function of cultivar and temperature. Depending on the cultivars being produced, foamflowers will reach full bloom in approximately four weeks when they are grown at 60° F (16° C) or three weeks at 65° F (18° C). Producing them at cooler temperatures only slightly increases the time to flower but will improve the overall color intensity of the foliage and flowers. To achieve the highest plant quality, I recommend growers produce foamflowers using cool temperatures, such as 55–60° F (13–16° C) and allow slightly more time (one to two weeks) to produce them.

References

Pilon, Paul. Perennial Solutions: "*Tiarella* 'Jeepers Creepers'." *Greenhouse Product News.* December 2003.

Veronica spicata

Speedwell

Veronica spicata, also referred to as speedwell, belongs to the Scrophulariaceae, or figwort, family, which contains numerous commercially produced perennials, including *Antirrhinum, Chelone, Digitalis, Linaria, Penstemon,* and *Verbascum.* The genus is named in honor of St. Veronica, as the markings on the flowers of certain species resemble the markings on the sacred handkerchief of St. Veronica. They are native to Europe and Asia and have become a very popular perennial in the American landscape.

Figure 12.57. *Veronica spicata* 'Royal Candles'

Veronica spicata is a clump-forming, bushy perennial often utilized in rock gardens, as border plants, and in mass plantings or containers. They are commonly used to attract butterflies into the garden and for cut flower production. *Veronica* plants prefer full sun, although in the South they perform best when grown under partial shade. They are both cold and heat hardy in USDA Hardiness Zones 3–8 and AHS Heat Zones 9–1. For more continuous blooming, deadhead old blooms.

Cultivars of *Veronica spicata* have many distinguishing characteristics, such as compact growth habits reaching 12–24 in. (30–61 cm) tall; clean, leafy foliage topped with numerous vertical flower spikes; and various shades of blue, pink, or white flowers lasting from late spring to midsummer. With these characteristics, speedwells are well suited for production in small container sizes and can be marketed alongside bedding plants.

Propagation

Depending on the cultivar, *Veronica spicata* is started from seed or vegetatively propagated by tip cuttings or division. The method of propagation is often dependent on cultivar, availability, and cost. Although using bare-root divisions provides for a quicker finishing time, it often leads to crop variability and lower survival rates. Many cultivars are patented, and unlicensed propagation is prohibited. To avoid illegal activities and the consequences, always check for plant patents before attempting to propagate *Veronica.*

Sow *Veronica* seeds in the intended plug flat, and do not cover the seed with germination mix or vermiculite. Light is required for germination, and covering reduces the amount of light reaching the seed and may adversely affect the germination process. The seed flats should be moistened and moved to a warm environment, where the temperatures can be maintained at 65–70° F (18–21° C) for germination. Many growers utilize germination chambers during this stage to provide uniform moisture levels and temperatures. Do not fertilize the seed flats prior to germination, as speedwells are sensitive to high salt levels during this stage.

The seeds should be germinated in seven to ten days. Following germination, reduce the moisture levels somewhat, allowing the growing medium to dry out slightly before watering to help promote rooting. The light levels can gradually be increased as the plug develops, from 500 f.c. (5.4 klux) following germination to 2,500 f.c. (27 klux) at the final stage. Fertilizers are usually applied once the

Commonly Grown *Veronica spicata* Cultivars

'Alba*	'Icicle'
'Blue Bouquet'*	'Pink Goblin'*
'Blue Carpet'*	'Red Fox'
'Blue Charm'	'Rose'
'Blue'*	'Royal Candles'
'Crater Lake Blue'	'Sightseeing'*

*Propagated from seed

true leaves are present, applying 100–150 ppm nitrogen every third irrigation using a balanced water-soluble source. When plugs are grown at 65° F (18° C), they are usually ready for transplanting in eight to ten weeks.

Tip cuttings should be taken while stock plants are vegetative. Cuttings that are already in bloom will take longer to root and have a lower survival rate than purely vegetative starting materials. Propagators have the most success when propagating during the early spring or late summer, when the plants are not flowering and are growing vegetatively. Tip cuttings should measure approximately 2 in. (5 cm) and contain several nodes. They are generally stuck into plug liners for rooting. The well-drained rooting medium should be moistened prior to sticking. The base of the cuttings can be dipped into a rooting hormone, such as a solution of indolebutyric acid (IBA), at rates of 750–1,000 ppm prior to sticking. *Veronica* can successfully root without rooting compounds but tend to root slightly faster and more uniformly when these treatments are provided.

Cuttings should be placed under low misting regimes for about the first two weeks of propagation. It is usually best to propagate under high humidity levels (90% relative humidity) with minimal misting. Prolonged exposure to mist may cause the leaves to rot during propagation. The misting can gradually be reduced as the cuttings form calluses and root primordia. They are usually rooted in approximately four to five weeks with soil temperatures at 68–73° F (20–23° C). For best results, the air temperature during rooting should be maintained above 60° F (16° C) and below 80° F (27° C). It is beneficial to begin constant liquid feeding with 200 ppm nitrogen at each irrigation beginning ten days from sticking.

Production

Veronica spicata perform best when grown in a moist, well-drained medium with a slightly acidic pH, 5.5–6.2. They perform well in many commercially available peat- or bark-based growing mixes, provided there is adequate drainage. They are well suited for commercial container growers and are commonly produced in 1 qt. (1 L) to 1 gal. (3.8 L) containers.

Speedwells are moderate feeders and perform best when either a constant liquid fertilization program is used, feeding at rates of 75–100 ppm nitrogen or a controlled-release fertilizer is incorporated at a rate equivalent to 1 lb. nitrogen per cubic yard (454 g/0.76 m^3) of growing medium. Plants grown under low fertility regimens will most likely appear chlorotic and exhibit a delay in flowering. They require frequent irrigation. When irrigation is necessary, I recommend watering thoroughly, allowing the soil to dry slightly between waterings. Providing high soil moisture and fertility levels will promote stem elongation and may delay flowering.

Veronica prefer to be grown at high light levels. Where possible, growers should place them in an environment with no less than 4,000 f.c. (43 klux) of light. Plants grown at low light levels tend to be of lower quality, becoming elongated and not flowering as profusely as those produced at ambient levels. During the winter months, providing 400–500 f.c. (4.3–5.4 klux) of supplemental lighting helps to improve the quality of *Veronica* crops.

Insects

Aphid	Southern root knot nematode
Caterpillar	Tarnished plant bug
Leaf miner	Thrips
Slug	Whitefly

Generally, speedwells are relatively insect free. Aphids, whiteflies, and thrips occasionally will become problematic. Of these insect pests, aphids are the most prevalent. A preventative application of imidacloprid (Marathon 60WP) as a drench will generally ensure aphid-free plants from spring planting until the plants are shipped. Other preventative strategies include monthly spray applications of systemic chemicals such as pymetrozine (Endeavor), thiamethoxam (Flagship), imidacloprid (Marathon II), or acetamiprid (TriStar). These chemicals will also prevent the occurrence of whiteflies.

Diseases

Botrytis	Powdery mildew
Downy mildew	Rust
Entyloma (leaf smut)	*Sclerotium*
Leaf gall	

Under the right circumstances, certain diseases may become prevalent when growing *Veronica spicata*. *Botrytis* is occasionally a problem on the lower foliage, where air movement is limited and the foliage often stays wet after irrigation for extended periods of time. Powdery mildew and downy mildew are also likely to be observed. At first, powdery mildew appears as small white, talcum-like colonies on the upper leaf surfaces, but under the right conditions they may engulf the plant, giving a powdery appearance. Downy mildew usually appears first on the undersides of the leaves as a mass of white or gray spores, and often the upper leaf surface (directly above where the spores are observed) will appear mottled, discolored, or blistered. To control these diseases, it is best to manage the environment by providing proper plant spacing and adequate air movement, controlling the humidity, and, if desired, following a preventative spray program using the appropriate chemicals.

Plant Growth Regulators

Controlling the plant height may be necessary when producing *Veronica spicata* under greenhouse conditions. Providing adequate spacing between the plants will reduce plant stretch caused by competition. Under certain growing conditions or under high plant densities, it may be necessary to use chemical PGRs to reduce or control plant height. In the northern parts of the country, I would recommend applying uniconazole-P at 5 ppm. Applying one or two applications seven days apart should provide adequate height control.

Forcing

Forcing *Veronica spicata* into bloom out of season is relatively easy, but there may be a few complications that could alter your ability to schedule blooming plants predictably. Although I have not seen research on all cultivars, I feel it is safe to make a few assumptions based from research conducted on particular cultivars of *Veronica spicata*. Growers should note that there are distinct differences between cultivars propagated by seed and cuttings in regard to their forcing requirements and performance. These differences are noted below where appropriate.

Vegetatively propagated *Veronica* have an obligate cold requirement in order to flower. I recommend vernalizing plugs or small containers of *Veronica* for a minimum of six weeks at

PGRs for *Veronica spicata* Cultivars

Trade Name	Active Ingredient	Northern Rate
A-Rest	Ancymidol	25 ppm
B-Nine or Dazide	Daminozide	2,500 ppm
Bonzi, Piccolo, or Paczol	Paclobutrazol	30 ppm
Sumagic	Uniconazole-P	5 ppm
B-Nine or Dazide + Cycocel	Daminozide + chlormequat chloride	2,000 ppm + 1,000 ppm

These are northern rates and need to be adjusted to fit the location where the crop is being produced. These rates are for spray applications and geared to growers using multiple applications.

Figure 12.58. ***Veronica spicata* 'Blue Carpet'**
Jelitto Perennial Seeds

35–44° F (2–7° C). When vernalizing plugs in the final container, it is best to bulk them up under short days for four to six weeks prior to the cold treatment.

Cuttings harvested from stock plants that have received a cold period will flower without receiving a cold treatment themselves. For example, you receive rooted cuttings from two suppliers in April. The first supplier had stock plants that never received vernalization, and the second supplier harvested cuttings from plants that were overwintered. If you are planning to produce flowering plants for sale two months after you receive your rooted cuttings, you will be able to produce flowering *Veronica* only from the second supplier. The first supplier's plants will grow to a shippable size, but will remain vegetative.

Growers should be aware of the following exceptions when producing *Veronica*. The straight species, *Veronica spicata,* and most seed-propagated cultivars will flower the first year without receiving a cold treatment. Growers should not pinch *Veronica* following the cold period. Pinching removes the inflorescences, which will cause the flowering to become sparse and non-uniform. Once stock plants have flowered, they will grow vegetatively until they are exposed to another cold period. This means growers can maintain non-flowering stock plants on a year-round basis if no cold treatment has been provided. Non-vernalized stock plants can be grown at any photoperiod to produce vegetative cuttings.

Speedwells are day-neutral plants and will flower under any photoperiod after the cold requirement has been achieved. The time to bloom after vernalization is a function of temperature. *Veronica spicata* cultivars grown at 68°F (20° C) will take six to eight weeks to reach flowering, while plants grown at 60°F (16° C) will flower in eleven weeks. *Veronica* grown under warmer temperatures will have smaller flowers than will plants grown under cooler temperature regimes. The best flower size is achieved by growing at temperatures averaging 60° F (16° C).

References

Enfield, A., E. Runkle, R. Heins, A. Cameron, and W. Carlson. "Herbaceous Perennials: Quick-cropping Part III." *Greenhouse Grower.* May 2002.

Pilon, Paul. Perennial Solutions: "*Veronica spicata* 'Royal Candles'." *Greenhouse Product News.* June 2003.

Index

*Page numbers with *f* indicate figures; page numbers with *t* indicate tables.

A

B

E

G

J

K

L

M

P

Q

R

S

T

U

V

W

X

Y

Z